W9-DHY-355

An Elementary Approach to Functions

An Elementary Approach to Functions

Second Edition

Henry R. Korn
Associate Professor of Mathematics
Westchester Community College

Albert W. Liberi
Associate Professor of Mathematics
Westchester Community College

McGRAW-HILL BOOK COMPANY
New York St. Louis San Francisco Auckland Bogotá Düsseldorf
Johannesburg London Madrid Mexico Montreal New Delhi
Panama Paris São Paulo Singapore Sydney Tokyo Toronto

AN ELEMENTARY APPROACH TO FUNCTIONS

1234567890 DODO 783210987

This book was set in Baskerville by Progressive Typographers.
The editors were A. Anthony Arthur and Shelly Levine Langman;
the designer was Merrill Haber;
the production supervisor was Charles Hess.
New drawings done by J & R Services, Inc.
R. R. Donnelley & Sons Company was printer and binder.

Library of Congress Cataloging in Publication Data

Korn, Henry R
An elementary approach to functions.

Includes index.
1. Functions. I. Liberi, Albert W., joint author.
II. Title.
QA331.3.K67 1978 515 77-8123
ISBN 0-07-035401-4

Contents

Preface to the Second Edition ix

Preface to the First Edition xi

1

The Straight Line 1

1.1 A Review of the Rectangular Coordinate System 1
1.2 The Symbol Δ and Absolute Value 4
1.3 Distance between Points 8
1.4 The Slope of a Line 13
1.5 Equation of a Line 17
1.6 Two Theorems 22
1.7 Systems of Linear Equations 26
1.8 The Perpendicular Distance from a Point to a Line 37

2

Functions 41

2.1 The Function Concept 41
2.2 Functional Notation 42
2.3 The Graph of a Function 46
2.4 Special Functions and Their Graphs 48

3

Quadratic Functions 56

3.1 Graphing Quadratic Functions 56
3.2 Roots 65
3.3 More on Roots; the Discriminant 71
3.4 Applications of Quadratic Functions 75

4

Inequalities 85

4.1 Introduction 85
4.2 Properties of Inequalities 86
4.3 Interval Notation 89
4.4 Linear Inequalities 90
4.5 Other Types of Inequalities 95
4.6 Absolute Value 101

5

Polynomial Functions 108

5.1 Polynomial Functions 108
5.2 Evaluating a Polynomial 109
5.3 The Division Algorithm 112
5.4 Two Theorems 114
5.5 Curve Sketching and the Roots of Polynomials 119
5.6 Complex Roots 130
5.7 More Aids to Locating Real Roots and Graphing 135
5.8 Irrational Roots 143
5.9 Practical Applications of the Third-Degree Polynomial 145

6

Rational Functions 153

6.1 The Rational Function 153
6.2 Horizontal and Vertical Asymptotes 154
6.3 Oblique Linear and Nonlinear Asymptotes 166
6.4 Partial Fractions 174

7

Constructing Functions 182

7.1 Algebraic Operations 182
7.2 Composite Functions 184
7.3 Inverse Functions 188
7.4 Parametric Equations 193
7.5 Algebraic Functions 195

8

Trigonometry 201

8.1 Introduction 201
8.2 The Right Triangle 201
8.3 Applying Trigonometry in Right Triangles 212
8.4 The General Angle 221
8.5 Oblique Triangles; the Law of Sines and Cosines 230
8.6 Vectors 240

9

Trigonometric Functions of Real Numbers 252

9.1 Introduction 252
9.2 The Winding Function 252
9.3 The Trigonometric Functions of Sine and Cosine 260
9.4 Some Important Trigonometric Formulas 263
9.5 Notation 269
9.6 Graphing: $f(x) = \cos x$ and $g(x) = \sin x$ 269
9.7 Further Graphing Techniques 273
9.8 Phase Shift: $f(x) = A \cos (Bx + C)$ and $g(x) = A \sin (Bx + C)$ 285
9.9 Inverse Cosine and Inverse Sine 290
9.10 Other Trigonometric Functions 296

10

Exponential Functions 301

10.1 Review 301
10.2 Exponential Functions 305
10.3 The Constant e 310
10.4 Logarithmic Functions 313
10.5 Computations 317

11

The Circle and the Parabola 330

11.1 Introduction 330
11.2 The Circle 331
11.3 Some Conditions That Determine a Circle 336
11.4 The Parabola 342
11.5 The Parabola with Vertex at $V(h, k)$ 346

12

The Ellipse and the Hyperbola 354

12.1 The Ellipse 354
12.2 The Ellipse with Center $C(h, k)$ 361
12.3 The Hyperbola 367
12.4 The Hyperbola with Center $C(h, k)$ 372
12.5 Rotation of Axes 378

Appendix: Tables 395

Answers to Odd-numbered Exercises 413

Index 455

Preface to the Second Edition

In this second edition the authors maintain the tenets of the first edition. The basic structure and tone of the original text remains the same. Two sections were deleted that proved to be of little use, and certain new materials were added. Notable among these changes are:

1 A new chapter on basic trigonometry, Chapter 8, has been added. This gives the user the option of a full course in numerical trigonometry; however, this chapter is independent of the others and may be omitted.

2 Systems of Linear Equations (in Chapter 1) has been expanded to include 3×3 systems. Determinants are introduced along with Cramer's rule.

3 The entire set of exercises in Chapter 2 has been reconstructed offering the student more problems starting at a slow pace and gradually increasing in difficulty.

4 Table 6 of the Appendix was shortened making its use more convenient.

5 Over 400 new problems were added throughout the book.

This text has a wide range of topics making it very flexible for many different meaningful courses. We recommend that individual instructors select topics most beneficial for their students.

Offered here are four tried course suggestions:

1 One-semester (no review of algebra and geometry; no analytic geometry); Chapters: 2, 4, 5, 6, 7, 9, and 10.

2 One-semester (emphasis on graphing techniques and analytic geometry); Chapters: 1, 2, 3, 5, 6, 11, and 12.

3 One-semester (algebra and analytic geometry); Chapters: 1, 3, 4, 11, and 12; Sections 6.4 and 10.5.

4 One-semester (algebra and trigonometry); Chapters: 1, 2, 3, 4, 8, and 10.

We wish to take this opportunity to thank all those who have expressed an interest in our text.

Henry R. Korn
Albert W. Liberi

Preface to the First Edition

The purpose of this text is to prepare students for an introductory course in the calculus. The approach is a concrete development motivated through problem solving and relevant applications. The student is led into a classical approach to mathematics, where a minimum amount of notation is employed and the reliance on set theory is omitted.

It is the opinion of the authors that a student entering the calculus should have a strong working knowledge of algebraic techniques and should also be familiar with such concepts as functions, absolute value, analytic geometry, and inequalities. Graphing techniques are used extensively throughout the text. The student is encouraged to draw information from algebraic procedures and apply the results in obtaining an illustrative graph.

An ample prerequisite for this text should be two years of high school mathematics (algebra and geometry). However, it is evident to most teachers that students with even stronger backgrounds often lack the mathematical maturity to bridge the gap between high school mathematics and the calculus. For this reason, there is a need for a one- or two-semester period of adjustment.

The examples in the text have been carefully selected and completely analyzed. Students should be encouraged to work through the examples with pencil and paper. Much of the theory and many useful algebraic techniques are exploited in the examples. The comments which are included should prove quite useful to the student since they were generated from actual classroom discussions.

There are over 300 examples with step-by-step solutions. The text also contains nearly 300 illustrations and 2000 exercises. The exercises in each chapter are of a varying degree of difficulty, and were chosen to illustrate as many facets of the topic covered as possible. Repetition has been avoided and algebraic manipulation emphasized.

The text contains adequate material for a meaningful two-semester course. However courses of this type are usually one semester in length and it is our recommendation that individual instructors select the topics most beneficial for their student.

Some course suggestions are as follows:

1 one-semester (no review of algebra and geometry; no analytic geometry); Chapters: 2, 4, 5, 6, 7, 8, and 9.

2 one-semester (emphasis on graphing techniques and analytic geometry); Chapters: 1, 2, 3, 5, 6, 10, and 11.

3 one-semester (algebra and analytic geometry); Chapters: 1, 3, 4, Sections: 6.5, 9.5, and Chapters: 10, and 11.

The authors wish to thank their wives, Susan and Jackie, for their endless hours of typing and constructive criticism, without which the manuscript would have certainly been disorganized and unfinished. We would also like to express special appreciation to Ms. Shelly Levine Langman, McGraw-Hill editing supervisor, not only for her enthusiasm and her professional abilities, but also for the extra attention she gave to us and the detailed analyses which transformed our manuscript into this text. We owe a debt of gratitude to Mr. Jack Farnsworth, McGraw-Hill editor, who gave us the incentive for completing this project by demonstrating his faith in our unproven abilities.

Finally, we would also like to extend our thanks to our students and fellow faculty members at Westchester Community College, who struggled through the original version of the text for three semesters.

HENRY R. KORN
ALBERT W. LIBERI

An Elementary Approach to Functions

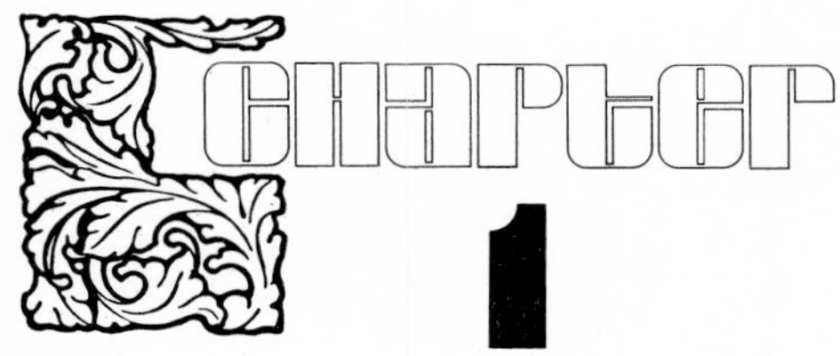

Chapter 1 The Straight Line

1.1 A REVIEW OF THE RECTANGULAR COORDINATE SYSTEM

In order to locate a point in a plane, we follow the *rectangular coordinate method*. In this system a plane is divided into four regions, called *quadrants*. The system contains two straight lines, called *axes*, one of which is horizontal and the other vertical. The horizontal line, which extends indefinitely to the left and right, is called the *x axis*. The vertical line, which extends indefinitely up and down (and intersects the horizontal line at right angles), is called the *y axis*. The point of intersection of the two lines is called the *origin* (see Diagram 1).

Although it is not necessary that the two axes be mutually perpendicular, we will use this frame of reference throughout this development. The four quadrants into which the axes divide the plane are conventionally labeled I, II, III, and IV (counterclockwise convention). Along the x axis, positive and negative units are to the right and left of the origin, respectively. Likewise, along the y axis, positive and negative units are above and below the origin, respectively (see Diagram 2).

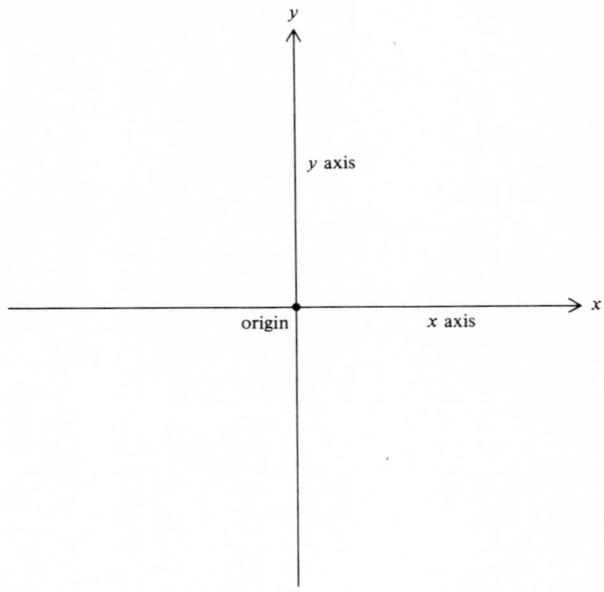

Diagram 1

To locate a point in this plane, we can choose two numbers. The first number will always indicate units to the right or left of the origin. The second number will always indicate units above or below the origin. The order in which we select these numbers is important in locating a point, and we write these numbers in parentheses. The first entry is called the *first coordinate,* and

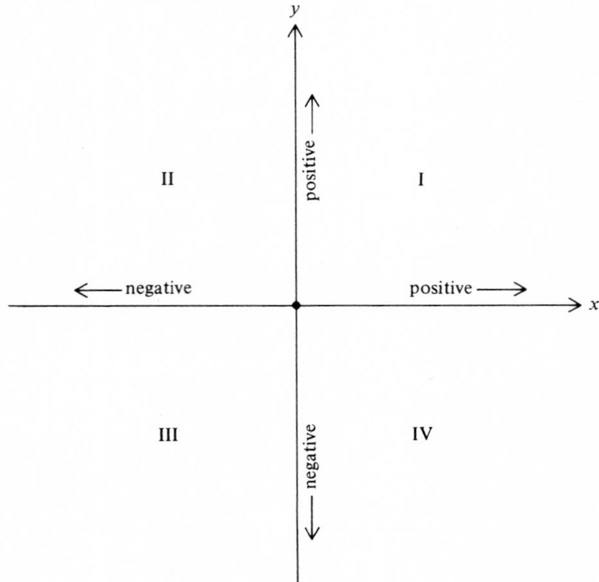

Diagram 2

the second entry is called the *second coordinate* of the point. The arrangement is called an *ordered pair.*

Example 1 (3, 1).
Here 3 is the first coordinate, and 1 is the second coordinate.

Example 2 (1, 3).
Here 1 is the first coordinate, and 3 is the second coordinate.

Let us locate these two points in Diagram 3. Note that each point in the plane has *two* coordinates and the order in which they are taken is important.

The first coordinate is referred to as the *abscissa* of the point, and the second coordinate is referred to as the *ordinate* of the point.

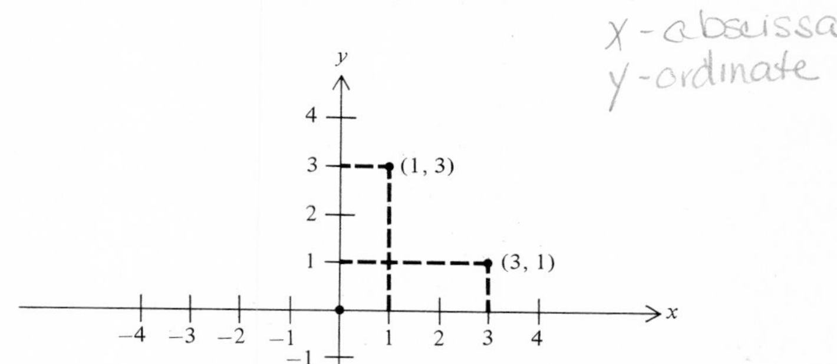

Diagram 3

DEFINITION 1 *Equality of ordered pairs* $(a, b) = (c, d)$ if and only if $a = c$ and $b = d$.

Example 3 Find a and b, given $(2, b) = (a + 1, 2)$.
From the definition for equality of ordered pairs $2 = a + 1$ and $b = 2$. Therefore, $a = 1$ and $b = 2$.

From Diagram 4, observe that the points P_1 through P_6 have associated with them the ordered pairs $P_1(2, 0)$, $P_2(0, -4)$, $P_3(4, 3)$, $P_4(-5, 4)$, $P_5(-3, -1)$, and $P_6(3, -2)$.

COMMENT It will be helpful to verify that the coordinates of points located in the different quadrants have the following signs:

Quadrant I:	(+, +)
Quadrant II:	(−, +)
Quadrant III:	(−, −)
Quadrant IV:	(+, −)

Points that lie on the x axis have the second coordinate zero, $(a, 0)$. See P_1 in Diagram 4. Points that lie on the y axis have the first coordinate zero, $(0, b)$. See P_2 in Diagram 4. There is no rule which demands using the same scale for both the x and y axes. A scale should be chosen which will be most adaptable.

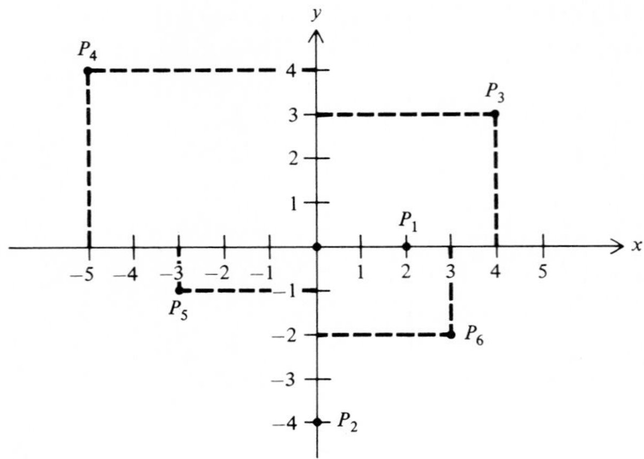

Diagram 4

Suppose that the point $P(-100, 1)$ is to be plotted. Note, in Diagram 5, that the scale of the x axis has been compressed to save space. In such cases the diagram should always be accurately labeled.

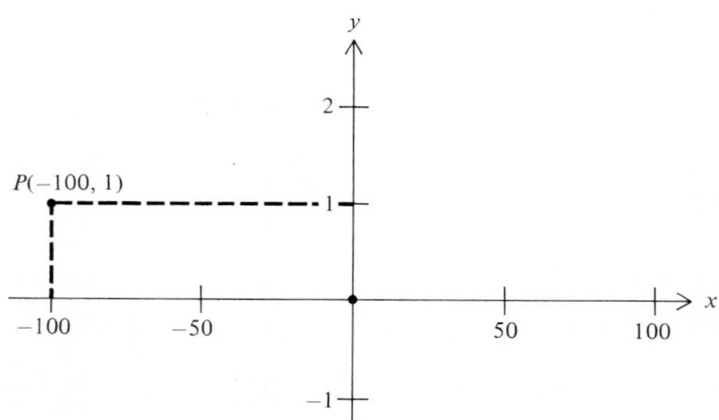

Diagram 5

1.2 THE SYMBOL Δ AND ABSOLUTE VALUE

If we select an *initial point* $A(x_1, y_1)$ and travel to a new point $B(x_2, y_2)$, which we call the *terminal point,* the value of the abscissa will change from x_1 to x_2 and the value of the ordinate will change from y_1 to y_2.

A symbol is used to represent such changes: Δx (read "delta x," not delta times x) and Δy (read "delta y," not delta times y).

$$\Delta x = x \text{ of the terminal point } - x \text{ of the initial point}$$

and

$$\Delta y = y \text{ of the terminal point } - y \text{ of initial point}$$

DEFINITION 1 If a particle moves from an initial point $A(x_1, y_1)$ to a terminal point $B(x_2, y_2)$, the *increments* Δx and Δy are given by

$$\Delta x = x_2 - x_1 \qquad \text{and} \qquad \Delta y = y_2 - y_1$$

COMMENT It follows from the above definition that both Δx and Δy can be positive, negative, or zero.

Example 1 Consider a particle moving from $A(-1, 4)$ to $B(2, -3)$ (see Diagram 6). Find Δx and Δy.

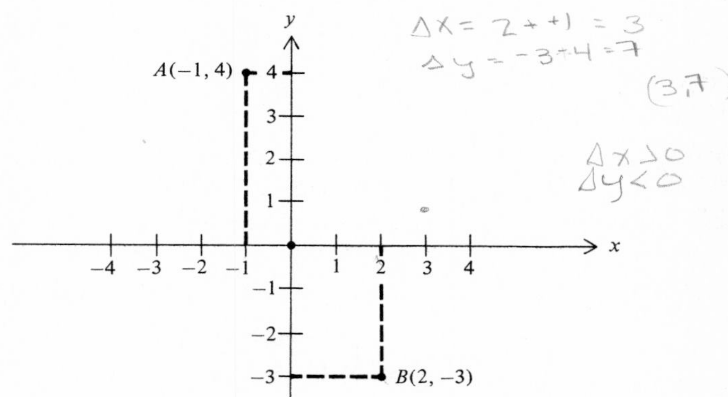

Diagram 6

Solution

Using Definition 1

$$\Delta x = 2 - (-1) = 3 \qquad \Delta y = -3 - (4) = -7$$

Note that x increases from A to B. Hence, $\Delta x > 0$. Since y decreases from A to B, $\Delta y < 0$.

Example 2 Consider a particle moving from $C(3, 2)$ to $D(-2, 2)$ (see Diagram 7). Find Δx and Δy.

Solution

$$\Delta x = (-2) - (3) = -5 \qquad \Delta y = (2) - (2) = 0$$

Example 3 Consider a particle moving from $E(-1, -2)$ to $F(-1, 0)$ (see Diagram 7). Find Δx and Δy.

Solution

$$\Delta x = (-1) - (-1) = 0 \qquad \Delta y = 0 - (-2) = 2$$

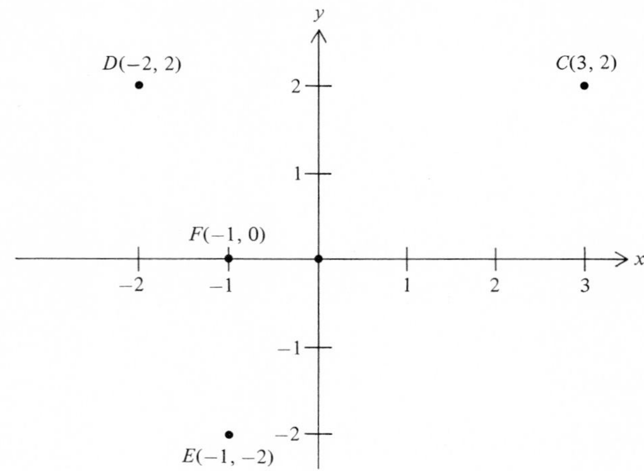

Diagram 7

We now discuss the concept of distance. In order to do this, we will define absolute value.

DEFINITION 2 The *absolute value* of x is denoted by $|x|$, where

$$|x| = \begin{cases} x & \text{if } x > 0 \\ 0 & \text{if } x = 0 \\ -x & \text{if } x < 0 \end{cases}$$

Example 4

$$|3| = 3$$
$$|0| = 0$$
$$|-3| = -(-3) = 3$$

Note that the absolute value of a negative number is never a negative number and, in general, $|x| = |-x|$.

This leads to the following theorem.

Theorem 1 $|x_2 - x_1| = |x_1 - x_2|$.

The proof of this theorem will be considered in Chapter 4. Let us look at a few examples to see what is implied by this theorem.

Example 5 If $x_1 = 4$ and $x_2 = 3$, then

$$|3 - 4| = |-1| = -(-1) = 1 \qquad |4 - 3| = |1| = 1$$

Example 6 If $x_1 = -2$ and $x_2 = 6$, then

$$|-2 - 6| = |-8| = -(-8) = 8 \qquad |6 - (-2)| = |8| = 8$$

Example 7 If $x_1 = -4$ and $x_2 = -1$, then

$$|-1 - (-4)| = |-1 + 4| = |3| = 3$$
$$|-4 - (-1)| = |-4 + 1| = |-3| = -(-3) = 3$$

DEFINITION 3 The *distance* between two points on a line when the ordinates are equal, that is, a line parallel to the x axis, is given by

$$|\Delta x| = |x_1 - x_2|$$

DEFINITION 4 The *distance* between two points on a line when the abscissas are equal, that is, a line parallel to the y axis, is given by

$$|\Delta y| = |y_1 - y_2|$$

Example 8 Find the distance from $A(-3, 2)$ to $B(2, 2)$ using Definition 3 since the ordinates are equal.

Solution $$|\Delta x| = |-3 - 2| = |-5| = 5$$

Example 9 Find the distance from $C(2, 4)$ to $D(2, -4)$ using Definition 4 since the abscissas are equal.

Solution $$|\Delta y| = |4 - (-4)| = |8| = 8$$

(See Diagram 8.)

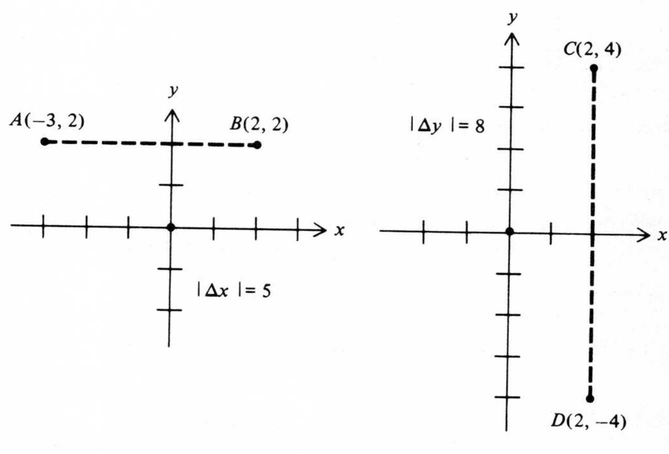

Diagram 8

COMMENT The distance between two points parallel to an axis is the absolute value of the difference of the unequal coordinates. Therefore, the order in which this difference is computed does not matter since

$$AB = |-3 - 2| = |2 - (-3)| = 5 \qquad CD = |4 - (-4)| = |-4 - 4| = 8$$

This means that distance is a nonnegative number.

EXERCISES

1 Locate the following points in the coordinate plane:
(a) $(6, 0)$ **(b)** $(-6, 0)$ **(c)** $(0, -6)$ **(d)** $(0, 6)$
(e) $(4, -3)$ **(f)** $(-3, -2)$ **(g)** $(-2, 5)$ **(h)** $(\frac{1}{2}, 4)$
(i) $(-\frac{3}{4}, \frac{5}{2})$ **(j)** $(500, 2)$ **(k)** $(-\frac{7}{3}, 1000)$ **(l)** $(\pi, -\pi)$
(m) $(2, -3)$ **(n)** $(1.8, -6.3)$ **(o)** $(0, 0)$

2 Solve for a and b:
(a) $(a + 3, b - 2) = (4, -3)$ **(b)** $(6, 3 - b) = (a + \frac{1}{2}, -2)$
(c) $\left(\frac{1}{a}, \frac{1}{b}\right) = (4, 3)$ **(d)** $\left(\frac{1}{a + b}, \frac{b}{1 - b}\right) = (1, 3)$
(e) $(\sqrt{a}, 4) = (2, 2\sqrt{b})$

3 **(a)** A particle moves from $A(-4, -5)$ to $B(-4, 6)$. Determine Δx and Δy.
(b) A particle moves from $A(-4, 6)$ to $B(-4, -5)$. Determine Δx and Δy.

4 **(a)** A particle moves from $A(-4, 2)$ to $B(2, 2)$. Determine Δx and Δy.
(b) A particle moves from $A(2, 2)$ to $B(-4, 2)$. Determine Δx and Δy.

5 Find $|\Delta x|$ and $|\Delta y|$ for Exercise 3.

6 Find $|\Delta x|$ and $|\Delta y|$ for Exercise 4.

7 Verify Theorem 1 for the following values of x_1 and x_2:

x_1	-3	4	-17	2
x_2	5	$-\frac{2}{3}$	-1	2

8 Find the distance between the following points:
(a) $(3, -8)$ and $(16, -8)$ **(b)** $(-\frac{5}{2}, \frac{1}{2})$ and $(-\frac{5}{2}, 3)$
(c) $(2.5, 3.76)$ and $(-2.5, 3.76)$ **(d)** $(-4, -\sqrt{2})$ and $(-3.08, -\sqrt{2})$
(e) $(2a, b)$ and $(2a, 6 - b)$

1.3
DISTANCE BETWEEN POINTS

So far we only have been finding the distance between two points parallel to one of the axes. Let us now find a general formula for the distance d between *any* two points in the coordinate plane.

Consider the points $A(x_1, y_1)$ and $B(x_2, y_2)$. Construct a right triangle with line segment AB as the hypotenuse (see Diagram 9).

PLAN In order to construct a right triangle with line segment AB as hypotenuse, through B, drop a perpendicular to the x axis and through A, drop a perpendicular to the y axis. Let C be the intersection of the two perpendiculars. [Do you see that the coordinates of C are (x_2, y_1)?]

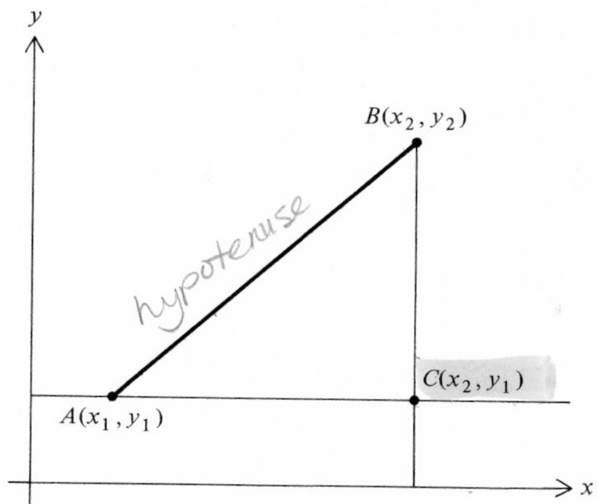

Diagram 9

Find the distance from A to C and from B to C. (These will be the legs of the right triangle.)

Use the *pythagorean theorem* to find the distance from A to B.

$$(AB)^2 = (AC)^2 + (BC)^2$$
$$AC = |x_2 - x_1| \qquad BC = |y_2 - y_1|$$

(See Diagram 10.) Then

$$(AB)^2 = |x_2 - x_1|^2 + |y_2 - y_1|^2$$

Note that $\quad |x_2 - x_1|^2 = (x_2 - x_1)^2 \quad$ and $\quad |y_2 - y_1|^2 = (y_2 - y_1)^2$

(See Property 7*b*, page 101.) Then

$$(AB)^2 = (x_2 - x_1)^2 + (y_2 - y_1)^2 \qquad \text{and} \qquad AB = \sqrt{(x_2 - x_1)^2 + (y_2 - y_1)^2}$$

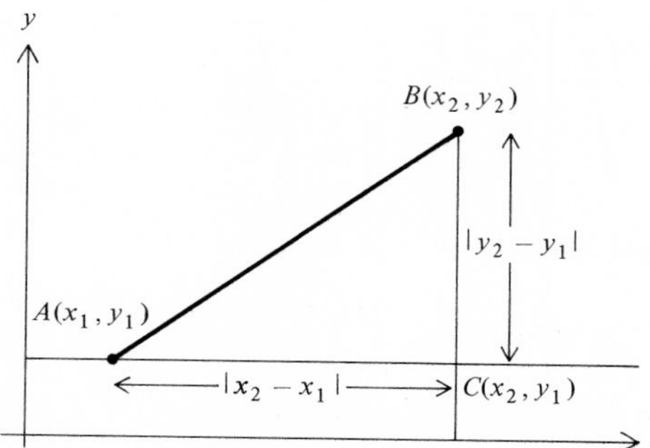

Diagram 10

This result can be summarized in the following theorem.

Theorem 1 The distance d between two points $A(x_1, y_1)$ and $B(x_2, y_2)$ in the coordinate plane is given by the formula

$$d = \sqrt{(x_2 - x_1)^2 + (y_2 - y_1)^2}$$

Example 1 Find the distance between $A(-1, -1)$ and $B(2, 3)$.

Solution

$$d = \sqrt{[2 - (-1)]^2 + [3 - (-1)]^2} \qquad = \sqrt{3^2 + 4^2}$$
$$= \sqrt{9 + 16} \qquad = \sqrt{25} \qquad = 5$$

If the order is reversed,

$$d = \sqrt{(-1 - 2)^2 + (-1 - 3)^2} \qquad = \sqrt{(-3)^2 + (-4)^2}$$
$$= \sqrt{9 + 16} \qquad = \sqrt{25} \qquad = 5$$

Example 2 Given that the distance from a point $A(-1, 4)$ to a point B whose ordinate is -8 is 13; find the abscissa of B (see Diagram 11).

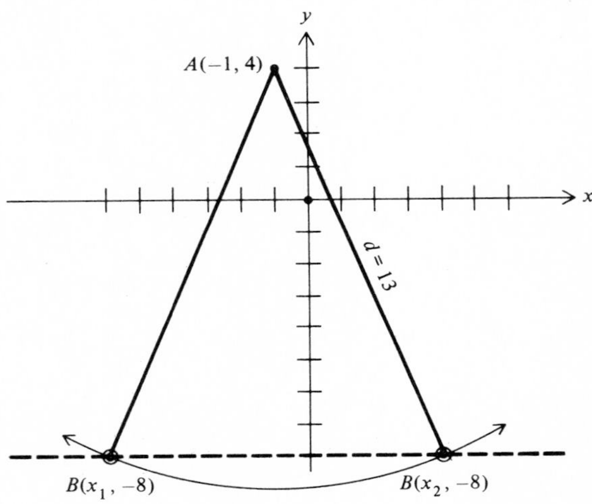

Diagram 11

Solution Draw a line 8 units below and parallel to the x axis and consider the line segment AB as though it were a pendulum arm. It becomes evident that the tip of this arm cuts the line at two positions. Hence, we must expect two solutions for the abscissa. Using the distance formula and letting x be the unknown abscissa, we obtain

$13 = \sqrt{[x - (-1)]^2 + (-8, -4)^2}$	
$= \sqrt{(x + 1)^2 + (-12)^2}$	Simplifying parentheses
$169 = (x + 1)^2 + 144$	Squaring both sides
$169 = x^2 + 2x + 145$	Squaring $x + 1$
$0 = x^2 + 2x - 24$	Subtracting 169 from both sides
$0 = (x + 6)(x - 4)$	Factoring

This implies that

$$x + 6 = 0 \quad \text{or} \quad x - 4 = 0$$
$$x = -6 \qquad\qquad x = 4$$

(Because if $ab = 0$, then $a = 0$ or $b = 0$.) Since $(x_1, -8)$ is the point in the third quadrant,

$$x_1 = -6$$

And, since $(x_2, -8)$ is the point in the fourth quadrant,

$$x_2 = 4$$

Example 3 Let us now use the distance formula to prove that

$$M\left(\frac{x_1 + x_2}{2}, \frac{y_1 + y_2}{2}\right)$$

is the midpoint of the line segment joining $A(x_1, y_1)$ and $B(x_2, y_2)$ (see Diagram 12).

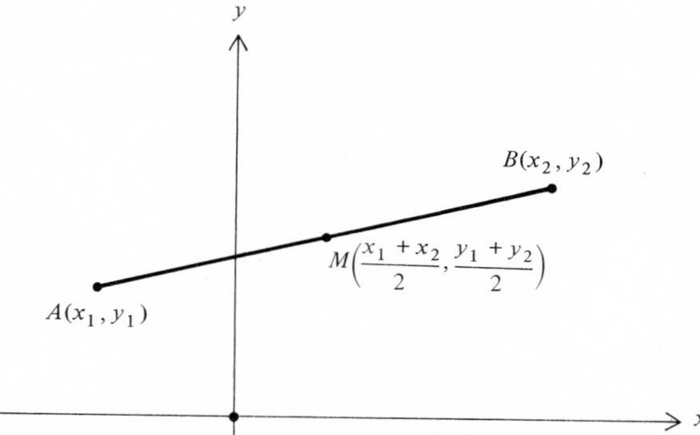

Diagram 12

Proof Our plan is to show that the distance from A to M is one-half the distance from A to B; that is,

$$AM = \tfrac{1}{2}(AB)$$

Show that the distance from B to M is also one-half the distance from A to B; that is,

$$BM = \tfrac{1}{2}(AB)$$

If we follow this plan, M will be on the segment containing A and B, since

$$AM + BM = \tfrac{1}{2}(AB) + \tfrac{1}{2}(AB) = AB$$

Find the distance from A to M.

$$AM = \sqrt{\left(\frac{x_1 + x_2}{2} - x_1\right)^2 + \left(\frac{y_1 + y_2}{2} - y_1\right)^2}$$

$$= \sqrt{\left(\frac{x_2 - x_1}{2}\right)^2 + \left(\frac{y_2 - y_1}{2}\right)^2} \qquad \text{Simplifying the parentheses}$$

$$= \sqrt{\frac{(x_2 - x_1)^2}{4} + \frac{(y_2 - y_1)^2}{4}} \qquad \text{Squaring denominators}$$

$$= \tfrac{1}{2}\sqrt{(x_2 - x_1)^2 + (y_2 - y_1)^2} \qquad \text{Factoring out the square root of 4 in the denominator}$$

Since $$AB = \sqrt{(x_2 - x_1)^2 + (y_2 - y_1)^2}$$

it follows that $$AM = \tfrac{1}{2}(AB)$$

In order to complete the proof, it must now be shown that

$$BM = \tfrac{1}{2}(AB)$$

We leave this demonstration to you.

Example 4 Find the midpoint of the line segment whose endpoints are $A(8, -2)$ and $B(-4, 3)$.

Solution

$$M = \left(\frac{8 + (-4)}{2}, \frac{(-2) + 3}{2}\right) = (2, \tfrac{1}{2})$$ From the midpoint formula

Example 5 If M is the midpoint of line segment AB, and if the coordinates of M are (2, 3) and the coordinates of B are $(-1, 2)$, find the coordinates of A.

Solution

$$2 = \frac{-1 + x}{2} \qquad 3 = \frac{2 + y}{2}$$

$$4 = -1 + x \qquad 6 = 2 + y$$

$$5 = x \qquad 4 = y$$

Therefore, the coordinates of A are (5, 4).

EXERCISES

Find the distance between the given points in Exercises 1 to 9.

1 $(-2, -3)$ and $(4, 5)$

2 $(3, -4)$ and $(6, -1)$

3 $(5, 4)$ and $(8, -2)$

4 $(\frac{2}{3}, -\frac{3}{2})$ and $(-\frac{1}{2}, \frac{2}{3})$

5 $(\sqrt{3}, \sqrt{2})$ and $(\sqrt{2}, -\sqrt{3})$

6 $(\sqrt{2}, \sqrt{8})$ and $(\sqrt{50}, 0)$

7 $(1 + x^2, -x)$ and $(1 - x^2, x)$

8 $(1, \pi)$ and $(-\pi, 1)$

9 (a, b) and $(-2a, 3b)$

Given the end points of a line segment, find the midpoint of each segment.

10 $(2, 4)$ and $(3, 6)$

11 $(-3, 4)$ and $(4, -4)$

12 $(-7, 12)$ and $(3, -8)$

13 $(0, -3)$ and $(-1, -1)$

14 $(\frac{2}{3}, -\frac{3}{2})$ and $(-\frac{1}{4}, -\frac{4}{3})$

15 $(1.8, -0.6)$ and $(-2.4, 0.5)$

16 (a, b) and $(-2a, 3b)$

17 $(a + 2b, 2a - b)$ and $(b - 3a, -3b)$

18 $(\sqrt{2}, \sqrt{8})$ and $(-\sqrt{50}, 0)$

19 Given a triangle joining the points $(0, -2)$, $(4, -6)$, and $(\frac{3}{2}, -\frac{9}{2})$. Show that the triangle is isosceles but not equilateral.

In Exercises 20 and 21, if P is the midpoint of QR, determine point R. For:

20 $P(3, 5)$ and $Q(-5, -3)$

21 $P(1, -3)$ and $Q(-2, 1)$

22 Show that $(-3, -4)$ is equidistant from $(4, -8)$ and $(-2, 4)$.

23 Into what ratio does $(-2, 3)$ divide the segment joining $(-5, 1)$ and $(4, 7)$?

24 Use the pythagorean theorem to show that the triangle with vertices $(-1, 4)$, $(2, -1)$, and $(-3, -4)$ is a right triangle.

25 Given that the distance from a point $A(-2, 2)$ to a point B, whose abscissa is 13, is 17, find the ordinate of B.

26 The coordinates of the point $C(x, y)$ which divide the line segment joining $A(x_1 y_1)$ and $B(x_2 y_2)$ into the ratio R/S are:

$$x = \frac{Sx_1 + Rx_2}{R + S} \qquad \text{and} \qquad y = \frac{Sy_1 + Ry_2}{R + S}$$

Using the above information, find the trisection points of the line segment joining the points $(-4, 2)$ and $(2, -2)$. *Hint:* A trisection point divides a segment into the ratio 1:2 (or 2:1).

27 Using the theorem in Exercise 26, find the coordinates of C on line segment AB, where $A(-1, -2)$ and $B(2, 2)$ such that C divides the segment into the ratio **(a)** 1:1 (midpoint of segment); **(b)** 1:3.

1.4 THE SLOPE OF A LINE

Two distinct points determine a straight line. In order to describe the steepness, or *slope,* of a line, let $A(x_1, y_1)$ and $B(x_2, y_2)$ be two points on the line (see Diagram 13). Recall from Section 1.2 that in moving from A to B,

$$\Delta x = x_2 - x_1 \qquad \text{and} \qquad \Delta y = y_2 - y_1$$

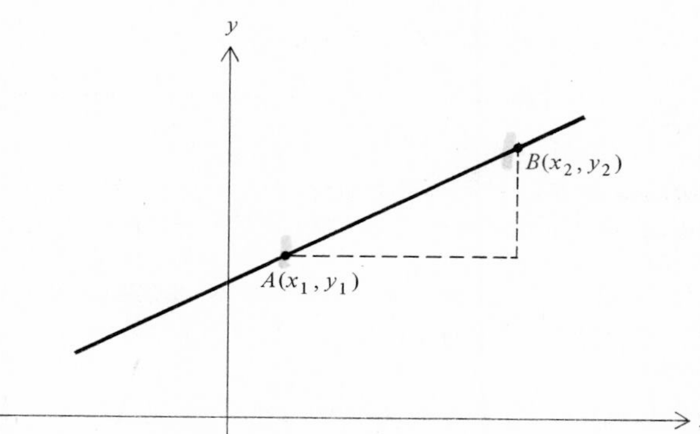

Diagram 13

DEFINITION 1 If (x_1, y_1) and (x_2, y_2) are two distinct points on a line not parallel to the y axis, then the *slope m* of the line is the ratio of Δy to Δx.

$$m = \frac{\Delta y}{\Delta x} = \frac{y_2 - y_1}{x_2 - x_1} \text{ or } \frac{y_1 - y_2}{x_1 - x_2}$$

Example 1 Find the slope of the line containing the points $(-2, -1)$ and $(0, 3)$.

Solution

$$m = \frac{3 - (-1)}{0 - (-2)} = \frac{4}{2} = 2 \qquad \text{Using Definition 1}$$

or, by reversing the initial point and the terminal point,

$$m = \frac{-1 - 3}{-2 - 0} = \frac{-4}{-2} = 2$$

Suppose $(-1, 1)$ is also a point on the line. If we now choose $(-1, 1)$ and $(0, 3)$ to compute the slope, then

$$m = \frac{3 - 1}{0 - (-1)} = \frac{2}{1} = 2$$

(See Diagram 14.) Why were lines parallel to the y axis excluded in the definition?

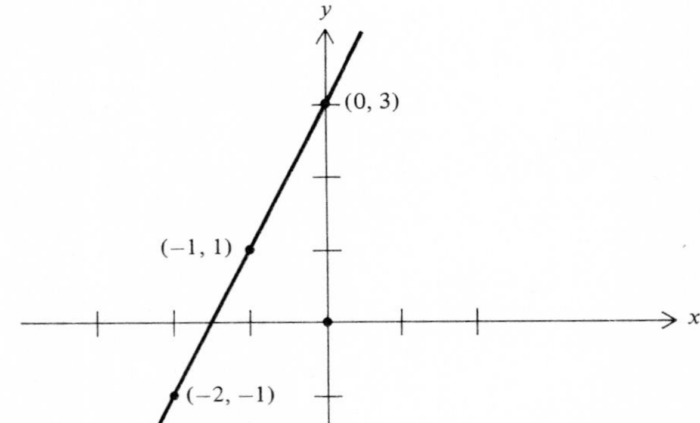

Diagram 14

On all vertical lines, the abscissas are equal ($x_1 = x_2$), and since $\Delta x = x_2 - x_1$, then $\Delta x = 0$ (see Diagram 15).

According to the definition of the slope of a line,

$$m = \frac{y_2 - y_1}{x_2 - x_1} = \frac{\Delta y}{\Delta x} = \frac{\Delta y}{0}$$

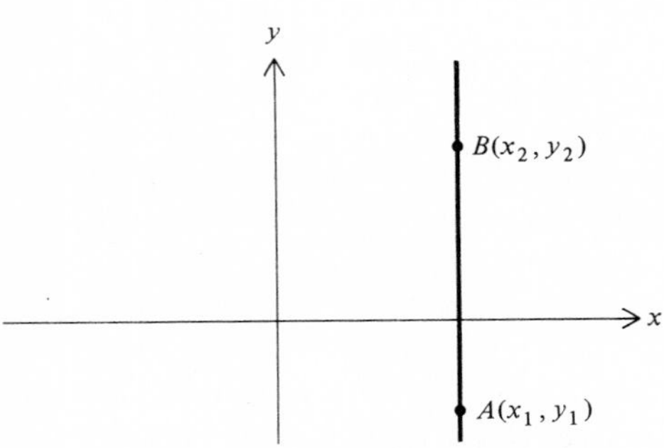

Diagram 15

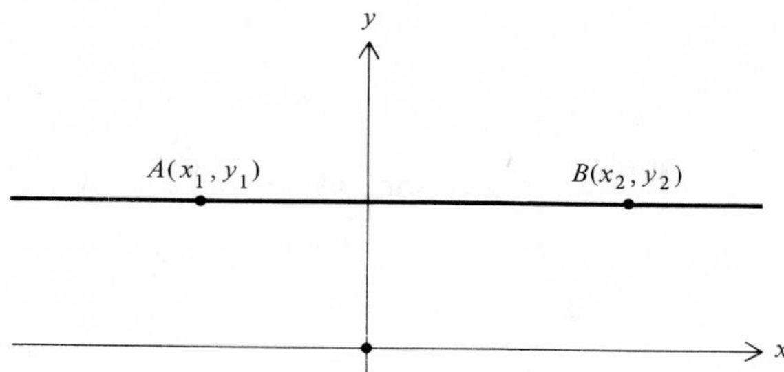

Diagram 16

Since division by 0 is undefined, the slope of a vertical line is also undefined. (It has *no slope*.)

COMMENT Each line has one and only one slope. Therefore, any two distinct points on the line can be used to determine its slope.

If a line is horizontal (see Diagram 16), the ordinates are equal ($y_1 = y_2$), and since $\Delta y = y_2 - y_1$, $\Delta y = 0$. According to the definition of the slope of a line,

$$m = \frac{\Delta y}{\Delta x} = \frac{0}{\Delta x} = 0$$

The slope of a horizontal line is zero.

The slope m may be positive, negative, or zero. The sign of m tells us whether the line rises or falls as we go from left to right.

Δy	Δx	$m = \frac{\Delta y}{\Delta x}$
+	+	+
+	−	−
−	+	−
−	−	+

When m is positive (+), the line rises (see Diagram 17*b*). When m is negative (−), the line falls (*from left to right*) (see Diagram 17*a*).

Example 2 Graph a line having a slope of $-\frac{3}{2}$ and passing through the point $P(4, -2)$.

Since the slope is negative, the line falls from left to right. The slope, $-\frac{3}{2}$, can also be interpreted as $3/-2$.

By referring to Diagram 18, $\Delta y = 3$ and $\Delta x = -2$, or $\Delta y = -3$ and $\Delta x = 2$.

COMMENT One fact which should be emphasized in closing this section is that a line with *zero* slope and a line with *no* slope are different.

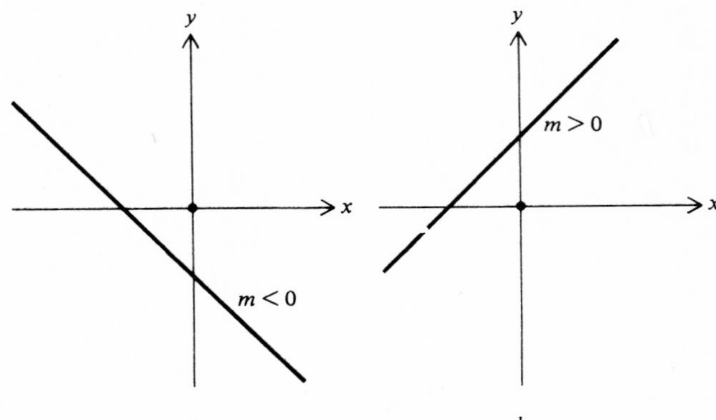

Diagram 17

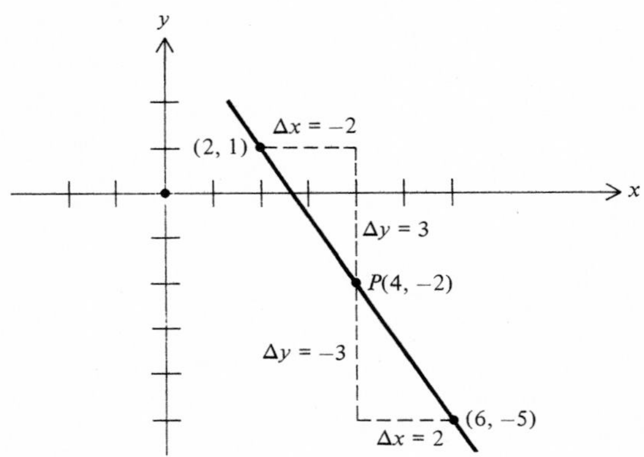

Diagram 18

EXERCISES

In Exercises 1 to 9, find the slope of the line passing through the given points.

1 $(4, 6)$ and $(6, 10)$

2 $(2, 3)$ and $(-1, 6)$

3 $(-4, 5)$ and $(-6, 3)$

4 $(2, -3)$ and $(-4, 3)$

5 $(5, 2)$ and $(-3, -4)$

6 $(2, -1)$ and $(6, -2)$

7 $(\frac{1}{2}, \frac{1}{2})$ and $(\frac{1}{3}, \frac{1}{4})$

8 $(\frac{2}{3}, -\frac{1}{2})$ and $(-\frac{3}{4}, \frac{1}{3})$

9 $(4, 6)$, $(4, 3)$, and $(4, 7)$

10 Let $A(2, 4)$, $B(-3, 2)$, and $C(-1, -1)$ be the vertices of triangle ABC. Show that the slope of the line BC is equal to the slope of the line segment containing the midpoints of AB and AC.

In Exercises 11 to 15 determine by means of *slope* whether the given points are on the same straight line.

11 $(-1, 3)$, $(4, -7)$, and $(0, 1)$

12 $(3, 2)$, $(4, 3)$, and $(-6, -12)$

13 $(1, -\frac{9}{2})$, $(-2, 3)$, and $(-\frac{1}{2}, -\frac{3}{4})$

14 $(-1, -2)$, $(1, 1)$, $(3, 4)$, and $(0, -\frac{1}{2})$

15 $(-4, 3)$, $(2, \frac{3}{2})$, $(-2, 2)$, and $(1, \frac{5}{4})$

16 Graph a line that has a slope of $\frac{3}{4}$ and passes through the point $P(-3, -2)$.
17 Graph a line that has a slope of $-\frac{3}{4}$ and passes through the point $(-3, -2)$.

1.5 EQUATION OF A LINE

Let us assume that each ordered pair of numbers (x, y) corresponds to a unique point in the plane, and conversely, that to each point in the plane there corresponds a unique ordered pair of numbers (x, y). Suppose we examine only those pairs whose coordinates satisfy a given relationship between x and y.

For example, if $x = y$, we would be interested only in those ordered pairs whose abscissa and ordinate are equal, such as (1, 1), (2, 2), (−4, −4), ... If $y = 2x$, we would be interested in ordered pairs satisfying this relationship, such as (1, 2), (2, 4), (−2, −4), ...

This relationship between the two variables x and y is called the *equation of the line.* The set of all ordered pairs that satisfy an equation of the line is called the *graph of the equation.*

Example 1 Find an equation of a line containing the points $A(2, -3)$ and $B(-3, 7)$.

The slope of the line through A and B is

$$m = \frac{7 - (-3)}{-3 - 2} = \frac{10}{-5} = -2$$

If $C(x, y)$ is any other point on this line (see Diagram 19), then the slope of line segment AC also equals -2.

$$m = \frac{y - (-3)}{x - 2} = -2 \qquad \text{and} \qquad \frac{y + 3}{x - 2} = -2$$

$$\begin{aligned} y + 3 &= -2(x - 2) && \text{Multiplying both sides by } (x - 2) \\ &= -2x + 4 \\ y &= -2x + 1 && \text{Solving for } y \end{aligned}$$

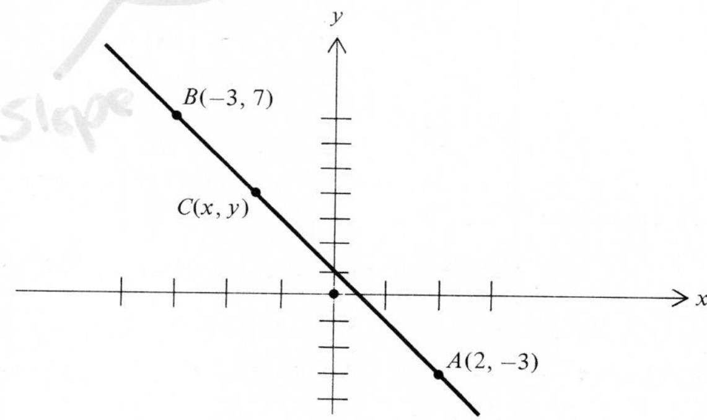

Diagram 19

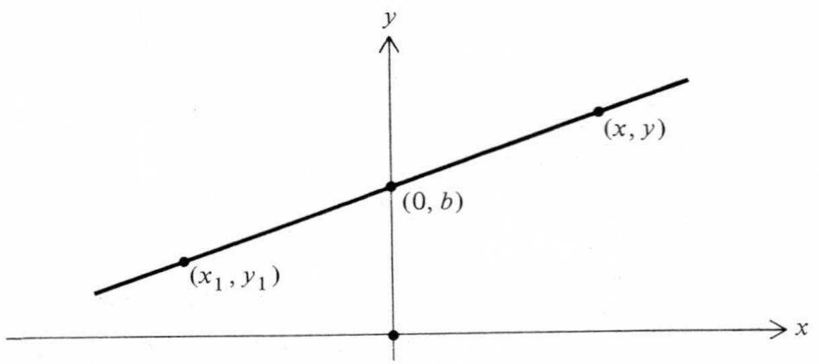

Diagram 20

This is an equation of the line passing through the two points $A(2, -3)$ and $B(-3, 7)$. *All* points on this line must satisfy this equation.

We now develop some procedures for finding the equation of a line under various conditions.

Since nonvertical lines intersect the y axis in exactly one point (see Exercise 25), select a line which crosses the y axis at the point $(0, b)$ (see Diagram 20).

The ordinate b is called the y intercept of the line. If (x, y) is any other point on the line, then using the definition of slope,

$$m = \frac{y - b}{x - 0} \qquad \text{or} \qquad m = \frac{y - b}{x}$$

(Note that since the y intercept is unique, the value of x for the other point cannot be zero.)

Multiplying both sides of the equation by x,

$$mx = y - b \qquad \text{or} \qquad y = \underset{\substack{\uparrow \\ \text{slope}}}{m}x + \underset{\substack{\uparrow \\ y \text{ intercept}}}{b}$$

This is the equation of a line whose slope is m and whose y intercept is b. It is called the *slope-intercept form.*

Referring to Diagram 20 again, suppose (x_1, y_1) is also a point on the line. Using the definition of slope,

$$\frac{y - y_1}{x - x_1} = m$$

$$y - y_1 = m(x - x_1) \qquad \text{Multiplying both sides by } (x - x_1)$$

This is the equation of a line which has slope m and contains the point (x_1, y_1). It is called the *point-slope form.*

Example 2 Using each of the two forms developed above, find the equation of the line passing through the points $(0, -1)$ and $(2, 2)$ (see Diagram 21).

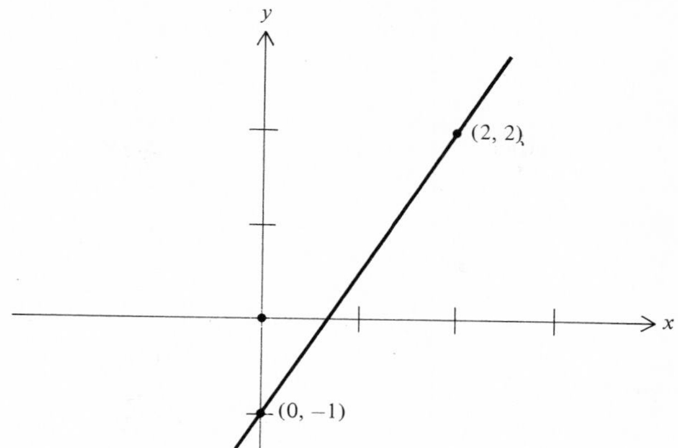

Diagram 21

SLOPE-INTERCEPT FORM Find the slope m.

$$m = \frac{2 - (-1)}{2 - 0} = \frac{3}{2}$$

Find the y intercept, b.

$$(0, -1) = (0, b) \Longrightarrow b = -1$$

$$y = \tfrac{3}{2}x - 1 \qquad \text{Slope-intercept form}$$

POINT-SLOPE FORM We already know the slope is $\frac{3}{2}$. Using either point $(0, -1)$ or $(2, 2)$, we have

$$y - (-1) = \tfrac{3}{2}(x - 0) \qquad \text{or} \qquad y - (2) = \tfrac{3}{2}(x - 2)$$

Note that $y = \frac{3}{2}x - 1$, $y + 1 = \frac{3}{2}x$, and $y - 2 = \frac{3}{2}(x - 2)$ are all equations of the same line.

Example 3 Find the equation of the line, L_1, parallel to the y axis and passing through the point (a, b).

Since every point on line L_1 has the same x coordinate, a, it is clear from Diagram 22 that points whose x coordinate do not equal a cannot be on line L_1.

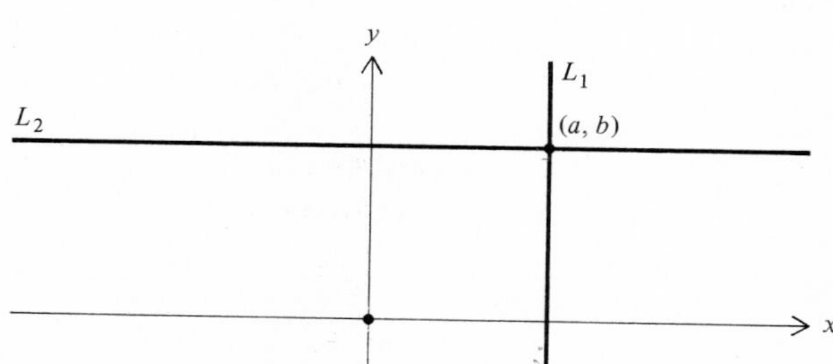

Diagram 22

Hence, $x = a$ is the description of *all* the points on line L_1. It should be noted that this line has *no* slope.

Example 4 Find the equation of the line, L_2, parallel to the x axis and passing through the point (a, b).

Since every point on line L_2 has the same y coordinate, b, it is clear from Diagram 22 that points whose y coordinate do not equal b cannot be on line L_2.

Hence, $y = b$ is the description of *all* the points on line L_2. It should be noted that this line has a *zero* slope.

It should be noted that a line can also be expressed by the equation $Px + Qy + R = 0$, where P, Q, and R are constants (P and Q are not both equal to zero).

Example 5 A line passes through the points $A(2, 3)$ and $B(-3, -1)$. Express an equation of the line in the form $Px + Qy + R = 0$.

Solution

$$m = \frac{3 - (-1)}{2 - (-3)} = \frac{4}{5} \qquad \text{Finding slope}$$

$$y - 3 = \tfrac{4}{5}(x - 2) \qquad \text{Point-slope form using point (2, 3)}$$

$$5y - 15 = 4x - 8 \qquad \text{Multiplying both sides by 5}$$

$$-4x + 5y - 7 = 0 \qquad \text{Expressing in the form } Px + Qy + R = 0$$

where $P = -4$, $Q = 5$, and $R = -7$.

SUMMARY

There are several forms for an equation of a line. Each form gives certain information about the line.

1 If line L is the y axis or parallel to the y axis, then

$$x = a \qquad \text{line parallel to the y axis}$$

2 If line L is the x axis or parallel to the x axis, then

$$y = b \qquad \text{line parallel to the x axis}$$

3 If (x_1, y_1) is a point on line L and m is the slope of the line, then

$$y - y_1 = m(x - x_1) \qquad \text{point-slope form}$$

4 If m is the slope of line L and b is the y intercept, then

$$y = mx + b \qquad \text{slope-intercept form}$$

5 If P, Q, and R are constants, and both P and Q are not zero, the general equation of any straight line

$$Px + Qy + R = 0 \qquad \text{general equation}$$

EXERCISES

In Exercises 1 to 12 find an equation of the line:

1 Through $(-2, 3)$ with slope 3
2 Through $(\frac{2}{3}, -\frac{1}{2})$ with slope -2
3 Through $(2, -4)$ with slope $-\frac{3}{2}$
4 Through $(2, -2\sqrt{2})$ with slope $\sqrt{2}$
5 With slope -4 and y intercept 2
6 With slope $\frac{1}{2}$ and x intercept -4
7 x intercept -2 and y intercept 3
8 Through $(\frac{3}{2}, -\frac{1}{2})$ and $(-\frac{2}{3}, 2)$
9 Through $(0, 32)$ and $(100, 212)$
10 Through $(2, -3)$ and $(4, 2)$
11 Through $(2, -3)$ and $(2, 5)$
12 Through $(-2, 4)$ and $(-1.994, 4.006)$

13 Determine the slope and y intercept of each of the following lines:
(a) $y = 2x + 4$
(b) $2y = 3x - 5$
(c) $3x + 2y - 4 = 0$
(d) $x - 2y + 4 = 0$
(e) $-x - 2y + 4 = 0$
(f) $\frac{x}{2} + \frac{y}{3} = 1$

14 The line $y = mx + b$ passes through $(-1, 2)$ and $(3, 5)$. Find m and b.

15 We define the x intercept of a line as the abscissa of the point at which the line intersects the x axis. Using this definition, show that $x/A + y/B = 1$ is an equation of a straight line such that A and B are the x and y intercepts, respectively. This is known as the *intercept form* of an equation of a straight line.

16 For lines that pass through the origin, the intercept form cannot be used. Why?

17 Express each of the following equations in intercept form; then graph each line:
(a) $3x + 2y = 6$
(b) $6x + 3y = 15$
(c) $4x - 3y = 12$
(d) $3x - 2y = 4$
(e) $3y - 2x = 0$

18 The line through $(2, -3)$ and $(1, -1)$ intersects the y axis at $(0, B)$ and the x axis at $(A, 0)$.
(a) Find A and B.
(b) Find the area formed by the line and the coordinate axes.

19 Find the smallest integral values of P, Q, and R, where $P > 0$, for the line that passes through $(-2, 1)$ and $(-4, 3)$ and has an equation of the form $Px + Qy + R = 0$.

20 A line contains the points $(1, 0)$ and $(0, \frac{1}{2})$. Write an equation of this line in the form $PX + Qy + R = 0$.

21 Find an equation of the line containing $(-2, -5)$ and $(4, 10)$. Does this line pass through the origin?

22 The diagram shows the graph of a line L. Interpret the necessary information from the graph and answer the questions below:

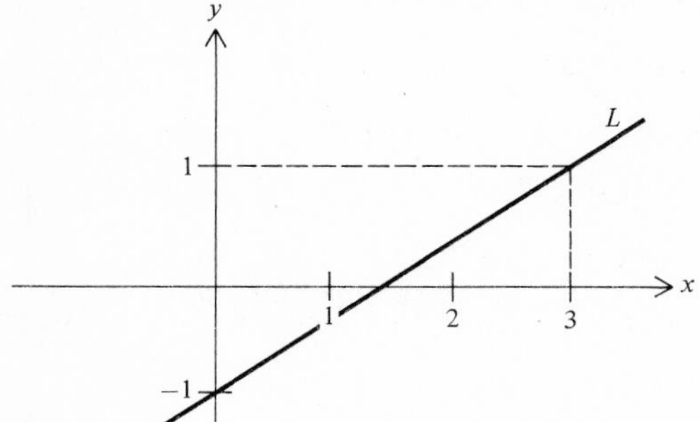

(a) Write the equation of L in the point-slope form.
(b) Write the equation of L in the slope-intercept form.
(c) Represent L in the form $Px + Qy + R = 0$.
(d) Determine A and B such that $x/A + y/B = 1$ is the equation of L.

23 Find the slope of the line not passing through the origin whose x intercept is:
(a) Equal to the y intercept.
(b) Twice the y intercept.
(c) Half the y intercept.
(d) The negative of the y intercept.

24 Find a point on the line $y = 9x/5 + 32$ such that the abscissa x equals the ordinate y.

25 Prove that: nonvertical lines intersect the y axis at one and only one point.

1.6 TWO THEOREMS

In plane geometry, a parallelogram is defined as a quadrilateral with opposite sides parallel. Using this fact, we prove the following theorem.

Theorem 1 If two lines are parallel, their slopes are equal.

Proof In Diagram 23, line L_1 passes through the origin $Q(0, 0)$. Line L_2 passes through the y axis at $R(0, b)$. Also, L_1 is parallel to L_2.

Line segment ST is drawn parallel to line segment QR, forming parallelogram $QRST$. Opposite sides of a parallelogram are equal; therefore $RQ = b$ and $ST = b$.

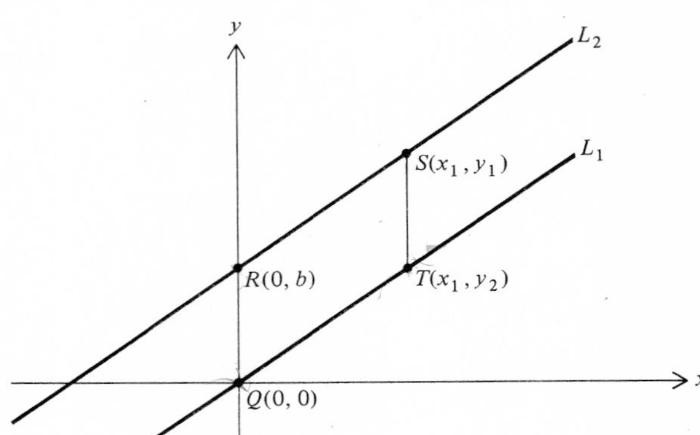

Diagram 23

$$b = y_1 - y_2$$

$$y_2 = y_1 - b \qquad \text{Solving for } y_2$$

$$\text{Slope of } L_2 = \frac{y_1 - b}{x_1 - 0} \qquad \text{slope of } L_1 = \frac{y_2 - 0}{x_1 - 0}$$

$$\text{Slope of } L_1 = \frac{y_1 - b - 0}{x_1} \qquad \text{Substituting } y_2 = y_1 - b$$

which is the *same* as the slope of L_2.

Converse of Theorem 1 If the slopes of two lines are equal, the two lines are parallel.

Theorem 2 If two lines are perpendicular, the product of their slopes is -1.

Proof In Diagram 24, lines L_1 and L_2 are drawn perpendicular and intersect at the origin $C(0, 0)$. $B(x_1, y_1)$ is a point on L_1 and $A(x_2, y_2)$ is a point on L_2. The slope of line segment BC is y_1/x_1, and the slope of line segment AC is y_2/x_2.

Note that since L_1 is perpendicular to L_2, Diagram 24 shows that line segment BC is perpendicular to line segment AC.

We must prove that

$$\frac{y_1}{x_1}\frac{y_2}{x_2} = -1$$

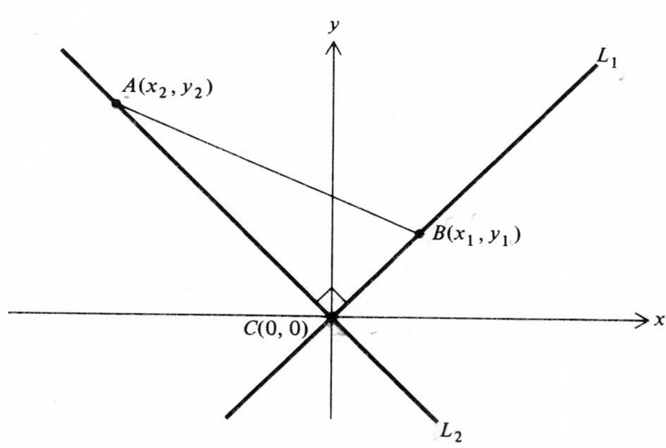

Diagram 24

Using the distance formulas,

$$BC = \sqrt{x_1^2 + y_1^2} \qquad AC = \sqrt{x_2^2 + y_2^2} \qquad AB = \sqrt{(x_2 - x_1)^2 + (y_2 - y_1)^2}$$

By the pythagorean theorem

$$\begin{aligned}
(AB)^2 &= (AC)^2 + (BC)^2 \\
(x_2 - x_1)^2 + (y_2 - y_1)^2 &= x_2^2 + y_2^2 + x_1^2 + y_1^2 \\
x_2^2 - 2x_1x_2 + x_1^2 + y_2^2 - 2y_1y_2 + y_1^2 &= x_2^2 + y_2^2 + x_1^2 + y_1^2 \\
-2x_1x_2 - 2y_1y_2 &= 0 \qquad \text{Simplifying} \\
-x_1x_2 &= y_1y_2 \\
-1 &= \frac{y_1y_2}{x_1x_2} \qquad \text{Dividing both sides by } x_1x_2
\end{aligned}$$

Line segment BC is perpendicular to line segment AC and

$$(\text{Slope of line segment } BC)\ (\text{slope of line segment } AC) = \frac{y_1y_2}{x_1x_2} = -1$$

Since the slope of line segment BC equals the slope of line L_1 and the slope of line segment AC equals the slope of line L_2,

$$(\text{Slope of } L_1)\ (\text{slope } L_2) = -1$$

the theorem is proved.

Converse of Theorem 2 If the product of the slopes of two lines is -1, the two lines are perpendicular.

Example 1 Find an equation of the perpendicular bisector L_1 of the line segment joining $A(-1, -3)$ and $B(5, 1)$.

STEP 1 Find the midpoint P of line segment AB:

$$P = \left(\frac{5 - 1}{2}, \frac{1 - 3}{2}\right) = (2, -1)$$

STEP 2 Find the slope of line segment AB

$$m = \frac{1 - (-3)}{5 - (-1)} = \frac{4}{6} = \frac{2}{3}$$

Therefore, the slope of the perpendicular bisector of line segment AB is $-\frac{3}{2}$ since $\frac{2}{3}(-\frac{3}{2}) = -1$.

STEP 3 Using the point-slope form of a line, we get

$$\begin{aligned}
y - (-1) &= -\tfrac{3}{2}(x - 2) \qquad \text{Point-slope form} \\
y + 1 &= -\tfrac{3}{2}x + 3 \\
y &= -\tfrac{3}{2}x + 2 \qquad \text{Slope-intercept form}
\end{aligned}$$

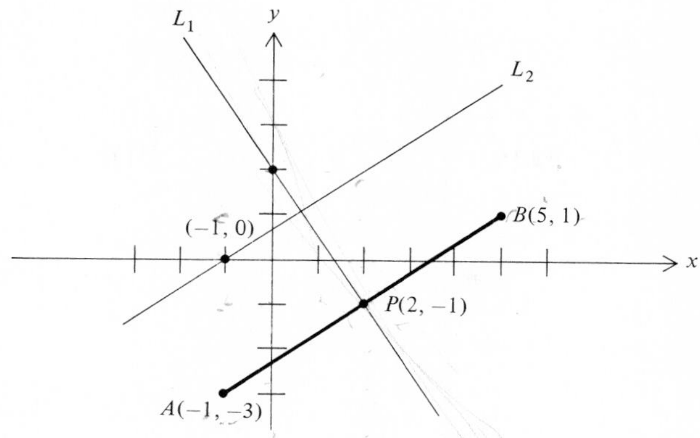

Diagram 25

See Diagram 25.

Example 2 Find an equation of the line parallel to the line segment in Example 1 that passes through the point $(-1, 0)$.

Solution Since the slope of line segment AB equals $\frac{2}{3}$, this must be the slope of the parallel line.

Using the point-slope form gives

$$y - 0 = \tfrac{2}{3}(x + 1) \qquad \text{or} \qquad y = \tfrac{2}{3}x + \tfrac{2}{3}$$

See Diagram 25.

EXERCISES

1 Find an equation of the perpendicular bisector of the line segment joining $A(-2, 4)$ and $B(3, 2)$.

2 Find an equation of the line passing through the origin and parallel to $4x - 3y - 2 = 0$.

3 Find an equation of the line passing through $(-4, 3)$ and parallel to $2x/3 - 3y/4 = 1$.

4 Find an equation of the line passing through $(3, -2)$ and perpendicular to the line passing through $(-4, 2)$ and $(2, 6)$.

5 Find an equation of the line passing through $(-4, 2)$ and perpendicular to $x + 2y = 4$.

6 Find an equation of the line in the form $Ax + By + C$ passing through $(6, 1)$ and parallel to $2x - y + 7 = 0$.

7 In triangle ABC, $A(4, -2)$, $B(2, 1)$, $C(-3, -3)$, find:

(a) An equation of the perpendicular bisector of side AC.

(b) An equation of the altitude to side BC.

8 Is there any general relationship between

$$Ax + By = C \qquad \text{and} \qquad Ay - Bx = D \qquad \begin{matrix} A \neq 0 \\ B \neq 0 \end{matrix}$$

9 Given: $A(-1, 4)$. Find two points M and N such that line segments AM and AN are parallel and equal in length to the segment joining $(-1, -2)$ and $(2, 2)$.

10 Find an equation of the line which has the same y intercept as $3x - 4y = 6$ and is perpendicular to the line.

11 Find an equation of the line parallel to $2x/BC + Cy/B = 1$ passing through $(C, -1/C)$.

12 Find an equation of the line passing through $(A, -2A)$ and perpendicular to the line $Ay + x = 1$.

13 Find an equation of the line perpendicular to $3/(x - 2) = 2/(y + 4)$ and passing through the x intercept of $x/2 - y/3 = 6$.

14 Find an equation of the two lines which have slopes of $-\frac{3}{2}$ and form with the coordinate axes a triangle of area 12 square units.

15 For what value of A will $2x - Ay = 3$ and $x + 2y = 1$ be **(a)** parallel and **(b)** perpendicular?

16 If $2y - 4x = 3$ and $Ax + y = 4$ are **(a)** parallel, find A, **(b)** perpendicular, find A.

17 Find T so that $P(-2, T)$ is on the line which is the perpendicular bisector of the segment joining $A(4, -1)$ $B(8, -7)$.

18 Find an equation of the line passing through $(-2, 3)$ and parallel to the line passing through $(3, 2)$ and $(-1, 6)$.

19 A line through $(-2, -3)$ and $(1, 2)$ is perpendicular to a line through $(-1, C)$ and $(2, -1)$. Determine C.

20 Given triangle ABC: $A(-5, 6)$, $B(-1, -4)$, and $C(3, 2)$. Find the equations of the three altitudes of the triangle.

21 Find the vertex of an isosceles triangle whose base is the line segment joining $A(-2, 2)$ and $B(4, -2)$ and:

(a) Whose side has a length of $\sqrt{13}$.

(b) Whose altitude has a length of $\sqrt{13}$ (two solutions).

1.7 SYSTEMS OF LINEAR EQUATIONS

Geometrically two straight lines may satisfy one of three conditions:

1 The two lines intersect at exactly one point.
2 The two lines have *all* their points in common and therefore coincide.
3 The two lines have *no* points in common and therefore are parallel.

Refer to Diagram 26.

One linear equation such as $2x - y = 0$ imposes one condition on x and y; that is, y is equal to 2 times x. Each ordered pair that satisfies this equation will be of the form (x, y), which is equal to $(x, 2x)$.

A linear equation such as $x + y = 3$ imposes a different condition on x and y; namely, y is equal to $3 - x$. Each ordered pair that satisfies this second equation will be of the form (x, y), which is equal to $(x, 3 - x)$.

If we impose *both* conditions simultaneously on x and y we form what is called a *system of linear equations*. Any ordered pair satisfying both these conditions is called a *solution of the system*.

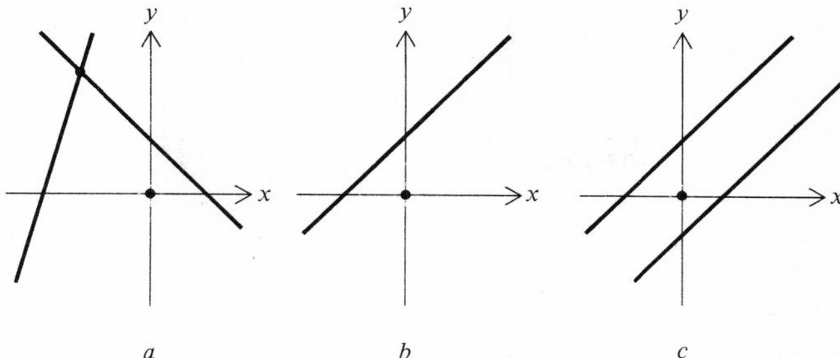

Diagram 26

We now illustrate three methods for obtaining the solution of a system of linear equations. The first method is graphical and is suggested only when an *approximate* solution is desired.

METHOD 1: GRAPHICAL METHOD

Solve the system (See Diagram 27.)

$$\begin{aligned} 2x - y &= 0 \\ x + y &= 3 \end{aligned}$$

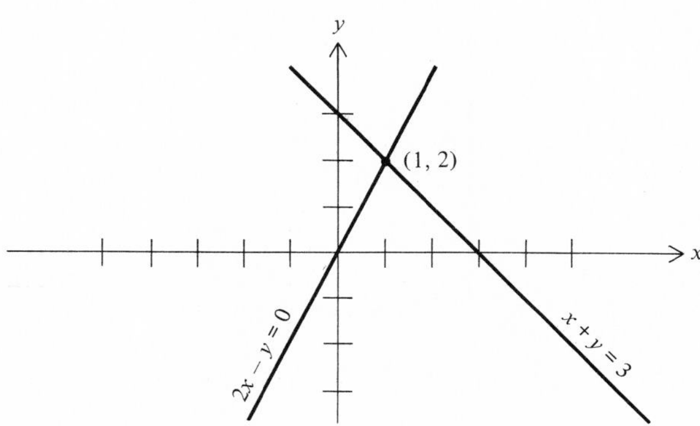

Diagram 27

The system happens to have an integer solution (1, 2), and so the method *appears* to be accurate. However, it is possible for a system to have a solution $(\frac{17}{7}, -\frac{8}{14})$. It is unlikely that the graphical method would yield this exact solution.

To illustrate Methods 2 and 3, we solve the system

$$\begin{aligned} x + 2y &= 4 \qquad (1) \\ -3x + 2y &= 12 \qquad (2) \end{aligned}$$

METHOD 2: SIMULTANEOUS EQUATIONS

Express both equations in the form $Px + Qy + R = 0$

$$\begin{aligned} x + 2y - 4 &= 0 \qquad (1) \\ -3x + 2y - 12 &= 0 \qquad (2) \end{aligned}$$

Eliminate *one* of the variables. If you want to eliminate y, multiply Equation (2) by -1 and *add*:

$$\begin{array}{rcrcrcl} x & + & 2y & - & 4 & = & 0 \\ 3x & - & 2y & + & 12 & = & 0 \\ \hline 4x & & & + & 8 & = & 0 \\ x & & & & & = & -2 \end{array}$$

If you want to eliminate x, multiply the Equation (1) by 3 and *add*:

$$\begin{array}{rcrcrcl} 3x & + & 6y & - & 12 & = & 0 \\ -3x & + & 2y & - & 12 & = & 0 \\ \hline & & 8y & - & 24 & = & 0 \\ & & y & & & = & 3 \end{array}$$

Once we determine either $x = -2$ or $y = 3$, we substitute into *either* equation.

$$\begin{aligned} (-2) + 2y - 4 &= 0 \qquad x = 2 \text{ from (1)} \\ 2y &= 6 \\ y &= 3 \end{aligned}$$

$$\begin{aligned} -3x + 2(3) - 12 &= 0 \qquad y = 3 \text{ from (2)} \\ -3x &= 6 \\ x &= -2 \end{aligned}$$

METHOD 3: SUBSTITUTION

In this method, we select either equation and solve for one variable in terms of the other. For this example, choose Equation (1) and solve for x in terms of y.

$$\begin{aligned} x + 2y - 4 &= 0 \\ x &= -2y + 4 \end{aligned}$$

$$\begin{aligned} -3(-2y + 4) + 2y &= 12 \qquad \text{Substituting for } x \text{ in (2)} \\ 6y - 12 + 2y &= 12 \\ 8y &= 24 \\ y &= 3 \end{aligned}$$

Since $x = -2y + 4$,

$$x = -2(3) + 4 = -2 \qquad \text{Substituting } y = 3$$

Hence, the solution of the system is $(-2, 3)$.

If we want to find a general formula to solve a system of two linear equations, let us examine the system consisting of a first equation

$$a_1x + b_1y = c_1 \tag{1}$$

and a second equation

$$a_2x + b_2y = c_2 \tag{2}$$

We can solve this general system by multiplying Equation (1) by b_2 and Equation (2) by $-b_1$, obtaining the new *equivalent* system

$$\begin{aligned} a_1b_2x + b_1b_2y &= b_2c_1 \quad (3)\\ -a_2b_1x - b_1b_2y &= -b_1c_2 \quad (4)\end{aligned}$$

Adding Equations (3) and (4), we obtain

$$a_1b_2x - a_2b_1x = b_2c_1 - b_1c_2$$

and

$$(a_1b_2 - a_2b_1)x = b_2c_1 - b_1c_2 \qquad \text{Factoring out an } x$$

$$x = \frac{b_2c_1 - b_1c_2}{a_1b_2 - a_2b_1} \qquad \text{Solving for } x \qquad (5)$$

Using the same method of elimination [this time multiplying Equation (1) by a_2 and Equation (2) by $-a_1$], we can solve for y:

$$y = \frac{a_1c_2 - a_2c_1}{a_1b_2 - a_2b_1} \qquad (6)$$

Formulas (5) and (6) can be used to solve any 2×2 system, when $a_1b_2 - a_2b_1 \neq 0$. Consider the system

$$\begin{aligned} 2x - 3y &= 14\\ 3x + y &= 10\end{aligned}$$

Using the general formula as a model, $a_1 = 2$, $a_2 = 3$, $b_1 = -3$, $b_2 = 1$, $c_1 = 14$, $c_2 = 10$. Then

$$x = \frac{(1)(14) - (-3)(10)}{(2)(1) - (3)(-3)} = \frac{44}{11} = 4$$

and

$$y = \frac{(2)(10) - (3)(14)}{11} = \frac{-22}{11} = -2$$

Instead of memorizing these formulas, we will introduce a *determinant*. A determinant is a square array of numbers that has a unique value. A 2×2 determinant is written

$$\begin{vmatrix} a & b \\ c & d \end{vmatrix}$$

The value of the determinant is defined to be $ad - bc$.

This definition suggests the following rule: to evaluate a 2×2 determinant, multiply the entries in the first diagonal and then subtract the product of the entries in the second diagonal.

$$\begin{vmatrix} a & b \\ c & d \end{vmatrix} = ad - bc$$

second diagonal

first diagonal

For example,

$$\begin{vmatrix} 2 & -3 \\ 4 & 1 \end{vmatrix} = (2)(1) - (-3)(4) = 14$$

Let us reexamine the general system defined by (1) and (2).

$$\begin{aligned} a_1x + b_1y &= c_1 \\ a_2x + b_2y &= c_2 \end{aligned}$$

Let $D = \begin{vmatrix} a_1 & b_1 \\ a_2 & b_2 \end{vmatrix} = a_1b_2 - a_2b_1$ Denominator in (5) and (6)

and $N_x = \begin{vmatrix} c_1 & b_1 \\ c_2 & b_2 \end{vmatrix} = b_2c_1 - b_1c_2$ Numerator of (5)

and $N_y = \begin{vmatrix} a_1 & c_1 \\ a_2 & c_2 \end{vmatrix} = a_1c_2 - a_2c_1$ Numerator of (6)

COMMENT N_x is formed by replacing the x coefficients by the constants c_1 and c_2; N_y is formed by replacing the y coefficients by the constants c_1 and c_2.

CONCLUSION The solution to the general 2×2 linear system is given by the formula

$$x = \frac{N_x}{D} \quad \text{and} \quad y = \frac{N_y}{D} \quad \text{provided } D \neq 0$$

This formula is known as *Cramer's rule.*†

Example 1 Find the solution of the linear system

$$\begin{aligned} 3x - 8y &= 4 \\ 4x + 6y &= 7 \end{aligned}$$

Using Cramer's rule, we have

$$D = \begin{vmatrix} 3 & -8 \\ 4 & 6 \end{vmatrix} = 18 - (-32) = 50$$

$$N_x = \begin{vmatrix} 4 & -8 \\ 7 & 6 \end{vmatrix} = 24 - (-56) = 80$$

$$N_y = \begin{vmatrix} 3 & 4 \\ 4 & 7 \end{vmatrix} = 21 - 16 = 5$$

Therefore

$$x = \frac{N_x}{D} = \frac{80}{50} = \frac{8}{5} \qquad y = \frac{N_y}{D} = \frac{5}{50} = \frac{1}{10}$$

† In 1750, the Swiss physicist Gabriel Cramer gave the rule bearing his name for expressing the solution of n linear equations in n unknowns in terms of determinants.

For systems of equations containing three variables, we can use either method 2 (simultaneous equations) or an expansion of Cramer's rule. Let us determine the solution of the following system, using a combination of method 2 *and* Cramer's rule. Consider

$$\begin{aligned} x - 5y + 3z &= 9 && (7)\\ 2x - y + 4z &= 6 && (8)\\ 3x - 2y + z &= 2 && (9) \end{aligned}$$

$$\begin{array}{rl} -2x + 10y + 6z = -18 & \text{Multiplying (7) by } -2 \text{ and} \\ 2x - y + 4z = 6 & \text{combining with (8)} \\ \hline 9y - 2z = -12 & (10) \\ -3x + 15y - 9z = -27 & \text{Multiplying (7) by } (-3) \text{ and} \\ 3x - 2y + z = 2 & \text{combining with (9)} \\ \hline 13y - 8z = -25 & (11) \end{array}$$

Let us now solve the resulting 2×2 system using Cramer's rule.

$$D = \begin{vmatrix} 9 & -2 \\ 13 & -8 \end{vmatrix} = -72 - (-26) = -46$$

$$N_y = \begin{vmatrix} -12 & -2 \\ -25 & -8 \end{vmatrix} = -225 - (-156) = -69$$

$$N_z = \begin{vmatrix} 9 & -12 \\ 13 & -25 \end{vmatrix} = 96 - 50 = 46$$

Therefore,

$$y = \frac{N_y}{D} = \frac{46}{-46} = -1 \qquad \text{and} \qquad z = \frac{N_z}{D} = \frac{-69}{-46} = \frac{3}{2}$$

To find x, select *one* of the original equations, for example, (8), $2x - y + 4z = 6$, and substitute $y = -1$ and $z = \frac{3}{2}$. Then

$$2x - (-1) + 4(\tfrac{3}{2}) = 6 \qquad \text{and} \qquad \begin{aligned} 2x + 1 + 6 &= 6 \\ 2x &= -1 \\ x &= -\tfrac{1}{2} \end{aligned}$$

The final solution to the problem is $x = -\frac{1}{2}$, $y = -1$, $z = \frac{3}{2}$. This can be written as an ordered triple $(x, y, z) = (-\frac{1}{2}, -1, \frac{3}{2})$.

COMMENT In general the solution to a system of three linear equations in x, y, and z is the ordered triple (x, y, z) that satisfies *all three equations.*

Systems of three linear equations involving three unknowns can be solved directly using determinants. Cramer's rule can now be extended to solve such a system. A third-order determinant has three rows and three columns,

$$\begin{vmatrix} a_1 & b_1 & c_1 \\ a_2 & b_2 & c_2 \\ a_3 & b_3 & c_3 \end{vmatrix}$$

By definition, the value of this determinant is given by

$$(a_1b_2c_3 + b_1c_2a_3 + c_1a_2b_3) - (c_1b_2a_3 + a_1c_2b_3 + b_1a_2c_3)$$

To remember this definition, it will be advisable to rewrite the first two columns to the right of the determinant.

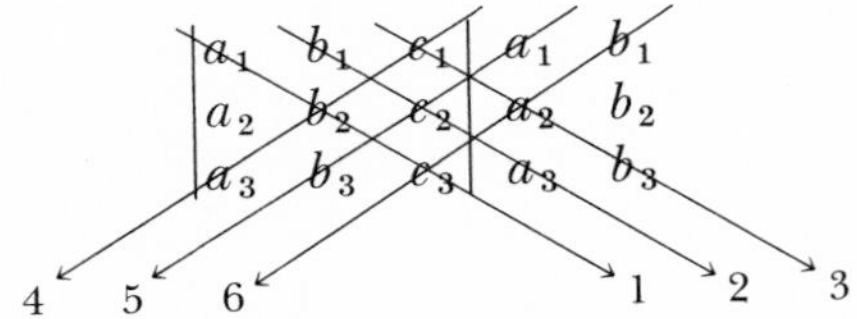

Find the products of each of the diagonals 1, 2, and 3 and add these three products together. Call this sum S_1. Then find the products of each of the reverse diagonals 4, 5, and 6 and add these three products together. Call this sum S_2. The value of the determinant is then $S_1 - S_2$.

Example 2 Evaluate the 3 × 3 determinant

$$\begin{vmatrix} 3 & -2 & 4 \\ 6 & 5 & -1 \\ -3 & 1 & 2 \end{vmatrix}$$

Solution

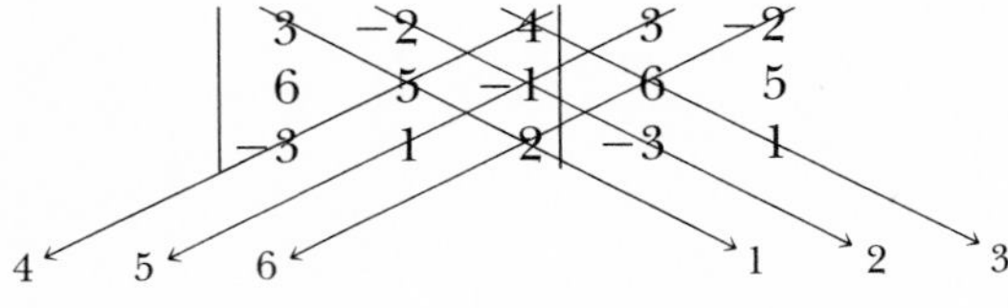

Repeating the first 2 columns

$$\begin{aligned} \text{Diagonal 1} &= 30 \\ \text{Diagonal 2} &= -6 \\ \text{Diagonal 3} &= 24 \\ \hline S_1 &= 48 \end{aligned} \qquad \begin{aligned} \text{Diagonal 4} &= -60 \\ \text{Diagonal 5} &= -3 \\ \text{Diagonal 6} &= -24 \\ \hline S_2 &= -87 \end{aligned}$$

$$S_1 - S_2 = 48 - (-87) = 135$$

To solve three equations in the form

$$\begin{aligned} a_1x + b_1y + c_1z &= d_1 \\ a_2x + b_2y + c_2z &= d_2 \\ a_3x + b_3y + c_3z &= d_3 \end{aligned}$$

the determinant solution is written

$$D = \begin{vmatrix} a_1 & b_1 & c_1 \\ a_2 & b_2 & c_2 \\ a_3 & b_3 & c_3 \end{vmatrix} \qquad x = \frac{\begin{vmatrix} \overset{\downarrow}{d_1} & b_1 & c_1 \\ d_2 & b_2 & c_2 \\ d_3 & b_3 & c_3 \end{vmatrix}}{D} \qquad y = \frac{\begin{vmatrix} a_1 & \overset{\downarrow}{d_1} & c_1 \\ a_2 & d_2 & c_2 \\ a_3 & d_3 & c_3 \end{vmatrix}}{D} \qquad z = \frac{\begin{vmatrix} a_1 & b_1 & \overset{\downarrow}{d_1} \\ a_2 & b_2 & d_2 \\ a_3 & b_3 & d_3 \end{vmatrix}}{D}$$

COMMENT Notice that all the denominators are the same. The numerator for x is formed by replacing the x coefficients a_1, a_2, and a_3 by the three constants d_1, d_2, and d_3. The numerator for y is formed by replacing the y coefficients b_1, b_2, b_3 by the three constants d_1, d_2, d_3. The numerator for z is formed in a similar manner.

Example 3 Solve the system

$$\begin{aligned} 3x - 2y - 3z &= -1 \\ 6x + y + 2z &= 7 \\ 9x + 3y + 4z &= 9 \end{aligned}$$

$$D = \left|\begin{array}{rrr|rr} 3 & -2 & -3 & 3 & -2 \\ 6 & 1 & 2 & 6 & 1 \\ 9 & 3 & 4 & 9 & 3 \end{array}\right| = (12 - 36 - 54) - (-27 + 18 - 48)$$

$$= (-78) - (-57) = -21$$

COMMENT If $D \neq 0$, the system has a unique solution (x, y, z).

Solution

$$x = \frac{\left|\begin{array}{rrr|rr} -1 & -2 & -3 & -1 & -2 \\ 7 & 1 & 2 & 7 & 1 \\ 9 & 3 & 4 & 9 & 3 \end{array}\right|}{-21}$$

$$= \frac{(-4 - 36 - 63) - (-27 - 6 - 56)}{-21}$$

$$= \frac{(-103) - (-89)}{-21} = \frac{-14}{-21} = \frac{2}{3}$$

$$y = \frac{\left|\begin{array}{rrr|rr} 3 & -1 & -3 & 3 & -1 \\ 6 & 7 & 2 & 6 & 7 \\ 9 & 9 & 4 & 9 & 9 \end{array}\right|}{-21}$$

$$= \frac{(84 - 18 - 162) - (-189 + 54 - 24)}{-21}$$

$$= \frac{(-96) - (-159)}{-21} = \frac{63}{-21} = -3$$

$$z = \frac{\left|\begin{array}{rrr|rr} 3 & -2 & -1 & 3 & -2 \\ 6 & 1 & 7 & 6 & 1 \\ 9 & 3 & 9 & 9 & 3 \end{array}\right|}{-21}$$

$$= \frac{(27 - 126 - 18) - (-9 + 63 - 108)}{-21}$$

$$= \frac{(-117) - (-54)}{-21} = \frac{-63}{-21} = 3$$

The solution is written as an ordered triple (x, y, z) or $(\frac{2}{3}, -3, 3)$

EXERCISES

Solve the linear systems in Exercises 1 to 4 by graphing.

1 $2x - y = 7$, $4x + y = 8$

2 $x + 2y = 5$, $3x - y = 1$

3 $x + y = -1$, $3x + y = 3$

4 $x - 3y + 4 = 0$, $2x + y + 10 = 0$

Solve the linear systems in Exercises 5 to 12 by algebraic methods.

5 $2x - 3y = 9$, $4x - y = 8$

6 $4x - y = -5$, $3x + 2y = 10$

7 $2x - 5y = 19$, $3x + 4y = -6$

8 $8x - 4y = 9$, $x - 2y = 9$

9 $x - 9y = 2$, $3x - 3y = -10$

10 $ax + by = c$, $bx + ay = d$

11 $1.6x - 2.05y = 0.39$, $5.2x + 4.1y = 3.42$

12 $cx + dy = c$, $c^2x + d^2y = c^2$

Find A so that $y = Ax - 1$ will pass through the intersection of:

13 $y = 2x + 2$, $3y + 2x - 2 = 0$

14 $5y = 9x + 160$, $9y = 5x - 160$

15 Find an equation of the line passing through the intersection of $2x + y = 3$ and $3x - 2y = 8$ and the point (3, 2).

16 Find an equation of the line perpendicular to $2x - y = 7$ passing through the intersection of

$$\frac{x}{2} + \frac{y}{3} = 5 \qquad \text{and} \qquad \frac{x}{4} - \frac{4y}{3} + 2 = 0$$

17 For what value(s) of B will $Bx + 2y = 2$ and $x + By = 2$ not intersect?

18 For what values of A and B will $2y - Ax = 1$ and $3x - By + A = 0$ have an *infinite* number of points of intersection?

19 Find the point on the line $y = 2x - 3$ equidistant from $(1, 0)$ and $(-1, 4)$.

20 Find an equation of the line passing through the intersection of $2x - y = 3$ and $3x + 2y = 4$ and:

(a) Parallel to the x axis.

(b) Parallel to the y axis.

(c) Perpendicular to $x + 2y = 3$.

(d) Parallel to the line containing $(-3, 2)$ and $(1, 4)$.

21 The line L_1 passes through $B(2, -1)$ with slope $-\frac{3}{4}$ and cuts the y axis at $(0, A)$. Find the point C such that CA is perpendicular to L_1 *and* another line L_2 contains points B and C and is parallel to the y axis.

22 Given triangle ABC, $A(2, 4)$, $B(3, -1)$, and $C(-5, 3)$.

(a) Find the coordinates of the intersection of the *medians* of the triangle. (This point is called the *centroid*.)

(b) Find the distance from A to the centroid.

(c) Compare the distance in part (*b*) with the length of the median from A.

23 Given triangle ABC, $A(-2, 1)$, $B(6, 3)$, and $C(2, -3)$. Find the coordinates of the intersection of the *perpendicular bisectors of the sides* of the triangle. (This point is called the *circumcenter* and is the center of the circumscribed circle.)

24 Given triangle ABC, $A(5, 3)$, $B(2, -3)$, and $C(-1, 1)$. Find the coordinates of the intersection of the *altitudes* of the triangle. (This point is called the *orthocenter.*)

25 Find A and B so that the graph of $Ax + By - 8 = 0$ contains the points $(2, 4)$ and $(-10, -4)$.

26 Find A and B so that the graph of $2x + Ay + B = 0$ contains the points $(5/2, 4/3)$, $(-1, -1)$.

27 A new corporate tax law specifies that the state tax is 20 percent of the amount that remains after paying the federal tax, while the federal tax is 50 percent of the amount that remains after paying the state tax. A corporation has $450,000 in taxable income. Find the state and federal tax.

28 If $(-2, -2)$, $(4, 2)$, and $(0, 4)$ are three vertices of a parallelogram, find the fourth vertex. (Find *all* solutions.)

29 In triangle ABC: $A(0, 0)$, $B(15, 20)$, and $C(15, 0)$. A *Nagel line* of a triangle is a line connecting a vertex and the point on the triangle halfway around the perimeter from that vertex.

(a) Find the equation of the three Nagel lines.

(b) Show that these lines are concurrent.

30 Given: two sets of collinear points: set 1 = $P_1(0, 0)$, $P_3(4, 0)$, and $P_5(2, 0)$, and set 2 = $P_2(2, 6)$, $P_4(8, 12)$, and $P_6(-2, 2)$.

(a) Find the equations of the lines passing through (1) P_1P_6, (2) P_3P_4, (3) P_1,P_2, (4) P_4,P_5, (5) P_2P_3, (6) P_5P_6.

(b) Find the point of intersection of the following pairs of lines computed from part (*a*): (1) and (2), and (4), and (5) and (6).

(c) Show that the three points of intersection in part (*b*) are collinear.

(d) Find the equation of the line passing through the three collinear points. This is called the *Pascal line.*

PROBLEMS INVOLVING THE USE OF DETERMINANTS

Evaluate the following determinants:

31 $\begin{vmatrix} 4 & 3 \\ 2 & 1 \end{vmatrix}$ **32** $\begin{vmatrix} -1 & 3 \\ -2 & 6 \end{vmatrix}$ **33** $\begin{vmatrix} 8 & -2 \\ 3 & 3 \end{vmatrix}$

34 $\begin{vmatrix} 5 & 14 \\ 2 & -3 \end{vmatrix}$ **35** $\begin{vmatrix} 5 & 1 & -1 \\ -3 & -2 & 2 \\ 2 & -3 & -3 \end{vmatrix}$ **36** $\begin{vmatrix} 1 & 1 & 1 \\ 2 & 3 & -1 \\ 6 & -2 & -3 \end{vmatrix}$

37 $\begin{vmatrix} 5 & 4 & 11 \\ 6 & -4 & 2 \\ 1 & 3 & 5 \end{vmatrix}$ **38** $\begin{vmatrix} 1 & -3 & 5 \\ 0 & 2 & -1 \\ 3 & 1 & 2 \end{vmatrix}$ **39** $\begin{vmatrix} 7 & 14 & -4 \\ -9 & -27 & 17 \\ 1 & 6 & -5 \end{vmatrix}$

40–47 Solve Exercises 5 to 12 using determinants.

Solve the linear systems in Exercises 48 to 51 by determinants.

48 $5x - 2y - 14 = 0$
$2x + 3y + 3 = 0$

49 $\dfrac{4x}{3} - \dfrac{3y}{5} = 1$
$x - y = \dfrac{-1}{6}$

50 $\dfrac{1}{x} + \dfrac{1}{y} = 8$
$\dfrac{1}{x} - \dfrac{1}{y} = 2$ *Hint:* Let $\dfrac{1}{x} = a$, $\dfrac{1}{y} = b$

51 $\dfrac{3}{x} - \dfrac{1}{y} = -4$
$\dfrac{1}{x} + \dfrac{2}{y} = 5$

52 Find the linear equation of the form $y = ax + b$ if $y = 2$ when $x = 5$ and if $y = 4$ when $x = 6$.

53 The sum of two numbers is 12, and their difference is 28. Find the numbers.

54 If 20 units of gold and 10 units of silver *or* 10 units of gold and 35 units of silver can be bought for \$2400, find the price of each unit.

Solve the 3×3 linear systems in Exercises 55 to 62 using determinants.

55 $\begin{aligned} 2x + y + 2z &= 1 \\ x + 2y - 3z &= 4 \\ 3x - y + z &= 0 \end{aligned}$

56 $\begin{aligned} x + y + z &= -1 \\ 2x + 3y + 2z &= -5 \\ 3x - 2y + 9z &= 0 \end{aligned}$

57 $\begin{aligned} x + 2y - z &= -1 \\ x - y + 2z &= 8 \\ 2x - y - z &= 5 \end{aligned}$

58 $\begin{aligned} 3x - 2y + z &= 1 \\ x + y + z &= 0 \\ y - 2z &= 2 \end{aligned}$

59 $\begin{aligned} 3x + 2z &= 0 \\ 4x + 3y &= 5 \\ 5y - z &= -2 \end{aligned}$

60 $\begin{aligned} 6x - 5y - 2z &= 2 \\ 4x + y + 3z &= 10 \\ 5x + 3y + 7z &= 13 \end{aligned}$

61 $\begin{aligned} 6x - 8y + 3z &= -5 \\ 9x + 20y - 4z &= 25 \\ 15x + 12y + 7z &= 12 \end{aligned}$

62 $\begin{aligned} \frac{3}{x} + \frac{1}{y} - \frac{4}{z} &= 0 \\ \frac{1}{x} + \frac{2}{y} - \frac{2}{z} &= -1 \\ \frac{4}{x} + \frac{4}{z} &= 3 \end{aligned}$

63 Given $s = at^2 + bt + c$. Determine a, b, and c if $s = 10$ when $t = 1$; $s = 22$ when $t = 2$; and $s = 50$ when $t = 3$.

64 Three boys have 35 marbles altogether. If the first gives one-fourth of his marbles to the third, the third will then have as many marbles as the second. However if the second gives one-half of his marbles to the third, the third will have one-half as many marbles as the first. How many marbles has each?

65 One angle of a triangle is three-sevenths the difference of the other two angles and one-fifth of their sum. Find the angles.

SPECIAL PROJECT

This problem must be accompanied by an illustrative diagram, utilizing a sensible color scheme, and of appropriate size to demonstrate the results of the problem visually. All pertinent data must be shown on the diagram.

Given the triangle ABC, where $A(-6, -8)$, $B(6, 4)$, and $C(-6, 10)$,

1 Find the equations of **(a)** medians, **(b)** altitudes, and **(c)** perpendicular bisectors of the sides.

2 Find the coordinates of **(a)** centroid G intersection of the medians, **(b)** orthocenter H intersection of the altitudes, and **(c)** circumcenter C intersection of the perpendicular bisectors of the sides.

3 Prove that G, H, and C are collinear and G divides the line HC in a 2:1 ratio.

4 Find **(a)** the question of line HC (*Euler line*), **(b)** the center of line segment HC, and **(c)** the length of HC.

Theorem In any triangle the midpoints of the sides, the feet of the altitudes, and the points halfway between each vertex and the orthocenter H are nine concyclic points (they lie on the same circle). The center of this circle is the center of the *Euler line.*

---

5 Determine **(a)** the set of nine concyclic points, **(b)** the radius of the nine-point circle, **(c)** the equation of the nine-point circle.

1.8 THE PERPENDICULAR DISTANCE FROM A POINT TO A LINE

In Section 1.3, we developed a formula for the distance between two points. Suppose we are now asked to find the distance between a point and a line.

Example 1 Find the distance from the point $A(1, 4)$ to the line

$$4x - 3y - 2 = 0 \tag{1}$$

STEP 1 Represent the information graphically.

Refer to Diagram 28. What we wish to calculate is the distance AB, which is the length of the *perpendicular segment* from $A(1, 4)$ to $4x - 3y - 2 = 0$. Line segment AC will not be considered, since it is *not* perpendicular to the line.

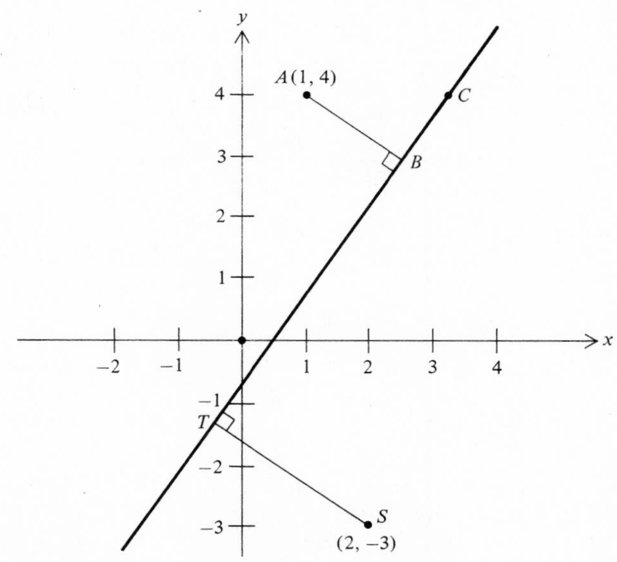

Diagram 28

Our plan is to first find an equation for the line through A perpendicular to $4x - 3y - 2 = 0$ and then solve the two equations simultaneously to obtain the coordinates of B. Then we must find the distance from A to B using the formula for the distance between two points.

STEP 2 Find the equation of the line through A perpendicular to $4x - 3y - 2 = 0$. First find the slope of $4x - 3y - 2 = 0$. Expressing the equation in slope-intercept form

$$-3y = -4x + 2$$
$$y = \tfrac{4}{3}x - \tfrac{2}{3}$$

We see that
$$m = \tfrac{4}{3}$$

Therefore, the slope of the line through A and B is $-\frac{3}{4}$ since line segment AB is perpendicular to $4x - 3y - 2 = 0$.

Next find the equation of line through A and B. Substituting $-\frac{3}{4}$ and $(1, 4)$ in the point-slope form

$$y - 4 = -\tfrac{3}{4}(x - 1)$$

Express the equation in the form $Px + Qy + R = 0$.

$$4y - 16 = -3x + 3$$
$$3x + 4y - 19 = 0 \qquad (2)$$

STEP 3 Solve the system

$$4x - 3y - 2 = 0 \qquad (1)$$
$$3x + 4y - 19 = 0 \qquad (2)$$
$$9x + 12y - 57 = 0 \quad \text{Multiplying (2) by 3} \qquad (3)$$
$$16x - 12y - 8 = 0 \quad \text{Multiplying (1) by 4} \qquad (4)$$

$$\begin{array}{r} 9x + 12y - 57 = 0 \\ 16x - 12y - 8 = 0 \\ \hline 25x \qquad\quad - 65 = 0 \end{array} \quad \text{Summing (3) and (4)}$$
$$x = \tfrac{65}{25} = \tfrac{13}{5}$$

$$4(\tfrac{13}{5}) - 3y - 2 = 0 \quad \text{Substituting } x = \tfrac{13}{5} \text{ in (1)}$$
$$\tfrac{52}{5} - 3y - 2 = 0$$
$$-3y = -\tfrac{52}{5} + \tfrac{10}{5} = -\tfrac{42}{5}$$
$$y = \tfrac{14}{5}$$

The coordinates of B are $(\frac{13}{5}, \frac{14}{5})$.

STEP 4 Find the distance from $A(1, 4)$ to $B(\frac{13}{5}, \frac{14}{5})$.
Using the distance formula,

$$d = \sqrt{\left(\frac{13}{5} - 1\right)^2 + \left(\frac{14}{5} - 4\right)^2} = \sqrt{\left(\frac{8}{5}\right)^2 + \left(-\frac{6}{5}\right)^2}$$

$$= \sqrt{\frac{64}{25} + \frac{36}{25}} = \sqrt{\frac{100}{25}} = \sqrt{4} = 2$$

Therefore the distance from $A(1, 4)$ to the line $4x - 3y - 2 = 0$ is 2.

To review the procedure and the mechanics of the process, you may want to show that the distance from $S(2, -3)$ to the line $4x - 3y - 2 = 0$ is 3 (see Diagram 28).

Because the above procedure is lengthy and involved, we now introduce a theorem (without proof) that will help simplify the mechanics.

Theorem 1 The distance from a point $A(x_1, y_1)$ to a line $Px + Qy + R = 0$ is given by the formula

$$d = \frac{|Px_1 + Qy_1 + R|}{\sqrt{P^2 + Q^2}}$$

Verification 1 Using point $A(1, 4)$ and line $4x - 3y - 2 = 0$ in Theorem 1

$$P = 4 \qquad Q = -3 \qquad R = -2 \qquad x_1 = 1 \qquad y_1 = 4$$

Substituting in the above formula,

$$d = \frac{|(4)(1) + (-3)(4) - 2|}{\sqrt{4^2 + 3^2}} = \frac{|4 - 12 - 2|}{\sqrt{25}} = \frac{10}{5} = 2$$

Verification 2 Using point $S(2, -3)$ and line $4x - 3y - 2 = 0$ in Theorem 1,

$$d = \frac{|(4)(2) + (-3)(-3) - 2|}{\sqrt{4^2 + 3^2}} = \frac{|8 + 9 - 2|}{5} = \frac{15}{5} = 3$$

EXERCISES

1 (a) Find an equation of the line perpendicular to $3x - 4y = 8$ passing through the point $(-\frac{1}{2}, \frac{3}{4})$.
(b) Find the point of intersection (label A) of the two lines above.
(c) Which of the two lines is closer to the point $(\frac{3}{2}, 1)$?
(d) Which point, $(-\frac{1}{2}, \frac{3}{4})$ or A, is closer to $(\frac{3}{2}, 1)$?

2 Find the distance from $P(-2, 2)$ to the line $3y = 4x + 5$.

3 Find the distance between the parallel lines $3x - 4y + 8 = 0$ and $3x - 4y - 4 = 0$.

4 Find the distance from (4, 5) to the line containing (2, −2) and (−1, 3).

5 Given the triangle ABC, $A(2, 6)$, $B(8, 2)$, and $C(-2, -8)$. Find the length of the altitude to side BC and the area of the triangle, using the formula $A = \frac{1}{2}bh$.

6 A line through (2, −3) is perpendicular to $3x - 4y + 6 = 0$. How close does this line come to the point (2, 2)?

7 Find the area of the triangle whose sides have equations

$$4y - 2x = 13 \qquad y + 4x = 10 \qquad \text{and} \qquad 11y + 8x = 2$$

8 Consider the point $(\frac{3}{16}, -2)$ and the line $6x - 8y = -4$. What is the distance from the point to the line?

9 Find C so that $P(4, 2)$ is $\sqrt{13}$ units from the line $2x + 3y = C$.

10 Find the distance from $(-3, \frac{3}{2})$ to $3y = 4x + 4$.

11 **(a)** Find the equation of the line passing through (−3, 5) and perpendicular to $x + 2y - 2 = 0$.

(b) Find the intersection of the two lines in part (*a*).

(c) Using the distance formula, find the distance from (−3, 5) to $x + 2y - 2 = 0$.

(d) Using the formula for distance from a point to a line, show that the answer to part (*c*) is correct.

12 A railroad has a problem with 1 mile of track lying in a flat desert. They suspect the trouble to be track-width expansion. Tracks are supposed to be 4.46 feet apart. A surveyor gets a fixed point of reference and observes that the tracks for the 1 mile run behave according to the equations $y = x + 3$ and $4y = 4x - 13$. To the nearest $\frac{1}{100}$ foot, what width correction must be made? (Use $\sqrt{2} = 1.414$.)

13 Find *all* the points on the y axis which are 3 units from $3x + 4y = 12$, and find all the points on the x axis which are 3 units from $3x + 4y = 12$.

14 Referring to Exercise 13, join all the points (consecutively) and form a plane geometric figure. Identify the figure.

15 Find the length of the three altitudes for the following triangles, and determine the area using the formula $A = \frac{1}{2}bh$:

(a) (2, 1), (3, −2), and (−4, −1) **(b)** (2, 4), (3, −1), and (−5, 3)

Functions

2.1 THE FUNCTION CONCEPT

In Chapter 1 we examined the linear equation quite thoroughly. Recall the equation $y = mx + b$. Upon fixing a value for the slope m and the y intercept b, we observed that for each value of x selected, a value of y was determined. This simple notion of using $mx + b$ as a rule, selecting an x, and generating a y describes the *function concept.*

For example, $y = 2x - 1$ represents a function. The rule is of the form $mx + b$, where m and b have been fixed ($m = 2$, $b = -1$) and y is used to denote the value of the rule for any value x.

Therefore when we refer to a *function,* we will mean that for each value of the variable x there is associated one and only one value of the variable y. Since in this context the value of y depends upon the choice of x, we say that y is the *dependent variable* and x is the *independent variable.* Furthermore the collection of all values that can replace the independent variable is called the *domain* of the function, and the collection of all the corresponding values of the dependent variable y is called the *range* of the function.

From this discussion we observe that a function is a rule and has a domain and a range.

DEFINITION 1 A *function* is a rule which assigns to each element in its domain one and only one element in its range.

Usually functions are specified by mathematical formulas, such as our first example, $y = 2x - 1$. Some other examples of functions are $y = 1/x$, $y = x^2$, $v = 16t$, and $C = 2\pi r$. Note that variables are only symbols: t and v or r and C can be used just as conveniently as x and y.

A function is not properly defined unless a domain is explicitly stated. *Hence, when no domain is specified, the domain of a function is understood to be the largest collection of real numbers for which the formula is meaningful.*

Example 1 Find the domain of $y = \dfrac{1}{x - 1}$.

Since division by zero is undefined, $x - 1 \neq 0$ or simply $x \neq 1$. Therefore the domain of this function is all real numbers except 1.

Example 2 Find the domain of $y = \sqrt{x}$.

Since the square root of a negative number is not real, we must limit the domain to all nonnegative real numbers.

Example 3 Find the domain of $y = \dfrac{1}{\sqrt{x - 1}}$

Using the ideas of Examples 1 and 2, $x \neq 1$ in order to avoid division by zero, and x cannot be less than 1 in order to avoid taking the square root of a negative number. Hence, the domain of the function is all real numbers greater than 1.

A function need not be in terms of a mathematical formula, nor do the domain and range have to be collections of real numbers. We end this section with such an example.

Example 4 Consider the collection of the members of the U.S. Senate and the collection of the states of the United States. To each senator there is associated the state he represents. One and only one state is determined each time we choose a senator, thus the state is a function of the senator. The domain of this function is a collection of people, and the range is a collection of states.

2.2 FUNCTIONAL NOTATION

Today in mathematics you will find many ways of expressing a function. It has already been shown that we can make use of a formula like $y = 2x - 1$

or $y = \sqrt{x}$. In this text we use a symbol like $f(x)$ and write the above functions as

$$f(x) = 2x - 1 \qquad \text{or} \qquad f(x) = \sqrt{x}$$

Here the dependent variable y has been replaced by the symbol $f(x)$, read "f of x" not "f times x." In writing $f(x) = 2x - 1$ or $f(x) = \sqrt{x}$, the symbol $f(x)$ is twofold: (1) it represents the function itself, and (2) it represents the value of the function associated with any particular value x, x being in the domain of the function. Besides f any other letter can be used for this notation. Hence you will see $g(x)$, $F(x)$, and so on being used.

Example 1 If $f(x) = 2x - 1$, then

$$\begin{aligned} f(2) &= 2(2) - 1 = 3 \\ f(t+1) &= 2(t+1) - 1 = 2t + 1 \\ f(\pi) &= 2\pi - 1 \end{aligned}$$

Example 2 If $g(x) = x^2$, then

$$g(x+h) - g(x) = (x+h)^2 - x^2 = x^2 + 2xh + h^2 - x^2 = 2xh + h^2$$

Example 3 If $f(x) = \sqrt{x}$, then

$$f(0) = \sqrt{0} = 0 \qquad f(4) = \sqrt{4} = 2$$

$f(-1)$ is meaningless since -1 is not in the domain of the function.

COMMENT $\sqrt{x}$ stands for the principal root of x. Hence the square roots of 4 are 2 or -2, but $\sqrt{4}$ is just 2.

Example 4 If $f(x) = 1/(x - 1)$, then

$$[f(-2)]^2 = [f(-2)][f(-2)] = (-\tfrac{1}{3})(-\tfrac{1}{3}) = \tfrac{1}{9}$$

Example 5 If $f(x) = 2x - 1$, then

$$f(a+b) = 2(a+b) - 1 = 2a + 2b - 1$$

and $$f(a) + f(b) = (2a - 1) + (2b - 1) = 2a + 2b - 2$$

Observe that $f(a + b) \neq f(a) + f(b)$.

EXERCISES

1 Define each of the following:

(a) Function **(b)** Domain
(c) Range **(d)** Independent variable
(e) Dependent variable **(f)** $f(x)$
(g) $\sqrt{Z}$

2 Rewrite using functional notation:

(a) $y = x^2 + \frac{1}{x}$ **(b)** $v = 16t + 5$ **(c)** $s = 16t^2$

(d) $C = 2\pi r$ **(e)** $A = \pi r^2$

3 When determining the domain of a function, what do we look for to keep the formula meaningful?

4 Find the domain of each function:

(a) $y = 3x$ **(b)** $y = \frac{3}{x}$ **(c)** $y = \sqrt{x + 1}$

(d) $y = \frac{1}{x} + \frac{1}{x - 1}$ **(e)** $y = \frac{1}{\sqrt{x}}$ **(f)** $y = x^4$

(g) $y = \sqrt[4]{x}$ **(h)** $y = 1$ **(i)** $y = \frac{x}{x}$

(j) $y = \sqrt{1 - x}$

5 $f(x)x^2 - 1$ and $g(x) = (x - 1)^2$. **(a)** Compute:
(1) $f(1)$ and $g(1)$ **(2)** $f(0)$ and $g(0)$ **(3)** $f(-1)$ and $g(-1)$
(b) From part (*a*) can you formulate a rule concerning $a^2 - b^2$ and $(a - b)^2$?

6 $f(x) = \sqrt{x + 9}$ and $g(x) = \sqrt{x} + \sqrt{9}$. **(a)** Compute:
(1) $f(0)$ and $g(0)$ **(2)** $f(16)$ and $g(16)$ **(3)** $f(40)$ and $g(40)$
(b) From part (*a*) can you formulate a rule concerning $\sqrt{a + b}$ and $\sqrt{a} + \sqrt{b}$?

7 $f(x) = x^3 - 1$ and $g(x) = (x - 1)^3$. **(a)** Compute:
(1) $f(0)$ and $g(0)$ **(2)** $f(1)$ and $g(1)$
(3) $f(2)$ and $g(2)$ **(4)** $f(-1)$ and $g(-1)$
(b) From part (*a*) can you formulate a rule concerning $x^3 - 1$ and $(x - 1)^3$?

8 Let $f(x) = \frac{x - 1}{1 - x}$. **(a)** Find the value of each of the following:
(1) $f(0)$ **(2)** $f(-1)$ **(3)** $f(-2)$
(4) $f(1)$ **(5)** $f(2)$ **(6)** $f(100)$
(b) What conclusion can be reached concerning $f(x)$?

In Exercises 9 to 18, given the functions

$$f(x) = -x^2 \qquad g(x) = 1 + |x| \qquad h(x) = 1 - |x|$$
$$j(x) = x - |x| \qquad k(x) = \pi$$

find the value of each of the following:

9 $f(-4)$ **10** $f(4)$ **11** $g(1)$ **12** $g(-1)$
13 $h(1)$ **14** $h(-1)$ **15** $j(2)$ **16** $j(-2)$
17 $k(3)$ **18** $k(-3)$

In Exercises 19 to 30 given the functions:

$$f(x) = x \qquad g(x) = \frac{x + 2}{x} \qquad h(x) = \sqrt{3 - x}$$

$$s(x) = |5 - x| \qquad t(x) = x^3 - 1$$

find the value of each of the following:
19 $f(\pi)$ **20** $g(-2)$ **21** $g(100)$ **22** $h(-1)$

23 $h(4)$ **24** $s(10)$ **25** $t(-2)$ **26** $t(0)$

27 $[t(-2)]^2$ **28** $f(x+h)$ **29** $g(x+c)$ **30** $t(x+h)$

31 Consider a function $y = f(x)$ that has a domain of all real numbers. What incorrect assumption is made in the computation $[f(-3)]^2 = f^2(-3)^2 = 9f^2$?

32 Consider $f(x) = 3x + 2$, where a and b are real numbers.
(a) $f(a) =$ **(b)** $f(b) =$
(c) $f(a) - f(b) =$ **(d)** $f(a - b) =$
(e) Does $f(a - b) = f(a) - f(b)$?

33 Does $y = \sqrt{-x^2 - 1}$ define a function? Give a good reason for your answer.

34 What is the domain of $f(x) = \sqrt{-x^2}$?

For each of the following functions:
(a) Find $f(c)$
(b) Find $\dfrac{f(x) - f(c)}{x - c}$ $x \neq c$ (reduce to simplest algebraic form)
(c) If $c = x$, evaluate the result of part *(b)*

35 $f(x) = x - 1$ **36** $f(x) = x^2 - 1$ **37** $f(x) = (x - 1)^2$

For each of the following functions find:
(a) $f(x + h)$
(b) $\dfrac{f(x+h) - f(x)}{h}$ $h \neq 0$ (reduce to simplest algebraic form)
(c) If $h = 0$, evaluate the result of part *(b)*

38 $f(x) = x$ **39** $f(x) = x^2$ **40** $f(x) = x^2 - 1$

41 $f(x) = x^2 - x$ **42** $f(x) = x^3$ **43** $f(x) = ax^2 + bx + c$

44 $f(x) = \dfrac{1}{x}$ **45** $f(x) = \dfrac{1}{x^2}$

46 $f(x) = \sqrt{x}$ *Hint:* rationalize the numerator

47 Consider the function $f(x) = 2x + 3$. **(a)** Find:

(1) $\dfrac{f(x+1) - f(x)}{1}$ **(2)** $\dfrac{f(x+0.1) - f(x)}{0.1}$ **(3)** $\dfrac{f(x+0.01) - f(x)}{0.01}$

(b) From the results to part *(a)* can you guess any significance to these results?
(c) For $f(x) = mx + b$, recompute part *(a)*. Does this result help your conclusion in part *(b)*?

48 Consider $f(0) = 1$. For *all* other x belonging to $y = f(x)$ we define $f(x - 1) + 3 = f(x)$. Suppose now we want to find $f(2)$. First we must calculate $f(1)$, since by definition $f(1) = f(1 - 1) + 3 = f(0) + 3 = 1 + 3 = 4$. From this result it follows that $f(2) = f(2 - 1) + 3 = f(1) + 3 = 4 + 3 = 7$. Using this procedure, compute the value of $f(5)$.

49 Using the technique in Exercise 48, if $f(1) = 2$ and $f(x - 1) - 1 = f(x)$ for *all* other x belonging to $y = f(x)$, compute the value of $f(4)$.

50 Consider $f(x) = x(x - 1)(x - 2)$, show that

$$\frac{f(a-1)}{a-1} = \frac{f(a-2)}{a-4}$$

2.3
THE GRAPH OF A FUNCTION

We begin with a definition.

DEFINITION 1 Let a be an element of the domain of a function $y = f(x)$. The *graph* of $y = f(x)$ is the totality of all points $(a, f(a))$.

Now we can consider drawing the graph of $f(x) = 2x - 1$, where x is any real number, and $g(x) = 2x - 1$, where x is a nonnegative real number; that is, $x \geq 0$ (see Diagram 1). Note the difference in the graphs of $y = f(x)$ and $y = g(x)$. The graph of $y = f(x)$ is a line, whereas the graph of $y = g(x)$ is a ray, whose end point is $(0, -1)$.

It is reasonable to suspect that these functions are not equal to each other. Without further discussion let us define the equality of two functions.

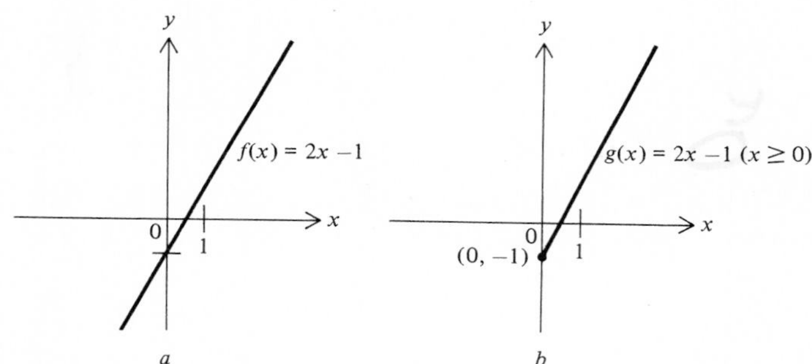

Diagram 1

DEFINITION 2 We say that two functions $y = f(x)$ and $y = g(x)$ are equal if the following conditions are satisfied:

(a) Their domains are equal.

(b) For each value a contained in each domain, $f(a) = g(a)$.

We can now conclude with certainty that $f(x) = 2x - 1$ does not equal $g(x) = 2x - 1$, $x \geq 0$, since their domains are different.

We can write a function in terms of an equation, that is, $y = 2x - 1$; however, does every equation supply us with a satisfactory rule to be a function? To answer this question let us examine the equation $y^2 = x^2$. Any ordered pair (x, y) that satisfies this equation will be a point of its graph. If we determine enough of these points, we can get a graph of $y^2 = x^2$. You can verify that the following points work: $(0, 0)$, $(1, 1)$, $(1, -1)$, $(-1, 1)$, $(2, 2)$, $(2, -2)$, $(-2, 2)$, $(-1, -1)$, $(-2, -2)$, ... The pattern is clear. We can generate all ordered pairs of the form $(a, \pm a)$. The graph of this equation is two lines (see Diagram 2*a*). Let us now draw a vertical line, say $x = 2$ (see Diagram 2*b*).

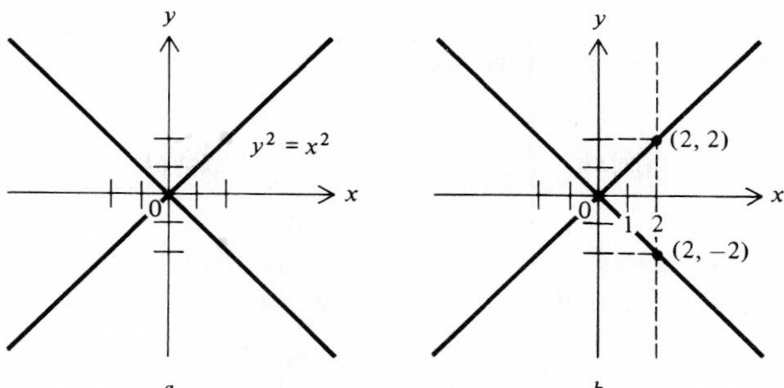

Diagram 2

This line intersects the graph of $y^2 = x^2$ at *two points,* namely, (2, 2) and (2, −2). This tells us that for one value of x two different values of y are obtained. This does not comply with the definition of a function. In general, equations that relate x to y can have more than one value of y for each given value of x. Therefore not every equation represents a function.

The procedure just discussed is known as the *vertical-line test.* In simple geometric terms, this test implies a vertical line can, at most, have one point in common with the graph of a function.

Example 1 By the vertical-line test in Diagram 3 graphs a, b, and c represent functions, while d, e, and f do not. Note that the y axis can be used in

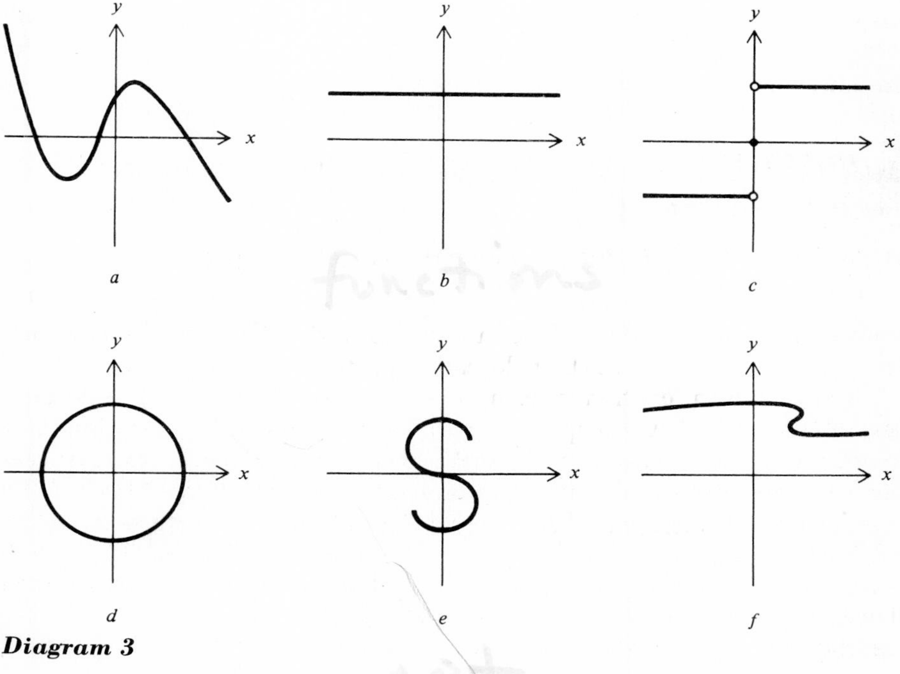

Diagram 3

this test. In graph c only point (0, 0) is common to the graph and this vertical line. For graph f where would you draw a vertical line?

2.4 SPECIAL FUNCTIONS AND THEIR GRAPHS

Constant Functions A function that has a range of exactly one element is called a *constant function.* In general we write such a function as

$$f(x) = c \qquad c = \text{real number}$$

A typical example if $f(x) = 2$. The domain of this function is all real numbers, and its range is the single number 2. Hence $f(0) = 2, f(1) = 2, f(-1) = 2$, and so on. The totality of all points $(x, 2)$, where x belongs to the domain of this function, defines its graph. It is a horizontal line which has a y intercept of 2 (see Diagram 4).

The Identity Function This function is of the form $f(x) = x$, where x can be any real number. Since in this function $y = x$, the totality of all points (x, x) defines this function graph. Its graph is a line that passes through the origin and makes equal angles with the coordinate axes. The range of this function is the same as its domain, namely, all real numbers (see Diagram 5).

Linear Functions The general formula is given by $F(x) = mx + b$. This function has a domain of all real numbers and is so named because its graph is a straight line. We have already seen specific examples of this function, that is, $f(x) = 2x - 1$ (Diagram 1) and $f(x) = x$ (Diagram 5). The slope of the line is m, and b is its y intercept. For $m \neq 0$ the range of this function is all real numbers.

Power Functions These are functions of the form $f(x) = x^n$, where n is some fixed positive integer. Some examples are

$$f(x) = x^2 \qquad f(x) = x^3 \qquad f(x) = x^{100}$$

In general, these functions have a domain of all real numbers. Depending on whether n is odd or even, their range will be all real numbers or

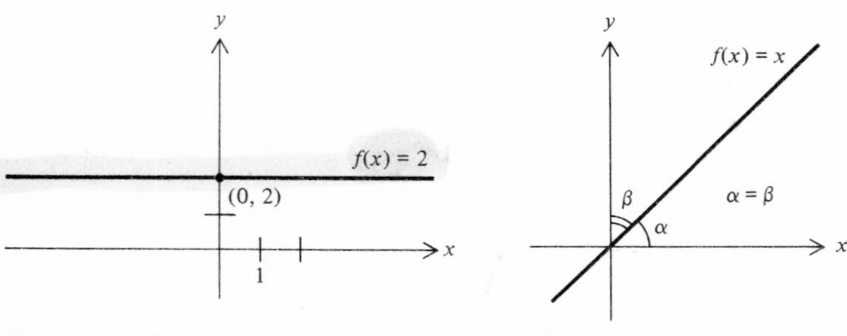

Diagram 4

Diagram 5

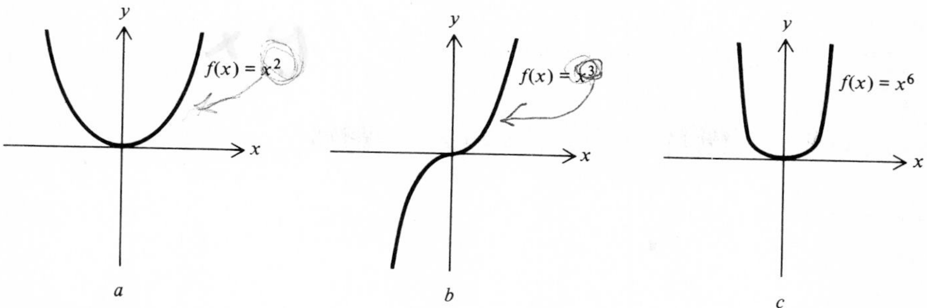

Diagram 6

all nonnegative real numbers, respectively. Hence $f(x) = x^3$ has a range of all reals whereas $f(x) = x^2$ and $f(x) = x^{100}$ have a range of all nonnegative reals, that is, $y \geq 0$.

Some special cases of the power function will be discussed in Chapters 3 and 5 (also see Exercise 3, page 52). Therefore we will not go into detail here on methods of obtaining their graphs. Just to give an idea of how they look we illustrate several examples in Diagram 6.

The Absolute-Value Function This function is of the form $g(x) = |x|$ and is defined for all real numbers. The definition of absolute value was given in Chapter 1. For convenience we repeat it:

$$|x| = \begin{cases} x & \text{if } x \text{ is nonnegative} \\ -x & \text{if } x \text{ is negative} \end{cases}$$

In terms of the idea of function, the absolute-value rule assigns to each real number x the nonnegative number $|x|$.

Functionally, $g(x) = |x|$ agrees with the identity function $f(x) = x$ for all nonnegative values of x ($x \geq 0$); however, for all negative values of x ($x < 0$), $g(x) = -f(x)$. This means that for *negative* x, the absolute-value function is just a reflection (mirror image) about the x axis of the identity function (see Diagram 7).

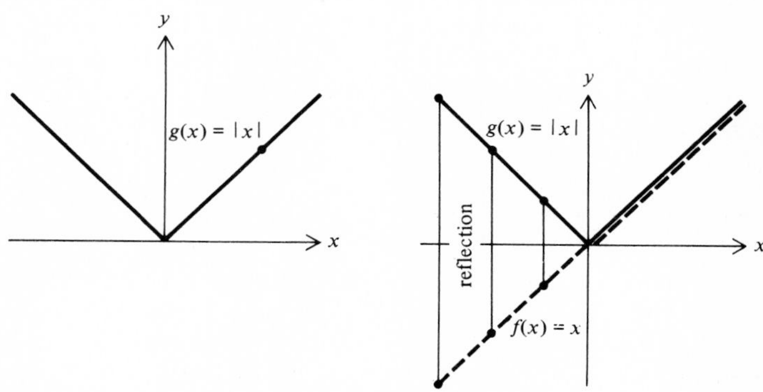

Diagram 7

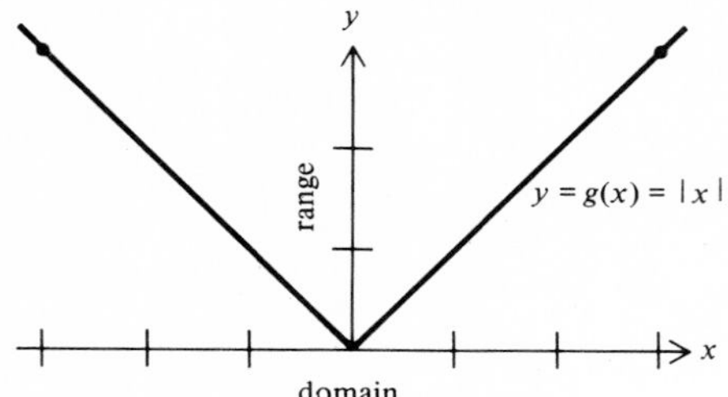

Diagram 8

Summarizing, we can see (Diagram 8) that the absolute-value function has a domain of all real numbers and a range of all nonnegative real numbers. Furthermore this is the first example of a function that can be defined by two separate rules over two different intervals, that is,

$$g(x) = \begin{cases} x & \text{for } x \geq 0 \\ -x & \text{for } x < 0 \end{cases}$$

The Cube-Root Function We write this function as $f(x) = \sqrt[3]{x}$ and define it for all real numbers. Every real number has a unique real cube root, and we define this root as follows.

DEFINITION 1 Let a and b be any real number; b is said to be the *cube root* of a, denoted $b = \sqrt[3]{a}$, if and only if $b \cdot b \cdot b = b^3 = a$.

Hence -2 is the cube root of -8 since $(-2)(-2)(-2) = -8$, and we can write $\sqrt[3]{-8} = -2$.

We can obtain a graph of this function by generating points of the form $(x, \sqrt[3]{x})$. Let us tabulate some of these points:

x	0	$\frac{1}{8}$	$-\frac{1}{8}$	1	-1	2	-2	3	-3	4	-4	8	-8
$f(x)$	0	$\frac{1}{2}$	$-\frac{1}{2}$	1	-1	1.26	-1.26	1.44	-1.44	1.59	-1.59	2	-2

Plotting these points gives a fair idea of how the graph of the cube-root function looks (see Diagram 9*a*). We could continue to plot more and more points, but this is not necessary. This function is defined for all real numbers and is *well behaved;* that is, for small changes in the value of x we obtain a corresponding small change in the value of $f(x)$. Hence we can correctly assume the graph to be a smooth unbroken curve through the points we generated (see Diagram 9*b*).

The Cost Function Consider the cost of mailing light packages up to 4 ounces in weight. We define *cost* as a function of the weight w as follows:

$$C(w) = \begin{cases} 5 & \text{if } w \text{ is between 0 and 1 including 1} \\ 10 & \text{if } w \text{ is between 1 and 2 including 2} \\ 15 & \text{if } w \text{ is between 2 and 3 including 3} \\ 20 & \text{if } w \text{ is between 3 and 4 including 4} \end{cases}$$

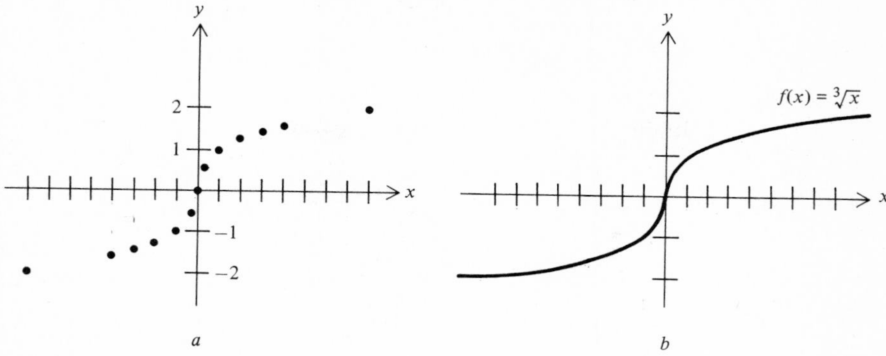

Diagram 9

$y = C(w)$ is a *cost function* defined by four separate rules over four different partial domains. Note that for each partial domain, $C(w)$ represents a constant function. w is the independent variable, and we can see that for each w between 0 and 1, $C(w) = 5$; for each w between 1 and 2, $C(w) = 10$ and so on. If we take care with our end points, the graph is easy to obtain (see Diagram 10).

COMMENT The axes are not labeled x and $f(x)$ but w and $C(w)$. $C(1) = 5$ and the open circle denotes exclusion of $(1, 10)$. The same argument applies to $C(0), C(2), C(3)$. Note that the range contains four and only four elements, namely, 5, 10, 15, and 20.

We have only scraped the surface in our discussion of functions. Throughout the text you will encounter many different kinds of functions.

We just saw that in the cost function, separate rules were defined for different intervals in the domain. Let us now consider one final example which defines an algebraic function in a similar manner. Draw a graph of

$$f(x) = \begin{cases} x + 2 & -3 \le x < 1 \\ 2 & x = 1 \\ (x - 2)^2 & 1 < x \le 4 \end{cases}$$

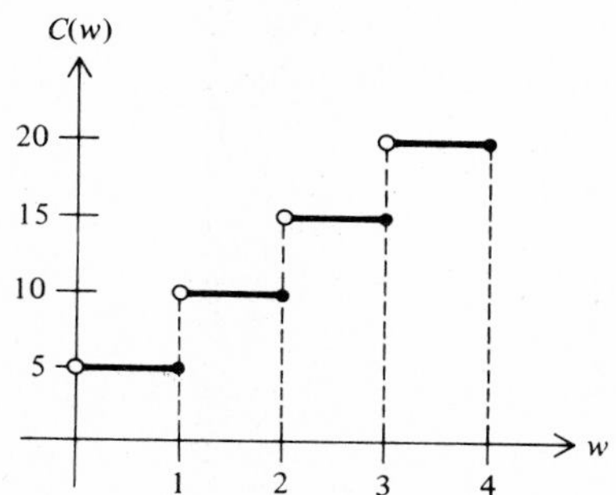

Diagram 10

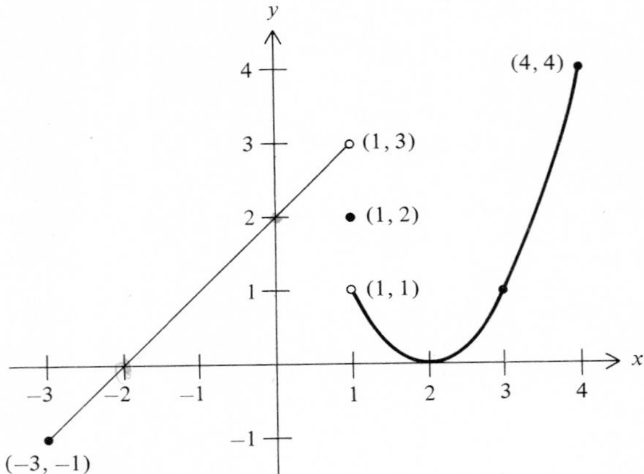

Diagram 11

The function in Diagram 11 is defined over the interval from $x = -3$ to $x = 4$ inclusive. Therefore the graph of the function will exist *only* in this interval. This is the *domain* of $f(x)$.

In the interval from -3 to 1, the graph of $f(x)$ is the straight line $f(x) = x + 2$. However, since $x = 1$ is *not* included in the given rule, we indicate the point (1, 3) with a circle. At $x = 1$, the rule states that $f(x) = 2$, hence a single dot is located at (1, 2). For the remainder of the domain, the graph of $f(x)$ is the second-degree polynomial $f(x) = (x - 2)^2$. Note, again, the circle at (1, 1) since the rule does *not* include $x = 1$.

By examining Diagram 11 it can also be seen that the *range* of this three-part function is all values of $f(x)$ from $y = -1$ to $y = 4$ inclusive.

EXERCISES

1 If you print each letter of the English alphabet in block-capital form so that the base of each letter lies on the x axis of a coordinate plane, how many letters can be a graph of a function?

2 State the domain of each function:
(a) $f(x) = \sqrt[4]{x}$ **(b)** $f(x) = \sqrt[5]{x}$ **(c)** $f(x) = \sqrt{x^2}$

3 Without plotting points make a quick sketch of the following power functions:
(a) $y = x^6$ **(b)** $y = x^7$ **(c)** $y = x^8$

In Exercises 4 to 11, draw a graph of each of the functions. Specify the domain and range.

4 $f(x) = \dfrac{1}{x}$ **5** $f(x) = -x^2$ **6** $f(x) = \dfrac{-1}{x^2}$

7 $f(x) = 2$ **8** $f(x) = \sqrt{x - 1}$ **9** $f(x) = \sqrt{1 - x}$

10 $f(x) = \dfrac{x}{|x|}$ **11** $f(x) = x|x|$

12 Express the area of a square as a function of **(a)** the length of its side, **(b)** the length of a diagonal, **(c)** its perimeter.

13 Express the volume of a cube as a function of **(a)** the length of its edge, **(b)** its surface area, **(c)** the *sum* of the lengths of its edges, and **(d)** the length of a diagonal.

14 A piece of wire 20 inches long is bent into the shape of a rectangle. If one of the sides is x inches, express the area of the rectangle as a function of x.

15 Consider a square and a circle, such that the diameter of the circle is equal to a side of the square. Express their combined area as a function of the radius of the circle.

16 Consider a square and a circle *inscribed* in the square. Express the area outside of the circle and inside the square as a function of: **(a)** the radius of the circle, **(b)** the side of the square, **(c)** the circumference of the circle.

17 Consider a circle and a square *inscribed* in the circle. Express the area outside of the square and inside of the circle as a function of **(a)** the radius of the circle, **(b)** the side of the square, **(c)** the diagonal of the square.

18 Express the volume of a sphere as a function of its diameter.

19 The volume of a cone is given by the formula $V = \frac{1}{3}\pi r^2 h$, where r is the radius of the base and h is the height. For a given cone, the height is 5 times the radius. Express the volume of this cone as a function of **(a)** its diameter and **(b)** its height.

20 The area of a trapezoid is given by the formula $A = \frac{1}{2}h(b_1 + b_2)$, where h represents its height with b_1 and b_2 the bases. Suppose a certain trapezoid has one base equal to half the length of the other and the height equal to two-thirds of the shorter base. Express the area of the trapezoid as a function of **(a)** the shorter base, **(b)** the longer base, **(c)** the height.

21 Construct a function that has an infinite domain and a range of **(a)** exactly one element and **(b)** exactly two elements.

22 Find a function $y = f(x)$ such that:

(a) $f(x) = f(-x)$ **(b)** $-f(x) = f(-x)$

(c) $f(x) \neq f(-x)$ **(d)** $-f(x) \neq f(-x)$

23 Draw a graph of the function. Specify the domain and the range.

$$g(x) = \begin{cases} 4 + x & x \le 1 \\ 4 - x & x > 1 \end{cases}$$

24 Draw a graph of the function. Specify the domain and the range.

$$h(x) = \begin{cases} x - 1 & x \le 0 \\ 0 & x \text{ between } 0 \text{ and } 1 \\ x & x \ge 1 \end{cases}$$

25 Draw a graph of the function sgn(x), the signum function. Specify the domain and the range.

$$\operatorname{sgn}(x) = \begin{cases} 1 & x > 0 \\ 0 & x = 0 \\ -1 & x < 0 \end{cases}$$

26 Draw a graph of the function. Specify the domain and the range.

$$f(x) = \begin{cases} x^2 & x \text{ between } 0 \text{ and } 2 \\ 3 & x = 2 \\ -x & x > 2 \end{cases}$$

27 Draw a graph of the function. Specify the domain and the range.

$$f(x) = \begin{cases} 3 & -2 < x \leq 0 \\ 3 - x & 0 < x \leq 3 \\ (x - 3)^2 & x > 3 \end{cases}$$

28 Draw a graph of the function. Specify the domain and the range.

$$g(x) = \begin{cases} -(x + 1) & -2 \leq x \leq -1 \\ x + 1 & -1 < x \leq 0 \\ (1 - x) & 0 < x < 1 \\ x - 1 & 1 \leq x \leq 2 \end{cases}$$

29 Draw a graph of the function. Specify the domain and the range.

$$f(x) = \begin{cases} x^2 & -2 \leq x < 1 \\ -|x - 2| & 1 \leq x < 3 \\ 2 & x = 3 \\ x - 4 & 3 < x \leq 5 \end{cases}$$

30 Referring to the example described in diagram 11, for what value(s) of x in the domain of the function is **(a)** $f(-2) = f(x)$ and **(b)** $f(1) = f(x)$?

31 A running back in a semipro football league gets paid a bonus for net yards rushed in a game. Rushing up to but not including 30 yards, he receives \$10. From 30 yards up to but not including 60 yards he receives \$20. From 60 yards up to but not including 100 yards he receives \$30. As an extra incentive, he gets \$50 for 100 yards gained *and* then \$10 for each additional yard gained beyond 100. Describe a function that will predict his exact bonus for a single game, where x denotes the number of yards gained. Sketch a graph of the function and specify the domain and the range of the function.

32 From the definition of absolute value, $f(x) = |x| = x$ if x is positive and $f(x) = |x| = -x$ if x is negative. Following this definition, draw a graph of the functions:

(a) $f(x) = |x| + x$ **(b)** $f(x) = |x| - x$ **(c)** $f(x) = x - |x|$

33 Suppose $f(x) = |x - 2|$.

(a) For what value of x does $f(x) = 0$? For what values of x is $f(x) > 0$, and for what values of x is $f(x) < 0$?

(b) Keeping in mind the above information and the definition of absolute value, rewrite $f(x)$ according to the scheme:

$$f(x) = \begin{cases} ? & x < 2 \\ ? & x = 2 \\ ? & x > 2 \end{cases}$$

(c) Draw a graph of $f(x)$.

34 Following the directions in Exercise 33, rewrite $f(x) = |2 - x|$ in a similar manner.

35 The displacement, s, of an object is given by

$$s(t) = \frac{a}{2}t^2 + bt + c$$

where a = acceleration
b = initial velocity
c = initial displacement
t = time

The velocity, v, of this object at any time $t > 0$ is given by $v(t) = at + b$. Suppose $s(0) = 0$ and $v(0) = 60$ feet per second. We are told that the acceleration is 32 feet per second per second.

(a) Determine a, b, and c and rewrite $s(t)$ and $v(t)$.
(b) Draw a graph of $v(t) = at + b$, for t between 0 and 5 seconds inclusive.
(c) Determine $s(2)$, $s(3)$, and $s(5)$.
(d) Find the area bounded by the graph of $v(t) = at + b$, the coordinate axes and the line:
(1) $t = 2$ **(2)** $t = 3$ **(3)** $t = 5$
Hint: Area of a trapezoid equals $\frac{1}{2}h(b + b')$.
(e) Compare parts (*c*) and (*d*). What is your conclusion?

36 An employee of a particular company receives $4 per hour for each hour he works in a regular 6-hour day. If he works more than 6 hours in any one day, he receives time and a half for the next 3 hours and double time for the 3 hours that follow the first overtime period. Describe a function that will predict his exact wage for a single day at any time t between 0 and 12 hours, inclusive.

37 Given

$$f(x) = \begin{cases} 1 & \text{if } x \text{ is rational} \\ -1 & \text{if } x \text{ is irrational} \end{cases}$$

Answer true or false:

(a) $f(\frac{22}{7}) = -1$ **(b)** $f(0.666 \cdots) = 1$ **(c)** $f(0) = 1$
(d) $f(\pi) = 1$ **(e)** $f(\sqrt{0.4}) = -1$ **(f)** $f(\sqrt[3]{1000}) = 1$

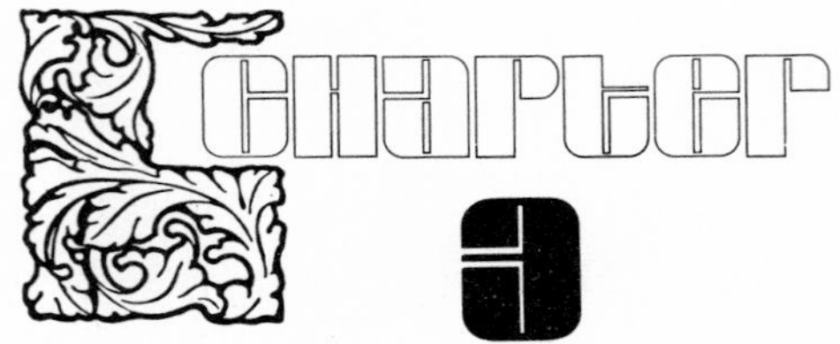

Quadratic Functions

3.1 GRAPHING QUADRATIC FUNCTIONS

A function described as $f(x) = ax^2 + bx + c$, where a, b, and c are real numbers, $a \neq 0$, is called a *quadratic function*.

Our problem in this chapter is to analyze the graphical characteristics of a quadratic and examine its usefulness in problem solving. We will sketch the graphs of the functions and classify them in special categories.

We begin by considering the simple function $f(x) = x^2$ and its graph. In the following table, several ordered pairs have been generated for the function $y = x^2$.

x	-3	-2	-1	0	1	2	3
y	9	4	1	0	1	4	9

In Diagram 1 these points have been plotted. If additional points are plotted, the pattern is as shown in Diagram 2. This curve is called a *parabola*. It is *symmetric* with respect to the y axis.

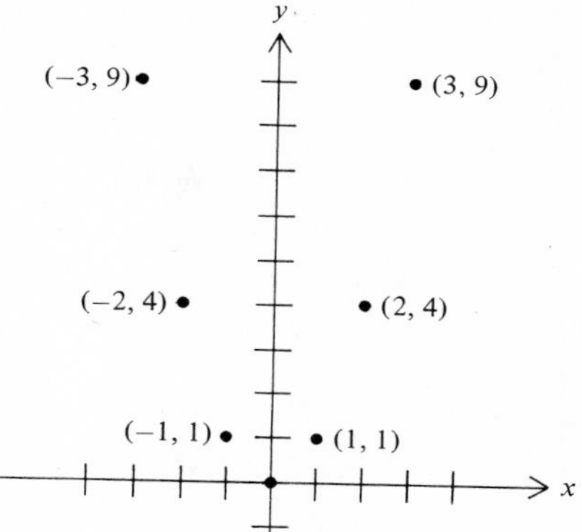

Diagram 1

If a function has the property that for every ordered pair (x, y) that belongs to the function, $(-x, y)$ also belongs to the function, then the graph of $y = f(x)$ is symmetric with respect to the y axis. The y axis for this example is called the *axis of symmetry*.

CASE 1

$$f(x) = ax^2 \qquad \begin{array}{l} a \text{ a real number} \\ a \neq 0 \end{array}$$

PROBLEM Analyze the following quadratic functions:

$$f(x) = x^2 \qquad a = 1$$

$$f(x) = 2x^2 \qquad a = 2$$

$$f(x) = \frac{x}{2} \qquad a = \tfrac{1}{2}$$

$$f(x) = -2x^2 \qquad a = -2$$

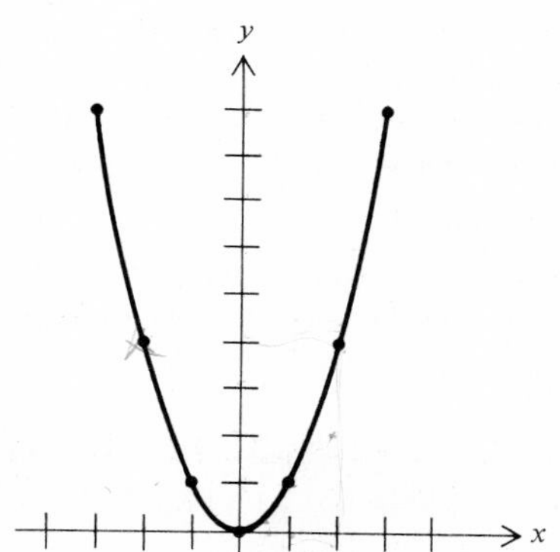

Diagram 2

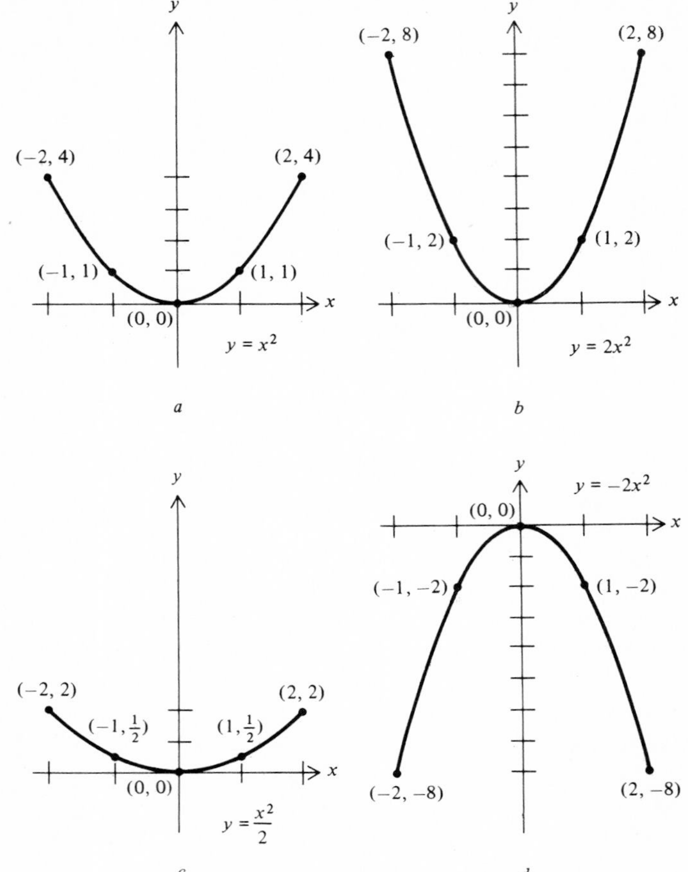

Diagram 3

In Diagram 3, each curve is sketched with a selection of various points indicated on the graph.

COMMENTS

When $a > 0$, the curve opens upward.

When $a < 0$, the curve opens downward.

Each curve is symmetric about the y axis. (For all the ordered pairs when the ordinates are the same, the abscissas are the negatives of each other.)

Each curve has a *turning point.*

In Diagram 3*a, b,* and *c* the point (0, 0) lies below every other point on the graph of $f(x)$ and is called the *minimum point.* In Diagram 3*d,* the point (0, 0) lies above every other point on the graph of $f(x)$ and is called the *maximum point.* This point is also referred to as the *vertex* of the graph.

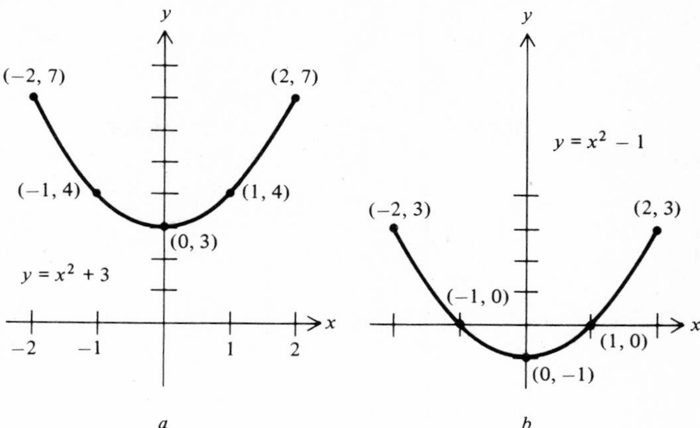

Diagram 4

CASE 2

$$f(x) = ax^2 + c \qquad \begin{array}{l} a \text{ and } c \text{ real numbers} \\ a \neq 0, c \neq 0 \end{array}$$

The only variation in Case 2 is the addition of the constant c. (Let us consider only $a > 0$; hence, the curve will open upward and possess a minimum point.) Consider

$$f(x) = x^2 + 3 \qquad f(x) = x^2 - 1$$

Determine the effect of adding 3 to the graph of $y = x^2$ (see Diagram 4*a*) and the effect of subtracting 1 from the graph of $y = x^2$ (see Diagram 4*b*). The graph of $y = x^2 + 3$ is the same as the graph of $y = x^2$ except that the vertex is *3 units above* the origin. The graph of $y = x^2 - 1$ is the same as the graph of $y = x^2$ except that the vertex is *1 unit below* the origin.

Diagram 5 gives a graphical summary of the effect of adding a constant c to the function $f(x) = ax^2$.

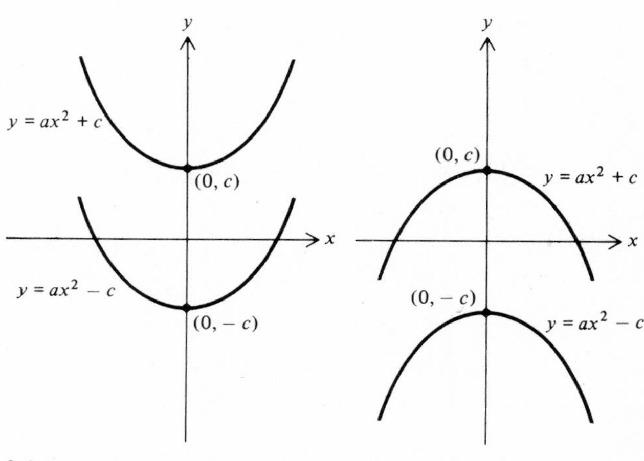

Diagram 5 $[a > 0; c > 0]$ $[a < 0; c > 0]$

CASE 3

$$f(x) = ax^2 + bx + c \qquad \begin{array}{l} a, b, c \text{ real numbers} \\ a \neq 0, b \neq 0 \end{array}$$

The General Quadratic Function In this case, we have the added effect of the bx term. Since this case is the general form of *all* quadratic functions of x, we must carefully analyze the results obtained. Suppose we generate ordered pairs which satisfy

$$f(x) = x^2 - 4x + 4 \qquad \text{and} \qquad g(x) = -x^2 + 4x - 3$$

x	-1	0	1	2	3	4	5
$f(x)$	9	4	1	0	1	4	9

See Diagram 6*a*.

x	-1	0	1	2	3	4	5
$g(x)$	-8	-3	0	1	0	-3	-8

For graphical analysis these functions are more easily handled by writing them in the following form:

$$y = a(x - H)^2 + K$$

All quadratic functions can be transformed from the form

$$f(x) = ax^2 + bx + c$$

into the form above. (This is explained in more detail in Example 4.)

Therefore let us write

$$f(x) = (x - 2)^2 \qquad \text{and} \qquad g(x) = -(x - 2)^2 + 1$$

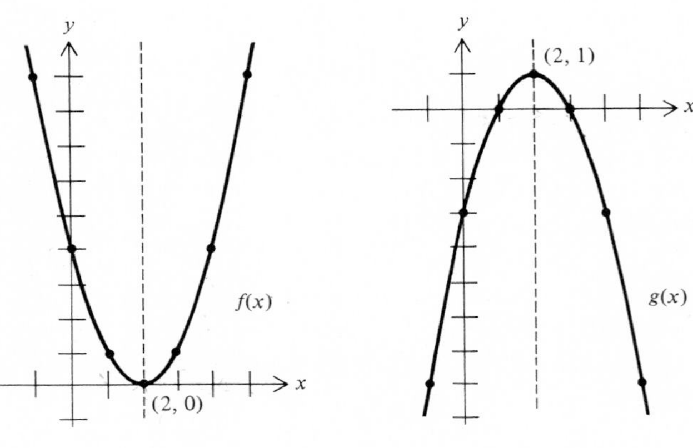

Diagram 6

The graph of $f(x) = (x - 2)^2$ is the same as the graph of $y = x^2$ except that it is 2 units to the right. The axis of symmetry of the graph of $y = x^2$ is $x = 0$; the axis of symmetry of the graph of $y = (x - 2)^2$ is $x = 2$. The vertex of the graph of $y = x^2$ is (0, 0); the vertex of the graph of $y = (x - 2)^2$ is (2, 0).

The line $x = 2$ is also the axis of symmetry of the graph of $y = -(x - 2)^2 + 1$. The graph of $y = -(x - 2)^2 + 1$ is the same as the graph of $y = -x^2$ except that it is 2 units to the right and 1 unit above. The vertex of the graph of $y = -x^2$ is (0, 0); the vertex of the graph of $y = -(x - 2)^2 + 1$ is (2, 1).

SUMMARY

1 The graph of $y = a(x - H)^2 + K$, $a \neq 0$, is the same as the graph of $y = ax^2$, except that the vertex is H units to the *right* (if $H > 0$) or H units to the *left* (if $H < 0$) *and* K units *above* the origin (if $K > 0$) or K units *below* the origin (if $K < 0$).

2 The graph is symmetric to the line $x = H$ (axis of symmetry), and the coordinates of the vertex are (H, K).

3 When $a > 0$, the vertex is the minimum point.

4 When $a < 0$, the vertex is the maximum point.

In Examples 1 to 3 analyze each quadratic function in the form

$$y = a(x - H)^2 + K$$

Example 1 $y = (x - 4)^2 + 2$; $a = 1, H = 4, K = 2$.

This is the graph of $y = x^2$, 4 units to the right and 2 units above the origin. $x = 4$ is the equation of the axis of symmetry. (4, 2) are the coordinates of the vertex (minimum point).

Example 2 $y = -2(x + 2)^2 - 3$; $a = -2, H = -2, K = -3$.

This is the graph of $y = -2x^2$, 2 units to the left and 3 units below the origin. $x = -2$ is the equation of the axis of symmetry. $(-2, -3)$ are the coordinates of the vertex (maximum point).

Example 3 $y = \frac{1}{2}(x - 3)^2 - 4$; $a = \frac{1}{2}, H = 3, K = -4$.

This is the graph of $y = \frac{1}{2}x^2$, 3 units to the right and 4 units below the origin. $x = 3$ is the equation of the axis of symmetry. $(3, -4)$ are the coordinates of the vertex (minimum point).

If we have a function in the form $f(x) = a(x + H)^2 + K$, we can sketch this graph easily by knowing how to graph $f(x) = ax^2$.

The function $y = (x - 2)^2 - 1$ can also be written $y = x^2 - 4x - 3$. In this form, the graph of the function cannot be recognized so easily.

How can we write a quadratic function $f(x) = ax^2 + bx + c$ in the form $f(x) = a(x - H)^2 + K$ for easy graphing? The method used is called *completing the square.*

Example 4 Express the function $f(x) = 2x^2 - 8x + 5$ in the form $f(x) = a(x - H^{\ 2} + K$.

STEP 1 Write the equation with y *and* the constant term to the left:

$$y - 5 = 2x^2 - 8x$$

STEP 2 Factor out the coefficient of the x^2 term so that the coefficient of the x^2 term in the parentheses is always 1

$$y - 5 = 2(x^2 - 4x)$$

STEP 3 Find one-half the coefficient of the x term (from step 2) and square the result:

$$\tfrac{1}{2}(-4) = -2 \qquad \text{then} \qquad (-2)^2 = 4$$

STEP 4 Add this term inside the parentheses $(x^2 - 4x + 4)$, which is now a *perfect square,* $(x - 2)^2$.

$$\overbrace{y - 5 + 8}^{} = \overset{2(4)\,=\,8}{2(x^2 - 4x + 4)} \qquad \text{then} \qquad y + 3 = 2(x - 2)^2$$

Add 8 (under the $+8$)

STEP 5 Write an equation with y as the only term on the left.

$$y = 2(x - 2)^2 - 3$$

CONCLUSION $f(x) = 2x^2 - 8x + 5 = 2(x - 2)^2 - 3$. This is the graph of $y = 2^2$, 2 units to the right and 3 units below the origin. The equation of the axis of symmetry is $x = 2$. The vertex is the point $(2, -3)$ and is the *minimum point* (see Diagram 7).

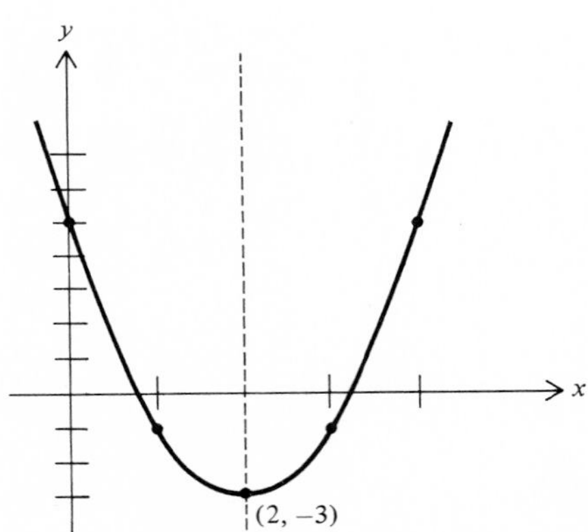

Diagram 7

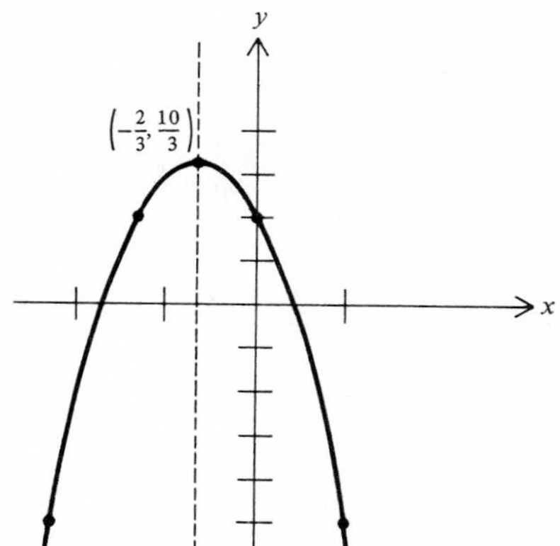

Diagram 8

Example 5 Express the function $f(x) = -3x^2 - 4x + 2$ in the form $f(x) = a(x - H)^2 + K$ and analyze.

STEP 1

$y - 2 = -3x^2 - 4x$

STEP 2

$y - 2 = -3(x^2 + \frac{4}{3}x)$

STEP 3

$\frac{1}{2}(\frac{4}{3}) = \frac{2}{3} \qquad (\frac{2}{3})^2 = \frac{4}{9}$

STEP 4

$$y - 2 - \frac{4}{3} = -3(x^2 + \frac{4}{3}x + \frac{4}{9})$$

Add $-\frac{4}{3}$ $\qquad -3(\frac{4}{9}) = -\frac{4}{3}$

and $y - \frac{10}{3} = -3(x + \frac{2}{3})^2$

STEP 5

$y = -3(x + \frac{2}{3})^2 + \frac{10}{3}$

This is the graph of $y = -3x^2$, $\frac{2}{3}$ unit to the left and $\frac{10}{3}$ units above the origin. The equation of the axis of symmetry is $x = -\frac{2}{3}$. The vertex is the point $(-\frac{2}{3}, \frac{10}{3})$, which is the *maximum point* (refer to Diagram 8).

Example 6 Find the equation of the axis of symmetry and the coordinates of the vertex for the graph of *any* quadratic function

$$f(x) = ax^2 + bx + c \qquad a \neq 0$$

STEP 1

$y - c = ax^2 + bx$

STEP 2

$y - c = a\left(x^2 + \frac{b}{a}x\right)$

STEP 3

$$\frac{1}{2}\frac{b}{a} = \frac{b}{2a} \qquad \left(\frac{b}{2a}\right)^2 = \frac{b^2}{4a^2}$$

STEP 4

$$y - c + \frac{b^2}{4a} = a\left(x^2 + \frac{b}{a}x + \frac{b^2}{4a^2}\right)$$

↑

Add $\frac{b^2}{4a}$ since $a\frac{b^2}{4a^2} = \frac{b^2}{4a}$

and $y - c + \dfrac{b^2}{4a} = a\left(x + \dfrac{b}{2a}\right)^2$

STEP 5

$$y = a\left(x + \frac{b}{2a}\right)^2 + c - \frac{b^2}{4a} = a\left(x + \frac{b}{2a}\right)^2 + \frac{4ac - b^2}{4a}$$

Comparing this result to the general form, we have

$$y = a(x - H)^2 + K$$

$$H = -\frac{b}{2a} \qquad K = \frac{4ac - b^2}{4a}$$

CONCLUSION The equation of the axis of symmetry is $x = -b/2a$, and the coordinates of the vertex are $(-b/2a, (4ac - b^2)/4a)$. The vertex is a minimum point if $a > 0$ or a maximum point if $a < 0$.

Example 7 Find the equation of the axis of symmetry and the coordinates of the vertex for $f(x) = 2x^2 + 3x - 2$, where $a = 2$, $b = 3$, $c = -2$.

The axis of symmetry is

$$x = \frac{-3}{2(2)} = -\frac{3}{4}$$

The coordinates of the vertex are

$$\left(-\frac{3}{4}, \frac{4(2)(-2) - 9}{4(2)}\right) \quad \text{or} \quad \left(-\frac{3}{4}, -\frac{25}{8}\right)$$

Since $a > 0$, this is a *minimum point*.

EXERCISES

In Exercises 1 to 12 graph both functions in the same coordinate plane.

1 $f(x) = 4x^2$; $g(x) = \frac{1}{4}x^2$

2 $f(x) = -3x^2$; $g(x) = \frac{1}{3}x^2$

3 $f(x) = \frac{3}{2}x^2$; $g(x) = -\frac{3}{2}x^2$

4 $f(x) = -2x^2$; $g(x) = -\frac{1}{2}x^2$

5 $f(x) = x^2 + 4$; $g(x) = x^2 - 2$

6 $f(x) = -x^2 - 1$; $g(x) = -x^2 + 3$

7 $f(x) = 2x^2 - 4$; $g(x) = -2x^2 + 3$

8 $f(x) = (x + 2)^2$; $g(x) = -(x + 2)^2 + 1$

9 $f(x) = 2(x - 1)^2 - 4$; $g(x) = 2(x + 1)^2 - 4$

10 $f(x) = -2(x + 1)^2 + 3$; $g(x) = -2(x - 1)^2 - 3$

11 $f(x) = \frac{1}{2}(x + 3)^2 + 2$; $g(x) = -\frac{1}{2}(x + 3)^2 - 2$
12 $f(x) = 3(x - \frac{5}{2})^2 + \frac{3}{2}$; $g(x) = \frac{1}{3}(x - \frac{5}{2})^2 + \frac{3}{2}$

Express the functions in Exercises 13 to 25 in the form $y = a(x - H)^2 + K$ and determine. **(a)** the equation of the axis of symmetry and **(b)** the coordinates of the vertex.

13 $f(x) = x^2 - 2x + 2$
14 $f(x) = -x^2 - 4x - 4$
15 $f(x) = -2x^2 + 4x - 1$
16 $f(x) = 2x^2 + 7x + 5$
17 $f(x) = -3x^2 + 4x - 2$
18 $f(x) = 2x^2 + 6x$
19 $f(x) = -x^2 - 8x$
20 $f(x) = -\dfrac{x^2}{2} - 4x + 8$
21 $f(x) = \frac{3}{2}x^2 - 3x + \frac{25}{6}$
22 $f(x) = \dfrac{x^2}{3} + \dfrac{4x}{3} - \dfrac{5}{3}$
23 $f(x) = -\frac{2}{3}x^2 + 2x - \frac{21}{3}$
24 $f(x) = 12x^2 + 7x - 12$
25 $f(x) = -(2 - 3x)^2 - (2x + 3)^2$

In Exercises 26 to 28 determine the value of K so that the graph of the given functions contain the given point.

26 $f(x) = 2(x - 1)^2 + K$, $(-1, 3)$
27 $f(x) = -\frac{1}{2}(x + 2)^2 + K$, $(-3, 1)$
28 $f(x) = 3(x - 2)^2 + K$, $(4, -2)$
29 If $(R, 0)$ and $(S, 0)$, $R \neq S$, are points on the graph of the function $f(x) = a(x - H)^2 + K$, show that

$$H = \frac{R + S}{2}$$

30 Compare the graphs of

$$f(x) = x^2 - 4 \quad \text{and} \quad g(x) = |x^2 - 4|$$

31 Compare the graphs of

$$y = x^2 - 3x - 4 \quad \text{and} \quad y = 2x^2 - 6x - 8$$

3.2 ROOTS

In Section 3.1 we defined a quadratic function as $f(x) = ax^2 + bx + c$. When $f(x) = 0$, we form a *quadratic equation*, $ax^2 + bx + c = 0$.

DEFINITION 1 All values of x such that $f(x) = 0$ are called *roots* of the equation [and *zeros* of the function $f(x)$].

A quadratic equation with real coefficients can have

Two different real roots
Two equal real roots
No real roots.

In order to perform a thorough analysis of a quadratic equation we introduce a four-step method (see Diagram 9).

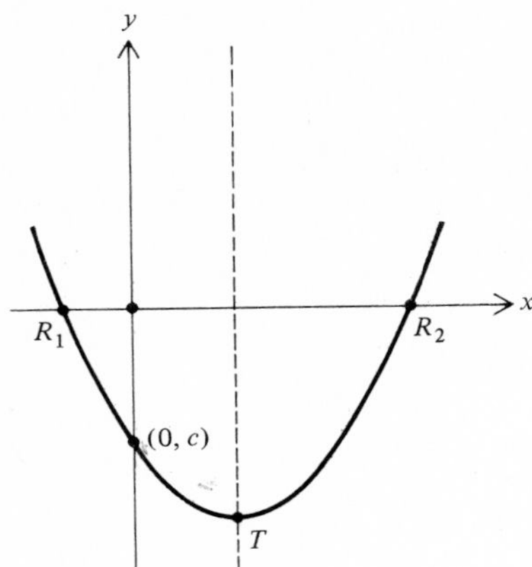

Diagram 9

STEP 1 Determine the real roots of $f(x) = 0$ (if they exist) labeled R_1 and R_2.

STEP 2 Find the y intercept of the graph of $f(x)$. The y intercept is determined by finding $f(0)$.

$$f(x) = ax^2 + bx + c \qquad f(0) = a(0)^2 + b(0) + c = c$$

Locate the point $(0, c)$.

STEP 3 Find the axis of symmetry of the graph of $f(x)$.

STEP 4 Find the coordinates of the vertex of the graph of $f(x)$ (labeled T).

Example 1 Analyze the function $f(x) = 2x^2 - 3x - 2$.

Since the graph of $f(x)$ crosses the x axis where $y = 0$, the roots are determined by finding the values of x for which $f(x) = 0$.

$$0 = 2x^2 - 3x - 2$$

STEP 1 *If* the quadratic can be expressed as the product of two factors, the equation can be solved by setting each factor equal to zero.

$$0 = (2x + 1)(x - 2)$$

$$2x + 1 = 0 \qquad\qquad x - 2 = 0$$

$$x = -\tfrac{1}{2} \qquad\qquad x = 2$$

The graph of $f(x)$ crosses the x axis at the points $(-\frac{1}{2}, 0)$ and $(2, 0)$. Therefore

$$x = -\tfrac{1}{2} \quad \text{and} \quad x = 2$$

are the roots of the quadratic equation.

JUSTIFICATION Since $AB = 0$ if and only if $A = 0$ or $B = 0$, it follows that $(x - R_1)(x - R_2) = 0$ if and only if $x - R_1 = 0$ or $x - R_2 = 0$. And solving for x, $x = R_1$ and $x = R_2$ are the roots.

STEP 2 The y intercept is $f(0) = -2$. Therefore, $(0, -2)$ is the point where the graph of $f(x)$ crosses the y axis.

STEP 3 Express the function in the form

$$y = a(x - H)^2 + K = 2x^2 - 3x - 2$$

$$y + 2 = 2x^2 - 3x$$

$$y + 2 + \tfrac{9}{8} = 2\left(x^2 - \frac{3x}{2} + \frac{9}{16}\right)$$

$$\uparrow \quad 2(\tfrac{9}{16}) = \tfrac{9}{8}$$

$$y + \tfrac{25}{8} = 2(x - \tfrac{3}{4})^2$$

$$y = 2(x - \tfrac{3}{4})^2 - \tfrac{25}{8}$$

STEP 4 The equation of the axis of symmetry is $x = \frac{3}{4}$.

STEP 5 The coordinates of the turning point are $(\frac{3}{4}, -\frac{25}{8})$, which is a minimum point since $2 > 0$ (see Diagram 10).

Suppose that in Example 1 the quadratic equation could *not* be factored simply. How could we find the roots? The process of completing the square can be used.

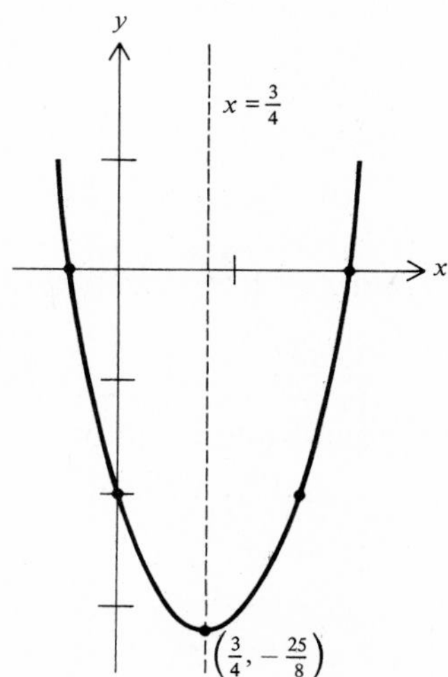

Diagram 10

Example 2 Find the roots of $3x^2 - 4x - 2 = 0$.

STEP 1 Write the quadratic equation in the form $ax^2 + bx + c = 0$:

$$3x^2 - 4x - 2 = 0$$

STEP 2 Transfer the constant term to the other side:

$$3x^2 - 4x = 2$$

STEP 3 Divide both sides of the equation by the coefficient of the x^2 term:

$$x^2 - \tfrac{4}{3}x = \tfrac{2}{3}$$

STEP 4 Find one-half the coefficient of the x term and square the result:

$$\tfrac{1}{2}(-\tfrac{4}{3}) = -\tfrac{2}{3} \qquad (-\tfrac{2}{3})^2 = \tfrac{4}{9}$$

STEP 5 Add this number to both sides of the equation:

$$x^2 - \tfrac{4}{3}x + \tfrac{4}{9} = \tfrac{2}{3} + \tfrac{4}{9}$$

STEP 6 The left side of the equation is a perfect square:

$$(x - \tfrac{2}{3})^2 = \tfrac{10}{9}$$

STEP 7 Take the square root of both sides of the equation:

$$x - \frac{2}{3} = \pm \frac{\sqrt{10}}{3}$$

Note that in order to find *all* the real roots (if any) we take both the positive *and* negative square roots of 10.

STEP 8 Solve for x

$$x = \frac{2}{3} \pm \frac{\sqrt{10}}{3} \quad \Longrightarrow \quad x = \frac{2 + \sqrt{10}}{3} \quad \text{or} \quad x = \frac{2 - \sqrt{10}}{3}$$

Therefore, the graph of the quadratic function $f(x)$ crosses the x axis at

$$\left(\frac{2 - \sqrt{10}}{3}, 0\right) \quad \text{and} \quad \left(\frac{2 + \sqrt{10}}{3}, 0\right)$$

Theorem The roots of the general quadratic equation

$$ax^2 + bx + c = 0$$

are given by the formula

$$x = \frac{-b \pm \sqrt{b^2 - 4ac}}{2a}$$

Proof

Use the eight-step procedure in the previous example.

STEP 1

$$ax^2 + bx + c = 0$$

STEP 2

$$ax^2 + bx = -c \qquad \text{Subtracting } c \text{ from both sides}$$

STEP 3

$$x^2 + \frac{bx}{a} = -\frac{c}{a} \qquad \text{Dividing by } a$$

STEP 4

$$\frac{1}{2}\frac{b}{a} = \frac{b}{2a} \qquad \left(\frac{b}{2a}\right)^2 = \frac{b^2}{4a^2}$$

Taking one-half the coefficient of the x term, then squaring

STEP 5

$$x^2 + \frac{bx}{a} + \frac{b^2}{4a^2} = \frac{b^2}{4a^2} - \frac{c}{a}$$

Adding the result of step 4 to both sides

STEP 6

$$\left(x + \frac{b}{2a}\right)^2 = \frac{b^2 - 4ac}{4a^2}$$

Factoring the left-hand side and combining the side as one term

STEP 7

$$x + \frac{b}{2a} = \pm\frac{\sqrt{b^2 - 4ac}}{2a}$$

Taking the square root of both sides and demanding both positive *and* negative roots

STEP 8

$$x = \frac{-b}{2a} \pm \frac{\sqrt{b^2 - 4ac}}{2a}$$

Adding $-b/2a$ to both sides, thereby solving for x

or
$$x = \frac{-b \pm \sqrt{b^2 - 4ac}}{2a}$$

COMMENT In order to use the formula, we need only determine the values of the constants a, b, and c. Direct substitution into the formula yields the required roots.

Example 3 Find the roots of $2x^2 - 4x - 7 = 0$ using the quadratic formula *and* graph the function $f(x) = 2x^2 - 4x - 7$, $a = +2$, $b = -4$, $c = -7$.

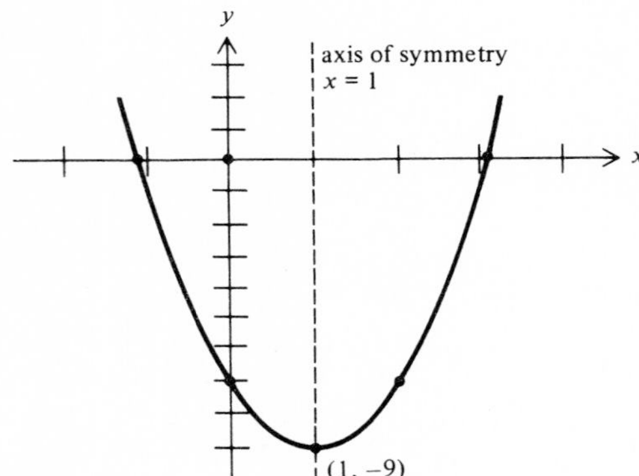

Diagram 11

Solution

$$x = \frac{-(-4) \pm \sqrt{(-4)^2 - 4(2)(-7)}}{2(2)} = \frac{4 \pm \sqrt{16 + 56}}{4}$$

$$= \frac{4 \pm \sqrt{72}}{4} = \frac{4 \pm 6\sqrt{2}}{4} = \frac{2 \pm 3\sqrt{2}}{2} \qquad \text{Since } \sqrt{72} = \sqrt{(36)(2)} = 6\sqrt{2}$$

The roots are

$$\frac{2 + 3\sqrt{2}}{2} \quad \text{or} \quad \frac{2 - 3\sqrt{2}}{2}$$

Find the y intercept:

$$f(0) = -7$$

Rewriting the equation in the form $y = a(x - H)^2 + K$, we obtain $y = 2(x - 1)^2 - 9$.

The equation of the axis of symmetry is $x = 1$. The coordinates of the vertex are $(1, -9)$, which is a minimum point. We can now graph the function (see Diagram 11).

EXERCISES

Analyze the quadratic functions in Exercises 1 to 9.

1 $f(x) = x^2 + 6x + 8$
2 $g(x) = 2x^2 + 5x - 3$
3 $h(x) = -2x^2 + 9x - 9$
4 $k(x) = 9x^2 - 6x + 1$
5 $f(x) = -(x - 1)^2 + 9$
6 $g(x) = -4x^2 + 4x - 1$
7 $h(x) = 8x^2 + 2x - 3$
8 $k(x) = 3x^2 + 14x - 8$
9 $f(x) = -(3x + 4)^2$

Find the roots of the quadratic equations in Exercises 10 to 24 by completing the square.

10 $x^2 - 6x + 8 = 0$
11 $x^2 + x - 6 = 0$
12 $x^2 - 3x - 4 = 0$
13 $x^2 + 4x + 2 = 0$

14 $3x^2 + x - 4 = 0$
15 $6x^2 + 13x - 5 = 0$
16 $9x^2 - 12x + 4 = 0$
17 $16x^2 + 40x + 25 = 0$
18 $2x^2 + 9x - 5 = 0$
19 $x^2 - 2x - 1 = 0$
20 $3x^2 + 8x + 5 = 0$
21 $5x^2 - 4x - 12 = 0$
22 $4x^2 - 12x + 7 = 0$
23 $x^2 + 2x - 2 = 0$
24 $3x^2 + 8bx - 3b^2 = 0$

Find the roots of the quadratic equations in Exercises 25 to 33 by using the quadratic formula and express the answer in the simplest form.

25 $4x^2 + 12x + 7 = 0$
26 $x^2 + 4x - 41 = 0$
27 $6x^2 - 19x + 15 = 0$
28 $3x^2 + 10x + 2 = 0$
29 $9x^2 + 24x + 16 = 0$
30 $4x^2 - 28x + 13 = 0$
31 $10x^2 - 11x - 6 = 0$
32 $3x^2 + 15x - 15 = 0$
33 $12x^2 - 9x - 5 = 0$

34 **(a)** Show that

$$x = \frac{2c}{-b \pm \sqrt{b^2 - 4ac}}$$

is a form of the quadratic formula.

(b) Using this form, find the roots of $4x^2 - 4x - 1 = 0$.

35 Prove:

(a) If $1/k$ and k are both roots of $ax^2 + bx + c = 0$ $(k \neq 0)$, then $a = c$.

(b) Verify the above statement for $6x^2 - 13x + 6 = 0$.

3.3 MORE ON ROOTS; THE DISCRIMINANT

For the quadratic equation $x^2 + 1 = 0$, $x^2 = -1$, but since the square of a real number cannot be negative, this equation has *no real solution.*

In order for $x^2 + 1 = 0$ to have a solution, we must have $x^2 = -1$ and $x = \pm\sqrt{-1}$. The number corresponding to $\sqrt{-1}$ is called an *imaginary number.*

The imaginary number i is defined as $i^2 = -1$. Therefore, $i = \sqrt{-1}$ (and $-i = -\sqrt{-1}$). By using the symbol i to represent $\sqrt{-1}$, we can write $\sqrt{-9}$ as

$$\sqrt{9(-1)} = i\sqrt{9} = 3i$$

In general, if $a \geq 0$, then

$$\sqrt{-a} = \sqrt{(-1)a} = i\sqrt{a}$$

Consider the quadratic equation

$$x^2 - 4x + 5 = 0$$

This equation has *no real solution:*

$$x = \frac{4 \pm \sqrt{16 - 4(5)}}{2} = \frac{4 \pm \sqrt{-4}}{2} = \frac{4 \pm 2i}{2}$$

$$= 2 \pm i$$

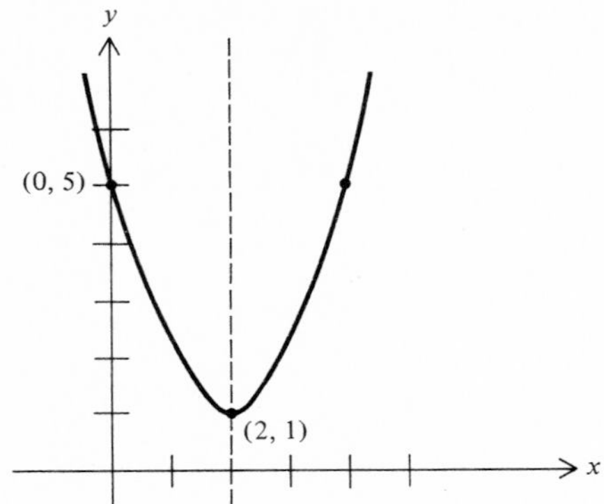

Diagram 12

The number $2 + i$ is called a *complex number*. It consists of a *real part*, 2, and an *imaginary part*, i.

Example 1 Graphically analyze $f(x) = x^2 - 4x + 5$.

Rewrite the equation in the form $y = a(x - H)^2 + K = (x - 2)^2 + 1$. The axis of symmetry is $x = 2$. The coordinates of the vertex are (2, 1), which is a minimum point. The y intercept is 5, since $f(0) = 5$ (see Diagram 12).

COMMENT If a quadratic equation has *no real roots*, the graph of the quadratic *does not intersect* the x axis.

Example 2 Consider the graphs of the functions

$$\begin{aligned} f(x) &= -3x^2 + 4x - 2 \\ g(x) &= 2x^2 + 3x - 5 \\ h(x) &= x^2 + 4x + 4 \\ k(x) &= -2x^2 + 4x - 1 \end{aligned}$$

Express each function in the form $y = (x - H)^2 + K$

$$\begin{aligned} f(x) &= -3(x - \tfrac{2}{3})^2 - \tfrac{2}{3} \\ g(x) &= 2(x + \tfrac{3}{4})^2 - \tfrac{49}{8} \\ h(x) &= (x + 2)^2 \\ k(x) &= -2(x - 1)^2 + 1 \end{aligned}$$

The equations of axes of symmetry are

$$\begin{aligned} f(x)&: \quad x = \tfrac{2}{3} \\ g(x)&: \quad x = -\tfrac{3}{4} \\ h(x)&: \quad x = -2 \\ k(x)&: \quad x = 1 \end{aligned}$$

The turning points are

$f(x)$:	$(\frac{2}{3}, -\frac{2}{3})$	maximum point
$g(x)$:	$(-\frac{3}{4}, -\frac{49}{8})$	minimum point
$h(x)$:	$(-2, 0)$	minimum point
$k(x)$:	$(1, 1)$	maximum point

and the roots are

$f(x)$:	$\dfrac{-2 + i\sqrt{2}}{3}$	and	$\dfrac{-2 - i\sqrt{2}}{3}$	two complex roots
$g(x)$:	$-\frac{1}{2}$	and	1	two real roots (rational)
$h(x)$:		-2		one double root
$h(x)$:	$\dfrac{2 - \sqrt{2}}{2}$	and	$\dfrac{2 + \sqrt{2}}{2}$	two real roots (irrational)

See Diagram 13.

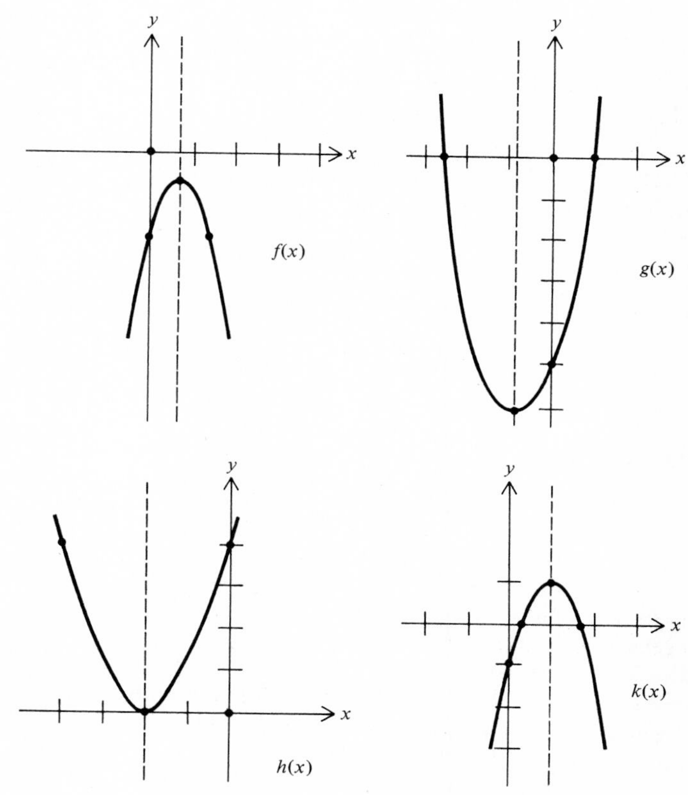

Diagram 13

The value of the expression $b^2 - 4ac$ determines the *type* of roots of a quadratic equation.

1 If $b^2 - 4ac > 0$, then $\sqrt{b^2 - 4ac} > 0$. Therefore, $ax^2 + bx + c = 0$ has *two real, unequal roots.*

2 If $b^2 - 4ac = 0$, then $\sqrt{b^2 - 4ac} = 0$. Therefore, $ax^2 + bx + c = 0$ has *two real, equal roots* (a double root).

3 If $b^2 - 4ac < 0$, then $\sqrt{b^2 - 4ac}$ is imaginary and $ax^2 + bx + c = 0$ has *no real roots.*

The term $b^2 - 4ac$ is called the *discriminant* of the quadratic function.

SUMMARY

Function	Number of x intercepts	Value of the discriminant	Roots
$f(x)$	0	-8	Two complex
$g(x)$	2	49	Two real, unequal (rational)
$h(x)$	1	0	Two real, equal (rational)
$k(x)$	2	8	Two real, unequal (irrational)

EXERCISES

Find the roots of the quadratic equations in Exercises 1 to 6.

1 $4x^2 + 4x + 5 = 0$ **2** $3x^2 + 8x + 7 = 0$ **3** $4x^2 - 16x + 25 = 0$

4 $x^2 + x + 1 = 0$ **5** $2x^2 - 4x + 5 = 0$ **6** $5x^2 - 4x + 3 = 0$

For each of the functions in Exercises 7 to 17 find:

(a) the equation of the axis of symmetry
(b) coordinates of the turning point
(c) the roots

7 $f(x) = 2x^2 + 6x$ **8** $g(x) = -x^2 - 8x$

9 $h(x) = -2x^2 + 8x - 5$ **10** $k(x) = 3x^2 - 4x + 2$

11 $f(x) = -4 + 4x - x^2$ **12** $g(x) = 2x^2 - 36x + 14$

13 $h(x) = \frac{x^2}{2} - 2x + \frac{7}{2}$ **14** $k(x) = -\frac{x^2}{2} - 4x + 8$

15 $f(x) = \frac{x^2}{3} + \frac{4x}{3} - \frac{5}{3}$ **16** $g(x) = -x^2 + \frac{4}{3}x + \frac{8}{9}$

17 $h(x) = -\frac{2}{3}x^2 + 2x - 7$

Using the discriminant, determine the type of roots in the quadratic equations in Exercises 18 to 21.

18 $2x^2 - 6x + 3 = 0$ **19** $6x^2 - 5x - 6 = 0$

20 $3x^2 - 8x + 6 = 0$ **21** $4x^2 - 36x + 81 = 0$

22 Examine (*a*) and (*b*). Comment on the results.

(a) $(\sqrt{-2})(\sqrt{-2}) = \sqrt{4} = 2$
(b) $(\sqrt{-2})(\sqrt{-2}) = (i\sqrt{2})(i\sqrt{2}) = i^2(2) = -2$

3.4 APPLICATIONS OF QUADRATIC FUNCTIONS

In many problems a quadratic function is used as a model which satisfies the conditions of the problem. We first have a description of the problem. Then we translate it from English into symbolic mathematics and formulate an equation which accurately expresses the quantity to be analyzed. This is a formidable task. Since we will concentrate entirely on quadratic functions (the subject of this chapter), we naturally follow the techniques presented in the preceding sections.

Example 1 A uniform border is placed around a rectangular picture 15 by 23 inches. If the area of the border is 315 square inches, find the width of the border (see Diagram 14).

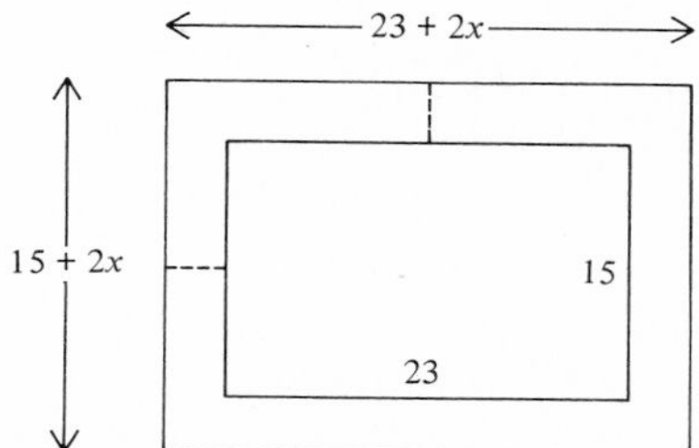

Diagram 14

We translate the information into symbols as follows:

$$(15 + 2x)(23 + 2x) = \text{area of picture} + \text{area of border}$$

$$\text{Area of picture} = 23(15) = 345 \qquad \text{Area of border} = 315$$

$$(15 + 2x)(23 + 2x) = 345 + 315 = 660$$

$$345 + 76x + 4x^2 = 660 \qquad \text{Multiplying left-side factors}$$

$$4x^2 + 76x - 315 = 0 \qquad \text{Subtracting 660 from both sides}$$

$$(2x - 7)(2x + 45) = 0 \qquad \text{Factoring}$$

$$x = \tfrac{7}{2} \quad \text{or} \quad x = -\tfrac{45}{2} \qquad \text{Solving for } x$$

The second solution is rejected as unrealistic, and therefore the width of the border is $\frac{7}{2}$ inches.

Example 2 In a simple electric circuit, two resistors can be placed in series or in parallel (see Diagram 15). The effective resistance in a series cir-

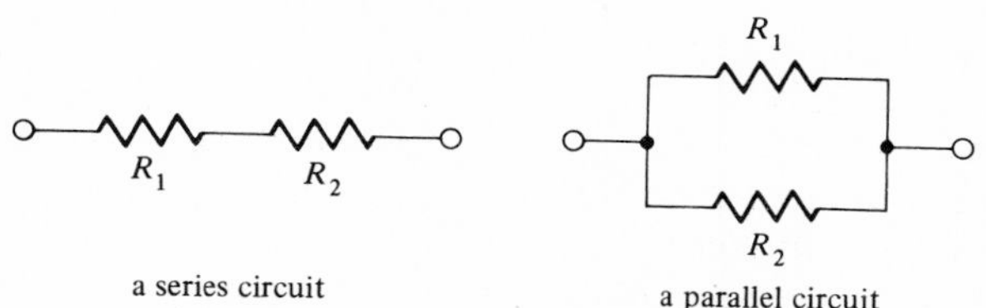

Diagram 15

cuit is given by the formula $R = R_1 + R_2$, and the effective resistance in a parallel circuit is given by the formula

$$R = \frac{R_1R_2}{R_1 + R_2}$$

Given, two resistors, R_1 and R_2; placed in series, their effective value is 18 and placed in parallel, their effective value is 4. Find R_1 and R_2.

Given

$$18 = R_1 + R_2 \qquad 4 = \frac{R_1R_2}{R_1 + R_2}$$

Solve for R_2 in the second equation.

$$4 = \frac{R_1R_2}{18} \qquad R_1R_2 = 72 \qquad R_2 = \frac{72}{R_1}$$

Substitute the value for R_2 in the first equation.

$$18 = R_1 + \frac{72}{R_1} \qquad 18R_1 = R_1^2 + 72$$

Solve your new equation for R_1.

$$R_1^2 - 18R_1 + 72 = 0$$
$$(R_1 - 6)(R_1 - 12) = 0$$
$$R_1 = 6 \qquad R_1 = 12$$

Therefore the answer is that when $R_1 = 6$, $R_2 = \frac{72}{6} = 12$.

In many practical situations we are interested in finding the most efficient or profitable method of performing a task. Corporations investigate methods which will yield the highest (maximum) profit and lowest (minimum) cost. A farmer may wish to enclose the largest amount of grazing area with a given amount of fencing. A beer-can manufacturer may wish to construct his 12-ounce aluminum can using the least amount of metal. A ballistics expert may wish to find the maximum height attained by a certain cannon shell.

Even nature itself provides a spectrum of best-value, that is, maximum or minimum problems. Light always seeks the path of minimum time. Electrons always seek the path of least resistance. In fact, physical phenomena in nature seem to tend toward a minimum principle.

In everyday life, problems in maxima and minima appear repeatedly, and mathematicians have developed efficient techniques for solving them. How should a rocket be designed to minimize air resistance? What advice should an economist give on the retail price of a new automobile to yield the greatest profit?

Although most of these problems are beyond the scope of our development, many of them can be solved. It is our intention to present a systematic approach which will yield a simple, direct, and meaningful solution.

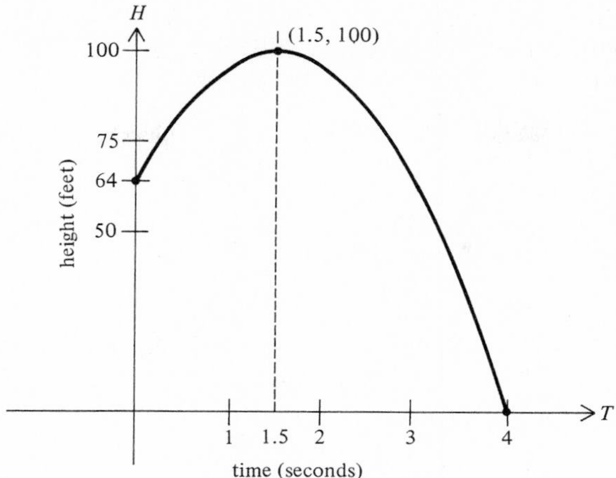

Diagram 16

Example 3 An object is projected vertically upward from a tower 64 feet high with an initial velocity of 48 feet per second. The height H of the object above the ground at any time T in seconds is described by the function

$$H(T) = -16T^2 + 48T + 64 \tag{1}$$

Find the height of the object when $T = 0$ (see Diagram 16).

Since the coefficient of the T^2 term is negative, the graph of $H(T)$ possesses a maximum point. The domain of the function is restricted to $T \geq 0$ (and will also be bounded from above by the time T it takes the object to return to earth, which at the present is unknown). When $T = 0$, $H(0) = 64$, which is the height of the object at time zero.

Example 4 When does the object in Example 3 return to the earth?

When its height above the ground is $H = 0$, or

$$0 = -16T^2 + 48T + 64 = T^2 - 3T - 4 = (T - 4)(T + 1)$$
$$T = 4 \text{ seconds} \quad T = -1 \text{ second}$$

The second solution must be rejected because it is not in the domain of the function. Therefore the amount of time the object remains in the air is 4 seconds.

Example 5 Find the maximum height attained by the object.

Since the axis of symmetry passes through the highest point,

$$T = -\frac{b}{2a} = -\frac{48}{2(-16)} = \tfrac{3}{2} \text{ seconds}$$

$$H(\tfrac{3}{2}) = -16(\tfrac{3}{2})^2 + 48(\tfrac{3}{2}) + 64 \qquad \text{Substituting } \tfrac{3}{2} \text{ for } T \text{ in (1)}$$
$$= -16(\tfrac{9}{4}) + 24(3) + 64 = 100 \text{ feet}$$

The maximum height is 100 feet. Diagram 16 represents the results of the problem graphically. Study the diagram and convince yourself of the importance of a graphical analysis as well as an algebraic analysis.

Example 6 How long does it take the object to pass the top of the tower, *when it is* on the way down?

You can check the diagram for this, or set Equation (1) equal to 64 (height of tower).

$$-16T^2 + 48T + 64 = 64$$
$$-16T^2 + 48T = 0 \qquad \text{Subtracting 64 from both sides}$$
$$-16T(T - 3) = 0 \qquad \text{Factoring}$$

$$T = 0 \text{ seconds} \qquad T = 3 \text{ seconds}$$

Example 7 A farmer has 800 feet of fencing and wants to build a rectangular enclosure divided into three sections by erecting two parallel fences within the enclosure (Diagram 17). What should the dimensions of the enclosure be for the farmer to enclose the *maximum area* with 800 feet of fencing?

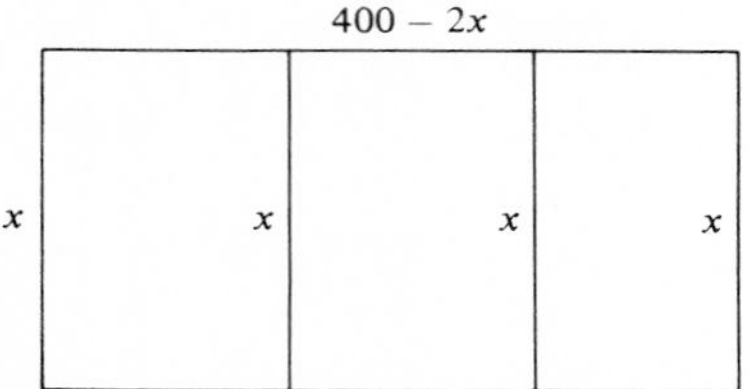

Diagram 17

If x represents the width of the rectangular enclosure, this leaves $800 - 4x$ feet of fence for the remaining two sides. This means $(800 - 4x)/2$ or $400 - 2x$ feet of fence is available for each of the other two sides. Since we are interested in maximizing the area, we must develop a formula for the area $A(x)$ in terms of x.

$$A(x) = x(400 - 2x) = 400x - 2x^2$$

$$x = \frac{-400}{2(-2)} = 100 \text{ feet} \qquad \text{Axis of symmetry}$$

$$A(100) = 100(400 - 200) = 20{,}000 \text{ square feet} \qquad \text{Maximum area}$$

The answer 100 by 200 feet (with two interior fences of 100 feet) gives a *maximum area* of 20,000 square feet (Diagram 18).

Example 8 Suppose the farmer had a natural 80-foot stone barrier and wanted to use 120 feet of fence to construct a simple rectangular enclo-

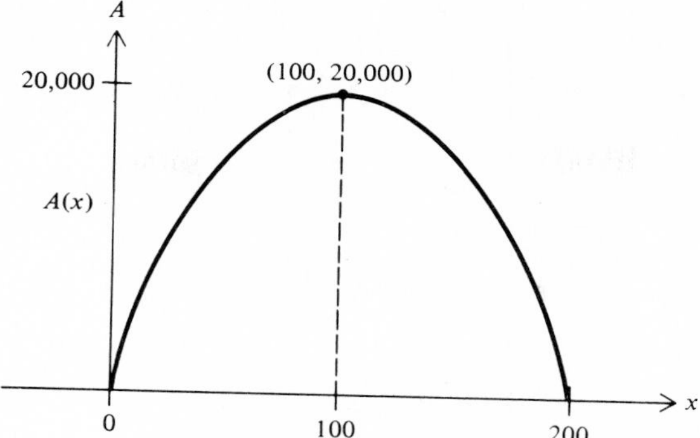

Diagram 18

sure of maximum area. In finding what dimensions maximize the area, we see how this problem differs from the previous one.

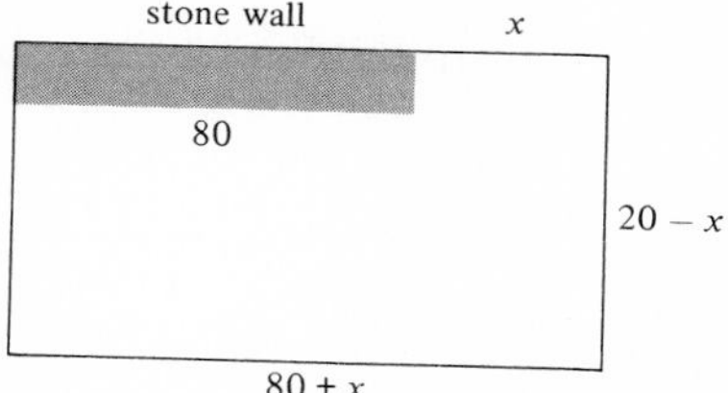

Diagram 19

From Diagram 19 we see that x is the amount of fence added to the 80-foot stone wall. This means we have accounted for $x + 80 + x$ or $2x + 80$ feet of fence, leaving $120 - (2x + 80)$ or $40 - 2x$ feet for the remaining two equal sides. This means each of the other sides is $(40 - 2x)/2 = 20 - x$ feet.

$$\begin{aligned} A(x) &= (20 - x)(80 + x) \\ &= 1600 - 60x - x^2 \qquad \text{Multiplying} \end{aligned}$$

$$x = \frac{60}{2(-1)} = -30 \qquad \text{Axis of symmetry}$$

The answer seems impossible, -30 feet?

Let us investigate the graph of $A(x)$ and discover why the problem has arisen (Diagram 20).

When $x = 0$ (in other words, using the stone wall as one of the natural sides of the rectangle) the 120 feet of fence will be partitioned as $80 - 20 - 20$ for an area of 80 by $20 = 1600$ square feet. Why?

From the diagram it can be seen that $(-30, 2500)$ is the graphical maximum. However, $x = -30$ is not in the acceptable realistic domain of the problem. The *least* amount of fence that can be added to the stone wall is 0 feet. We must examine the graph in the region $0 \leq x \leq 20$ to find the correct answer.

This is an example where the solution occurs at a point within the domain

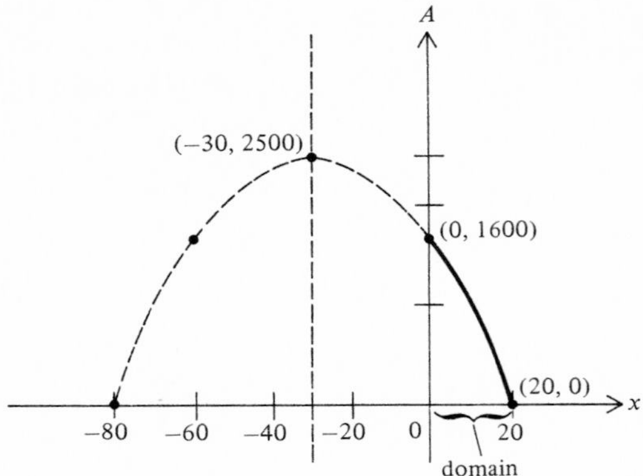

Diagram 20

which is not necessarily the highest point on the graph. Hence, when using mathematics to represent the conditions of a practical problem, *always* determine the realistic domain of the problem, otherwise you may accept a meaningless solution.

EXERCISES

omit for Exam

PART 1

1 The distance from $A(-2, 1)$ to $B(4, y)$ is $3\sqrt{5}$; determine the value of y.

2 The distance from $(3, 5)$ to $(2, 1)$ is the same as from $(3, 5)$ to $(0, x)$; determine x.

3 Show that the line through $(-1, 5)$ on the parabola $y = -x^2 + 2x + 8$ which is parallel to the line $y = 4x - 2$ intersects the parabola at exactly one point.

4 Two cubes of different sizes are on a table, the smaller cube atop the larger. The sum of their heights is 6 inches, and their *total exposed area* is $\frac{441}{4}$ square inches. Find the dimensions of the cubes.

5 If $P(2, -3)$ and $(c, -1/3)$ are on the graph of $y = ax^2$, find c.

6 An open box is to be constructed from a square piece of cardboard by cutting 2-inch squares from each corner and turning up the sides. If the volume of the box is to be 128 cubic inches, what should the size of the initial piece of cardboard be?

7 A picture frame is 9 by 12 inches and has a uniform border inside the edges. Find the width of the border if one-half the area in the frame is taken up by the picture. Find the width of the border if three-fourths of the area is taken up by the picture (approximate answer).

8 Given: two resistors, R_1 and R_2. In series their effective value is 6, and in parallel their effective value is $\frac{9}{8}$. Find R_1 and R_2.

9 Under what conditions will the effective value of two resistors in series be 4 times the effective value of the same two resistors in parallel?

10 Show that it is impossible for the effective value of two resistors in series to be double the effective value of the same two resistors in parallel.

11 **(a)** A circular arena 16 yards in diameter is surrounded on the outside by a uniform circular border which is one-half the area enclosed in the arena. Find the width of this border.

(b) If the circular border is placed on the *inside* the arena, so that it divides the arena into two sections of equal area, find the width of the border.

12 An artist designs a rectangular op-art poster to conform to specific conditions. The ratio of the width to the length must equal the ratio of the length to the sum of the length and the width. If the poster has a perimeter of 32 feet, find the dimensions of the poster.

13 The rear wheel of an experimental racing car has a circumference which is 2 feet more than the circumference of the front wheel. In a 1-mile race, the rear wheel revolves 256 times less than the front wheel. Find the circumference of each wheel.

14 A space capsule is in the shape of the frustum of a cone (refer to the diagram below). The volume of the solid is given by the formula.

$$V = \frac{\pi}{3} H(a^2 + ab + b^2)$$

where $\pi = \frac{22}{7}$
H = height of frustum
a = radius of top
b = radius of bottom

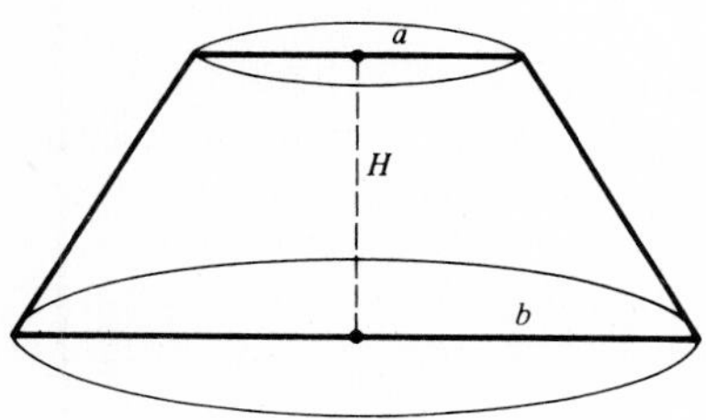

If the capsule contains 3696 cubic feet, is 14 feet high, and has a 24-foot diameter for the bottom base, find the diameter of the upper base.

15 A piece of wire 40 inches long is bent into the shape of a right triangle. If the hypotenuse is 17 inches, find the length of the longer side.

16 A rectangular block has a height which is 2 feet less than the width and a length which is 5 feet longer than the height. The sum of the areas of the faces of the block (total surface area) is 100 square feet. Find the dimensions of the block.

17 A uniform walk is built on the outside of a pool 24 by 16 feet. The area of the walk is 225 square feet. Find the width of the walk.

18 The sum of the squares of three consecutive integers is 149. Find the numbers.

19 Consider two congruent rectangles each having a perimeter of 30 inches. Suppose both are used to form a block letter L (refer to the diagram). The common intersection forms a square. If the resulting exposed area is 63 inches, find the dimensions of the rectangles.

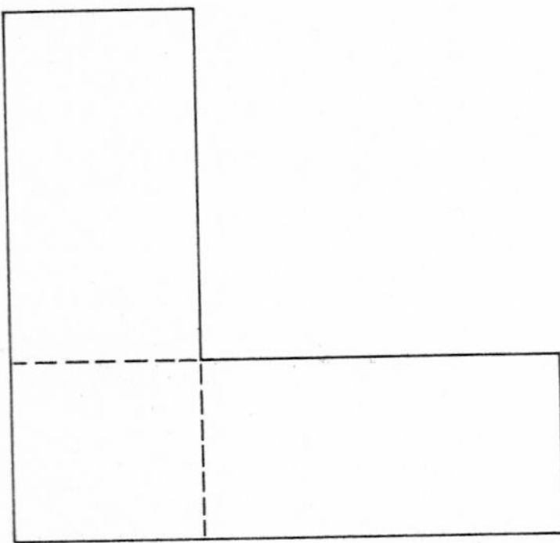

20 Given a pool 36 by 32 feet. A footbridge 4 feet wide is constructed over the pool (see diagram below). Find the area of the footbridge.

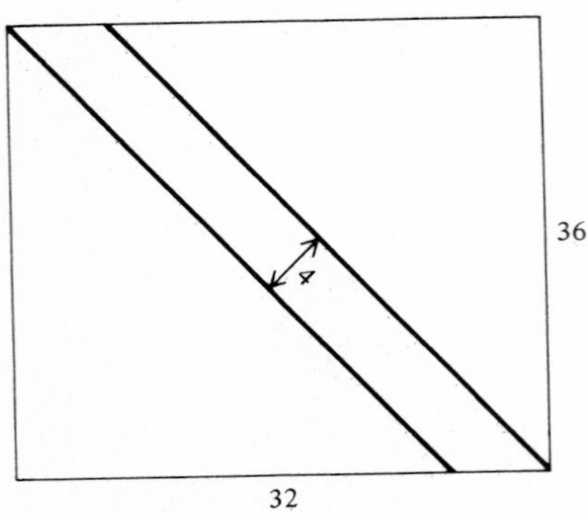

PART 2: MAXIMUM AND MINIMUM PROBLEMS

21 Given $S = 128T - 16T^2$ and $V = 128 - 32T$, $0 \leq T \leq 8$.
- **(a)** Sketch a graph of S versus T.
- **(b)** Sketch a graph of V versus T.
- **(c)** For what value of T is S a maximum?
- **(d)** Find V when S is a maximum.

22 A stock manipulator estimates that if he sells his MBI stock it will be worth \$2 per share. However, if he waits, his present holding of 150 shares will increase by 30 shares per week but the market price will also drop 20 cents per share each week.
- **(a)** Express as a quadratic function of x (the number of weeks) the income from the stock under these specific conditions.
- **(b)** In how many weeks should he sell the stock to realize maximum return?

23 A rectangle is in the first quadrant with one vertex at the origin, two sides along the axes, and the fourth vertex on $x + 2y - 6 = 0$. Find the maximum area of the rectangle.

24 Find two integers whose sum is 24 and whose product is a maximum.

25 Find two numbers whose difference is 8 and whose product is a minimum.

26 A Florida citrus grower can ship 12 tons of oranges to New York now at a profit of \$50 per ton. If he waits, he can add 3 tons per week, but his profit per ton will be reduced by \$5. When should he ship the oranges for maximum profit?

In Exercises 27 to 33 a farmer has 1200 feet of fence and wishes to enclose a rectangular area. Find the dimensions of the rectangle that will maximize the area enclosed under each of the stated conditions.

27 Two sections are to be formed by erecting one parallel fence within the enclosure.

28 Three sections are to be formed by erecting two parallel fences within the enclosure.

29 Four sections are to be formed by erecting three parallel fences within the enclosure.

30 One side of the enclosure is bounded naturally by a river and hence only three sides of fencing are required.

31 There already is a 200-foot stone wall as part of the rectangular enclosure.

32 There already is a 400-foot stone wall as part of the rectangular enclosure.

33 There already is a 600-foot stone wall as part of the rectangular enclosure.

34 An object is projected upward from a tower 400 feet high with an initial velocity of 120 feet per second. The height above the ground at any time t is given by the function $h(t) = 400 + 120t - 16t^2$. The velocity of the object at any time t is given by $v(t) = 120 - 32t$. Find:

(a) The maximum height attained by the object.
(b) The time it takes the object to hit the ground.
(c) The velocity of the object at impact.
(d) The velocity of the object after 2.5 seconds.
(e) How high the object was 1 second before impact.

35 An object is projected upward from a mountain 900 feet above the surface of Venus with an initial velocity of 105-feet per second. The height above the ground at any time t is given by the function $h(t) = -15t^2 + 105t + 900$. Find: **(a)** the maximum height attained by the object and **(b)** the time it takes the object to return to the surface of Venus.

36 An object is projected upward from a mountain 1200 feet above the surface of the moon, with an initial velocity of 40-feet per second. The height above the ground at any time t is given by the function

$$h(t) = -\frac{8t^2}{3} + 40t + 1200$$

The velocity of the object at any time t is given by

$$v(t) = -\frac{16t}{3} + 40$$

Find:

(a) The maximum height attained by the object
(b) The time it takes the object to return to the surface of the moon
(c) The speed of the object at impact
(d) The height of the object one second before impact
(e) The speed of the object after 6 seconds

37 An object is projected upward from a height of 720 feet above the surface of Saturn with an initial velocity of 180-feet per second. The height above the ground

at any time t is given by the function $h(t) = -20t^2 + 180t + 720$. Find:
(a) The maximum height attained by the object
(b) The time it takes the object to return to the surface of Saturn

38 An object is projected upward from a height of 1 mile above the surface of a new planet with an initial velocity of 400 feet per second. The height above the ground at any time t is given by the function $h(t) = -80t^2 + 400t + 5280$. Find **(a)** the maximum height attained by the object and **(b)** the time it takes the object to return to the surface of the planet.

39 The formula for the output P of a battery is given by $P = VI - RI^2$, where I is the current, R the resistance, and V the voltage. Find the current I which corresponds to a maximum value of P for a battery where the voltage is 12 volts and $R = \frac{1}{2}$ ohm.

40 The perimeter of a rectangular sheet is 40 inches. There is a 1-inch border on three of the sides and a 2-inch border on the bottom, within which is the printed area. Find the dimensions of the paper if the printed area is to be a maximum.

41 A company produces a product and wants to decide how much to market during the upcoming quarter to maximize profits. If x is the number of items to be produced, then $P(x) = I(x) - C(x)$, where P is the profit on x items, I is the income from the sale of x items, and C is the total cost of producing the x items. For a company with the following $I(x)$ and $C(x)$:

$$I(x) = -x^2 + 40x \qquad C(x) = 400 - 10x \qquad 10 \le x \le 30$$

(a) Sketch graphs of $C(x)$, $I(x)$, $P(x)$.
(b) What amount x minimizes $C(x)$?
(c) What amount x maximizes $I(x)$?
(d) What amount x maximizes $P(x)$?

42 Find the area of the largest rectangle that can be inscribed in a right triangle whose dimensions are 60 by 20 units (see diagram). *Hint:* Find a formula for the area of $RSTC$ in terms of x alone by using the property of similar triangles $ASR \sim ABC$.

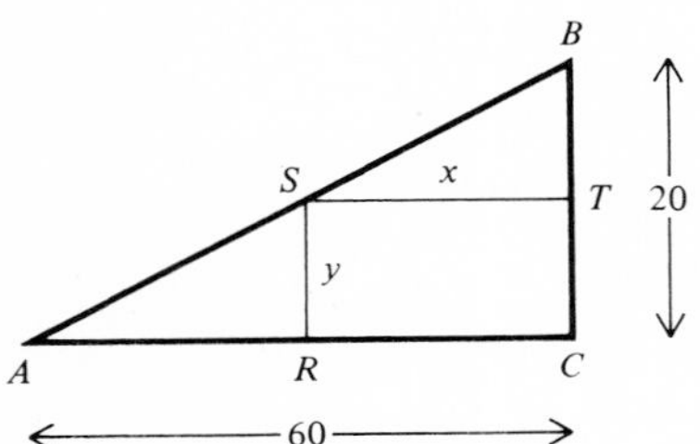

Inequalities

4.1 INTRODUCTION

The ability to analyze different types of inequalities is essential in higher mathematics. In the calculus, precise definitions are written in terms of such statements, and in practical applications of mathematics, inequalities play just as important a role as equations. For example, when we state that the diameter d of the earth is 7900 miles we really mean d is some number between 7850 and 7950; that is, $7850 < d < 7950$.

Generally, whenever we measure quantities such as the weight of a body, distance, velocity, electrical pressure, and so on, the data collected will be only as accurate as the instruments used. Hence, the inequality will appear in a very natural way.

Both from a theoretical and practical viewpoint inequalities are essential in mathematics.

4.2 PROPERTIES OF INEQUALITIES

We have already used the symbols $<$ and $>$ sparingly. These symbols are called *less than* and *greater than,* respectively. We now define what we mean when we say that one real number is less than another real number.

DEFINITION 1 For real numbers a and b, $a < b$ if and only if $b - a$ is positive. Further, $b > a$ if and only if $a < b$.

Example 1 $3 < 5$ because $5 - 3 = 2$ is positive.

Example 2 $\frac{1}{3} < \frac{1}{2}$ because $\frac{1}{2} - \frac{1}{3} = \frac{1}{6}$ is positive.

Example 3 $5 > 3$ since $3 < 5$.

The statement $a < b$ is called an *inequality.* Before we state properties of inequalities we need one more definition.

DEFINITION 2 If $a < b$ and $c < d$, we say that these inequalities have the *same sense;* if $a < b$ and $c > d$, we say these inequalities have the *opposite sense.*

PROPERTY 1 If $a < b$ and $b < c$, then $a < c$.

Example 4 $-3 < -1$ and $-1 < 2$, and so $-3 < 2$. We can show this result on a real-number line (see Diagram 1).

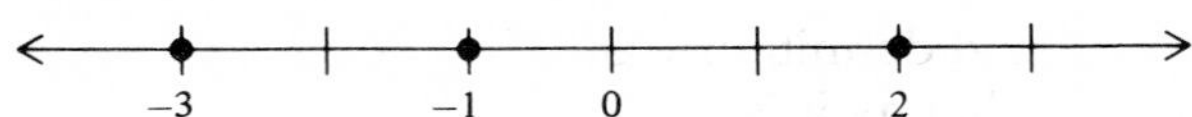

Diagram 1

Geometrically if $a < b$, a can be represented on a real-number line as a point to the left of the point that corresponds to b. Likewise if $b < c$, point c will be to the right of b, making point a necessarily to the left of c. This gives a graphic verification of Property 1 known as the *transitive law* (see Diagram 2).

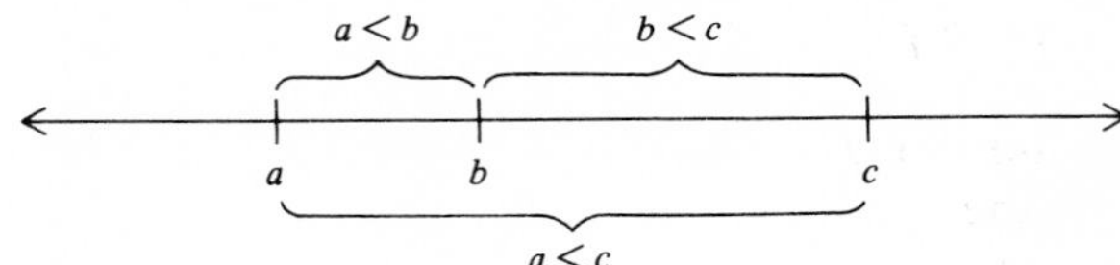

Diagram 2

PROPERTY 2 The sense of an inequality is not changed when the same number is added to both sides of the inequality. That is, if $a < b$, then $a + c < b + c$, for any real number c.

Example 5 If $3 < 5$, then $3 + 2 < 5 + 2$.

Example 6 If $2 < 3$, then $2 + (-10) < 3 + (-10)$.

In general, if $a < b$, then a is to the left of b on the real-number line. If we add any number c to a and b, $a + c$ is still to the left of $b + c$ (see Diagram 3).

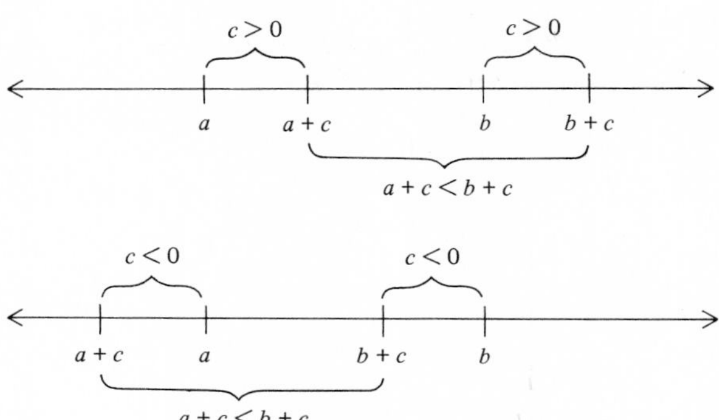

Diagram 3

COMMENT If $c = 0$, we simply have $a < b$.

PROPERTY 3 The sense of an inequality is not changed by multiplying or dividing both sides of the inequality by the same positive number. That is,

(a) If $a < b$, then $ac < bc$ for any $c > 0$.
(b) If $a < b$, then $a/c < b/c$ for any $c > 0$.

Perhaps you read definition 1 and thought it trivial. Let us verify Property 3 and see how handy it is.

Proof

PART (a) $a < b$ means that $b - a > 0$ by definition. Since the product of two positive numbers is again a positive number, we obtain $c(b - a) = cb - ca$, which is positive. Again by Definition 1, $cb - ca > 0$ means $ca < cb$ and our proof of the part (a) is complete.

PART (b) When we divide a number a by c, this is the same as multiplying the number a by the reciprocal of c; that is, $a/c = (1/c)(a)$, and so part (b) of this property is a consequence of part (a).

Example 7 If $3 < 4$, then $10(3) < 10(4)$.

Example 8 If $-2 < 4$, then $-\frac{2}{2} < \frac{4}{2}$.

We simply state the next two properties and leave the proofs as exercises.

PROPERTY 4 The sense of an inequality is reversed when multiplying or dividing both sides of the inequality by the same negative number. That is,
(a) If $a < b$, then $ac > bc$ for any $c < 0$,
(b) If $a < b$ then $a/c > b/c$ for any $c < 0$.

Example 9 If $-4 < 6$, then

$$(-2)(-4) > (-2)(6) \qquad \text{Multiplying both sides by } -2$$
$$8 > -12 \qquad \text{(note change in sense)}$$

Example 10 If $3 < 9$, then

$$\frac{3}{-3} > \frac{9}{-3} \qquad \text{Dividing both sides by } -3 \text{ (note change in sense)}$$

$$-1 > -3$$

PROPERTY 5 The sense of an inequality is reversed when each side of the inequality is inverted provided its members agree in sign. That is,
(a) If $a < b$, $a > 0$, and $b > 0$, then $1/a > 1/b$,
(b) If $a < b$, $a < 0$, and $b < 0$, then $1/a > 1/b$.

Example 11 If $2 < 3$, then $\frac{1}{2} > \frac{1}{3}$.

Example 12 If $-5 < -2$, then $-\frac{1}{5} > -\frac{1}{2}$.

COMMENT At first glance $-\frac{1}{2}$ might seem larger than $-\frac{1}{5}$, but keep in mind these are negative quantities. $-\frac{1}{2}$ is to the left of $-\frac{1}{5}$, and one quick glance at their positions on the real-number line clears up the situation (see Diagram 4).

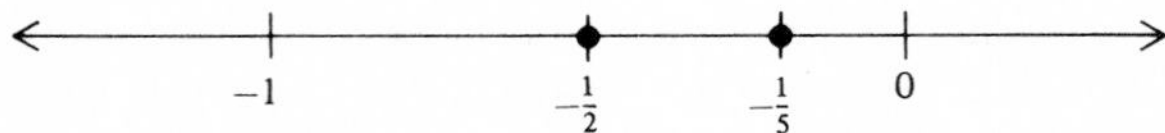

Diagram 4

Two other convenient symbols are $\leq$ and $\geq$, called, respectively, *less than or equal to* and *greater than or equal to.*

DEFINITION 3 For two real numbers a and b, $a \leq b$ means that "$a < b$ or $a = b$," and $b \geq a$ means that $a \leq b$.

PROPERTY 6 The sense of an inequality is not changed if the same positive root is taken of both members or if both members are raised to the same positive power provided the members are nonnegative. That is,
(a) If $a < b$ and $a \geq 0$, then $\sqrt[n]{a} < \sqrt[n]{b}$.
(b) If $a < b$ and $a > 0$, $b \geq 0$, then $a^n < b^n$.

Example 13 If $4 < 9$, then

$$\sqrt{4} < \sqrt{9} \qquad \text{Taking square root of both sides}$$
$$2 < 3$$

Example 14 If $6 < 10$, then

$$6^3 < 10^3 \qquad \text{Cubing both sides}$$
$$216 < 1000$$

COMMENT Suppose we did not stipulate that both members be positive; for example, $-4 < 3$, which is a true statement. Now let us square both sides of the inequality:

$$(-4)^2 > (3)^2$$
$$16 > 9$$

We are forced to change the sense of the inequality. However, if we take $-4 < 3$ and cube both sides of the inequality

$$(-4)^3 < (3)^3$$
$$-64 < 27$$

the inequality retains its original sense. Why? Obviously no general property could be stated without paying an inordinate amount of attention to signs. Also, to take the square root of both members of $-4 < 3$ would be to take the square root of a negative number, which causes some problems. (How do we compare imaginary numbers and real numbers?) Therefore, be careful to observe all stipulations in Property 6 before using it in a problem.

4.3 INTERVAL NOTATION

In the following discussion it is convenient to use the real-number line. In doing so we will use the words *point* and *real number* interchangeably. Hence, we will speak of "the point x" rather "the point corresponding to the real number x."

If $a < b$, then between a and b lies a continuous collection of points. Suppose x is a point between a and b; then $a < x$ and $x < b$. We can write this concisely as $a < x < b$.

DEFINITION 1 The *open interval* from a to b is the collection of all points x such that $a < x < b$ and is denoted by (a, b). The points a and b are called *end points* of the interval.

Note here that the end points are not included in the open interval. If we include the end points, we have the *closed interval.*

Diagram 5

DEFINITION 2 The *closed interval* from a to b is the collection of all points x such that $a \leq x \leq b$ and is denoted by $[a, b]$.

Let us illustrate the open and closed intervals (see Diagram 5).

The interval is *half-open* if it includes all the points between a and b and exactly *one* end point.

DEFINITION 3 The *interval half-open on the left* from a to b is the collection of all points x such that $a < x \leq b$ and is denoted by $(a, b]$.

DEFINITION 4 The *interval half-open on the right* from a to b is the collection of all points x such that $a \leq x < b$ and is denoted by $[a, b)$.

We illustrate each of these intervals on the real-number line (see Diagram 6).

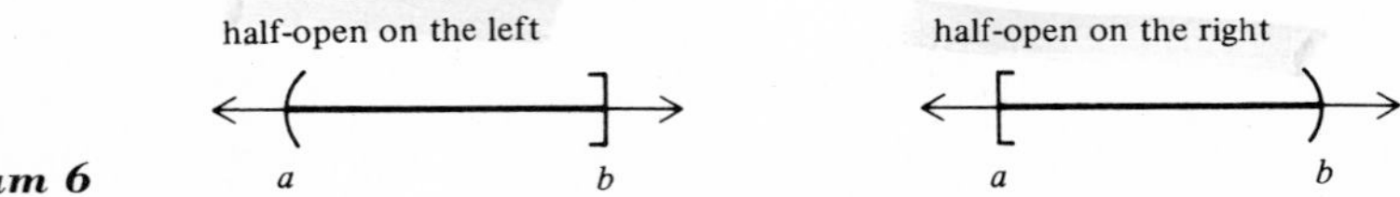

Diagram 6

We finish our discussion of intervals by discussing collections of real numbers of the form $x > a$ and $x \leq b$. On the real-number line these collections graph as *rays* (see Diagram 7). Since each of these intervals is *endless* or *infinite* in one direction, we use the symbols $+\infty$ and $-\infty$, read "plus infinity" and "minus infinity," respectively. Hence the collection of all real numbers x, $x > a$, will be denoted by $(a, +\infty)$ and the collection of all real numbers x, $x \leq b$ will be denoted by $(-\infty, b]$.

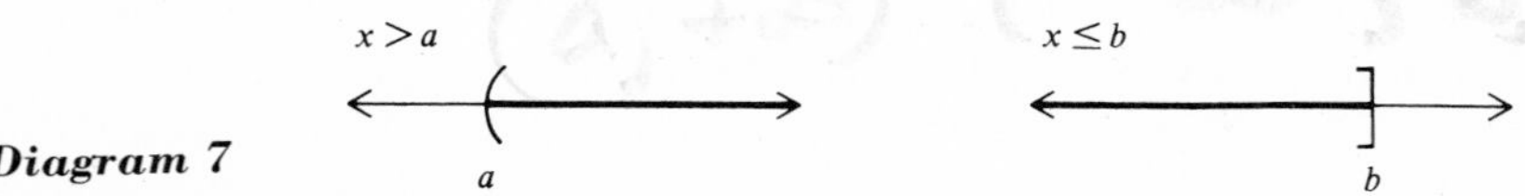

Diagram 7

Note that when the end point a is excluded, a parenthesis is used; in the second interval the end point b is included, and a bracket is used. Furthermore, $(-\infty, +\infty)$ denotes the collection of all real numbers. It should also be noted that we do not refer to such intervals as open or closed.

4.4 LINEAR INEQUALITIES

Now that we have stated some of the basic properties concerning inequalities we can work through some examples.

Example 1 Find all values of x such that $10x + 2 < 8 - 3x$.

Collect all variables (x's) on one side of the inequality and all constants on the other side. This is easily done by adding -2 and $3x$ to both sides of the inequality, $10x + 2 < 8 - 3x$:

$$13x < 6 \qquad \text{Property 2}$$
$$x < \tfrac{6}{13} \qquad \text{Dividing by 13; Property 3}$$

COMMENT We can also state the solution to be $(-\infty, \frac{6}{13})$. Graphically we can represent this interval on a real-number line (see Diagram 8).

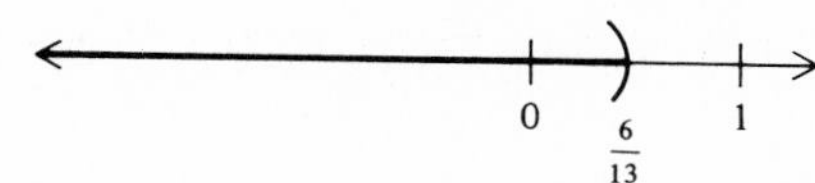

Diagram 8

Example 2 Find all the values of x that satisfy

$$3 - 2x \geq 9$$
$$-2x \geq 6 \qquad \text{Property 2}$$

Hence

$$x \leq -3 \qquad \text{Property 4}$$

or

$$(-\infty, -3]$$

The graphic solution is shown in Diagram 9.

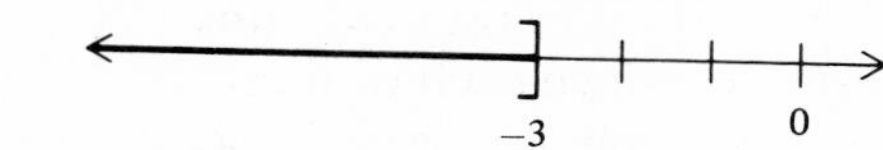

Diagram 9

COMMENT We can solve the problem and avoid the use of Property 4 as follows:

$$3 - 2x \geq 9$$
$$-6 \geq 2x \qquad \text{Property 2}$$
$$-3 \geq x \qquad \text{Property 3}$$

Note that $-3 \geq x$ is *identical to* $x \leq -3$.

Example 3 Solve $0 \leq 2x - 6 < 4$.

This example compounds two inequalities into one statement and can be solved by first adding 6 to *every* member:

$$0 \leq 2x - 6 < 4$$
$$6 \leq 2x < 10 \qquad \text{Property 2}$$
$$3 \leq x < 5 \qquad \text{Property 3}$$

or [3, 5). (See Diagram 10.)

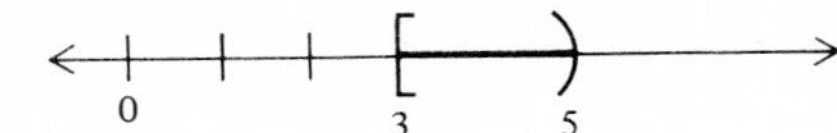

Diagram 10

We now examine a simple but interesting proof.

Example 4 Show that $\sqrt{2} + \sqrt{7} < \sqrt{3} + \sqrt{6}$.

Since

$$\sqrt{2} + \sqrt{7} > 0 \qquad \text{and} \qquad \sqrt{3} + \sqrt{6} > 0$$

let us assume that the inequality is true and use Property 6*b*:

$$(\sqrt{2} + \sqrt{7})^2 < (\sqrt{3} + \sqrt{6})^2$$

$$2 + 2\sqrt{2}\sqrt{7} + 7 < 3 + 2\sqrt{3}\sqrt{6} + 6$$

$$2\sqrt{14} < 2\sqrt{18} \qquad \text{Property 2}$$

$$\sqrt{14} < \sqrt{18} \qquad \text{Property 3}$$

$$14 < 18 \qquad \text{Property } 6b$$

It is clear that this last inequality is true. If we now reverse this process, it follows that the original inequality is true.

When combining two or more inequalities with the words *and* or *or*, particular care must be taken to arrive at a correct solution. We can clarify this with a few examples.

Example 5 Find all values of x that satisfy

$$x + 3 > 2 \qquad \text{and} \qquad 3 - 2x < x$$

We must find a collection of x values that satisfy *both* inequalities

$$\begin{aligned} x + 3 &> 2 & \qquad \text{and} \qquad 3 - 2x &< x \\ x &> -1 & -3x &< -3 \\ & & x &> 1 \end{aligned}$$

For $x > -1$, $x + 3 > 2$ is true; however, there are values of x, $x > -1$, that do not satisfy $3 - 2x < x$, namely, $-1 < x \leq 1$. When we exclude $-1 < x \leq 1$ from the interval $x > -1$, we have the collection of x values that satisfy *both* inequalities; that is, $x > 1$. Graphically, we can obtain the correct solution by simply observing where the graph of $x > -1$ overlaps the graph of $x > 1$ (see Diagram 11).

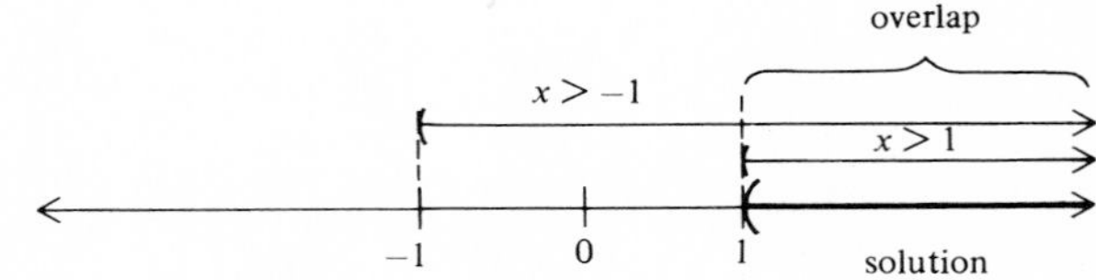

Diagram 11

Example 6 Find all values of x that satisfy

$$x + 3 > 2 \qquad \text{or} \qquad 3 - 2x < x$$

Since the two inequalities are connected by the word *or,* we simply must look for those values of x that satisfy either equation. From the previous example we have $x > -1$ and $x > 1$ as the collections of x that satisfy the respective inequalities. Our solution is therefore $x > -1$ since this interval includes all values of x satisfying $3 - 2x < x$ as well as those satisfying $x + 3 > 2$.

Therefore, graphing $x > -1$ will give us a picture of our solution (see Diagram 12).

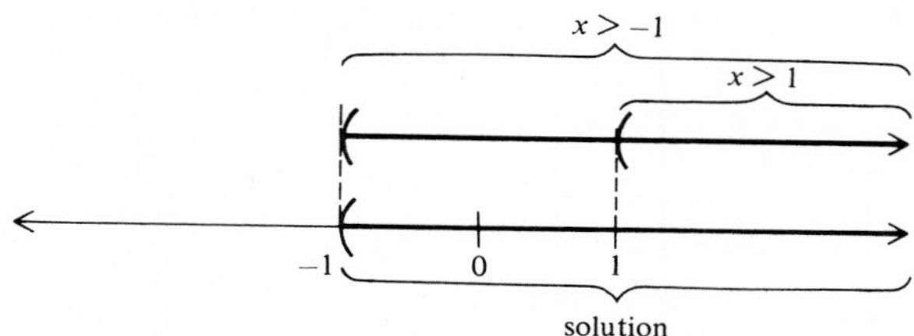

Diagram 12

Example 7 Find all values of x that satisfy

$$3x + 5 > 8 \qquad \text{and} \qquad 2x - 5 < 3$$

Solution

$$\begin{array}{rcr} 3x + 5 > 8 & \text{and} & 2x - 5 < 3 \\ 3x > 3 & & 2x < 8 \\ x > 1 & & x < 4 \end{array}$$

Hence only those values of x that are greater than 1 and less than 4 will satisfy *both* inequalities, that is, $1 < x < 4$.

Example 8 Find all values of x that satisfy

$$3x + 5 > 8 \qquad \text{or} \qquad 2x - 5 < 3$$

Since the inequalities involved are those of Example 7, we immediately obtain

$$x > 1 \qquad \text{or} \qquad x < 4 \tag{1}$$

However, this tells us that any real number will satisfy at least one of the two inequalities involved. Therefore, expressions (1) simplify to

$$-\infty < x < +\infty \tag{2}$$

which can also be written $(-\infty, +\infty)$. We can illustrate the solutions to Examples 7 and 8 graphically (see Diagram 13).

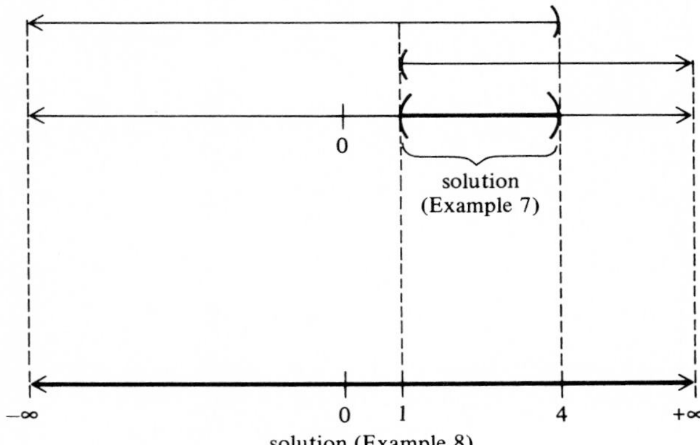

Diagram 13

Example 9 Solve $3x - 1 < 2x + 5 < x + 1$.

Since the variable x is present in all three members, we must separate the compound inequality into two inequalities connected by the word *and* and solve them:

$$3x - 1 < 2x + 5 \quad \text{and} \quad 2x + 5 < x + 1$$
$$x < 6 \quad \text{and} \quad x < -4$$

which is equivalent to $x < -4$.

EXERCISES

1 In each of the following state a conclusion:
(a) If $2 < 5$ and $5 < 7$, then . . .
(b) If $\sqrt{3} < \sqrt{5}$ and $\sqrt{5} < \sqrt{\pi^3}$, then . . .
(c) If $(a - b) < c$ and $c < (a + b)$, then . . .

2 Illustrate the following intervals:
(a) $(-2, 3)$ **(b)** $(-4, 0]$ **(c)** $[-3, 1]$
(d) $(1, +\infty)$ **(e)** $[\sqrt{2}, 2)$ **(f)** $(-\infty, -100]$
(g) $(-\infty, \frac{1}{2})$ **(h)** $[a, 2 + a)$

In Exercises 3 to 8, show that the following are true inequalities by making use of the properties presented in this section.

3 $\sqrt{6} + \sqrt{7} > \sqrt{3} + \sqrt{10}$ **4** $\sqrt{5} - \sqrt{2} < \sqrt{3}$
5 $\sqrt{7} + \sqrt{8} > \sqrt{5} + \sqrt{10}$ **6** $\sqrt{11} + \sqrt{15} - \sqrt{12} > \sqrt{13}$
7 $\frac{14}{32} < \frac{23}{53}$ **8** $\frac{1}{3} > 0.333$

In Exercises 9 to 14 show that each of the following inequalities is true. In each case the letters represent *unequal* positive numbers.

9 $\frac{1}{2}(x + y) > \sqrt{xy}$ **10** $a^2 + b^2 + c^2 < (a + b + c)^2$

11 $a^3 + b^3 > a^2b + ab^2$ **12** $\dfrac{a}{b} + \dfrac{b}{a} > 2$

13 $\frac{a}{b^2} + \frac{b}{a^2} > \frac{1}{a} + \frac{1}{b}$

14 $a^3 + b^3 + c^3 < (a + b + c)^3$

Solve the linear inequalities in Exercises 15 to 37.

15 $2x - 1 \le 3$

16 $-4 < 3x + 4$

17 $1 - x \ge 4$

18 $3x - 6 > 4x - 8$

19 $\frac{4}{3}x - 9 > 2x - 1$

20 $\frac{x}{2} + \frac{x-2}{3} < 2x - \frac{1}{12}$

21 $2.8x - 0.3 \ge 3.2x - 2.7$

22 $\frac{x}{2} - \frac{3}{4} > 12 - \frac{x}{3}$

23 $2x - 3 > \frac{8x}{3} - 21$

24 $-3x + 4 \le 5x - 8$

25 $1 \le \frac{3x}{4} - \frac{1}{2} < 4$

26 $-1 < \frac{4 - 3x}{5} \le 7$

27 $\frac{2x - 1}{3} < x - \frac{x - 1}{2}$

28 $3 + 4x \ge 2x + 1 > 3x - 5$

29 $6 \le 2 - 4x \le 10$

30 $3 < x + \frac{1}{2} < 4$

31 $x - 5 < 3x + 1 < 2 - x$

32 $2x - 1 \le 3$ and $1 - x \ge 4$

33 $2x - 1 \le 3$ or $1 - x \ge 4$

34 $3 - 2x \ge 5$ or $3 - 2x < -5$

35 $3x - 6 > 4x - 8$ and $2x - 1 \le 5 - x$ and $2x - 1 \ge -5 + x$

36 $-11 < 4 - 3x < 1$ and $-8 \le 2x - 6 < 0$

37 $-11 < 4 - 3x < 1$ or $-8 \le 2x - 6 < 0$

38 Prove Property 4.

39 Prove Property 5.

4.5 OTHER TYPES OF INEQUALITIES

Given two numbers, if a positive product is obtained, *both* factors must agree in sign. This gives us only two possibilities, + + or − −. If the product is negative, the two numbers disagree in sign and we have either + − or − +.

Theorem 1 For all real numbers a and b:
(a) $ab > 0$ if and only if either $a > 0$ and $b > 0$ or $a < 0$ and $b < 0$.
(b) $ab < 0$ if and only if either $a > 0$ and $b < 0$ or $a < 0$ and $b > 0$.

We can generalize this result for products of more than two factors. Suppose a product is made of n factors, where n is some integer greater than 2. We write this product

$$a_1 \cdot a_2 \cdot a_3 \cdot \cdot \cdot a_n$$

and have Theorem 2.

Theorem 2 Let $a_1 \cdot a_2 \cdot a_3 \cdot \cdot \cdot a_n$ be a product of n nonzero factors:
(a) $a_1 \cdot a_2 \cdot a_3 \cdot \cdot \cdot a_n > 0$ means that all the factors are positive or that there are an *even number of negative factors.*
(b) $a_1 \cdot a_2 \cdot a_3 \cdot \cdot \cdot a_n < 0$ means that there are an *odd number of negative factors.*

We illustrate Theorem 2 with some trivial examples.

Example 1 $(2)(4)(1)(-1)(-2)(-1) < 0$ since we have three negative factors.

Example 2 $\dfrac{(3)(4)}{(-6)(-2)} > 0$ since this quotient can be written as the product $(3)(4)(-\frac{1}{6})(-\frac{1}{2})$ having two negative factors.

Example 3 $(\sqrt{2})(\sqrt{3})(\sqrt{5})(-\sqrt{3}) < 0$ because this product has one negative factor.

Suppose we have a product $(x + 2)(x - 1)$. Then $x + 2$ and $x - 1$ are factors of this product. Say we draw a real-number line and indicate with a vertical broken line the point where $x + 2$ is zero. This broken line divides the real-number line into two intervals (see Diagram 14). We indicate with plus and minus signs the sign the factor has in each of these intervals (see Diagram 15).

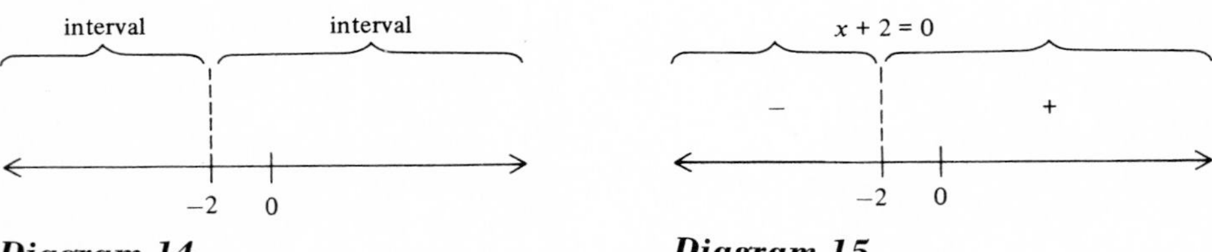

Diagram 14

Diagram 15

Diagram 15 then depicts what we call a *sign pattern.* Following the same procedure for the factor $x - 1$, we obtain its sign pattern (see Diagram 16).

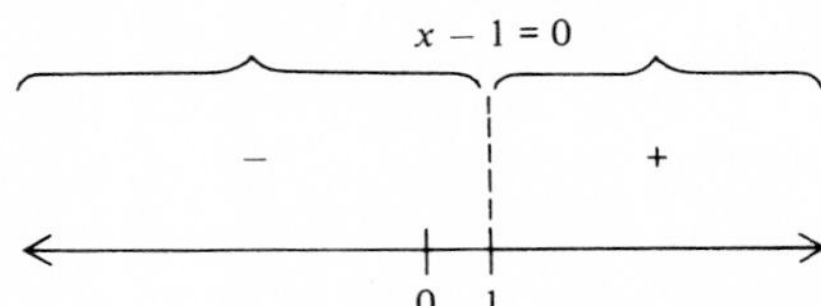

Diagram 16

We can now combine the patterns of both factors in tabular form, obtaining the *product sign pattern* (see Diagram 17). This geometrical device will be helpful in solving the problems that follow.

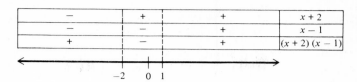

−	+	+	$x + 2$
−	−	+	$x - 1$
+	−	+	$(x + 2)(x - 1)$

−2 0 1

Diagram 17

Suppose we now are asked to find the solution for

$$x^2 < 1 \qquad x^2 > 2x + 3$$
$$4t - t^2 \geq 0 \qquad 2x^3 \leq 3x^2 + 5x$$
$$\frac{1}{x} \geq \frac{1}{3} \qquad \frac{x}{x + 1} < 1 - \frac{1}{x}$$

Our plan of attack is basically the same for all six problems:

1 Transpose terms so that one side of the inequality is zero.
2 Convert the nonzero side into a product.
3 Make up a product sign pattern.
4 Read the solution from step 3.

We now illustrate this procedure with examples. Keep in mind that we are looking for the collection of real numbers which makes each of the inequalities a true statement.

Example 4 $x^2 < 1$

STEP 1 Obtain zero member $x^2 - 1 < 0$.

STEP 2 Factor as $(x + 1)(x - 1) < 0$.
We will now look for those real numbers where the product is negative.

STEP 3 See Diagram 18 for the product sign pattern.

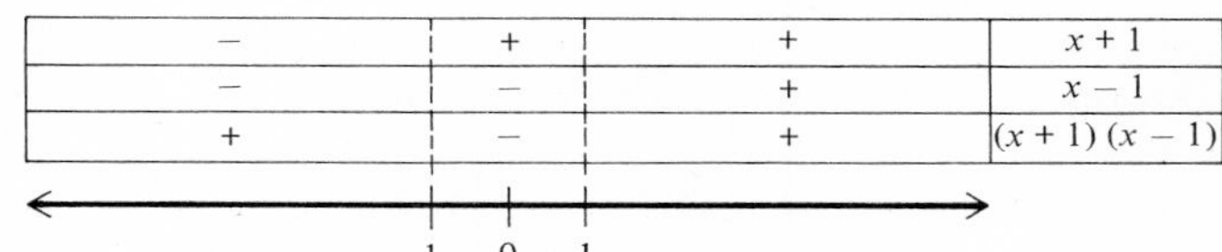

−	+	+	$x + 1$
−	−	+	$x - 1$
+	−	+	$(x + 1)(x - 1)$

Diagram 18

STEP 4 The bottom line of the table indicates the collection of real numbers for which the product is negative, namely, $-1 < x < 1$.

Example 5 $x^2 > 2x + 3$

STEP 1

$x^2 - 2x - 3 > 0$

STEP 2

$(x - 3)(x + 1) > 0$ Factoring

We will now look for those real numbers that make this product positive.

STEP 3

See Diagram 19.

STEP 4

$x < -1$ or $x > 3$

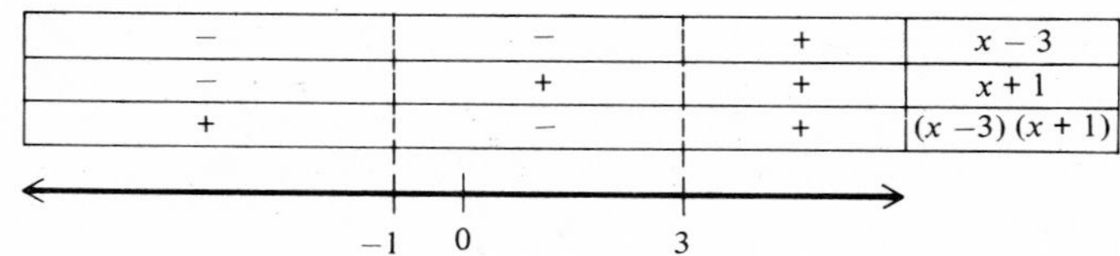

Diagram 19

Example 6 $4t - t^2 \geq 0$

STEP 1

Already done

STEP 2

$t(4 - t) \geq 0$ Factoring

We will now look for those real numbers that make this product positive or zero.

STEP 3

See Diagram 20.

STEP 4

$0 \leq t \leq 4$

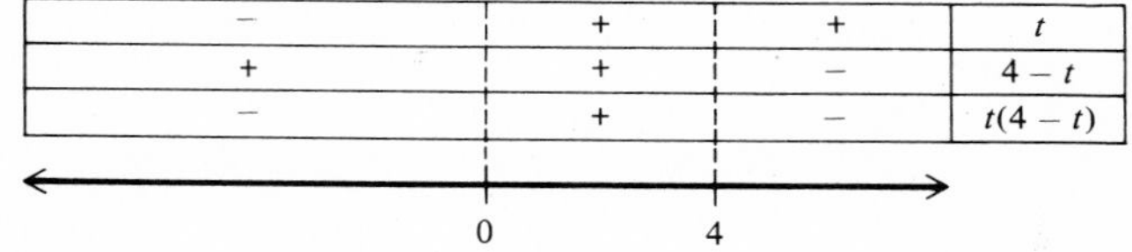

Diagram 20

COMMENT Note that the solution is in terms of t. Be cautious when constructing a sign pattern. *Do not* simply place a plus sign to the right of a dotted line and a minus sign to the left of it. The factor $4 - t$ did not behave this way!

Example 7 $2x^3 \leq 3x^2 + 5x$

STEP 1 $2x^3 - 3x^2 - 5x \leq 0$

STEP 2 $x(2x^2 - 3x - 5) = x(2x - 5)(x + 1) \leq 0$ Factoring

We will now look for those real numbers that make this product zero or negative.

STEP 3 See Diagram 21.

STEP 4 $x \leq -1$ or $0 \leq x \leq 2.5$

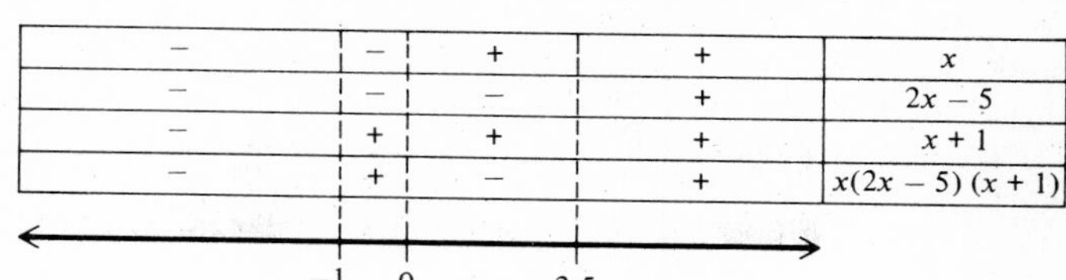

−	−	+	+	x
−	−	−	+	$2x - 5$
−	+	+	+	$x + 1$
−	+	−	+	$x(2x - 5)(x + 1)$

Diagram 21

Example 8

$$\frac{1}{x} \geq \frac{1}{3}$$

STEP 1

$$\frac{1}{x} - \frac{1}{3} \geq 0$$

STEP 2

$$\frac{3 - x}{3x} \geq 0 \qquad \text{Combining as one term}$$

Equivalently,

$$(1/3x)(3 - x) \geq 0,$$

STEP 3 Caution: The factor $1/3x$ has a variable in the denominator. In this example when $x = 0$, $1/3x$ is *meaningless*. However $1/3x$ does have a definite sign pattern to the left and right of this point (see Diagram 22).

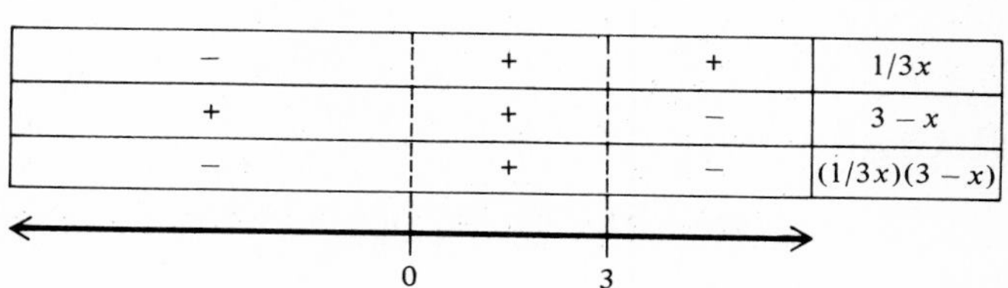

−	+	+	$1/3x$
+	+	−	$3 - x$
−	+	−	$(1/3x)(3 - x)$

Diagram 22

STEP 4

$$0 < x \leq 3$$

COMMENT The solution is a half-open interval excluding the end point on the left.

Example 9

$$\frac{x}{x + 1} < 1 - \frac{1}{x}$$

STEP 1

$$\frac{x}{x + 1} - 1 + \frac{1}{x} < 0 \qquad \text{Subtracting } 1 - \frac{1}{x} \text{ from both sides}$$

STEP 2 The least common denominator of the three terms of the left side of Step 1 is $x(x + 1)$. Expressing each term with $x(x + 1)$ as its denominator, we obtain

$$\frac{x^2}{x(x + 1)} - \frac{x(x + 1)}{x(x + 1)} + \frac{x + 1}{x(x + 1)} < 0$$

or

$$\frac{x^2 - x^2 - x + x + 1}{x(x + 1)} < 0 \qquad \text{Combining fraction with like denominators}$$

$$\frac{1}{x(x+1)} < 0 \qquad \text{Simplifying numerator}$$

Equivalently $$\frac{1}{x}\left(\frac{1}{x+1}\right) < 0$$

STEP 3 See Diagram 23.

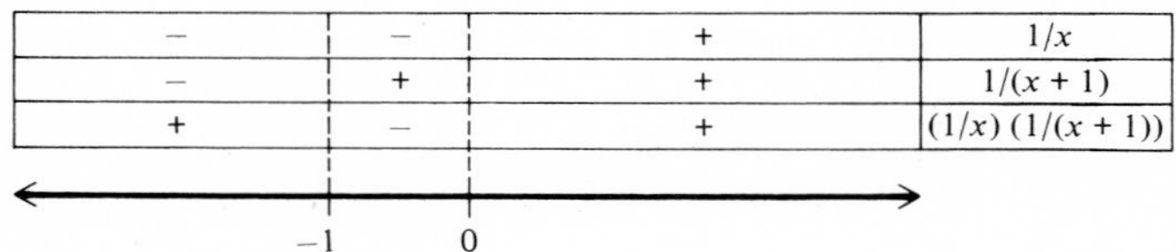

−	−	+	$1/x$
−	+	+	$1/(x+1)$
+	−	+	$(1/x)(1/(x+1))$

Diagram 23

STEP 4 $$-1 < x < 0$$

COMMENTS Review the algebra involved in this problem, carefully, and note that $1/(x+1)$ is meaningless when $x = -1$, and $1/x$ is meaningless when $x = 0$.

EXERCISES

Draw a sign pattern for each of the following:

1 **(a)** $x + 2$ **(b)** $x - 5$ **(c)** $3x - 2$ **(d)** $2 - 3x$

2 **(a)** $\frac{1}{2x}$ **(b)** $\frac{1}{x+2}$ **(c)** $\frac{1}{x-2}$ **(d)** $\frac{1}{2-x}$

3 Using the sign patterns derived from Exercise 2, solve the following inequalities:
(a) $x + 2 \geq 0$ **(b)** $3x - 2 < 0$ **(c)** $2 - 3x > 0$

Solve each of the inequalities in Exercises 4 to 42 making use of a product sign pattern.

4 $x^2 - x \leq 0$ **5** $t^2 > 2t$ **6** $(x + 2)(x - 3) > 0$
7 $(x - 1)(x - 4) \leq 0$ **8** $x^2 - x - 12 \geq 0$ **9** $2x^2 + 5x - 12 \leq 0$
10 $5x > 3 - 2x^2$ **11** $3x^2 < 17x + 6$ **12** $(x - 2)^2 > (2x + 1)^2$

13 $(2x - 1)^2 > (3x + 4)^2$ **14** $\frac{x}{x-1} < \frac{1}{4}$

15 $x^2 + 5x - 50 > -3x^2 + x - 2$ **16** $3x^2 - 8x + 7 > 2x^2 - 3x + 1$
17 $x^3 - 4x \geq 0$
18 $x^2 + x - 2 \leq 0$ and $x^2 - x - 6 \geq 0$
19 $x^2 + x - 2 \leq 0$ or $x^2 - x - 6 \geq 0$
20 $2x^2 + x - 6 \geq 0$ and $2x^2 - 3x - 2 < 0$
21 $2x^2 + x - 6 \geq 0$ or $2x^2 - 3x - 2 < 0$
22 $(x + 4)(x - 1)(x - 2) < 0$
23 $(x - 1)^2(x - 3) > 0$

24 $\dfrac{2x+1}{x-1} < 1$

25 $\dfrac{2x+1}{x-4} \leq 1$

26 $\dfrac{3}{x-2} > 0$

27 $\dfrac{x+2}{x-1} < \dfrac{8}{x}$

28 $3x^3 - 4x^2 - 20x < 0$

29 $\dfrac{x-1}{x+2} \geq 0$

30 $\dfrac{2}{x} > x$

31 $\dfrac{x}{x+1} \geq 1$

32 $\dfrac{1}{x} + \dfrac{2}{x} > \dfrac{4}{x}$

33 $x > x + \dfrac{1}{x}$

34 $\dfrac{(x+2)(x-4)}{2x-1} > 0$

35 $\dfrac{x^2-1}{x+2} > 0$

36 $\dfrac{x^2-4}{x+1} < -x - 2$

37 $\dfrac{x^3}{x^2-5x+4} \leq 0$

38 $\dfrac{x-2}{x-1} < \dfrac{1}{x}$

39 $\dfrac{(x+1)(x-4)}{(x+2)(x-2)} < 0$

40 $2 < \dfrac{1}{x-3} < 8$

41 $\dfrac{1}{x-4} > \dfrac{1}{x-3}$

42 $(x-1)^3(x-3)(x+2) < 0$

4.6 ABSOLUTE VALUE

In Chapter 1 we defined and made use of the concept of absolute value. Recall that the absolute value of x is denoted by $|x|$ and defined as

$$|x| = \begin{cases} x & \text{if } x \geq 0 \\ -x & \text{if } x < 0 \end{cases}$$

Recall also that $|x - a|$ is the distance between x and a. In Chapter 2 we revised this notion; however, we viewed it as a formula that defined a function. In this section we discuss some of the basic properties of absolute value and make use of these properties to solve equations and inequalities involving this concept.

PROPERTY 7 For any real number a:

(a) $|a| \geq 0$.
(b) $|a|^2 = a^2$.
(c) $\sqrt{a^2} = |a|$.

Proof

PART (*a*)

$$\begin{aligned} a < 0 \quad &\text{means that} \quad |a| = -a > 0 \\ a = 0 \quad &\text{means that} \quad |a| = a = 0 \\ a > 0 \quad &\text{means that} \quad |a| = a > 0 \end{aligned}$$

PART (b) For $a \geq 0$, $|a| = a$; hence

$$|a|^2 = |a||a| = a \cdot a = a^2$$

For $a < 0$, $|a| = -a$; hence

$$|a|^2 = |a||a| = (-a)(-a) = a^2$$

PART (c) Recall that $\sqrt{a^2}$ denotes the principal root of a^2; hence $\sqrt{a^2} \geq 0$. $a^2 \geq 0$; however, a could be positive, zero, or negative. Now

$$\sqrt{a^2} = \begin{cases} a & \text{whenever } a \geq 0 \\ -a & \text{whenever } a < 0 \end{cases}$$

but this is how we defined the absolute value of a. Therefore

$$\sqrt{a^2} = |a|$$

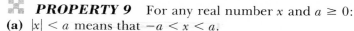

PROPERTY 8 For any real numbers a and b

$$|a - b| = |b - a|$$

Proof

If $a - b > 0$, then $b < a$, which means that $a - b > 0$ and

$$|a - b| = -(a - b) = b - a = |b - a|$$

If $a - b > 0$, then $b < a$, which means that $a - b > 0$ and

$$|a - b| = a - b = -(b - a) = |b - a|$$

If $a - b = 0$, then $a = b$, which means that $b - a = 0$ and

$$|a - b| = 0 = |b - a|$$

Example 1 Let $a = 2$ and $b = 3$; then

$$|2 - 3| = |-1| = 1 = |3 - 2|$$

PROPERTY 9 For any real number x and $a \geq 0$:

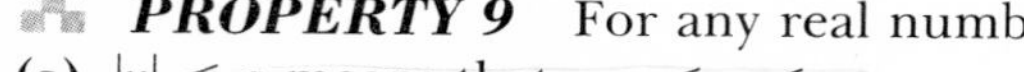

(a) $|x| < a$ means that $-a < x < a$.
(b) $|x| > a$ means that $x > a$ or $x < -a$.
(c) $|x| = a$ means that $x = a$ or $x = -a$.

Proof

We prove part (a), leaving parts (b) and (c) as exercises.

PART (a) Use Property 6a of Section 4.1: $|x| < a$ means that $|x|^2 < a^2$.

$$x^2 < a^2 \qquad \text{Property 7}b$$

and

$$x^2 - a^2 < 0 \qquad \text{Subtracting } a^2 \text{ from both sides}$$

$$(x - a)(x + a) < 0 \qquad \text{Factoring}$$

Hence when we set up a product sign pattern for these factors, it follows that $-a < x < a$.

PROPERTY 10 The absolute value of the product of any two real numbers a and b is equal to the product of their absolute values; that is,

$$|ab| = |a||b|$$

Proof

$$|ab| = \sqrt{(ab)^2} = \sqrt{a^2b^2} = \sqrt{a^2}\sqrt{b^2} = |a||b| \qquad \text{Property 7c}$$

Example 2

$$|2(-3)| = |-6| = -(-6) = 6$$
$$|2||-3| = 2[-(-3)] = 2(3) = 6$$

PROPERTY 11 The absolute value of the sum of any two real numbers a and b is less than or equal to the sum of their absolute values; that is,

$$|a + b| \leq |a| + |b|$$

Proof

$$\begin{aligned}
|a + b|^2 &= (a + b)^2 && \text{Property 7b} \\
&= a^2 + 2ab + b^2 && \text{Squaring} \\
&= |a|^2 + 2ab + |b|^2 && \text{Again by Property 7b} \\
&\leq |a|^2 + 2|ab| + |b|^2 && \text{Since } ab \leq |ab| \\
&\leq |a|^2 + 2|a||b| + |b|^2 && \text{Property 10} \\
&\leq (|a| + |b|)^2
\end{aligned}$$

Hence, by Property 6*b*, it follows that

$$|a + b| \leq |a| + |b|$$

Example 3 For $a = 2$ and $b = 3$

$$|2 + 3| = 5 = |2| + |3|$$

Example 4 For $a = -2$ and $b = -5$

$$|-2 + (-5)| = 7 = |-2| + |-5|$$

Example 5 For $a = 3$ and $b = -7$

$$|3 + (-7)| = 4$$

but $$|3| + |-7| = 10$$

Hence $$|3 + (-7)| < |3| + |-7|$$

Let us end the discussion with some other worked-out examples.

Example 6 Solve $|y| > -1$.
By Property 7*a*, $|y| \geq 0$ for all real y, so surely it is greater than -1 for all real numbers.

Example 7 Solve $\sqrt{(x)^2} < 1$.
From Property 7*c*, $\sqrt{x^2} = |x|$, so that we have $|x| < 1$; and by Property 9*a*, $|x| < 1$ means $-1 < x < 1$.

Example 8 Solve $|x + 2| = 3$.
From Property 9*c*, $|x + 2| = 3$ means

$$x + 2 = 3 \qquad \text{or} \qquad x + 2 = -3$$

so that $$x = 1 \qquad \text{or} \qquad x = -5$$

Example 9 Solve $|x + 2| < 3$.
From Property 9*a*, $|x + 2| < 3$ means $-3 < x + 2 < 3$. Subtracting 2 from each member, we obtain $-5 < x < 1$.

Example 10 Solve $|2x + 1| > 1$.
From Property 9*b*, $|2x + 1| > 1$ means that $2x + 1 > 1$ or $2x + 1 < -1$. Simplifying gives $2x > 0$ or $2x < -2$, and we obtain $x > 0$ or $x < -1$.

Example 11 Solve $|2x| > |x + 1|$.
Since $|2x| \geq 0$ and $|x + 1| \geq 0$, Property 6*b* is applicable:

$$|2x|^2 > |x + 1|^2$$

$$(2x)^2 > (x + 1)^2 \qquad \text{Property 7}b$$

$$4x^2 > x^2 + 2x + 1 \implies 3x^2 - 2x - 1 > 0 \qquad \text{Simplifying}$$

Hence $$(3x + 1)(x - 1) > 0 \qquad \text{Factoring}$$

It now follows from a product sign pattern that

$$x > 1 \qquad \text{or} \qquad x < -\tfrac{1}{3}$$

Example 12 Using absolute-value notation, write: The variable x is within 1 unit of some number a.
Since $|x - a|$ means the distance between a and x, we write $|x - a| < 1$. This further means that x is between $a + 1$ and $a - 1$ (see Diagram 24).

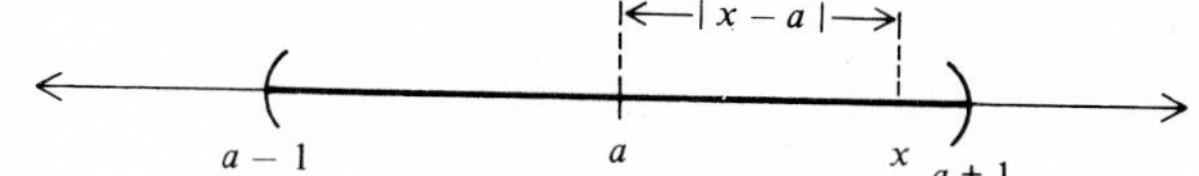

Diagram 24

Example 13 Using absolute-value notation, write: The variable x is within 1 unit of some number a but not equal to a.

We know that for any values of x and a, $|x - a| \geq 0$ (Property 7a). The sentence stipulates that $x \neq a$. This means that $|x - a| > 0$. Therefore adding this restriction to the preceding example, we obtain the desired results, namely, $0 < |x - a| < 1$ (see Diagram 25).

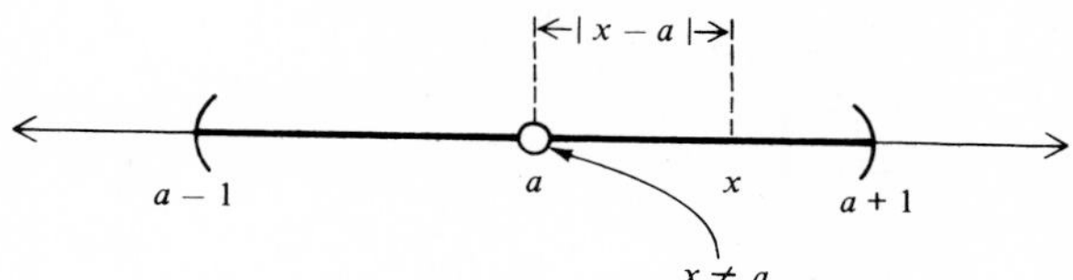

Diagram 25

Some important applications of absolute value can be studied in the following examples.

Example 14 Show that if $|x - 4| < 1$ and $y = 3x + 2$, then y is in the interval (11, 17).

$|x - 4| < 1$ means that $-1 < x - 4 < 1$, that is, $3 < x < 5$. Hence $9 < 3x < 15$ and $11 < 3x + 2 < 17$. Since $y = 3x + 2$, $11 < y < 17$, or, equivalently, y is in the interval (11, 17).

Example 15 For $f(x) = 2x$, find d such that $|f(x) - 6| < 1$ whenever $|x - 3| < d$.

Since $f(x) = 2x$,

$$|f(x) - 6| = |2x - 6| = 2|x - 3| < 1$$

From $2|x - 3| < 1$ it follows that $|x - 3| < \frac{1}{2}$. Hence, $d = \frac{1}{2}$.

Example 16 Solve Example 15 graphically.

First graph $f(x) = 2x$. Then on the y axis graph $|f(x) - 6| < 1$. Now trace two horizontal segments from the end points of this interval so that they meet the graph of $y = f(x)$. At the two points of intersection drop perpendiculars, which intersect the x axis, determining an interval about the point $x = 3$. We can now choose any value for d that will guarantee that the distance between d and 3 is within this interval (see Diagram 26).

For this example the maximum value of d is $\frac{1}{2}$. You should realize, however, that any value d, $0 < d < \frac{1}{2}$, will satisfy the condition that $|f(x) - 6| < 1$.

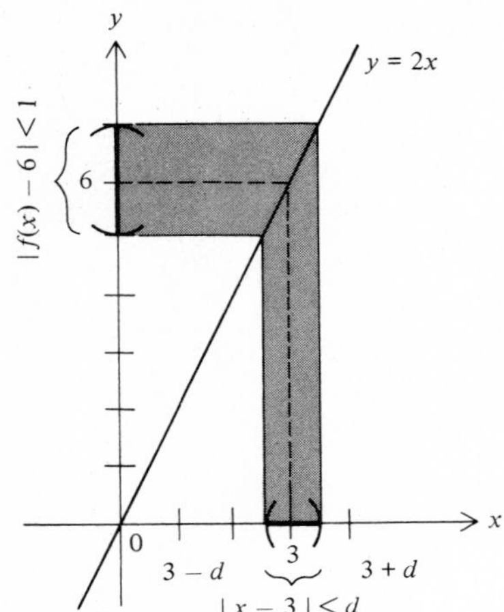

Diagram 26

EXERCISES

1 Depict graphically:
(a) $|x| < a, a > 0$ **(b)** $|x| > a, a > 0$ **(c)** $|x| = a, a > 0$

2 Express $|a + b|$ by using a radical.

3 Express $|a/b|$ using a radical.

4 Show that $|a^3| = a^2 \sqrt{a^2}$.

5 Given $|x - y| = 5$, what is the value of $|y - x|$?

Determine the solution to Exercises 6 to 44.

6 $|x - \frac{1}{3}| = \frac{3}{4}$ **7** $|2x - 1| = 4$ **8** $|3 - x| = -1$

9 $|3x - 1| < 5$ **10** $|x + \frac{1}{2}| > \frac{5}{2}$ **11** $|x - 3| < 1$

12 $|2 - 8x| \geq 6$ **13** $|x - 2| < 0.001$ **14** $|x - 2| > 0.001$

15 $|2x - 3| \leq |4x + 3|$ **16** $|x + 1| < |1 - 2x|$

17 $|2x - 1| > 3$ and $|x - 2| < 1$ **18** $|2x - 3| \leq 1$ or $|x + 2| < 3$

19 $|2x - 5| < |x + 4|$ **20** $|2x - 3| \leq 5$ and $|2x - 1| > 3$

21 $|2x - 3| \leq 5$ or $|2x - 1| > 3$ **22** $|2x - 3| < 9 - x$ and $4x - 3 > 6x - 7$

23 $|2x - 3| < 9 - x$ or $4x - 3 > 6x - 7$

24 $|x^2 + 5x - 19| = 5$ **25** $|x^2 - 3x - 7| < 3$

26 $|x|\,(2x - 3) < 0$ **27** $2|2x - 3| < |x + 10|$

28 $0 < |x - 2| \leq 4$ **29** $0 < |2x + 3| \leq 5$

30 $|4x^2 - 2x - 7| > 5$ **31** $0 < |x - 2| < 1$

32 $0 < |x - a| < b, b > 0$ **33** $|x^2 - 5x| = 6$

34 $|2x^2 - 1| = -1$ **35** $3 - |x| = x$ **36** $|\sqrt{x^2} - x| > 1$

37 $|x - 2| = 2 - x$ **38** $|x - 2||1 - x| = 0$

39 $|x| - |3x - 1| < 0$ **40** $\left|\dfrac{3x}{1 + x}\right| < 2$ **41** $\left|\dfrac{x - 2}{x + 1}\right| = 3$

42 $\left|\dfrac{1}{x} - 3\right| < 6$ **43** $\left|\dfrac{x + 3}{x - 1}\right| < 1$ **44** $\left|\dfrac{x - 1}{x - 2}\right| < 1$

45 Show that if $|x - 1| < 3$ and $y = 4 - 3x/2$, then y is in the interval $(-2, 7)$.
46 Show that if $|x - 3| < 2$ and $y = x^2 - 3x + 2$, then y is in the interval $(-12, 24)$.
47 Show that if $|x - 6.5| < 2.5$ and $y = x/2 + 1/\sqrt{x}$, then y is in the interval $(\frac{7}{3}, 5)$.
48 **(a)** For $f(x) = 3x - 1$, find d such that $|f(x) - 5| < 1$ whenever $|x - 2| < d$.
(b) Solve this problem graphically.
49 **(a)** For $f(x) = 2 - x/2$, find d such that $|f(x) - 1| < 1$ whenever $|x - 2| < d$.
(b) Solve this problem graphically.
50 Given $f(x) = x^2$ and $|f(x) - 9| < 1$ whenever $|x - 3| < d$. Find two values for d graphically such that the above statement is true.

16 $|x+1|^2 < |1-2x|^2$
$(x+1)^2 < (1-2x)^2$

3 distance form

Polynomial Functions

5.1 POLYNOMIAL FUNCTIONS

In Chapter 2 we analyzed the linear function $f(x) = mx + b$. In Chapter 3 we analyzed the quadratic function $f(x) = ax^2 + bx + c$. In this chapter we analyze the general polynomial function of the form

$$p(x) = a_n x^n + a_{n-1}x^{n-1} + a_{n-2}x^{n-2} + \ldots + a_1 x^1 + a_0$$

where $n \geq 0$ and is an integer and the coefficients $a_0, a_1, a_2, a_3, \ldots, a_n$ are *real numbers.*

If $a_n \neq 0$, the polynomial is said to be of *degree n* and the term a_n is called the *leading coefficient.* If all the coefficients of $p(x)$ are zero, then $p(x) = 0$ and $p(x)$ is called a *zero polynomial.* An important term a_0 is called the *constant term.*

Example 1 $p(x) = 2x^4 - 9x^2 + 17x + 3$.

Since $n = 4$, $p(x)$ is a fourth-degree polynomial (called a quartic). The leading coefficient is

$$a_n = a_4 = 2$$

Zero coefficients are equally important and must be accounted for.

$$\begin{aligned} a_{n-1} &= a_3 = 0 \\ a_{n-2} &= a_2 = -9 \\ a_{n-3} &= a_1 = 17 \\ a_{n-4} &= a_0 = 3 \qquad \text{constant term} \end{aligned}$$

Example 2 $r(x) = x - 3x^3 + \sqrt{2}x^2$.

$r(x)$ is a polynomial of degree 3 (called a cubic) with leading coefficient -3 and constant term 0. The order of the terms does not follow the proper convention and should be rewritten

$$r(x) = -3x^3 + \sqrt{2}x^2 + x$$

The expression

$$r(x) = -3x^3 + \sqrt{2}x^2 + \sqrt{x}$$

does not qualify as a polynomial function since $\sqrt{x} = x^{1/2}$ and the exponent 1/2 is not an integer. The expression

$$r(x) = -3x^3 + \sqrt{2}x^2 + \frac{1}{x}$$

also does not qualify as a polynomial function since $1/x = x^{-1}$ and the exponent is not a *positive* integer.

5.2 EVALUATING A POLYNOMIAL

Frequently we are required to evaluate the function $p(x)$ for particular values of the variable x. This can easily be done by direct substitution into the function. For example, if

$$p(x) = 2x^3 - 8x^2 + 5x + 7$$

find the value of the function when $x = 3$.

$$\begin{aligned} p(3) &= 2(3)^3 - 8(3)^2 + 5(3) + 7 = 2(27) - 8(9) + 15 + 7 \\ &= 54 - 72 + 15 + 7 = 76 - 72 = 4 \end{aligned}$$

When $x = -2$,

$$\begin{aligned} p(-2) &= 2(-2)^3 - 8(-2)^2 + 5(-2) + 7 = 2(-8) - 8(4) - 10 + 7 \\ &= -16 - 32 - 10 + 7 = -58 + 7 = -51 \end{aligned}$$

Although not very difficult, such methods may be cumbersome, and other techniques have been developed to evaluate such functions and reduce the numerical work. We investigate one such method, called *synthetic substitution,* using the above problem as our guide.

Example 1 Find $p(3)$ for

$$p(x) = 2x^3 - 8x^2 + 5x + 7$$

STEP 1 Factor the x^2 from the first two terms

$$p(x) = (2x - 8)x^2 + 5x + 7$$

STEP 2 Factor the x from the new first two terms

$$p(x) = [(2x - 8)x + 5]x + 7$$

STEP 3 To find $p(3)$, substitute 3 for x

$$\begin{aligned} p(3) &= \{[2(3) - 8](3) + 5\}(3) + 7 = [(6 - 8)(3) + 5](3) + 7 \\ &= [(-2)(3) + 5](3) + 7 = (-6 + 5)(3) + 7 = (-1)(3) + 7 = -3 + 7 \\ &= 4 \end{aligned}$$

If we follow the rules of algebra and evaluate the statement working from the inside out, we find that $p(3) = 4$.

We can summarize the mechanics:

Step	Operation	Result
1	Multiply the leading coefficient by 3	6
2	Add the result of step 1 to −8, the second coefficient	−2
3	Multiply the result of step 2 by 3 *again*	−6
4	Add the result of step 3 to 5, the third coefficient	−1
5	Multiply the result of step 4 by 3 *again*	−3
6	Add the result of step 5 to the constant term 7	4

This process, called *synthetic substitution*, can be summarized in tabular form

$$\begin{array}{r|rrrr} \text{from } p(3) & \multicolumn{4}{l}{\text{coefficients of } p(x) \textit{ in order}} \\ 3 & 2 & -8 & 5 & 7 \\ & & 6 & -6 & -3 \\ \hline & 2 & -2 & -1 & \boxed{4} \Longrightarrow p(3) = 4 \end{array}$$

To show the mechanical simplicity of this process, let us repeat the procedure to find $p(-2)$

$$\begin{array}{r|rrrr} -2 & 2 & -8 & 5 & 7 \\ & & -4 & 24 & -58 \\ \hline & 2 & -12 & 29 & \boxed{-51} \Longrightarrow p(-2) = -51 \end{array}$$

The results are identical to substituting directly into the function, but *beware* of the zero coefficients. They must be accounted for by using a zero entry.

Example 2 $f(x) = 2x^4 - 9x^2 + 17x + 3$.
Find $f(-3)$ by synthetic substitution

		↓ must be included			
−3	2	0	−9	17	3
		−6	18	−27	30
	2	−6	9	−10	33 ⟹ $f(-3) = 33$

If we neglected the second highest power, which happens to be missing, we would erroneously compile

−3	2	−9	17	3
		−6	45	−186
	2	−15	62	−183 incorrect

For a polynomial of degree n there should be $n + 1$ terms. Since in this example $n = 4$, there should have been five terms (not four).

EXERCISES

Using the method of synthetic substitution, evaluate the given polynomials in Exercises 1 to 12.

1 $P(x) = 2x^3 - 3x^2 + x - 4$
(a) $P(1)$ **(b)** $P(-2)$ **(c)** $P(4)$ **(d)** $P(-8)$

2 $f(x) = 3x^4 + 10x^3 - 20x^2 - 40x + 32$
(a) $f(-4)$ **(b)** $f(1)$ **(c)** $f(-3)$ **(d)** $f(\frac{2}{3})$

3 $F(x) = 6 - 4x - x^2 + x^3$
(a) $F(0)$ **(b)** $F(1)$ **(c)** $F(-1)$ **(d)** $F(-3) - F(3)$

4 $G(x) = x^3 - x^3 - 3x^2 + 4$
(a) $G(2)$ **(b)** $[G(0)][G(1) - G(-1)]$

5 $H(x) = x^4 - 3x^3 + x - 3$
(a) $\dfrac{H(-1) + H(4)}{100}$ **(b)** $\dfrac{H(-2) + H(4)}{H(1)}$

6 $R(x) = 4x^4 - 2x^2 + 1$
(a) $R(2)$ **(b)** $R(\sqrt{2})$ **(c)** $R(-\sqrt{2})$ **(d)** $R(-\frac{1}{2})$

7 $T(x) = x^3 - (a + b)x^2 + bx - 2ab$
(a) $T(a)$ **(b)** $T(b)$ **(c)** $T(a + b)$ **(d)** $T(1)$

8 If $P(x) = 4x^3 - x^2 + 3x + b$, find b so that:
(a) $P(-1) = 0$ **(b)** $P(\frac{1}{2}) = 1$

9 If $Q(x) = x^3 - 3x + 2c$, find c so that:
(a) $Q(1) = 1$ **(b)** $Q(-1) = 0$

10 If $R(x) = 2x^3 - 3x^2 + ax + b$, find the values of a and b so that

$$R(2) = 4 \quad \text{and} \quad R(-1) = 1$$

11 If $S(x) = x^4 - 2x^3 - 2x^2 + 3x + 2$, find each of the following:
(a) $S(4)$ **(b)** $S(\frac{3}{2})$ **(c)** $S(-\frac{2}{3})$ **(d)** $S(-4)$

12 Consider the cubic polynomial: $P(x) = ax^3 + bx^2 + cx + d$. Find a, b, c, and d so that

$$P(0) = -1 \qquad P(1) = 2 \qquad P(-1) = -10 \qquad P(3) = 38.$$

13 Consider the quadratic $Q(x) = ax^2 + bx + c$ and find a, b, and c so that

$$Q(-1) = 1 \qquad Q(1) = -5 \qquad Q(-\tfrac{1}{2}) = -2$$

5.3 THE DIVISION ALGORITHM

When dividing 324 by 24, we obtain a *quotient* of 13 and a *remainder* of 12. This result can also be written

$$324 = 13(24) + 12$$

In general, if we have a positive integer P (called the *dividend*) and a positive integer D (called the *divisor*), we should be able to find a nonnegative integer Q (called the *quotient*) and another nonnegative integer R (called the *remainder*) which is less than D such that

$$\frac{P}{D} = Q + \frac{R}{D} \qquad \text{or} \qquad P = QD + R$$

Division of polynomials leads to a polynomial quotient with a polynomial remainder. This property of polynomials can be expressed in a theorem, which we present without proof.

Theorem 1 The Division Algorithm. If $P(x)$ and $D(x)$ are polynomials, and if $D(x)$ is not the zero polynomial, then there exist unique polynomials $Q(x)$ and $R(x)$ such that:

(a) Either $R(x) = 0$ or the degree of $R(x)$ is less than the degree of $D(x)$ and $P(x) = Q(x)D(x) + R(x)$.

(b) If the degree of $D(x)$ is *greater* than that of $P(x)$, then $Q(x) = 0$ and $P(x) = R(x)$.

Example 1 Suppose $P(x) = 3x^4 - 4x^3 + 2x^2 + x - 1$ and $D(x) = 2x^2$. Find $Q(x)$ such that

$$3x^4 - 4x^3 + 2x^2 + x - 1 = Q(x) \cdot 2x^2 + R(x)$$

Either $R(x)$ will be 0 *or* $R(x)$ has a degree less than 2. It should be noted that $Q(x)$ must be of degree 2, since $Q(x) \cdot 2x^2$ is a polynomial of degree 4 [the degree of $P(x)$]. Divide $D(x)$ into $P(x)$:

$$\begin{array}{r|l}
 & \frac{3}{2}x^2 - 2x + 1 \\
\cline{2-2}
2x^2 & 3x^4 - 4x^3 + 2x^2 + x - 1 \\
 & 3x^4 \\
\cline{2-2}
 & \quad - 4x^3 + 2x^2 + x - 1 \\
 & \quad - 4x^3 \\
\cline{2-2}
 & \qquad\quad 2x^2 + x - 1 \\
 & \qquad\quad 2x^2 \\
\cline{2-2}
 & \qquad\qquad\quad x - 1
\end{array}$$

$$Q(x) = \tfrac{3}{2}x^2 - 2x + 1$$
$$R(x) = x - 1$$

Example 2 $P(x) = 4x^3 + 2x^2 + x - 2$ and $D(x) = 2x^2 - x + 1$. Find $Q(x)$ and $R(x)$ such that

$$4x^3 + 2x^2 + x - 2 = Q(x)[(2x^2 - x + 1)] + R(x)$$

In this example, $Q(x)$ must be of degree 1 for the product $Q(x)D(x)$ to yield a polynomial of degree 3. Divide $D(x)$ into $P(x)$:

$$\begin{array}{r|l}
 & 2x + 2 \\
\cline{2-2}
2x^2 - x + 1 & 4x^3 + 2x^2 + x - 2 \\
 & 4x^3 - 2x^2 + 2x \\
\cline{2-2}
 & \qquad 4x^2 - x - 2 \\
 & \qquad 4x^2 - 2x + 2 \\
\cline{2-2}
 & \qquad\qquad\quad x - 4
\end{array}$$

$$Q(x) = 2x + 2$$
$$R(x) = x - 4$$

Example 3 $P(x) = x$ and $D(x) = x^2 - 1$. Find $Q(x)$ and $R(x)$ such that

$$x = Q(x)(x^2 - 1) + R(x)$$

In this example, the degree of $D(x)$ is *greater* than that of $P(x)$. Then

$$x = 0(x^2 - 1) + x \qquad Q(x) = 0 \qquad R(x) = x$$

Suppose we consider the specific case of a polynomial $P(x)$ of degree n, $n \neq 0$, being divided by a *linear polynomial* of the form $x - c$. Then

$$P(x) = (x - c)Q(x) + R(x)$$

Since either $R(x) = 0$ or the degree of $R(x)$ is less than that of $x - c$, $R(x)$ *must be a constant.*

Example 4 Divide $P(x) = 2x^3 + x^2 - 8x - 4$ by $D(x) = x - 1$. Let us illustrate the method by *long division* first.

$$
\begin{array}{r}
2x^2 + 3x - 5 \\
x - 1 \overline{) 2x^3 + x^2 - 8x - 4} \\
\underline{2x^3 - 2x^2} \\
3x^2 - 8x - 4 \\
\underline{3x^2 - 3x} \\
- 5x - 4 \\
\underline{- 5x + 5} \\
- 9
\end{array}
$$

$$Q(x) = 2x^2 + 3x - 5 \qquad R(x) = -9$$

It should be noted that $P(1) = -9$. Using the method of synthetic substitution from Section 5.2, we can verify this fact.

$$
\begin{array}{r|rrrr}
1 & 2 & 1 & -8 & -4 \\
 & & 2 & 3 & -5 \\
\hline
 & 2 & 3 & -5 & -9
\end{array}
$$

$$P(1) = -9 \qquad P(1) = R(x)$$

When $P(x)$ is divided by $x - c$, the remainder $P(c)$ is the value of $P(x)$ at c.

EXERCISES

In Exercises 1 to 15, $P(x)$ is divided by $D(x)$. Determine $Q(x)$ and $R(x)$.

1 $P(x) = 9x^3 - 2x + 1$; $D(x) = 3x^2 + x - 2$
2 $P(x) = x^8 + x^2 + 1$; $D(x) = x^2 - 1$
3 $P(x) = x^8 - 1$; $D(x) = x^2 - 1$
4 $P(x) = 2x^3 + 8 + 2x^2$; $D(x) = 2x^2 + 2x + 4$
5 $P(x) = 3x^4 - 4x^3 + 7x^2 - x + 4$; $D(x) = x^2 - 6x + 1$
6 $P(x) = x^3$; $D(x) = x^2 - 1$
7 $P(x) = 3x^3 + 2x^2 - x + 1$; $D(x) = 2x^2 + x + 1$
8 $P(x) = 2x^4 - x^3 + x^2 + 4x - 1$; $D(x) = 3x^2$
9 $P(x) = 2x^4 + 3x^3 + x + 1$; $D(x) = 3x$
10 $P(x) = x^4 - 5x^3 + 3x^2 + x - 5$; $D(x) = x^2 - x - 1$
11 $P(x) = 16x^4 + 4x^2 + 1$; $D(x) = 4x^2 - 2x + 1$
12 $P(x) = 4x^2 - 9$; $D(x) = -2x + 3$
13 $P(x) = 3x^3 - 10x^2 + 21x - 26$; $D(x) = x - 2$
14 $P(x) = 2x^3 + 3x^2 - 4x + 5$; $D(x) = x + 1$
15 $P(x) = 2x^3 + 5x^2 + x - 13$; $D(x) = x + \frac{3}{2}$

5.4 TWO THEOREMS

Theorem 1 (The Remainder Theorem). If a polynomial $P(x)$ of nonzero degree n is divided by $x - c$, a constant remainder can be found by determining $P(c)$.

Since, by the division algorithm, $P(x) = Q(x)D(x) + R(x)$, this theorem implies that

$$P(x) = Q(x)(x - c) + k$$

k = constant remainder

When $x = c$,

$$P(c) = Q(c)(c - c) + k = k$$

Example 1 Determine the remainder on dividing $P(x) = 2x^3 + x^2 - 8x - 4$ by $D(x) = x - 3$.

$D(x)$ is in the form $x - c$. Then, $x - c = x - 3$ and $c = 3$.

Find $P(3)$ by synthetic substitution:

$$\begin{array}{r|rrrr} 3 & 2 & 1 & -8 & -4 \\ & & 6 & 21 & 39 \\ \hline & 2 & 7 & 13 & \boxed{35} \end{array}$$

$P(3) = 35$, or 35 is the remainder.

Example 2 If we divide $P(x)$ in Example 1 by $D(x) = x + 2$, then $x - c = x + 2$ and $c = -2$. Find $P(-2)$.

Solution

$$\begin{array}{r|rrrr} -2 & 2 & 1 & -8 & -4 \\ & & -4 & 6 & 4 \\ \hline & 2 & -3 & -2 & \boxed{0} \end{array}$$

$$P(-2) = 0$$

COMMENT The method of synthetic substitution is used for determining values of the polynomial and can be used to construct a table of values to graph $y = P(x)$. The last result, $P(-2) = 0$, is significant. -2 is called a *zero* of $P(x)$. In arithmetic, when 15 is divided by 5, the quotient is 3 and the remainder is 0. We say that 5 is a *factor* of 15. When $P(x) = 2x^3 + x^2 - 8x - 4$ is divided by $x + 2$, the remainder is also zero. Then we say that $x + 2$ is a factor of $P(x)$. Since $P(x)$ divided by $x - 3$ yields a nonzero remainder, $x - 3$ is *not* a factor of $P(x)$. In general, if $P(x) = Q(x)(x - c) + 0$, then $x - c$ is a factor of $P(x)$.

Theorem 2 (Factor Theorem). If c is a zero of the polynomial $P(x)$, then $x - c$ is a factor of $P(x)$.

Converse of Theorem 2 If $x - c$ is a factor of $P(x)$, then c is a zero of the polynomial $P(x)$.

Example 3 Show that $x + 2$ is a factor of $P(x) = 4x^4 - 15x^2 + 5x + 6$. If $x + 2$ is a factor, then -2 is a zero of $P(x)$. Show that $P(-2) = 0$:

$$\begin{array}{r|rrrrr} -2 & 4 & 0 & -15 & 5 & 6 \\ & & -8 & 16 & -2 & -6 \\ \hline & 4 & -8 & 1 & 3 & \boxed{0} \end{array}$$

QUESTION How would we test $2x - 3$ as a factor of $P(x)$?

ANSWER Express $2x - 3$ as $2(x - \frac{3}{2})$ and compare $x - \frac{3}{2}$ with $x - c$ so that $c = \frac{3}{2}$.

Example 4 Show that $2x - 3$ a factor of $P(x) = 4x^4 - 15x^2 + 5x + 6$?

$$\begin{array}{r|rrrrr} \frac{3}{2} & 4 & 0 & -15 & 5 & 6 \\ & & 6 & 9 & -9 & -6 \\ \hline & 4 & 6 & -6 & -4 & \boxed{0} \end{array}$$

Some interesting observations from *arithmetic* will be beneficial here.

The *prime factors* of 120 are $2 \cdot 2 \cdot 2 \cdot 3 \cdot 5$. We can divide 120 evenly by any one of these factors, and the new quotient will still be divisible by the remaining factors. For example,

$\frac{120}{5} = 24$	leaves	$2 \cdot 2 \cdot 2 \cdot 3$
$\frac{120}{2} = 60$	leaves	$2 \cdot 2 \cdot 3 \cdot 5$
$\frac{120}{3} = 40$	leaves	$2 \cdot 2 \cdot 2 \cdot 5$

It should be noted that 2 is a *triple* factor of 120. Therefore, if we can discover *one* factor of a number, to find the other factors we need only analyze the resulting quotient, not the original number again. Hence to find *all* the factors of 120,

$\frac{120}{3} = 40$	remainder 0	3 is a factor of 120
$\downarrow$		
$\frac{40}{5} = 8$	remainder 0	5 is a factor of 120
$\frac{8}{2} = 4$	remainder 0	2 is a triple factor of 120
$\frac{4}{2} = 2$	remainder 0	
$\frac{2}{2} = 1$	remainder 0	

Note that although 8 is a factor of 120, since 8 can be written as the product of three prime numbers $2 \cdot 2 \cdot 2$, we write 2 as a triple factor, rather than 8.

Referring back to $P(x) = 4x^4 - 15x^2 + 5x + 6$ and showing that $x + 2$ is a factor, let us revisit what is known as *long division,* from elementary algebra, and analyze the results very carefully.

$$\begin{array}{r l}
 & 4x^3 - 8x^2 + x + 3 \\
x + 2 \,\big) & 4x^4 \qquad\quad - 15x^2 + 5x + 6 \\
 & 4x^4 + 8x^3 \\
\hline
 & \quad - 8x^3 - 15x^2 \\
 & \quad - 8x^3 - 16x^2 \\
\hline
 & \qquad\qquad x^2 + 5x \\
 & \qquad\qquad x^2 + 2x \\
\hline
 & \qquad\qquad\quad 3x + 6 \\
 & \qquad\qquad\quad 3x + 6 \\
\hline
 & \qquad\qquad\qquad\quad 0 \quad \text{remainder}
\end{array}$$

Using the synthetic process

$$\begin{array}{r|rrrrr}
-2 & 4 & -0 & -15 & 5 & 6 \\
 & & -8 & 16 & -2 & -6 \\
\hline
 & 4 & -8 & 1 & 3 & \boxed{0} \\
 & \uparrow & \uparrow & \uparrow & \uparrow & \\
 & x^3 & x^2 & x & c &
\end{array}$$

The resulting constants in the synthetic process correspond to the coefficients of the quotient. This process is called *synthetic division* (a welcome substitute for the traditional long-division method shown previously). The quotient is, of course, another polynomial of degree *one less* than the original polynomial, and the new equation is referred to as a *depressed equation.*

In this example, the depressed equation is

$$Q(x) = 4x^3 - 8x^2 + x + 3$$

Suppose we now wish to test whether $2x - 3$ is a factor of the original $P(x)$. Instead of using the original fourth-degree polynomial, we can use $Q(x)$, the depressed equation.

From $2x - 3$, $c = \frac{3}{2}$. Then

$$\begin{array}{r|rrrr}
\frac{3}{2} & 4 & -8 & 1 & 3 \\
 & & 6 & -3 & -3 \\
\hline
 & 4 & -2 & -2 & \boxed{0}
\end{array} \qquad 4x^2 - 2x - 2 \qquad \text{New quotient}$$

In fact, now we can continue testing factors on the new quotients. For example, show that $2x + 1$ is a factor of $P(x)$:

$$2x + 1 = 2(x + \tfrac{1}{2}) \qquad c = -\tfrac{1}{2}$$

$$\begin{array}{r|rrr}
-\frac{1}{2} & 4 & -2 & -2 \\
 & & -2 & 2 \\
\hline
 & 4 & -4 & \boxed{0}
\end{array} \qquad 4x - 4 \qquad \text{New quotient}$$

And now show that $x - 1$ is also a factor:

$$x - 1 = x - c \qquad c = 1$$

$$\begin{array}{r|rr} 1 & 4 & -4 \\ & & 4 \\ \hline & 4 & \boxed{0} \end{array}$$

Some important time-saving devices can be discovered here. Since we reduced $P(x)$ eventually to a polynomial of degree 2, we could have employed the methods of Chapter 3 and found the resulting factors.

$$Q(x) = 4x^2 - 2x - 2 = 2(2x + 1)(x - 1)$$

Therefore $2x + 1$ and $x - 1$ are also factors.

COMMENTS Finding the linear factors of a polynomial $P(x)$ is the same as finding the *roots* of the polynomial equation $P(x) = 0$. If c is a zero of the polynomial $P(x)$, we can say that c is a *root* of the equation $P(x) = 0$. Referring to the previous example, $P(x) = 4x^4 - 15x^2 + 5x + 6$, we see that the factors of $P(x)$ are $2x - 3$, $2x + 1$, $x + 2$, and $x - 1$. When $P(x) = 0$, the roots are $\frac{3}{2}$, $-\frac{1}{2}$, -2, and 1.

Example 5 Using the factor theorem, find $P(x)$ whose zeros are 1, -2, and 3.

Since 1, -2, and 3 are zeros of $P(x)$, $x - 1$, $x - (-2)$, and $x - 3$ are factors of $P(x)$. Or

$$P(x) = (x - 1)(x + 2)(x - 3) = x^3 - 2x^2 - 5x + 6$$

Example 6 Show that $x - 1$ is a factor of $P(x) = 2x^3 - 3x^2 + 1$.

Solution

$$\begin{array}{r|rrrr} 1 & 2 & -3 & 0 & 1 \\ & & 2 & -1 & -1 \\ \hline & 2 & -1 & -1 & \boxed{0} \end{array}$$

Therefore, $P(x) = (x - 1)(2x^2 - x - 1)$. It is easy to show that $x - 1$ is also a factor of $2x^2 - x - 1$.

$$\begin{array}{r|rrr} 1 & 2 & -1 & -1 \\ & & 2 & 1 \\ \hline & 2 & 1 & \boxed{0} \end{array}$$

And $P(x) = (x - 1)^2(2x + 1)$. Since $x - 1$ is *not* a factor of $2x + 1$, we can say that $x - 1$ is a factor of *multiplicity* 2 of $P(x)$. In general, if

$P(x) = (x - c)^k Q(x)$, then $x - c$ is a factor of *multiplicity* k. Correspondingly, c is a root of multiplicity k of the equation $P(x) = 0$.

EXERCISES

In Exercises 1 to 9, use synthetic division to find the quotient $Q(x)$ and the remainder $R(x)$ by dividing the first polynomial by the second polynomial.

1 $3x^3 - 5x^2 - 16x + 12;\ x - 2$

2 $x^4 - 3x^2 + 5x + 8;\ x + 1$

3 $x^4 - 2x^3 - 24x^2 + 12x + 30;\ x + 4$

4 $2x^4 - 6x^2 - 5;\ x + 3$

5 $x^5 - 2x^4 - 3x^3 - 6x + 15;\ x - 3$

6 $x^3 - 1;\ x - 1$

7 $x^6 + 64;\ x + 2$

8 $8x^3 - 10x^2 - x + 3;\ 4x - 3$

9 $2x^3 - x^2 - 4x + 2;\ 2x + 3$

10 Using the factor theorem, show that $x - a$ is a factor of $x^n - a^n$ (n is positive integer).

11 Using the factor theorem, show that $x^3 + a^3$ is exactly divisible by $x + a$.

12 Using the factor theorem, show that $x^4 + a^4$ is *not* exactly divisible by $x + a$. Find the remainder.

13 When is $x + a$ a factor of $x^n - a^n$? (State conditions that must be satisfied by n.)

14 For what value of c is $3x^2 - cx + 4x + 2c$ exactly divisible by $x + 1$?

15 For what value of d does

$$\frac{x^4 - dx^2 - 6x + (d + 4)}{x - 2}$$

have a remainder of 2?

Find the *possible* multiplicity of factors in Exercises 16 to 19.

16 $x^4 - 5x^3 + 6x^2 + 4x - 8$; factor $x - 2$

17 $x^3 - 3x - 2$; factor $x + 1$

18 $9x^3 - 3x^2 - 8x + 4$; factor $3x - 2$

19 $x^4 + 2x^2 + 1$; factor $x + 1$

20 Find a and b so that $P(x) = x^4 + ax^3 + (b - a)x^2 - 4bx - 2$ has $x + 1$ as a *double factor.*

21 $P(x) = 2x^4 - 4x^3 + 3x^2 + 1$. Compute: $P(0)$, $P(1)$, $P(0.2)$, $P(0.4)$, $P(2)$, and $P(0.6)$.

22 When $2x^2 + 3x - 12$ is divided by $x + k$, the remainder is 8. Find k.

In Exercises 23 to 25 use the remainder theorem to find the remainder when:

23 $3x^{12} - 6x^3 + 5$ is divided by $x + 1$.

24 $x^{27} - 2x^{18} + 12$ is divided by $x - 1$.

25 $2x^{32} + 3x^{15} - 1$ is divided by $x + 1$.

26 Show that c is a root of $x^3 - x^2(1 + c) - x(c + 4c^2) + (4c^3 + 2c^2) = 0$.

In Exercises 27 to 29 find k so that the first polynomial is a factor of the second.

27 $x - 2;\ x^4 - 2x^2 - 3x + k$

28 $2x + 1;\ 2x^4 - 3x^3 - 7x^2 - 8x + k$

29 $x + 1;\ 4x^3 + kx + 1$

5.5 CURVE SKETCHING AND THE ROOTS OF POLYNOMIALS

In the terminology of Chapter 2, the domain of a polynomial function is the set of all real numbers. The range, however, depends on the characteristics of the function. Graphing these functions is of considerable importance to a basic understanding of the calculus and its practical applications. Our aim is to eliminate the popular brute-force methods used by most students and apply the methods previously discussed in this chapter to ensure maximum efficiency in curve sketching. Special theorems will help us attain this goal while using a minimum number of points.

To sketch a graph of a polynomial function, we first use synthetic substitution and generate a set of points for $P(x)$. Then we plot the points.

Example 1 Sketch the graph of $P(x) = x^3 - 2x^2 - 5x + 6$.

Solution

x	-3	-2	-1	0	1	2	3	4
$P(x)$	-24	0	8	6	0	-4	0	18

Notice that $P(-2) = 0$, $P(1) = 0$, and $P(3) = 0$. The real roots of $P(x) = 0$ are -2, 1, and 3. In Diagram 1 we locate the various points from the table.

Diagram 1 suggests that perhaps there are numerous polynomials which pass through these points. Plotting additional points will help to determine the shape of the graph more accurately. For example,

x	$-\frac{5}{2}$	$-\frac{3}{2}$	$-\frac{1}{2}$	$\frac{1}{2}$	$\frac{3}{2}$	$\frac{5}{2}$	$\frac{7}{2}$
$P(x)$	$-\frac{77}{8}$	$\frac{45}{8}$	$\frac{63}{8}$	$\frac{25}{8}$	$-\frac{21}{8}$	$-\frac{27}{8}$	$\frac{55}{8}$

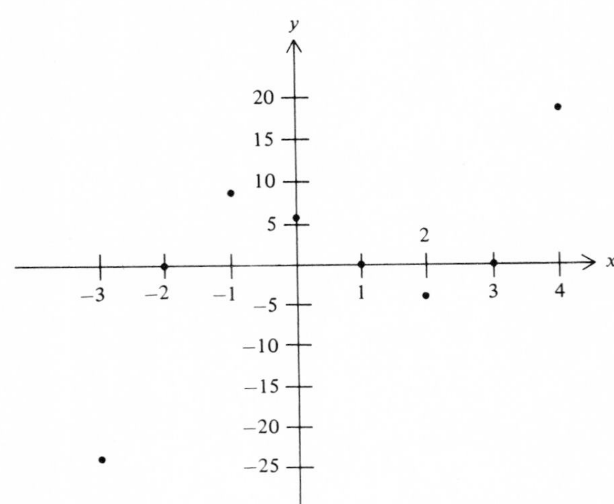

Diagram 1

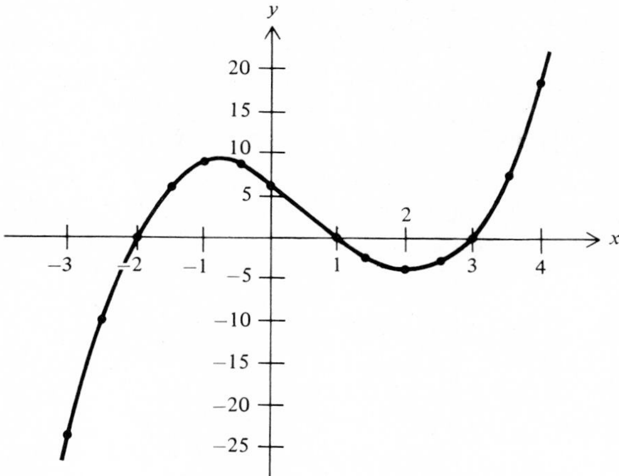

Diagram 2

For a better approximation see Diagram 2.

Without attempting to establish a rigorous set of rules for graphing polynomial functions, let us assume (1) that the graph of a polynomial function is a smooth and unbroken curve and (2) that although plotting points generally does not determine the *exact* shape of the graph, this method is acceptable.

Suppose $P(x) = x^3 + x^2 - 7x - 3$. We are able to write $P(x) = (x + 3)(x^2 - 2x - 1)$ as a consequence of the factor theorem; then -3 is a root of $P(x) = 0$. Also, any value of x such that $x^2 - 2x - 1 = 0$ is a root. Using the quadratic formula, we find that $x = 1 + \sqrt{2}$ and $x = 1 - \sqrt{2}$ are also roots.

In general, if $P(x)$ is a third-degree polynomial, it is impossible to have more than three zeros. Suppose $P(x)$ of degree 3 had four zeros, c_1, c_2, c_3, and c_4. Then, by the factor theorem,

$$P(x) = (x - c_1)(x - c_2)(x - c_3)(x - c_4)$$

The product of these factors would give us a *fourth*-degree polynomial, which is a contradiction.

How many roots can a polynomial equation of degree n have?

Theorem 1 If $P(x) = 0$ is a polynomial equation of degree n, it has exactly n roots.

COMMENTS Some equations have no real roots. In Chapter 3 an equation $x^2 + 1 = 0$ had two complex roots. (We examine nonreal roots in the next section.) Hence, the above theorem *does not imply n real* roots. Some of the roots may be repeated (of multiplicity k). Hence, the theorem *does not imply n different* roots. At this point in the study of polynomial equations, there is no way to determine how many of the roots are real and how many are nonreal.

Below are some examples of polynomial equations and their roots.

Example 2 $2x^4 - 3x^3 - 20x^2 + 27x + 18 = 0$.
The roots are 2, 3, $-\frac{1}{2}$, and -3. There are four different real roots, all rational.

Example 3 $18x^3 - 39x^2 + 8x + 16 = 0$.
The roots are $\frac{4}{3}$, $\frac{4}{3}$, and $-\frac{1}{2}$. There are three real roots, with two repeated rational roots.

Example 4 $2x^3 - 8x^2 - x + 4 = 0$.
The roots are 4, $\sqrt{2}/2$, and $-\sqrt{2}/2$. There are three real roots, one rational and two irrational.

Example 5 $2x^5 - 11x^4 + 14x^3 - 2x^2 + 12x + 9 = 0$.
The roots are 3, 3, $-\frac{1}{2}$, i, and $-i$. There are five roots: three real roots (two of them repeated) and two imaginary roots (nonreal).

The problem of discovering the roots of a polynomial equation is not as easy as the previous examples seem to indicate. Let us concentrate our efforts on polynomials with rational coefficients. In fact, if the coefficients are rational, such as

$$\frac{x^3}{2} + \frac{3x^2}{8} + 2x + \frac{3}{2} = 0$$

by multiplying both sides of the equation by 8, we can express the polynomial as an equation with only integral coefficients. That is,

$$4x^3 + 3x^2 + 16x + 12 = 0$$

If we convert polynomial equations with rational coefficients to polynomial equations with *integer* coefficients, we can examine the following theorem (called the *rational-root theorem*).

Theorem 2 If

$$P(x) = a_n x^n + a_{n-1}x^{n-1} + a_{n-2}x^{n-2} + \cdots + a_1x^1 + a_0 = 0$$

and *all* the coefficients $a_n, a_{n-1}, \ldots, a_1, a_0$ are *integers,* then the rational roots can be found by examining a restricted number of possibilities. More specifically, if P/Q is a rational root of $P(x) = 0$, then P must be a divisor of the constant term a_0 and Q must be a divisor of the leading coefficient a_n.

COMMENT If fractions of the form P/Q are possible roots of $P(x) = 0$, P/Q is expressed in *simplest reduced form.* If we find *all* possible roots of the form P/Q, positive *and* negative, the rational roots of $P(x) = 0$ *must* be on the list. If *all* the possibilities are properly tested and *none* of them is a root of $P(x) = 0$, then the polynomial possesses *no* rational roots. This means that the roots are either irrational or complex or both.

Example 6 $P(x) = 6x^4 - 13x^3 - 6x^2 + 5x + 2$. Find all the rational roots of $P(x) = 0$.

According to Theorem 1, we know there are exactly four roots. Now if there are any rational roots among these four, they are in the form P/Q. According to Theorem 2, P must be a divisor of the constant term 2 (that is, $+1$, -1, $+2$, or -2) and Q must be a divisor of the leading coefficient 6 (that is, $+1$, -1, $+2$, -2, $+3$, -3, $+6$, or -6), and it will be advisable to write this as

$$\frac{\pm(1, 2)}{\pm(1, 2, 3, 6)}$$

Now we compute all possible combinations of numerators and denominators. This will generate *all possible rational roots*

$$\pm\left(1, \tfrac{1}{2}, \tfrac{1}{3}, \tfrac{1}{6}, 2, \tfrac{2}{3}\right)$$

There is a total of 12 possibilities. Let us test some of them.

$$\begin{array}{r|rrrrr} 1 & 6 & -13 & -6 & 5 & 2 \\ & & 6 & -7 & -13 & -8 \\ \hline & 6 & -7 & -13 & -8 & \boxed{-6} \end{array} \qquad \text{no root}$$

$$\begin{array}{r|rrrrr} -1 & 6 & -13 & -6 & 5 & 2 \\ & & -6 & 19 & -13 & 8 \\ \hline & 6 & -19 & 13 & -8 & \boxed{10} \end{array} \qquad \text{no root}$$

Hence, $P(1) = -6\,P(-1) = 10$. See Diagram 3 for a graphical interpretation of this seemingly uninteresting result.

In order for the graph of $P(x)$ to go from $(-1, 10)$ to $(1, -6)$ *it seems logical* that the curve must cross the x axis somewhere between $x = -1$ and $x = 1$ [keep in mind that the graph of $P(x)$ is unbroken].

$$\begin{array}{r|rrrrr} \frac{1}{2} & 6 & -13 & -6 & 5 & 2 \\ & & 3 & -5 & -\frac{11}{2} & -\frac{1}{4} \\ \hline & 6 & -10 & -11 & -\frac{1}{2} & \boxed{\frac{7}{4}} \end{array} \qquad \text{no root}$$

$$\begin{array}{r|rrrrr} -\frac{1}{2} & 6 & -13 & -6 & 5 & 2 \\ & & -3 & 8 & -1 & -2 \\ \hline & 6 & -16 & 2 & 4 & \boxed{0} \end{array} \qquad \text{root}$$

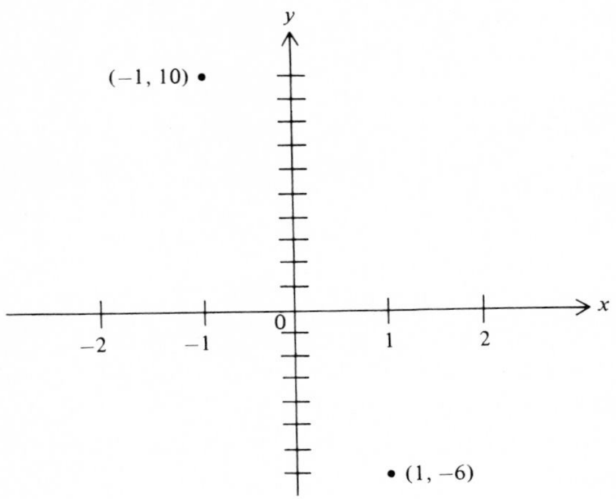

Diagram 3

Now that we have found one rational root, we can test the remaining possible roots on the depressed equation

$$6x^3 - 16x^2 + 2x + 4 = 0$$

and since 2 is a common factor of the coefficients, it is advisable to rewrite the polynomial as

$$3x^3 - 8x^2 + 2 = 0$$

At this point let us compute the y intercept, $P(0) = 2$, and take a look at Diagram 4. Notice, from Diagram 4 that there *seems* to be another root between $x = 0$ and $x = 1$.

$$\begin{array}{r|rrrr} \frac{2}{3} & 3 & -8 & 1 & 2 \\ & & 2 & -4 & -2 \\ \hline & 3 & -6 & -3 & \boxed{0} \end{array} \qquad \text{root}$$

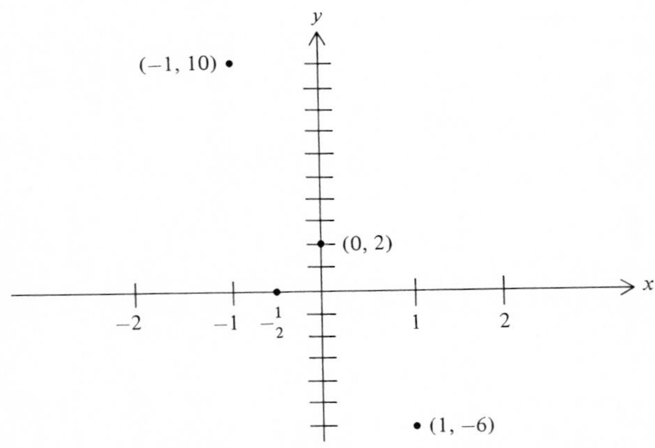

Diagram 4

The resulting depressed equation is

$$3x^2 - 6x - 3 = 0 \qquad \text{or} \qquad x^2 - 2x - 1 = 0$$

If we now apply the rational-root theorem to the equation $x^2 - 2x - 1 = 0$, the only possible rational roots are $+1$ and -1. However, we have previously tested these possibilities, and neither qualified as a root. We therefore conclude that $-\frac{1}{2}$ and $\frac{2}{3}$ are the *only* rational roots of the polynomial equation.

The fourth-degree polynomial has been reduced to a quadratic polynomial. Two roots are still unaccounted for. Even though the rational-root theorem helps us discover only rational roots, if we can reduce the original polynomial equation to a quadratic, the quadratic formula (Chapter 3) will yield the remaining two roots, whether they are irrational or complex. For $x^2 - 2x - 1 = 0$, then

$$x = \frac{2 \pm \sqrt{4 - 4(1)(-1)}}{2} = \frac{2 \pm \sqrt{4 + 4}}{2} = \frac{2 \pm \sqrt{8}}{2} = 1 \pm \sqrt{2}$$

There are two irrational roots (approximately 2.4 and -0.4).

When additional points are supplied, the graph of $P(x) = 6x^4 - 13x^3 - 6x^2 + 5x + 2$ is as shown in Diagram 5.

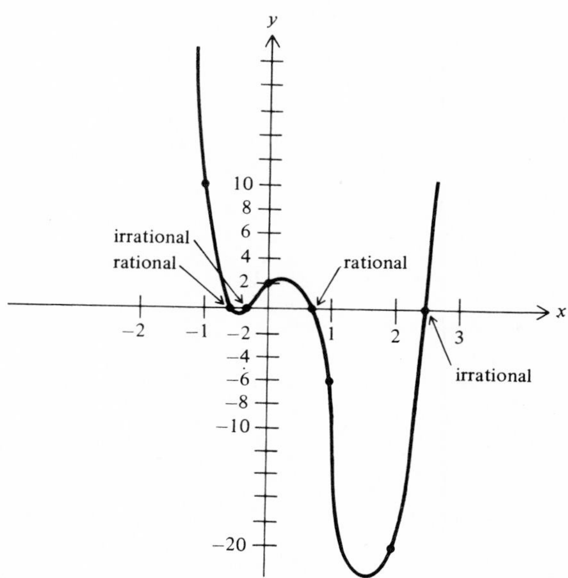

Diagram 5

Example 7 Find all the rational roots of $P(x) = x^3 - 3x + 2 = 0$.

Of the possible roots of the form P/Q, P must be a divisor of 2, and Q must be a divisor of 1.

$$\frac{\pm(1, 2)}{\pm 1}$$

Hence, the only possible *rational* roots are 1, -1, 2, and -2. Let us test the limited number of possibilities:

$$\begin{array}{r|rrrr} 2 & 1 & 0 & -3 & 2 \\ & & 2 & 4 & 2 \\ \hline & 1 & 2 & 1 & \boxed{4} \end{array} \qquad \begin{array}{r|rrrr} -1 & 1 & 0 & -3 & 2 \\ & & -1 & 1 & 2 \\ \hline & 1 & -1 & -2 & \boxed{4} \end{array}$$

$$\begin{array}{r|rrrr} -2 & 1 & 0 & -3 & 2 \\ & & -2 & 4 & -2 \\ \hline & 1 & -2 & 1 & \boxed{0} \end{array}$$

The result is the depressed equation

$$Q(x) = x^2 - 2x + 1 = 0 = (x - 1)(x - 1) = 0$$

The other two roots are $x = 1$ and $x = 1$. This is not surprising. We have seen a double root in Chapter 3. Let us examine its graphical significance (refer to Diagram 6).

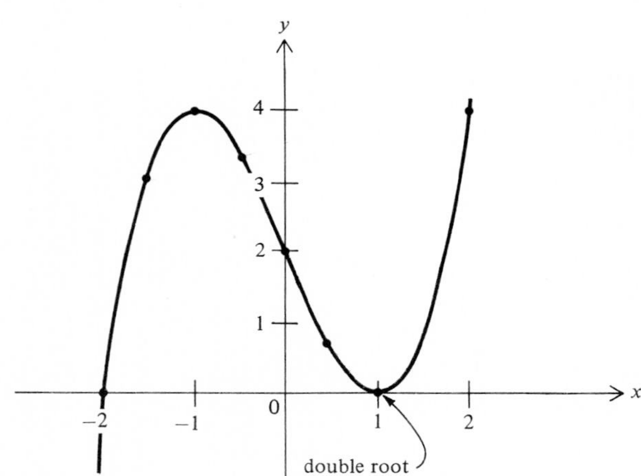

Diagram 6

Theorem 2 (the rational-root theorem) provides a *seemingly* abbreviated list of possible rational roots. In Example 6 the roots were in the form

$$\pm\left(\frac{1,\ 2}{1,\ 2,\ 3,\ 6}\right)$$

There was a total of 12 possible rational roots. Not bad. However, in many instances a_0 and a_n have so many factors that the list of possibilities is extremely long. For example, if $F(x) = 24x^4 - 42x^3 - 77x^2 + 56x + 60$, the rational-root theorem yields a total of 72 possible roots. It is possible to reduce this list significantly by using the following *special theorem.*

Theorem 3 If P/Q is a rational root (in lowest terms) of $f(x) = 0$, then
(a) $P - Q$ must divide $f(1)$, *and*
(b) $P + Q$ must divide $f(-1)$,
or in general

$$P - MQ \text{ must divide } f(M) \qquad M \text{ an integer}$$

COMMENT If $M = 0$, then according to this theorem P must divide $f(0)$. In other words, P must divide a_0 (this is part of the rational-root theorem). If $M = 2$, then $P - 2Q$ must divide $f(2)$; if $M = -3$, then $P + 3Q$ must divide $f(-3)$, and so on.

To illustrate this theorem, let us analyze

$$f(x) = 15x^4 - 7x^3 - 49x^2 + 21x + 12 = 0$$

Using the rational root theorem, there are 36 possible rational roots of the form P/Q:

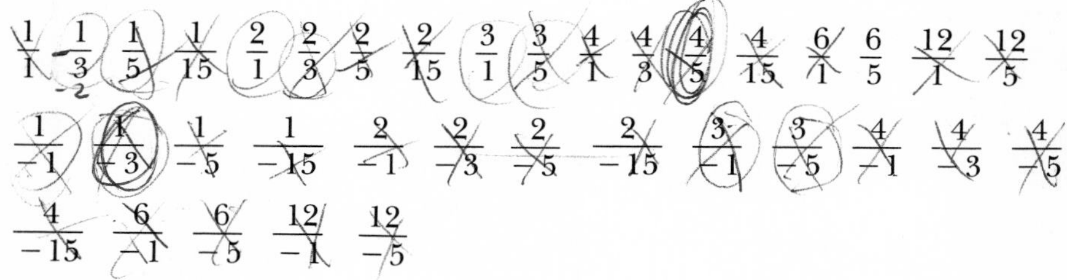

$$\frac{1}{1} \ \frac{1}{3} \ \frac{1}{5} \ \frac{1}{15} \ \frac{2}{1} \ \frac{2}{3} \ \frac{2}{5} \ \frac{2}{15} \ \frac{3}{1} \ \frac{3}{5} \ \frac{4}{1} \ \frac{4}{3} \ \frac{4}{5} \ \frac{4}{15} \ \frac{6}{1} \ \frac{6}{5} \ \frac{12}{1} \ \frac{12}{5}$$

$$\frac{1}{-1} \ \frac{1}{-3} \ \frac{1}{-5} \ \frac{1}{-15} \ \frac{2}{-1} \ \frac{2}{-3} \ \frac{2}{-5} \ \frac{2}{-15} \ \frac{3}{-1} \ \frac{3}{-5} \ \frac{4}{-1} \ \frac{4}{-3} \ \frac{4}{-5}$$

$$\frac{4}{-15} \ \frac{6}{-1} \ \frac{6}{-5} \ \frac{12}{-1} \ \frac{12}{-5}$$

NOTE The positive and negative possibilities are listed separately. For computational reasons, it will be desirable to associate the negative sign with the denominators.

Now
$$\begin{array}{r|rrrrr} 1 & 15 & -7 & -49 & 21 & 12 \\ & & 15 & 8 & -41 & -20 \\ \hline & 15 & 8 & -41 & -20 & -8 \end{array} \qquad f(1) = -8$$

$$\begin{array}{r|rrrrr} -1 & 15 & -7 & -49 & 21 & 12 \\ & & -15 & 22 & 27 & -48 \\ \hline & 15 & -22 & -27 & 48 & -36 \end{array} \qquad f(-1) = -36$$

Neither 1 nor -1 is a root of $f(x) = 0$. If P/Q is a root of $f(x) = 0$, then $P + Q$ must divide $f(-1)$. This means that the sum of P (the numerator) and Q (the denominator) must be a divisor of -36. The only fractions which satisfy this condition are

$$\frac{1}{3} \ \frac{1}{5} \ \frac{2}{1} \ \frac{3}{1} \ \frac{4}{5} \ \frac{1}{-3} \ \frac{1}{-5} \ \frac{2}{-1} \ \frac{2}{-3} \ \frac{2}{-5} \ \frac{3}{-1} \ \frac{3}{-5} \ \frac{4}{-1} \ \frac{4}{-3} \ \frac{4}{-5} \ \frac{6}{-5}$$

For example,

$\frac{1}{5}$ works	Since $1 + 5 = 6$ and 6 divides -36
$\frac{4}{15}$ does *not* work	Since $4 + 15 = 19$ and 19 does *not* divide -36
$\frac{6}{-5}$ works	Since $6 + (-5) = -1$ and -1 divides -36

At this point in the analysis, we have reduced the total number of possibilities from 36 to 16.

Now $P - Q$ must divide $f(1)$. This means that the numerator minus the denominator must be a divisor of -8. From the reduced list of 16, the fractions that satisfy this condition are

$$\frac{1}{3} \quad \frac{1}{5} \quad \frac{2}{1} \quad \frac{3}{1} \quad \frac{4}{5} \quad \frac{1}{-3} \quad \frac{3}{-1} \quad \frac{3}{-5}$$

For example,

$\frac{3}{1}$ works	Since $3 - 1 = 2$ and 2 divides -8
$\frac{4}{-1}$ does *not* work	Since $4 - (-1) = 5$ and 5 does *not* divide -8
$\frac{3}{-5}$ works	Since $3 - (-5) = 8$ and 8 divides -8

We now have two alternatives. We can test each of the eight remaining possibilities or recall from the theorem that $P - MQ$ must divide $f(M)$. Suppose we take the latter alternative and let $M = 2$, which means $P - 2Q$ must divide $f(2)$.

$$\begin{array}{r|rrrrr} 2 & 15 & -7 & -49 & 21 & 12 \\ & & 30 & 46 & -6 & 30 \\ \hline & 15 & 23 & -3 & 15 & \boxed{42} \end{array}$$

Hence $P - 2Q$ must divide 42. Testing each of the eight finalists, we obtain the new list of possibilities:

$$\frac{3}{1} \quad \frac{4}{5} \quad \frac{1}{-3}$$

For example,

$\frac{4}{5}$ works	Since $4 - 2(5) = -6$ and -6 divides 42
$\frac{3}{-5}$ does *not* work	Since $3 - 2(-5) = 13$ and 13 does *not* divide -42

Since the list has been substantially reduced, we will be better off testing each member of the new list. As a final solution we find that the rational roots are $\frac{4}{5}$ and $1/(-3)$ and the remaining two roots are $\pm\sqrt{3}$.

EXERCISES

In Exercises 1 to 10 find all the real roots of the polynomial equations and express irrational roots in the simplest radical form.

1 $3x^3 + 4x^2 - 7x + 2 = 0$

2 $2x^4 - 11x^3 + 17x^2 - 10x + 2 = 0$

3 $2x^4 - 3x^3 - 20x^2 + 27x + 18 = 0$

4 $18x^3 - 39x^2 + 8x + 16 = 0$

5 $9x^4 - 19x^2 - 6x + 4 = 0$

6 $6x^4 + 7x^3 - 13x^2 - 4x + 4 = 0$

7 $2x^3 - 8x^2 - x + 4 = 0$

8 $x^3 - 19x + 30 = 0$

9 $x^4 - 39x^2 + 108 = 0$

10 $(x - 1)(x - 2)(x + 3) = 12$

11 Find the negative root of $x^3 - 6x^2 - x + 30 = 0$.

12 Show that $10x^4 - 13x^3 + 15x^2 - 18x - 24 = 0$ has no rational roots.

13 Find the fractional root of $4x^3 + 3x^2 + 16x + 12 = 0$.

14 $6x^4 - 25x^3 - 21x^2 + 50x + 18 = 0$

15 $12x^4 - 40x^3 - 5x^2 + 45x + 18 = 0$

16 $12x^3 - 5x^2 - 11x + 6 = 0$

17 $6x^3 + 16x^2 - 3x - 8 = 0$

18 $8x^3 + 14x^2 + 7x + 1 = 0$

19 $9x^3 + 18x^2 + 11x + 2 = 0$

20 $4x^3 - 13x + 6 = 0$

21 $4x^4 - 12x^3 - 7x^2 + 24x - 9 = 0$

In Exercises 22 to 28 find *all* the roots of the polynomial equations using the *special theorem* (Theorem 3).

22 $24x^4 + 26x^3 - 53x^2 - 52x + 10 = 0$

23 $15x^4 + 23x^3 - 61x^2 + 17x + 6 = 0$

24 $15x^4 - 89x^3 + 52x^2 + 16x - 8 = 0$

25 $10x^4 - 41x^3 - 38x^2 + 82x + 36 = 0$

26 $12x^4 + 65x^3 + x^2 - 130x - 50 = 0$

27 $16x^4 + 6x^3 - 103x^2 - 44x + 35 = 0$

28 $20x^4 + 67x^3 + 6x^2 - 70x + 16 = 0$

29 Using the concept of the rational-root theorem, prove that $\sqrt{3} + \sqrt{2}$ is irrational. *Hint:* Let $x = \sqrt{3} + \sqrt{2}$, and through the process of squaring form a polynomial equation with integral coefficients.

30 On the same axis sketch $y = x^3$ and $3x - y - 2 = 0$. Use algebraic methods and the rational-root theorem to determine where the two curves intersect.

Using the techniques of this chapter and Chapter 4, solve the following problems:

31 $6x^3 - 17x^2 - 5x + 6 \leq 0$

32 $6 + 13x + x^2 - 2x^3 > 0$

33 $4x^3 - 3x^2 < 16x - 12$

34 $6x^3 + 7x^2 - 13x - 12 \geq 0$

SOME APPLICATIONS OF POLYNOMIAL FUNCTIONS

Exercises 35 to 38 can be solved by translating the description into a third-degree polynomial. The domain of each of the polynomials is restricted by the conditions of the problem. Set up the required polynomial and find a meaningful solution.

35 A rectangular sheet of tin 15 by 9 inches has equal squares cut from its four corners. The resulting sheet is folded up on the sides, forming a topless box. Find all the *different* dimensions of the cutout square such that the box will have a volume of 56 cubic inches.

36 The height of a closed box is 1 inch greater than the width, and the length is 2 inches greater than the height. The sum of the surface area and the volume is 180. Find the dimensions of the box.

37 The edges of a rectangular box are 2, 3, and 4 units. If each edge is increased by the same amount, the volume is 5 times as much. Find the amount added.

38 A rectangle has a perimeter of 32 units and is rotated about a vertical line connecting the midpoints of the longer parallel sides, forming a cylinder of volume 308. Find the dimensions of the rectangle. (Let $\pi = \frac{22}{7}$. The volume of a cylinder is $\pi r^2 h$.)

5.6 COMPLEX ROOTS

So far we have analyzed polynomial equations with real roots. However, the set of real numbers is not extensive enough to provide solutions for all polynomial equations.

In Chapter 3, we learned that the solution for a quadratic polynomial, $f(x) = ax^2 + bx + c$, depends on its *discriminant*, $b^2 - 4ac$. When $b^2 - 4ac \geq 0$, then $f(x)$ has two real zeros. However, when $b^2 - 4ac < 0$, then $f(x)$ has *no* real zeros.

Algebraically, we introduced the number i such that $i^2 = -1$. The squares of all real numbers are *nonnegative*. Since no real number c has the property such that $c^2 = -1$, i is *not* in the system of reals. i is called the *imaginary unit* (a new number whose square is -1).

In this section we briefly examine a number system capable of supplying solutions for *every* polynomial equation. This is known as the *complex number system.*

In general, a complex number is written in the form $a + bi$, where a and b are real numbers and i is the imaginary unit. For example,

$2 + 0i = 2$	real number
$0 + 2i = 2i$	pure imaginary number
$2 + 2i$	complex number (combination of a real and an imaginary part)

Example 1 Sketch and analyze $P(x) = x^3 - x + 6$.

Using the rational-root theorem, we see that the only possible rational roots of $P(x) = 0$ are $\pm(1, 2, 3, 6)$.

$$\begin{array}{r|rrrr} -2 & 1 & 0 & -1 & 6 \\ & & -2 & 4 & -6 \\ \hline & 1 & -2 & 3 & 0 \end{array}$$

$$x^2 - 2x + 3 = 0 \qquad \text{depressed equation}$$

$$x = \frac{2 \pm \sqrt{4 - 12}}{2} \quad \text{or} \quad x = \frac{2 \pm \sqrt{-8}}{2} \qquad \text{Using the quadratic equation}$$

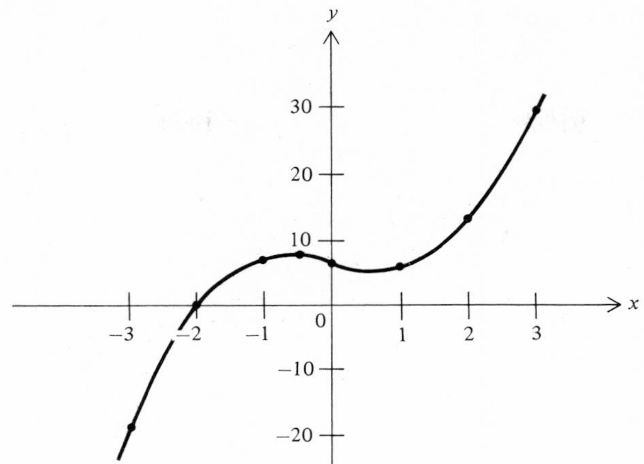

Diagram 7

Now $\sqrt{-8} = 2i\sqrt{2}$, and $x = 1 \pm i\sqrt{2}$, that is two complex roots in the form $a + bi$ and $a - bi$. Diagram 7 illustrates the graph of $P(x)$.

COMMENT The graph of $P(x)$ crosses the x axis at exactly one point, indicating *one* real root. The other two roots are complex: $1 + i\sqrt{2}$ and $1 - i\sqrt{2}$. The graph of $P(x)$ does *not* cross the x axis at any point corresponding to a complex root.

Example 2 Graph and analyze $Q(x) = x^3 - 6x^2 + 9x + 16$.

Solution

$$\begin{array}{r|rrrr} -1 & 1 & -6 & 9 & 16 \\ & & -1 & 7 & -16 \\ \hline & 1 & -7 & 16 & 0 \end{array}$$

$x^2 - 7x + 16 = 0$ depressed equation

$\frac{7}{2} \pm \frac{i\sqrt{15}}{2}$ roots

We determine some other points on the graph of $P(x)$:

x	−2	0	1	2	3	4
$P(x)$	−34	16	20	18	16	20

Referring to Diagram 8, we see that the graph intersects the x axis once at $(-1, 0)$. (The other two roots are complex, in the form $a + bi$ and $a - bi$.)

In each of the two polynomial equations just analyzed, one may wonder whether complex solutions *always* occur in *pairs* of the form $a + bi$ and $a - bi$. If the coefficients of the polynomial equation are real, the answer will be yes.

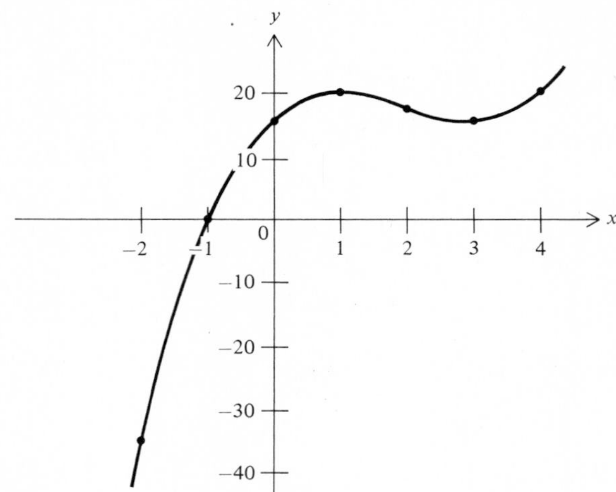

Diagram 8

Theorem 1 If a complex number $a + bi$ ($b \neq 0$) is a root of a polynomial equation with *real* coefficients, then $a - bi$ is also a root.

This implies that if $P(a + bi) = 0$, then $P(a - bi) = 0$. Complex numbers such as $2 + 3i$ and $2 - 3i$ which are expressed in the form $a + bi$ and $a - bi$ are called *conjugates* of each other.

Intuitively, we see that since complex roots of real polynomials occur only in conjugate pairs, a polynomial equation of degree 3 must possess *at least* one real root.

Theorem 2 All real polynomial equations of *odd* degree must have at least one real root.

The basic properties of addition and multiplication of complex numbers are necessary in analyzing polynomial equations.

We *add* two complex numbers by adding their real parts and their imaginary parts *separately*.

Example 3 $(a + bi) + (c + di) = (a + c) + (b + d)i.$

Example 4 $(-2 + i) + (3 - 4i) = 1 - 3i.$

In order to *multiply* two complex numbers, recall that $i(i) = -1$; that is, $i^2 = -1$.

The product of two binomials, in elementary algebra, is given by

$$(a + b)(c + d) = ac + ad + bc + bd$$
$$(a + b)(a - b) = a^2 - b^2$$

The product of two complex numbers is performed in the same manner.

Example 5

$$\begin{aligned}(a + bi)(c + di) &= ac + adi + bci + bdi^2 \\ &= ac + adi + bci + bd(-1) && \text{Substituting } -1 \text{ for } i^2 \\ &= (ac - bd) + (ad + bc)i && \text{Grouping terms} \\ (-2 + i)(3 - 4i) &= (-2)(3) + (-2)(-4i) + (i)(3) + (i)(-4i) \\ &= -6 + 8i + 3i - 4i^2 && \text{Recalling } -4i^2 = -4(-1) = 4 \\ &= -2 + 11i && \text{Grouping terms}\end{aligned}$$

It is much easier to multiply two complex conjugates.

Example 6

$$\begin{aligned}(a + bi)(a - bi) &= aa + a(-bi) + abi + (bi)(-bi) \\ &= a^2 - abi + abi - b^2i^2 \\ &= a^2 + b^2 && \text{Since } -b^2i^2 = -b^2(-1) = b^2 \\ (3 - 2i)(3 + 2i) &= 9 + 4 = 13\end{aligned}$$

Note that $a = 3$ and $b = 2$. We are now able to factor the sum of two squares, that is, $x^2 + 1 = (x + i)(x - i)$.

Example 7 Consider $P(x) = x^4 + x^2 - 2x + 6$. We have been informed that $1 + i$ is a root of $P(x) = 0$. How do we determine the remaining three roots?

Using synthetic substitution, we verify the given information. This requires the multiplication and addition of complex numbers.

$1 + i$	1	0	1	−2	6
		$1 + i$	$2i$	$-1 + 3i$	−6
$1 - i$	1	$1 + i$	$1 + 2i$	$-3 + 3i$	0
		$1 - i$	$2 - 2i$	$3 - 3i$	
	1	2	3	0	

$$Q(x) = x^2 + 2x + 3 = 0 \qquad x = \frac{-2 \pm \sqrt{4 - 4(1)(3)}}{2}$$

$$x = \frac{-2 \pm \sqrt{-8}}{2} = -1 \pm i\sqrt{2}$$

Generating some additional information, we can find

x	−2	−1	0	1	2
$P(x)$	30	10	6	6	22

(see Diagram 9).

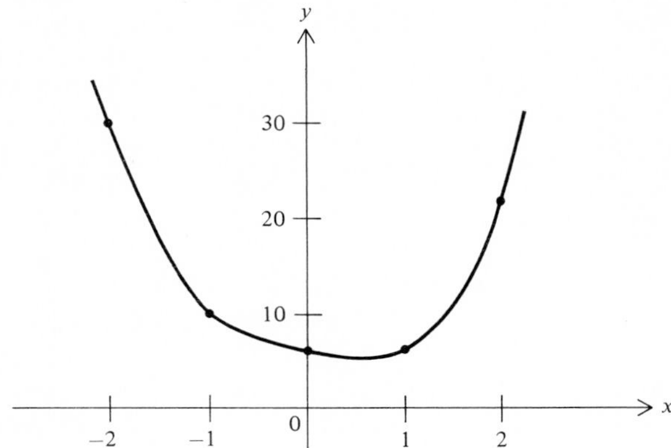

Diagram 9

COMMENT The polynomial equation has *two pairs* of complex conjugate roots

$$1 \pm i \qquad \text{and} \qquad -1 \pm i\sqrt{2}$$

Since there are *no* real roots, the graph of $P(x)$ never crosses the x axis. Using the factor theorem, if c is a zero of $P(x)$, then $x - c$ is a factor of $P(x)$. Then each complex pair $a \pm bi$ forms a real quadratic factor

$$(x - a - bi)(x - a + bi) = x^2 - 2ax + a^2 + b^2$$

or

$$P(x) = (x^2 - 2x + 2)(x^2 + 2x + 3)$$

If a real polynomial equation of *even* degree is given, it is possible that *all* the roots will be complex conjugates. Unless we can reduce equations of this type to the second degree, we require additional information or more advanced techniques to find *all* the roots.

EXERCISES

In Exercises 1 to 10 find all the roots.

1 $3x^4 + 2x^3 + 2x^2 + 2x - 1 = 0$

2 $4x^4 - 5x^2 - 6x + 7 = 0$

3 $8x^3 - 4x^2 + 6x + 5 = 0$

4 $4x^4 + x^2 - 3x + 1 = 0$

5 $4x^4 + 3x^3 - 32x - 24 = 0$

6 $2x^3 + 7x^2 + 6x - 5 = 0$

7 $2x^4 - 4x^3 + 3x^2 - 2x + 1 = 0$

8 $2x^3 - x^2 - 2x + 6 = 0$

9 $2x^3 - 7x^2 + 4x - 14 = 0$

10 $2x^4 - 5x^3 + 3x^2 + 4x - 6 = 0$

11 $2x^4 - 3x^3 - 7x^2 - 8x + 6 = 0$

12 $6x^4 + 5x^3 - 16x^2 - 9x - 10 = 0$

13 $(x + 1)(x - 2)(x + 3) = 24$

14 $6x^4 - 7x^3 + 5x^2 + x - 2 = 0$

In Exercises 15 to 21 find *all* the solutions for the polynomial equations (the *special theorem* in Section 5.5 will be very useful).

15 $24x^3 - 47x^2 + 190x + 8 = 0$

16 $16x^4 - 12x^3 - 128x^2 - 73x - 63 = 0$

17 $6x^4 - x^3 - 45x^2 + 92x - 70 = 0$
18 $40x^4 + 59x^3 + 32x^2 + 59x - 8 = 0$
19 $24x^3 - 19x^2 + 91x + 20 = 0$
20 $16x^4 - 34x^3 + 17x^2 - 68x - 30 = 0$
21 $18x^4 + 9x^3 + 67x^2 + 36x - 20 = 0$
22 If $1 + i$ is a root of $x^4 - 6x^3 + 11x^2 - 10x + 2 = 0$, show that $1 - i$ is also a root. Find the other two roots.
23 Show that two of the roots of $x^3 + x^2 + x + 1 = 0$ are the reciprocals of each other.
24 Sketch the graphs of each of the polynomial functions. (Refer to Exercises 1 to 10 for additional information.)
(a) $P(x) = 8x^3 - 4x^2 + 6x + 5$
(b) $P(x) = 4x^4 + x^2 - 3x + 1$
(c) $P(x) = 2x^3 + 7x^2 + 6x - 5$
(d) $P(x) = 2x^3 - x^2 - 2x + 6$
(e) $P(x) = 2x^4 - 5x^3 + 3x^2 + 4x - 6$
25 If $-1 + i$ is a root of $x^4 - 2x^3 + 4x + 12 = 0$, find the remaining roots.
26 **(a)** Express $P(x) = x^4 - 2x^3 + 4x + 12$ as the product of two real quadratic factors (refer to Exercise 25).
(b) Sketch a graph of $P(x)$.
27 Find the roots of $x^2 - 2ix - 2 = 0$. (Why are they *not* complex conjugates?)

5.7
MORE AIDS IN LOCATING REAL ROOTS AND GRAPHING

Location Principle If $P(x) = 0$ is a real polynomial equation and a and b are real numbers, then:
(a) If $P(a)$ and $P(b)$ have the *same sign,* there is either *no* root or an *even* number of real roots between $x = a$ and $x = b$.
(b) If $P(a)$ and $P(b)$ have *opposite signs,* there is either *one* real root or an *odd* number of real roots between $x = a$ and $x = b$.

Example 1 $P(x) = 2x^3 + x^2 - 8x - 4$.
The roots of $P(x) = 0$ are -2, $-\frac{1}{2}$, and 2, and $P(0) = -4$.

1	2	1	−8	−4
		2	3	−5
	2	3	−5	−9

−1	2	1	−8	−4
		−2	1	7
	2	−1	−7	3

3	2	1	−8	−4
		6	21	39
	2	7	13	35

−3	2	1	−8	−4
		−6	15	−21
	2	−5	7	−25

$P(1) = -9$ $\quad P(-1) = 3$ $\quad P(3) = 35$ $\quad P(-3) = -25$

(See Diagram 10.)

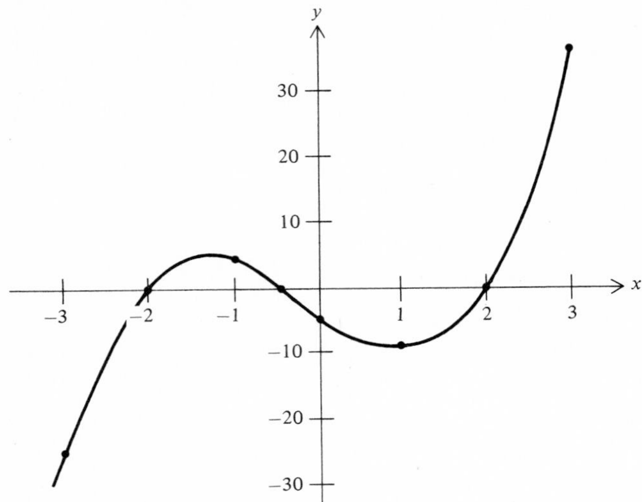

Diagram 10

1 Since $P(0)$ and $P(1)$ have the *same* sign, there are either *no* real roots between 0 and 1 *or* an *even* number of real roots between 0 and 1. An examination of Diagram 10 shows that there are *no* real roots in the interval.

2 $P(0)$ and $P(-1)$ have opposite signs, and there is an *odd* number of real roots between 0 and -1 (exactly *one* root at $x = -\frac{1}{2}$).

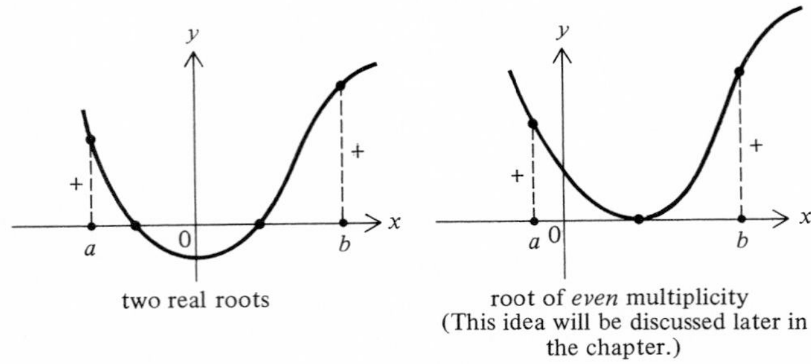

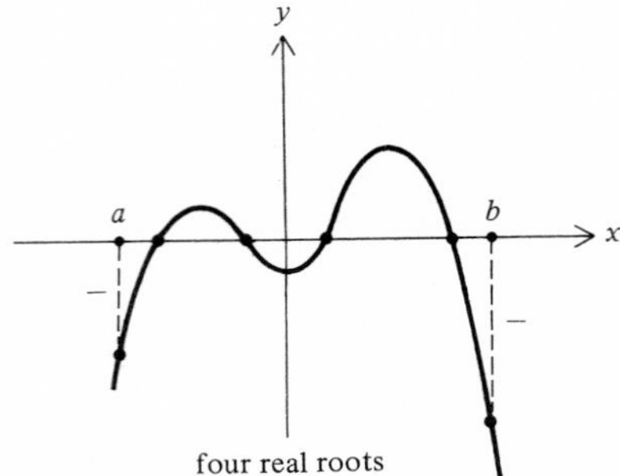

Diagram 11

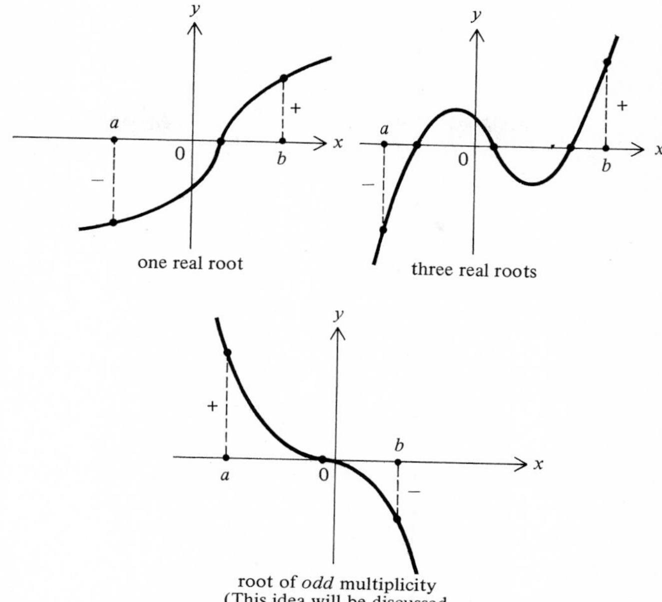

Diagram 12

3 $P(0)$ and $P(-3)$ have the same sign, and there are two real roots between 0 and -3 (an even number).

4 In fact, by examining the diagram, we can see that *all* roots of $P(x) = 0$ are between -3 and 3:

$$P(3) = 35 \qquad \text{and} \qquad P(-3) = -25$$

This is consistent with the example. They have opposite signs and hence an *odd* number of real roots between -3 and 3.

COMMENT If we assume that the graph of a polynomial function is a smooth unbroken curve *and* that it satisfies the definition of a function (see Chapter 2), then Diagrams 11 and 12 will give the *basic ideas* contained in the location principle.

If we refer to the graph of $P(x) = 2x^3 + x^2 - 8x - 4$ in Diagram 10, we can see that the three zeros of the function divide the x axis into four intervals (see Diagram 13).

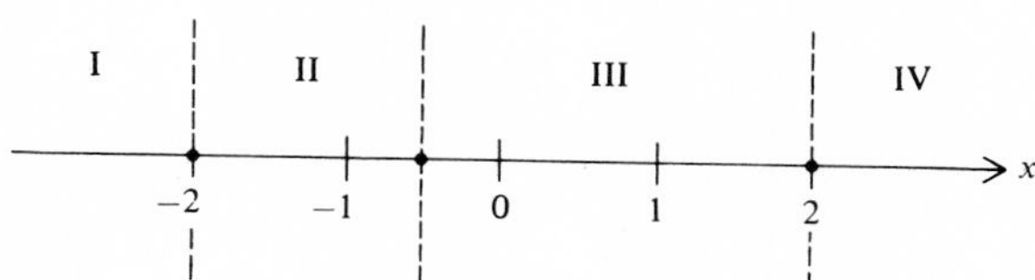

Diagram 13

We need only evaluate $P(x)$ at *any* point in a given interval. For example, if c is a point in an interval and $P(c) > 0$, the graph of the polynomial will be *entirely above* the x axis for that interval. If $P(c) < 0$, then the graph of the polynomial will be *entirely below* the x axis for that interval. The only exception is when we use an end point, which is a zero of the function. At this point, $P(c) = 0$, and the graph touches the x axis.

Example 2

For $P(-3) = -25$ $\quad P(x) < 0 \quad$ in the interval $(-\infty, -2)$

For $P(-1) = 3$ $\quad P(x) > 0 \quad$ in the interval $(-2, -\frac{1}{2})$

For $P(0) = -4$ $\quad P(x) < 0 \quad$ in the interval $(-\frac{1}{2}, 2)$

For $P(3) = 35$ $\quad P(x) > 0 \quad$ in the interval $(2, +\infty)$

Verify these conclusions in Diagram 10.

The following theorem introduces an additional technique for facilitating the rapid graphing of polynomial functions.

Theorem 1 If $(x - R)^c$ is a factor of a real polynomial $P(x)$, where c is a positive integer representing the *number of times* that $x - R$ is a factor of $p(x)$, then:

(a) If $c = 1$, the graph of $P(x)$ intersects the x axis *once* at $x = R$. R is called a *unique zero* of $P(x)$.

(b) If c is an *even* integer, the graph of $P(x)$ has its *turning point* on the x axis at $x = R$. R is called a *multiple zero* of even order c.

(c) If c is an *odd* integer and not equal to 1, the graph of $P(x)$ intersects the x axis *once* at $x = R$. R is called a *multiple zero* of odd order c.

Without the techniques of the calculus, it is difficult to distinguish between cases (*a*) and (*c*) in the theorem. We will try to simplify the difference by a careful analysis of selected examples. Diagram 14 illustrates, in general, the graphical interpretation of the theorem.

Example 3 $P(x) = x - 1$.

This is described in part (*a*) of the theorem. Since $x - 1$ is the only factor of $P(x)$, $c = 1$ and the graph of $P(x)$ intersects the x axis once at $x = 1$; 1 is a unique zero of $P(x)$ (see Diagram 15).

Example 4 $P(x) = (x + 1)^3$.

We refer to part (*c*) of the theorem. The graph of $P(x)$ intersects the x axis *once* at $x = -1$. However, -1 is a zero of multiplicity 3. In other words, $x = -1$ is a *triple root* of $P(x) = 0$ (see Diagram 16).

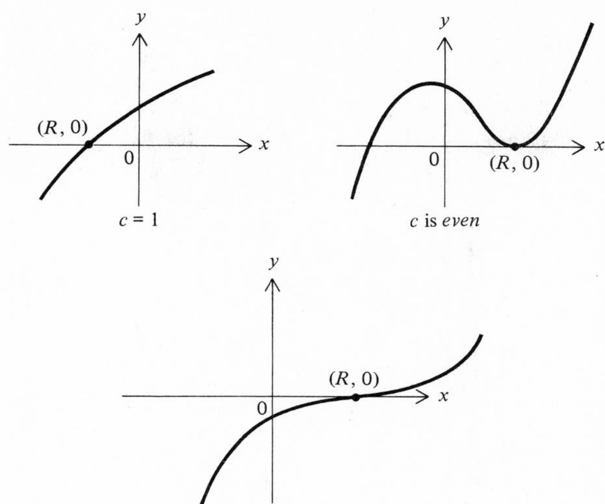

Diagram 14

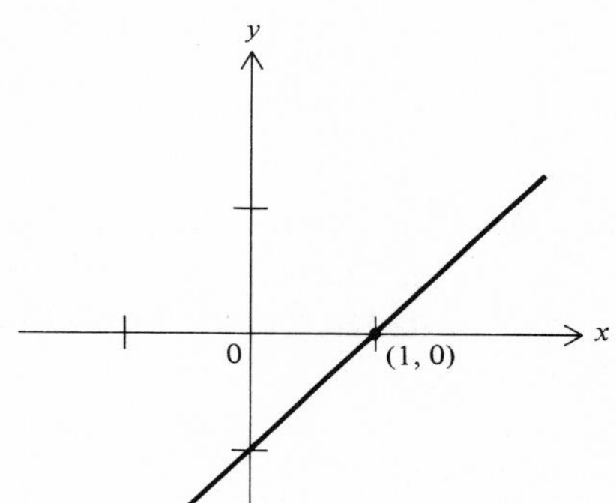

Diagram 15

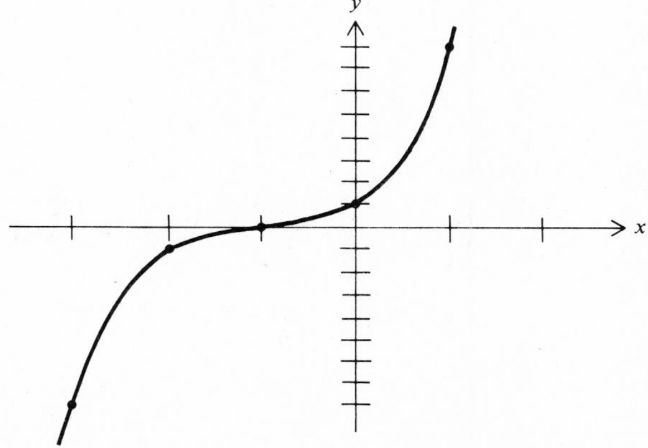

Diagram 16

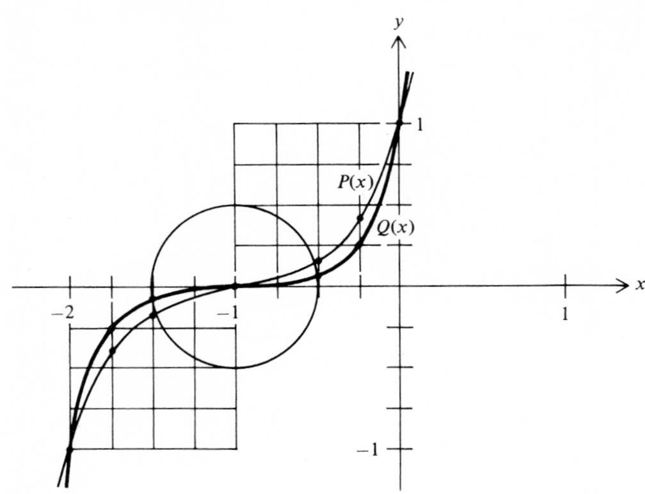

Diagram 17

Example 5 Compare the graphs of $P(x) = (x + 1)^3$ and $Q(x) = (x + 1)^5$ in the interval $[-2, 0]$.

Both graphs intersect the x axis *once* at $x = -1$. However, -1 is a zero of multiplicity 3 for $P(x)$, while -1 is a zero of multiplicity 5 for $Q(x)$. Is there any graphical significance to this result?

CONCLUSION From Diagram 17 it can be observed that as the degree of odd multiplicity of the zeros increases, a definite behavior of the graph occurs about the zero. Note that in the circular region about the zero $x = -1$, the graph of $Q(x)$ is *flatter* than the graph of $P(x)$.

In the interval $(-2, -1)$	$Q(x) > P(x)$
In the interval $(-1, 0)$	$Q(x) < P(x)$
In the interval $(0, +\infty)$	$Q(x) > P(x)$
In the interval $(-\infty, -2)$	$Q(x) < P(x)$

Example 6 $P(x) = (x - 2)^2$.

Referring to part (b) of the theorem, we see that the graph of $P(x)$ has its *turning point* on the x axis at $x = 2$. However, 2 is a zero of multiplicity 2. In other words, $x = 2$ is a *double root* of $P(x) = 0$ (see Diagram 18).

Example 7 Compare the graphs of

$$P(x) = (x - 2)^2 \quad \text{and} \quad Q(x) = (x - 2)^4$$

in the interval $[1, 3]$.

Both graphs have their turning point on the x axis at $x = 2$. However, 2 is a zero of multiplicity 2 for $P(x)$, while 2 is a zero of multiplicity 4 for $Q(x)$. Is there any graphical significance to this result? (See Diagram 19.)

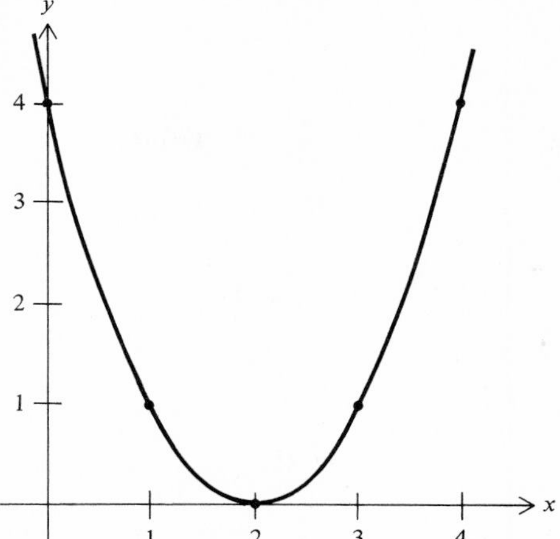

Diagram 18

CONCLUSION From Diagram 19 it can be observed that as the degree of even multiplicity of the zero increases, a definite behavior occurs about the zero. Note that in the circular region about the zero $x = 2$, the graph of $Q(x)$ is *flatter* than the graph of $P(x)$. In the interval $(1, 3)$, $Q(x) < P(x)$; outside the interval, $Q(x) \geq P(x)$.

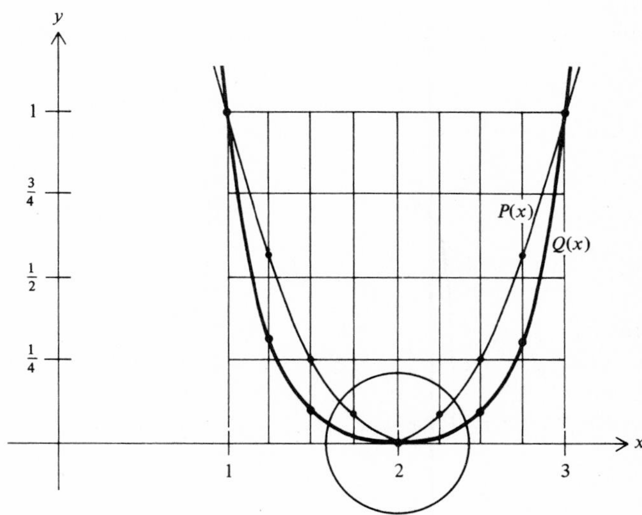

Diagram 19

Example 8 $P(x) = (x - 1)^3(x + 2)^2$.

Referring to parts (*b*) and (*c*) of the theorem, we should expect a turning point on the x axis at $x = -2$ *and* one intersection of the x axis at $x = 1$, where 1 is a zero of multiplicity 3. Because of the complexity of this problem, let us

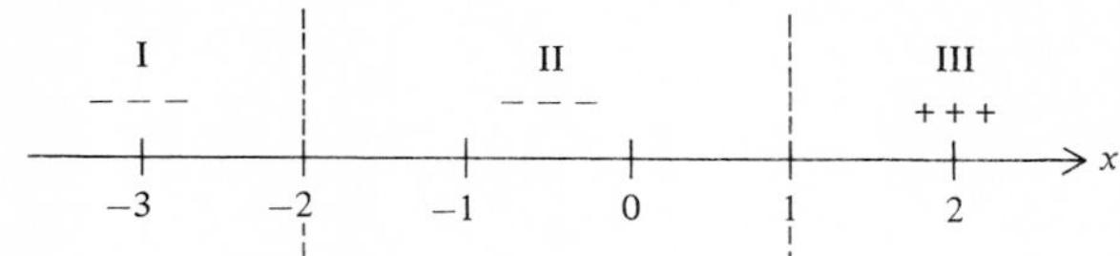

Diagram 20

examine $P(x)$ for *consistency of sign* in the three intervals determined by the zeros, $x = 1$ and $x = -2$ (see Diagram 20).

For $P(-3) = -64$ $\quad P(x) < 0$ $\quad$ in the interval $(-\infty, -2)$

For $P(0) = -4$ $\quad P(x) < 0$ $\quad$ in the interval $(-2, 1)$

For $P(2) = 16$ $\quad P(x) > 0$ $\quad$ in the interval $(1, +\infty)$

COMMENT Unless the turning point of the graph of $P(x)$ occurs at a multiple zero on the x axis, the graphical methods just presented do not generate other turning points on the graph of the polynomial. In Diagram 21, a turning point *seems* to occur at $x = -1$. The methods of the calculus show that the turning point *actually* occurs at $x = -\frac{4}{5}$. This, of course, is one serious limitation of precalculus graphical techniques.

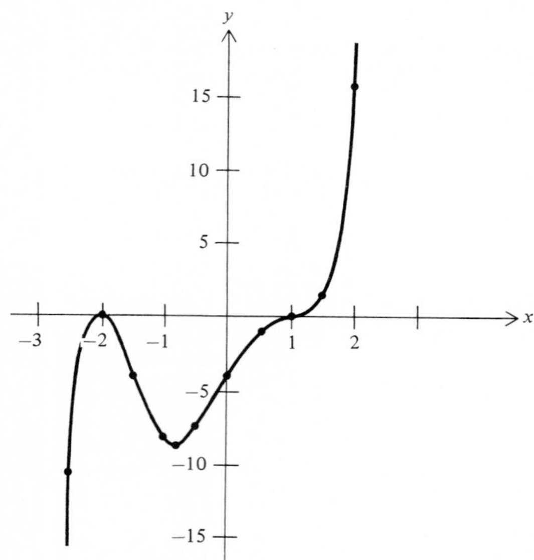

Diagram 21

EXERCISES

1 Between what two consecutive positive integers does a root of $x^3 - 5x - 3 = 0$ lie?
2 Between what two consecutive negative integers does a root of $x^3 + 2x^2 + 3 = 0$ lie?

3 Given: $f(x)$. If $f(-1) > 0$ and $f(-2) < 0$, and if $f(-1.1) > 0$ and $f(-1.2) < 0$, then $f(-1.15) > 0$. From this information, find a real root of $f(x) = 0$ to the nearest tenth.

Locate consecutive integers between which there are zeros of each of the functions in Exercises 4 to 7 in the interval $-4 \leq x \leq 4$.

4 $f(x) = x^3 - 2x^2 + x - 3$
5 $f(x) = 3x^4 - 4x^3 - 2x^2 + 1$
6 $f(x) = x^4 - x^3 - 5x^2 - 1$
7 $f(x) = x^3 + 3x^2 - 2x - 5$

In Exercises 8 to 22 sketch the graph.

8 $P(x) = 4x^3 + x^2 - 11x + 6$
9 $P(x) = x(x - 1)(x^2 - 1)$
10 $P(x) = 3x^4 - 4x^3 + 1$
11 $P(x) = x^4 - 12x^3 + 48x^2 - 64x$
12 $P(x) = 2x^3 - 3x^2 + 1$
13 $f(x) = (x - 2)^2(x + 1)^3$
14 $g(x) = (x + 2)^2(x - 3)^4$
15 $h(x) = (x - 4)(x + 1)^3$
16 $f(x) = x^4 - 8x^2 + 16$
17 $f(x) = x^3 - 6x^2 + 12x - 8$
18 $f(x) = (x + 3)^2(x - 2)^3(x + 1)$
19 $f(x) = (2 - x)^2(x + 1)^2$
20 $f(x) = x^6 + 3x^5 - 3x^4 - 11x^3 + 6x^2 + 12x - 8$
21 $f(x) = (x - 2)^5(2x + 1)^4$
22 $f(x) = (x - 1)^2(x + 2)(x - 3)$

5.8 IRRATIONAL ROOTS

Throughout this chapter we have concentrated on analyzing polynomial equations with at least one rational root. Unless we were able to reduce the original polynomial to a quadratic polynomial, thereby utilizing the quadratic formula, the location of irrational roots was never mentioned.

In introductory courses in computer programming, one of the most popular exercises is to find the irrational roots of real polynomial equations to a specified degree of accuracy. The method, called *binary search,* entails continual partitioning of regions until you find some number R such that $P(R) \approx 0$. The programmer himself directs the computer to continue the search until it comes within a specified tolerance of the actual root. A closeness of 0.00001 is not uncommon in such an exercise ($|P(R) - 0| < 0.00001$). This type of accuracy is difficult to attain in general.

Example 1 Find to the nearest hundredth the positive real root of

$$P(x) = x^3 + 3x^2 - 2x - 5 = 0$$

STEP 1 Using the location principle, we see that there is exactly one sign change in $P(x)$ between $x = 1$ and $x = 2$

$$P(1) = -3 \qquad P(2) = 11$$

Divide the interval from 1 to 2 into half and test $P(1.5)$ (see Diagram 22)

$$P(1.5) = 2.125$$

This result is positive and indicates (by the location principle) that a root is between $x = 1$ and $x = 1.5$ (Diagram 23).

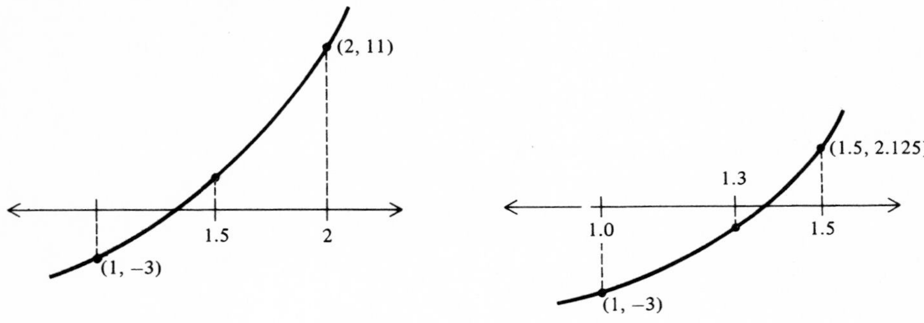

Diagram 22 **Diagram 23**

STEP 2 Again take approximately one-half the difference between 1.0 and 1.5 or 1.3:

1.3	1	3	−2	−5
		1.3	5.59	4.667
	1	4.3	3.59	−0.333

Diagram 23

STEP 3 Hence a root occurs between 1.3 and 1.5. Again take one-half the difference between 1.3 and 1.5 and find $P(1.4)$:

1.4	1	3	−2	−5
		1.4	6.16	5.824
	1	4.4	4.16	0.824

A root is now between 1.3 and 1.4.

STEP 4 In order to determine whether the root is closer to 1.3 or 1.4 we test $P(1.35)$. If $P(1.35) > 0$, the root is closer to 1.3, and if $P(1.35) < 0$, the root is closer to 1.4 (Diagram 24).

STEP 5 $P(1.35) \approx 0.23$, and since $P(1.35) > 0$, the root to the nearest tenth is 1.3.

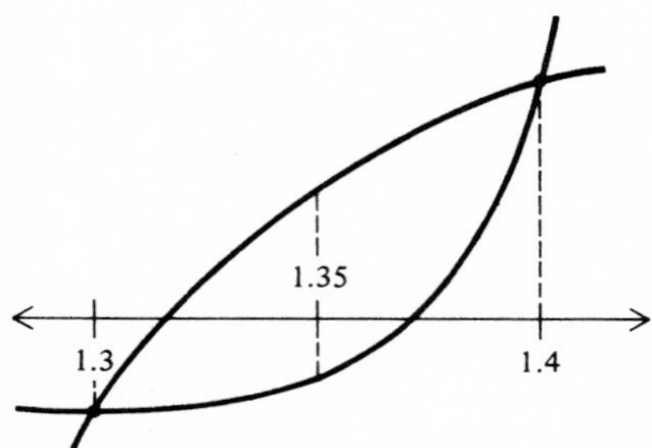

Diagram 24

COMMENT By now you have probably seen why the method is referred to as *binary search.* We partition our intervals into two parts in our test-and-retest procedure.

EXERCISES

1 Find the smallest positive root of $x^3 + 2x - 6 = 0$ to the nearest tenth.
2 Find *all* the real roots of $x^4 + 2x^2 - 8 = 0$ to the nearest tenth.
3 Find the positive root of $x^3 - 3x^2 + 3x - 4 = 0$ to the nearest hundredth.
4 Find the positive real root of $x^3 + x - 3 = 0$ to the nearest tenth.
5 Find the largest negative real root of $x^3 - 3x^2 - 3x + 18 = 0$.
6 Find the smallest positive real root of $x^3 - 2x^2 - 2x - 8 = 0$.
7 To find $\sqrt{3}$, we let $x = \sqrt{3}$; then $x^2 = 3$ and $P(x) = x^2 - 3 = 0$. Using this method, approximate $\sqrt[3]{18}$.
8 Find all the irrational roots of $f(x) = x^3 + 3x^2 - 2x - 5$.
9 Find all the irrational roots of $f(x) = x^3 - 3x^2 - 4x + 1$.
10 Find a root of $f(x) = x^4 - 3x^3 + 3x^2 - 2$ in the interval (1, 2).
11 If 1 foot is added to the side of a cube, the volume is doubled. Find the side of the original cube.
12 The problem of duplicating a cube originated in ancient Greece. The Delians were instructed by the gods that they must double the size of Appollo's cubical altar if they were to get rid of the plague. Assuming that the altar of Apollo was a 6-foot cube, find the dimensions of the altar which will double the volume. Explain why the instructions from the gods are generally impossible to carry out.

5.9 PRACTICAL APPLICATIONS OF THE THIRD-DEGREE POLYNOMIAL (*Optional*)

In Section 5.7 we mentioned that the turning points on the graphs of polynomial functions could be found by methods of calculus. If we restrict ourselves to third-degree polynomials, it is possible to present this method (without justification) and show how it is applied to practical problems. This presentation may give you a slight insight into one practical technique from the calculus involving polynomial functions.

Given a real polynomial function

$$f(x) = Ax^3 + Bx^2 + Cx + D$$

If the graph of the polynomial has turning points, then the equation $3Ax^2 + 2Bx + C = 0$ describes the *slope* of the line passing through the turning points and *parallel* to the x axis. (Recall that all lines parallel to the x axis have a slope of 0.) Diagram 25 gives a graphical interpretation of this statement.

The slopes of L_1 and L_2 equal 0; x_1 and x_2 are the abscissas of the turning points.

Since $3Ax^2 + 2Bx + C = 0$, using the quadratic formula, we can deter-

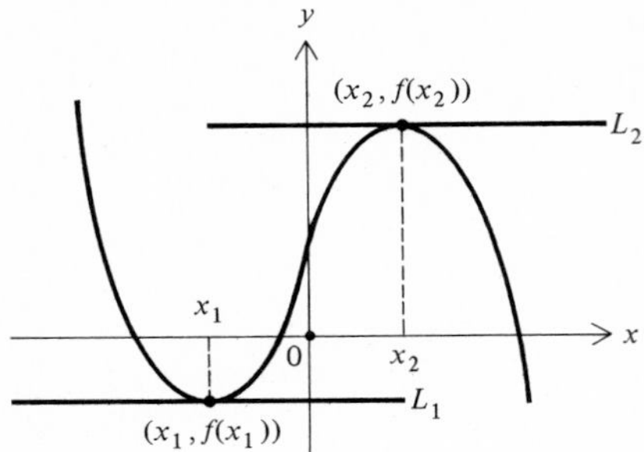

Diagram 25

mine these abscissas of the turning points:

$$x = \frac{-2B \pm \sqrt{4B^2 - 4(3A)C}}{6A} = \frac{-2B \pm \sqrt{4(B^2 - 3AC)}}{6A}$$

$$= \frac{-2B \pm 2\sqrt{B^2 - 3AC}}{6A} = \frac{-B \pm \sqrt{B^2 - 3AC}}{3A}$$

(where each x is a coordinate of the turning points)

COMMENT If $B^2 - 3AC \leq 0$ the graph of $f(x)$ has *no turning points.*

Example 1 A company wishes to construct an open-top container by cutting equal squares from a piece of cardboard 6×6 feet and folding up the sides. Find the size of the square which should be cut out to form a container of *maximum volume.*

First we outline the procedure we will follow.

STEP 1 Draw a sketch to represent the information given above (see Diagram 26)

STEP 2 Determine the quantity to be maximized: volume.

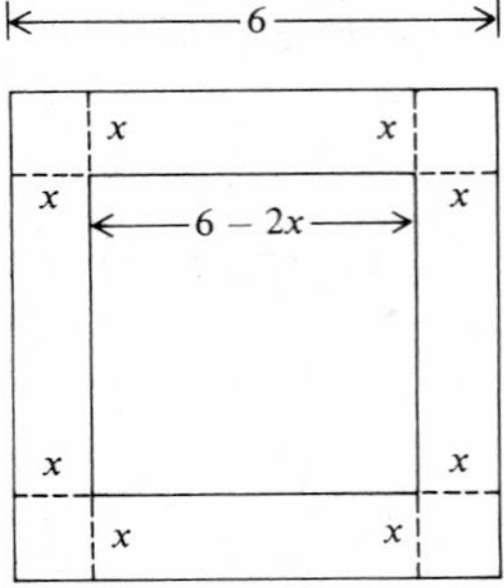

Diagram 26

STEP 3 Translate the problem into symbolic form. Let

$$x = \text{side of square}$$
$$6 - 2x = \text{side of required container}$$

(Again see Diagram 26.)

STEP 4 Find a formula which will relate the given information and the quantity to be maximized:

$$V(x) = x(6 - 2x)(6 - 2x)$$

STEP 5 Determine any restrictions on the variables involved.

$$V \geq 0$$
$$0 \leq 6 - 2x \leq 6$$
$$-6 \leq -2x \leq 0$$

or

$$0 \leq x \leq 3 \qquad \text{domain of } V(x)$$

ANALYSIS

$$V(x) = x(6 - 2x)(6 - 2x) = 4x^3 - 24x^2 + 36x \qquad 0 \leq x \leq 3$$

This function is now in the form

$$f(x) = Ax^3 + Bx^2 + Cx + D$$

Hence, $A = 4 \qquad B = -24 \qquad C = 36 \qquad D = 0$

Find the x coordinates of the turning points using the given formula:

$$x = \frac{24 \pm \sqrt{576 - 3(4)(36)}}{12} = \frac{24 \pm 12}{12} = 1 \text{ or } 3$$

For the graph of $V(x)$ see Diagram 27

$$V(3) = 0 \qquad \text{minimum volume}$$
$$V(1) = 16 \qquad \text{maximum volume}$$

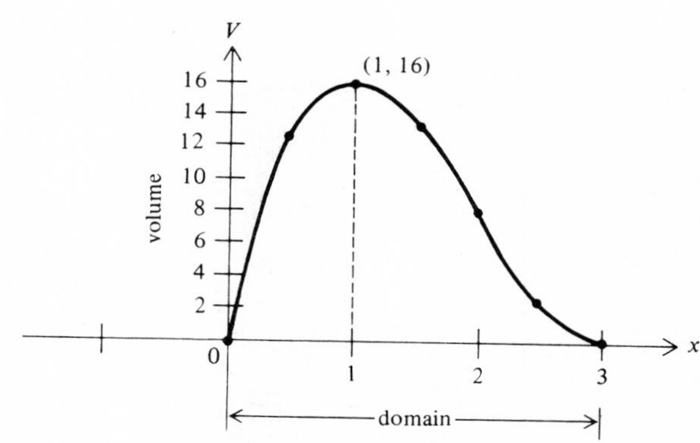

Diagram 27

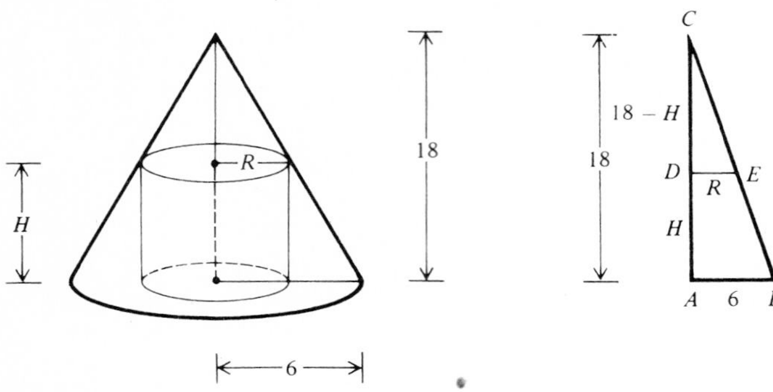

Diagram 28 *Diagram 29*

The answer therefore is to cut out a 1-inch square from each corner.

Example 2 Find the dimensions on the cylinder of maximum volume that can be inscribed in a right circular cone of radius 6 inches and height of 18 inches.

STEP 1 See Diagram 28 for a sketch of the information.

STEP 2 Maximize the *volume* of the cylinder.

STEP 3 Translate the problem into symbolic form (see Diagram 29)

R = radius of cylinder and H = height of cylinder
Radius of cone = 6 and height of cone = 18

STEP 4 The volume of the cylinder is $V = \pi R^2 H$. In order to satisfy the general form of a polynomial function, express the volume as a function of *one variable*. By examining Diagram 29, ΔCAB is similar to ΔCDE. From the properties of similar triangles in elementary geometry

$$\frac{18 - H}{18} = \frac{R}{6} \Longrightarrow 18R = 108 - 6H$$

Since R is squared in the formula for V, it may be easier to solve for H in terms of R.

$$H = \frac{108 - 18R}{6}$$

Now $$V = \pi R^2 \frac{108 - 18R}{6}$$

$$V(R) = -3\pi R^3 + 18\pi R^2$$ V now a function of one variable R

STEP 5 $V \geq 0$ domain $= 0 \leq R \leq 6$

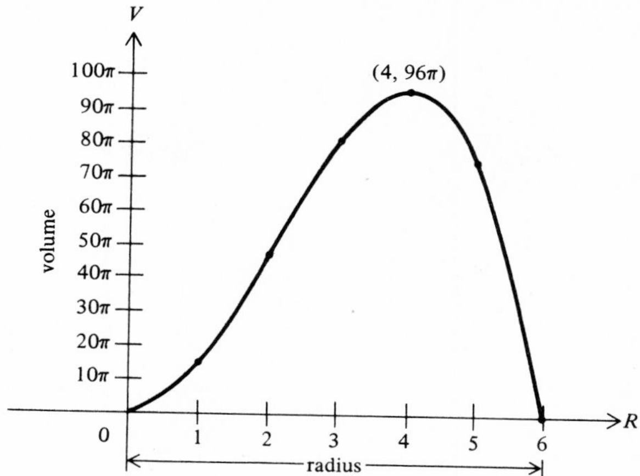

Diagram 30

ANALYSIS

$$R = \frac{-18\pi \pm \sqrt{(18\pi)^2 - 3(3\pi)(0)}}{-9\pi} \qquad \text{Substituting } A = -3\pi,\ B = 18\pi,\ C = 0$$

$$= \frac{-18\pi \pm 18\pi}{-9\pi} = 4 \text{ or } 0$$

For the graph of $V(R)$ see Diagram 30.

$$H = \frac{108 - 18(4)}{6} = 6 \qquad \text{Substituting } R = 4$$

The dimensions of the cylinder are therefore

$$R = 4 \qquad \text{and} \qquad H = 6$$

Example 3 Using the methods of this section, if $f(x) = x^3 + Bx^2 + Cx$, find B and C if $(2, -10)$ is a turning point, and find the other turning point.

Solution $\qquad 3Ax^2 + 2Bx + C = 0$

Since the coefficient of x^3, in $f(x)$, is 1, then $A = 1$.

$$3x^2 + 2Bx + C = 0$$
$$3(2)^2 + 2B(2) + C = 0 \qquad \text{Substituting } x = 2$$
$$12 + 4B + C = 0 \qquad \text{Simplifying} \qquad (1)$$

When $x = 2, f(2) = -10$, since $(2, -10)$ is a point on the graph of $f(x)$. Therefore, if $f(x) = x^3 + Bx^2 + Cx$,

$$(2)^3 + B(2)^2 + C(2) = -10$$
$$8 + 4B + 2C = -10$$
$$4B + 2C = -18$$
$$2B + C = -9 \qquad (2)$$

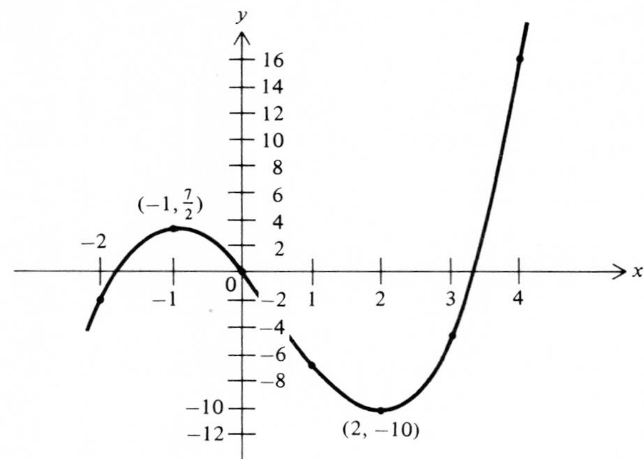

Diagram 31

$$\begin{aligned} 4B + C &= -12 \\ 2B + C &= -9 \\ \hline 2B \qquad &= -3 \\ B &= -\tfrac{3}{2} \end{aligned}$$ Solving (1) and (2) simultaneously

$$\begin{aligned} 2(-\tfrac{3}{2}) + C &= -9 \\ -3 + C &= -9 \\ C &= -6 \end{aligned}$$ Substituting $B = -\frac{3}{2}$ into (2)

Therefore, our equation is

$$f(x) = x^3 - \frac{3x^2}{2} - 6x$$

We find the other turning point as follows:

$$x = \frac{\frac{3}{2} \pm \sqrt{\frac{9}{4} - 3(1)(-6)}}{3} = \frac{\frac{3}{2} \pm \sqrt{\frac{81}{4}}}{3} = \frac{\frac{3}{2} \pm \frac{9}{2}}{3} = 2 \text{ or } -1$$

$$f(-1) = -1 - \tfrac{3}{2} + 6 = \tfrac{7}{2}$$ Using $x = -1$

Therefore the second turning point is $(-1, \frac{7}{2})$ (see Diagram 31).

EXERCISES

1 If $f(x) = x^3 + Bx^2 + Cx$:

(a) Find B and C if the point (2, 14) is a turning point.

(b) Find the other turning point.

(c) Sketch a graph of $f(x)$.

2 A company wishes to construct an open box from a piece of cardboard 8 inches wide and 15 inches long by cutting a square from each corner and folding up the sides. Find the dimensions of the box of largest volume.

In Exercises 3 to 17 find the coordinates of the turning points (if they exist) and sketch a graph of $f(x)$.

3 $f(x) = 2x^3 - 9x^2 + 12x$

4 $f(x) = x^3 - 6x^2 + 9x$

5 $f(x) = x^3 - 3x^2 - 9x + 11$

6 $f(x) = (x - 2)^2(x - 5)$

7 $f(x) = 2x^3 + 3x - 1$

8 $f(x) = x^3 + 3x^2 - 9x + 1$

9 $f(x) = x^3 - 3x + 1$

10 $f(x) = x^3 + 3x^2 - 1$

11 $f(x) = (x - 1)(x + 1)^2$

12 $f(x) = 2x^3 + 3x^2 + 12x - 4$

13 $f(x) = x^3 - 6x^2 + 17$

14 $f(x) = x^3 - 9x^2 + 15x - 5$

15 $f(x) = 2x^3 - x^2 - 4x + 2$

16 $f(x) = (x - 3)(x + 1)^2$

17 $f(x) = \dfrac{x^3}{4} - \dfrac{3x^2}{4} + 2$

18 Find the volume of the largest cone that can be inscribed in a sphere of radius 9. ($V = \frac{1}{3}\pi R^2H$.)

19 Find the volume of the largest cylinder that can be inscribed in a sphere of radius $3\sqrt{3}$. ($V = \pi R^2H$.)

20 A right triangle of hypotenuse $3\sqrt{3}$ is rotated about one of its legs, forming a cone. Find the cone of greatest volume.

21 Find the rectangle of greatest area with base on the x axis and its other two vertices on the parabola $y = -x^2 + 8x - 4$.

22 $f(x) = x^3 + Bx^2 + Cx + D$.

(a) Find B and C so that turning points will occur at $x = -1$ and $x = 3$.

(b) If the graph of $f(x)$ passes through the point $(-2, 2)$, determine D. Sketch the graph of $f(x)$.

23 A package is to be sent parcel post. The sum of the *girth* (distance around the package) and the length cannot exceed 84 inches. Find the dimensions of the largest package that can be sent if the base is square.

24 A large metropolitan city proposes a daily commuter tax on cars entering the mid-town section of the city. Economists predict that as the tax increases, the number of cars will decrease. However, if the tax is excessive, there will be little revenue for the city. The economists develop a formula to represent the relationship between x (the number of cars in thousands) and y (the tax per car in pennies): $y = 192 - 4x^2$ The total revenue R is therefore $R = xy$. Find the tax which should be imposed to yield maximum revenue for the city.

25 A rectangular box with a top is to be constructed from a square piece of cardboard by cutting along the broken lines. The cardboard is to be turned up, forming the four sides, and then the flap is folded over to form the top. Find the dimensions of the container of largest volume which can be formed from a cardboard sheet (see diagram below).

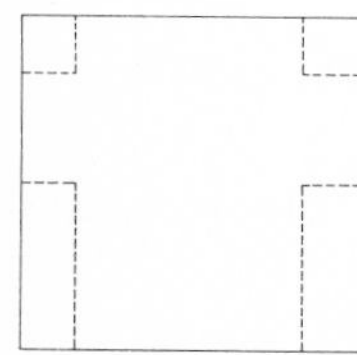

(a) 12 by 12 inches **(b)** 8 by 15 inches (squares cut from 8 inch side)

26 A rectangle of perimeter 24 is revolved about one of its sides, forming a cylinder.

Find the dimensions of the rectangle which generates the cylinder of maximum volume.

27 A company plans to market its product in an aluminum rectangular box 4 times as long as it is wide. The total amount of material to be used for each box is 600 square inches. Find the dimensions of the box which will hold the largest volume: **(a)** the box has no top; **(b)** with a top included.

28 Find the largest and smallest value of $f(x) = 4x^3 - 8x^2 + 5x$ in the interval $[0, 2]$; graph $f(x)$ in the given interval.

29 Find the largest and smallest value of $f(x) = x^3 - 3x + 2$ in the interval $[-3, \frac{3}{2}]$; graph $f(x)$ in the given interval.

30 Find the largest and smallest value of $f(x) = 4x^3 - 9x^2 - 12x$ in the interval $[-2, 1]$; graph $f(x)$ in the interval.

31 Find the largest and smallest value of $f(x) = x^2(x - 3)$ in the interval $[-1, 3]$; graph $f(x)$ in the interval.

32 Find the largest and smallest value of $f(x) = x^3 - 6x^2 + 9x + 1$ in the interval $[0, 2]$; sketch $f(x)$ in the interval.

33 Find the dimensions of the right circular cylinder of greatest volume if the total surface area is 96π.

34 Find the rectangle of greatest area that can be inscribed in $y = 16 - x^2$ with its base on the x axis.

35 Find the trapezoid of greatest area that can be inscribed in $y = 16 - x^2$ with its longer base on the x axis.

36 A piece of wire 60 inches long is cut into six sections, two of one length and four of another. Each of the two sections having the same length is bent to form a circle. The two circles are then braced together by the other four remaining sections, forming the frame of a right circular cylinder. How should the piece of wire be partitioned so that the volume of the cylindrical frame is a maximum?

37 A company plans to market its product in a rectangular box whose length is twice the width. The total amount of material to be used for the box is 384 square inches. Find the dimensions of the box which will hold the largest volume if the box has no top.

38 A company plans to market its product in a rectangular box whose length is 3 times the width. The total amount of material to be used for the box is 450 square inches. Find the dimensions of the box which will hold the largest volume if the box has a top.

Chapter 6 Rational Functions

6.1 THE RATIONAL FUNCTION

In Chapter 5 we saw a collection of functions which all fall into the same category. Had we attempted to divide one polynomial by another nonzero polynomial, we could have obtained an expression which would generally not comply with our understanding of polynomial expressions. The operation of division leads us into a new class of functions called the *rational functions,* which are developed and analyzed in this chapter.

DEFINITION 1 A *rational function* $R(x)$ is the quotient of polynomial functions $P(x)$ and $Q(x)$ such that

$$R(x) = \frac{P(x)}{Q(x)} \qquad Q(x) \neq 0$$

DEFINITION 2 The *domain of a rational function* $R(x) = P(x)/Q(x)$ is the collection of all real x such that $Q(x) \neq 0$.

Let us examine a selection of five rational functions and determine the domain of each.

Example 1 $R(x) = \dfrac{x + 1}{3x - 2}$.

When $3x - 2 = 0$, $R(x)$ is undefined. The domain of $R(x)$ is all real numbers, $x \neq \frac{2}{3}$.

Example 2 $f(x) = \dfrac{3}{x - 1}$.

When $x - 1 = 0$, $f(x)$ is undefined. The domain of $f(x)$ is all real numbers, $x \neq 1$.

Example 3 $g(x) = \dfrac{x^2}{x + 1}$.

When $x + 1 = 0$, $g(x)$ is undefined. The domain of $g(x)$ is all real numbers, $x \neq -1$.

Example 4 $h(x) = \dfrac{x - 1}{x^2 - x - 6}$.

When $x^2 - x - 6 = 0$, $h(x)$ is undefined. Since $x^2 - x - 6 = (x - 3)(x + 2)$, the domain of $h(x)$ is all real numbers with *two* restrictions, $x \neq 3, x \neq -2$.

It is also possible to have a rational function whose domain is *unrestricted.* Consider the following example.

Example 5 $k(x) = \dfrac{x + 1}{x^2 + 1}$.

When $x^2 + 1 = 0$, $k(x)$ is undefined. However, there is no real number which satisfies the condition $x^2 = -1$. Therefore $k(x)$ has a domain of *all real numbers.*

To get a deeper understanding of the behavior of rational functions, we resort to *graphing*. Graphing these functions, although similar in technique to that of polynomial functions, will offer more of a challenge. With practice, you should have little trouble in mastering the additional procedures necessary to obtain interesting and accurate results.

6.2 HORIZONTAL AND VERTICAL ASYMPTOTES

As you have witnessed in the first four examples given in Section 6.1, certain restrictions were placed on the domains. The zero(s) of the denominator in each case produced these restrictions. We define the zeros (of the denominator only) as *crucial values.* We can now ask: What is the behavior of a rational function, say $y = f(x)$, as x takes on values closer and closer to these crucial values? (Note that we do not talk of the function at a crucial value because it is not defined there.) It is convenient to use certain notation in discussing such questions. The notation needed is as follows:

1 $x \to +\infty$ is read "x increases without bound." For example, $x \to +\infty$, $1/x \to 0^+$ reads "as x increases without bound, $1/x$ approaches zero through positive values."

2 $x \to -\infty$ is read "x decreases without bound." For example, $x \to -\infty$, $1/x \to 0^-$ reads "as x decreases without bound, $1/x$ approaches zero through negative values."

3 $|x| \to \infty$ is read "x increases or decreases without bound" and is a combination of the two previous statements.

We are now ready to study the following example.

Example 1 What is the behavior of $f(x)$ as $x \to 2$ for

$$f(x) = \frac{3}{x - 2}$$

DISCUSSION $f(x)$ has one crucial value, namely 2. To study the behavior of $f(x)$ near the value 2 we test values slightly smaller and slightly larger than 2. In other words, we squeeze in on 2 from the left and from the right, as the following tables demonstrate.

x	1	1.5	1.9	1.99	1.999	$x \to 2^-$
$f(x)$	-3	-6	-30	-300	-3000	$f(x) \to -\infty$

x	3	2.5	2.1	2.01	2.001	$x \to 2^+$
$f(x)$	3	6	30	300	3000	$f(x) \to +\infty$

From these tables it becomes apparent that

1 $|f(x)| \to \infty \qquad$ as $x \to 2$

2 When $x < 2$, $f(x) < 0$ and when $x > 2$, $f(x) > 0$. Graphically $f(x)$ rises sharply as $x \to 2^+$ and drops sharply as $x \to 2^-$ (see Diagram 1).

Since $x \neq 2$ [2 is not in the domain of $f(x)$] *the graph of* $f(x)$ *will never intersect the line defined by the equation* $x = 2$. Such a line is an example of a *vertical asymptote*.

COMMENT In general, the graph of a function can *never* intersect one of its vertical asympotes, say $x = a$, since the function is undefined at a.

We now see our first new procedure in graphing rational functions, that is, search for a vertical asymptote. We can now formulate our first rule.

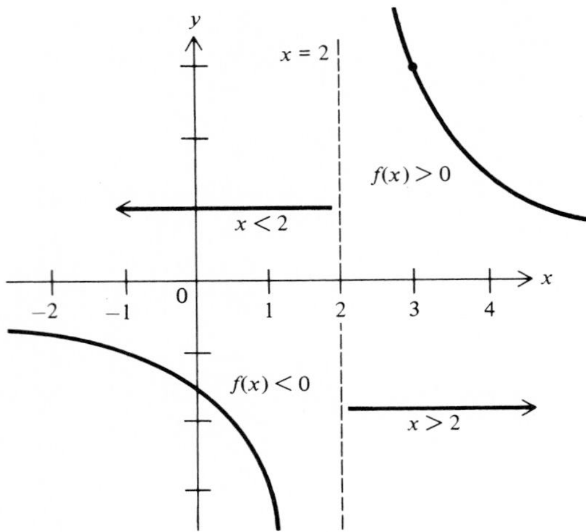

Diagram 1

Rule 1 If $f(x) = P(x)/Q(x)$ and there exists a number c such that $Q(c) = 0$ and $P(c) \neq 0$, then $x = c$ is a *vertical asymptote* of $y = f(x)$.

The second phase is to determine whether our function possesses horizontal asymptotes. This is done by observing the function's behavior as x grows both positively and negatively. If we can show that there exists a horizontal line $y = c$ such that our function tends to get closer and closer to as $|x| \to \infty$, then $y = c$ is a *horizontal asymptote* of $y = f(x)$.

Again let us make use of $f(x) = 3/(x - 2)$ to search for horizontal asymptotes.

Example 2 Find the horizontal asymptote of $f(x) = \dfrac{3}{x - 2}$.

DISCUSSION For this particular function we observe that the numerator is constant, namely, 3. The denominator of $f(x)$ is $x - 2$, and therefore as x grows large, so does $x - 2$. We know that the larger the denominator in comparison to the numerator, the smaller the value of the rational expression. Hence, $3/(x - 2)$ approaches zero as x grows.

Symbolically we can write this as follows:

$$\text{As } x \longrightarrow +\infty, \qquad \frac{3}{x - 2} \longrightarrow 0^+$$

Likewise, as x becomes more and more negative $3/(x - 2)$ again approaches zero. We can write this symbolically as follows:

$$\text{As } x \longrightarrow -\infty, \qquad \frac{3}{x - 2} \longrightarrow 0^-$$

Examine the tables of values to see what is actually occurring.

x	3	32	302	3002	$x \to +\infty$
$f(x)$	3	0.1	0.01	0.001	$f(x) \to 0^+$

x	-1	-28	-298	-2998	$x \to -\infty$
$f(x)$	-1	-0.1	-0.01	0.001	$f(x) \to 0^-$

Zero is the number we are looking for. Hence, $y = 0$ is the equation of the line which is the *horizontal asymptote.*

COMMENT Zero is the only value that works here. No value of x (regardless of how large it is) can make $3/(x - 2)$ equal to zero. Yet we can get as close to zero as we wish. For example, had we thought 1/1,000,000 was a small enough value, we could have shown that the function can be made less than this. If we let $x > 3{,}000{,}002$, then

$$\frac{3}{x - 2} < \frac{1}{1{,}000{,}000}$$

Rule 2 If $f(x) = P(x)/Q(x)$ and $f(x) \to c$, where c is some real number, as $|x| \to \infty$, $y = c$ is a *horizontal asymptote.*

Before we graph $f(x) = 3/(x - 2)$, using both vertical and horizontal asymptotes, let us mention two aids already discussed in Chapter 5, intervals and consistency of sign in the interval.

For rational functions we not only have zeros but crucial values to contend with. The *zeros and the crucial values* together determine how the horizontal axis is divided into intervals. For each interval, the function will be consistent in sign, that is have the same sign. You need only evaluate the function at one point (say a) in the given interval. If $f(a) > 0$, the graph of the function will be above the horizontal axis for that interval. The only exception is at the end points of the interval. If the end point is a zero of $f(x)$, then naturally at that single point the graph touches the horizontal axis. Likewise, if for any point, say b, in the interval $f(b) < 0$, the function $f(x)$ will be graphed below the horizontal axis.

If there are *m distinct* zeros in the *numerator* and *n distinct* zeros (crucial values) in the *denominator,* the *total number of intervals* will be $m + n + 1$. Taking all these facts into considerstion, let us now graph $f(x) = 3/(x - 2)$.

STEP 1 The domain is all real numbers x, $x \neq 2$.

STEP 2 $x = 2$ is the vertical asymptote.

STEP 3 $y = 0$ is the horizontal asymptote.

STEP 4 Since there is only one crucial value at $x = 2$, this divides the x axis into two intervals, $(-\infty, 2)$ and $(2, +\infty)$ (see Diagram 2).

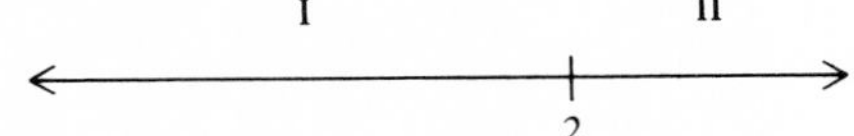

Diagram 2

STEP 5 For the interval $(-\infty, 2)$ choose any arbitrary x, say $x = 1$. $f(1) = -3 < 0$, and hence $f(x) < 0$ in this interval. For the interval $(2, +\infty)$ choose any arbitrary x, say $x = 3$. $f(3) = 3 > 0$, and hence $f(x) > 0$ in this interval.

STEP 6 For the y intercept set $x = 0$ and $f(0) = -\frac{3}{2}$. There is no x intercept.

This is all the information we need to make an accurate sketch of $f(x)$. Note how the graph of the function draws toward its asymptotes (Diagram 3). This is very important because one of the major reasons for discussing asymptotes is to obtain guidelines for our sketch.

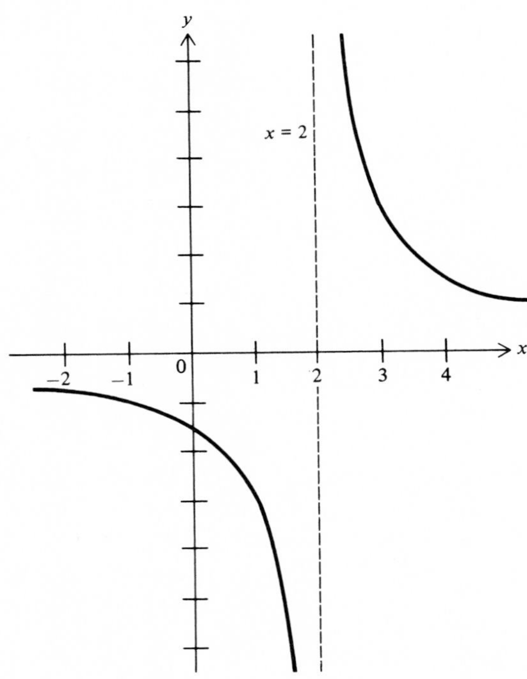

Diagram 3

Example 3 Analyze $f(x) = \dfrac{x}{x^2 - 1}$.

$$f(x) = \frac{x}{(x + 1)(x - 1)} \qquad \text{Factoring}$$

STEP 1 The domain is all real numbers x, $x \neq \pm 1$.

STEP 2 Since $f(x) = 0$ when $x = 0$, 0 is a zero of the function.

STEP 3 The vertical asymptotes are $x = 1$ and $x = -1$.

STEP 4 To find the horizontal asymptote follow this procedure very carefully. Although it is mechanical, it provides an elementary approach to horizontal asymptotes.

Take $x/(x^2 - 1)$ and divide the numerator and denominator by x^2 (the *highest* power of x in the expression):

$$\frac{x}{x^2 - 1} = \frac{x/x^2}{x^2/x^2 - 1/x^2} = \frac{1/x}{1 - 1/x^2}$$

This expression is equivalent to the original, $x \neq 0$, and since we are looking for the horizontal asymptote we are only interested in the behavior of the function as $|x| \to \infty$. Now, as $|x| \to \infty$, $1/x$ and $1/x^2 \to 0$.

Therefore $$\frac{1/x}{1 - 1/x^2} \longrightarrow 0 \qquad \text{as } |x| \longrightarrow \infty$$

This implies that $y = 0$ is a horizontal asymptote.

STEP 5 We have one zero and two crucial values; hence we have four intervals to consider $(-\infty, -1)$, $(-1, 0)$, $(0, 1)$, and $(1, +\infty)$ (see Diagram 4).

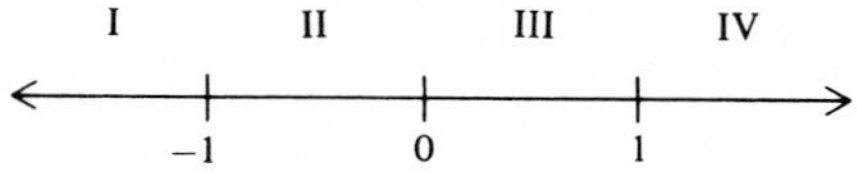

I II III IV

−1 0 1

Diagram 4

STEP 6 For $(-\infty, -1)$ choose $x = -2$; then $f(-2) = -\frac{2}{3} < 0$. Therefore

$$f(x) < 0 \qquad \text{for } (-\infty, -1)$$

For $(-1, 0)$ choose $x = -\frac{1}{2}$; then $f(-\frac{1}{2}) = \frac{2}{3} > 0$. Therefore

$$f(x) > 0 \qquad \text{for } (-1, 0)$$

For $(0, 1)$ choose $x = \frac{1}{2}$; then $f(\frac{1}{2}) = -\frac{2}{3} < 0$. Therefore

$$f(x) < 0 \qquad \text{for } (0, 1)$$

For $(1, +\infty)$ choose $x = 2$; then $f(2) = \frac{2}{3} > 0$. Therefore

$$f(x) > 0 \qquad \text{for } (1, +\infty)$$

STEP 7 Both the x and y intercepts are zero.

Diagram 5 uses all the information we have accumulated.

COMMENT The graph of $f(x)$ intersects its horizontal asymptote at (0, 0).

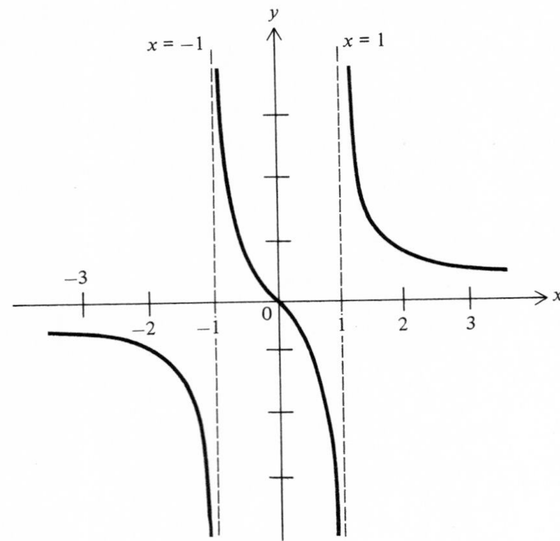

Diagram 5

Example 4 Analyze $g(x) = \dfrac{x + 1}{x + 2}$.

STEP 1 The domain is all real x, $x \neq -2$.

STEP 2 When $x = -1$, $g(x) = 0$; hence -1 is a zero of $g(x)$.

STEP 3 The vertical asymptote in $x = 2$.

STEP 4 To find the horizontal asymptote use the method from Example 3. Start by dividing the numerator and denominator by x, the highest power of the variable

$$\frac{x + 1}{x + 2} = \frac{1 + 1/x}{1 + 2/x} \qquad \text{Dividing through by } x$$

$$\text{As } |x| \longrightarrow \infty \qquad \frac{1 + 1/x}{1 + 2/x} \longrightarrow \frac{1}{1} = 1 \Longrightarrow y = 1 = \text{horizontal asymptote}$$

STEP 5 One zero and one crucial value give us three intervals: $(-\infty, -2)$, $(-2, -1)$, and $(-1, +\infty)$ (see Diagram 6).

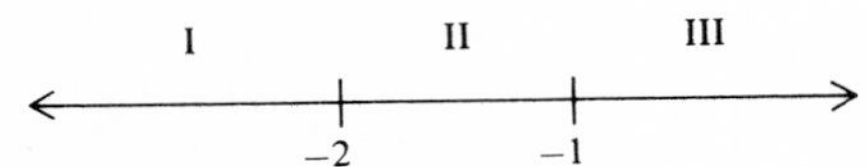

Diagram 6

STEP 6

$$g(x)\begin{cases} > 0 & \text{for } (-\infty, -2) \\ < 0 & \text{for } (-2, -1) \\ > 0 & \text{for } (-1, +\infty) \end{cases}$$

STEP 7 The x intercept is -1 since $g(-1) = 0$, and the y intercept is $\frac{1}{2}$ since $g(0) = \frac{1}{2}$. [See Diagram 7 for the graph of $g(x)$.]

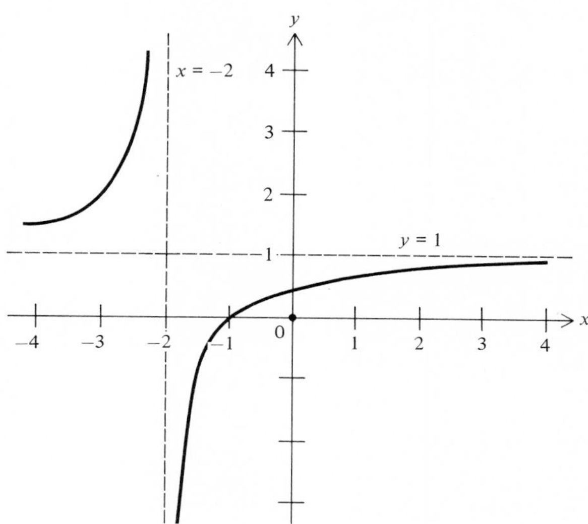

Diagram 7

COMMENT We should clear up one question: How do we know that the graph of $g(x)$ does not intersect $y = 1$, the horizontal asymptote? A method for finding the proper answer is as follows: if $g(x)$ intersects $y = 1$, then for some x in its domain

$$\frac{x+1}{x+2} = 1$$

Since $x \neq -2$ we can clear the rational expression by multiplying both sides of the equation by $x + 2$

$$x + 1 = x + 2$$
$$1 = 2 \qquad \text{false}$$

This leads to a contradiction, and therefore no value of x in the domain of $g(x)$ works; hence there is no intersection.

Whenever asymptotes other than vertical are present, you should get into the habit of checking for such intersection points. Learning of such strategic points changes a graph significantly. Let us take a look at such an example.

Example 5 Analyze $f(x) = \dfrac{3x^2 - 3x - 6}{2x^2}$.

Write $f(x)$ in factored form:

$$f(x) = \frac{3(x-2)(x+1)}{2x^2}$$

STEP 1 The domain is all real numbers x, $x \neq 0$.

STEP 2 Since $f(2) = 0$ and $f(-1) = 0$, the zeros are -1 and 2.

STEP 3 The vertical asymptote is $x = 0$.

STEP 4 We find the horizontal asymptote from

$$f(x) = \frac{3 - 3/x - 6/x^2}{2} \qquad \text{as } |x| \longrightarrow \infty, f(x) \longrightarrow \tfrac{3}{2}$$

Therefore the horizontal asymptote is $y = \frac{3}{2}$.

STEP 5 Does the graph of $f(x)$ intersect the horizontal asymptote?

$$\frac{3x^2 - 3x - 6}{2x^2} = \frac{3}{2}$$

Multiply by $2x^2$ to give

$$3x^2 - 3x - 6 = 3x^2 \quad \Longrightarrow \quad x = -2$$

The answer is yes, at $(-2, \frac{3}{2})$.

STEP 6 Two zeros and one crucial value will generate four intervals: $(-\infty, -1)$, $(-1, 0)$, $(0, 2)$, and $(2, +\infty)$ (see Diagram 8).

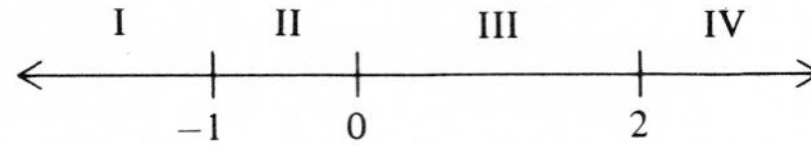

Diagram 8

STEP 7 For $(-\infty, -1)$ $f(x) > 0$; for $(-1, 0)$, $f(x) < 0$; for $(0, 2)$, $f(x) < 0$ and for $(2, +\infty)$, $f(x) > 0$.

STEP 8 The x intercepts are -1 and 2, since $f(-1) = 0$ and $f(2) = 0$. There is no y intercept since $f(0)$ is undefined. [See Diagram 9 for the graph of $f(x)$.]

COMMENT The most important observation to be made in this example occurs in the region of the point $(-2, \frac{3}{2})$. Note that to the left of the point, the graph of the function approaches the horizontal asymptote, $y = \frac{3}{2}$, from

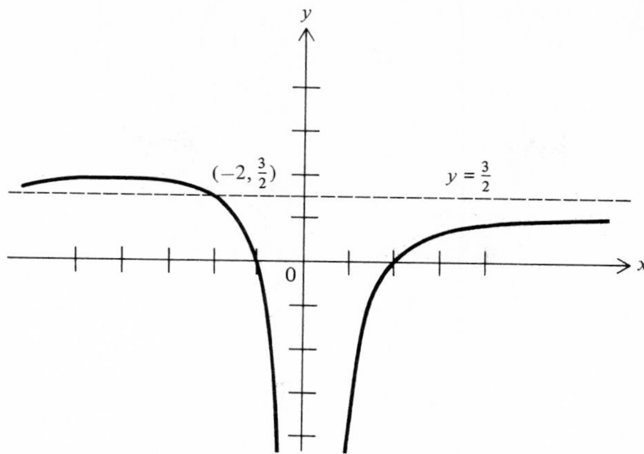

Diagram 9

above. We know this to be the case by virtue of the fact that there is one and only one point of intersection with $y = \frac{3}{2}$. Had we not found this point of intersection, $y = f(x)$ could have been graphed *incorrectly* (Diagram 10).

Diagram 10 is a logical graph but incorrect. With the point of intersection unknown, one could easily satisfy the ordinary condition here, namely, $f(x) > 0$ in the interval $(-\infty, -1)$, and the graph of $f(x)$ might just as well have tended toward the asymptote $y = \frac{3}{2}$ from below, as $x \rightarrow -\infty$.

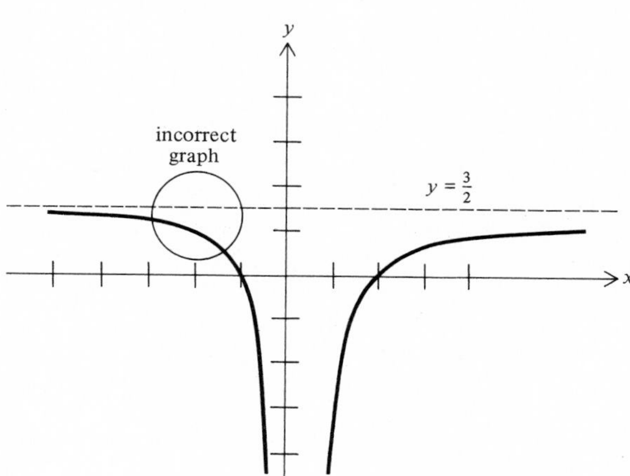

Diagram 10

Example 6 Analyze $h(x) = \dfrac{x^2 + 1}{x}$ for horizontal asymptotes.

Since $x = 0$ is not in the domain of $h(x)$, we can now write the function in the form

$$h(x) = \frac{x^2}{x} + \frac{1}{x} = x + \frac{1}{x} \qquad \text{as } |x| \longrightarrow \infty, \frac{1}{x} \longrightarrow 0$$

Therefore, $h(x)$ behaves like the function $h(x) = x$. Hence, $h(x)$ approaches x, *not* some finite number c. We do not have a horizontal asymptote (Rule 2).

Before ending this section, let us make some observations on the preceding five examples. By doing so, we can get a feeling for a theorem on horizontal asymptotes which will shorten the analysis.

In Examples 1 and 2, the degree of the numerator was *less* than that of the denominator. As $|x|$ grew larger, the quotients decreased, with the result that both functions had horizontal asymptotes at $y = 0$.

In Examples 3 and 4, the degree of the numerator and denominator is the *same*. As $|x|$ grew larger, the numerators and denominators in each case grew at approximately the same rate, so that the resulting horizontal asymptote corresponded to the *ratio of the leading coefficients* in each polynomial.

Function	Horizontal asymptote
$g(x) = \dfrac{x+1}{x+2}$	$y = 1$
$f(x) = \dfrac{3x^2 - 3x - 6}{2x^2}$	$y = \frac{3}{2}$

Finally, in Example 6, the degree of the numerator was *larger* than that of the denominator. Since the numerator grew more rapidly than the denominator, the quotient constantly grew and no horizontal asymptote existed. This trend is generally true and can be formulated in the following theorem.

Theorem 1 Consider a rational function of the form

$$f(x) = \frac{a_0x^m + a_1x^{m-1} + a_2x^{m-2} + \cdots}{b_0x^n + b_1x^{n-1} + b_2x^{n-2} + \cdots}$$

(a) If $n > m$, $y = 0$ is the horizontal asymptote of $f(x)$.
(b) If $n = m$, $y = a_0/b_0$ is the horizontal asymptote of $f(x)$.
(c) If $n < m$, no horizontal asymptote exists.

EXERCISES

Sketch and completely analyze each of the rational functions in Exercises 1 to 24.

1 $f(x) = \dfrac{2}{x+4}$

2 $f(x) = \dfrac{-3}{2x-1}$

3 $f(x) = \dfrac{4}{x^2+1}$

4 $f(x) = \dfrac{\sqrt{2}}{x^2-9}$

5 $f(x) = \dfrac{2x+1}{3x-2}$

6 $f(x) = \dfrac{2x^2-x-3}{x^2+x-2}$

7 $f(x) = \dfrac{2x^2}{(x+2)(x-1)}$

8 $f(x) = \dfrac{2x^2}{x^2-1}$

9 $f(x) = \dfrac{3x-2}{x^2-4}$

10 $f(x) = \dfrac{(x+1)^2}{x^2+1}$

11 $f(x) = \dfrac{4x+6}{x^2+4}$

12 $f(x) = \dfrac{x}{x^2-64}$

13 $f(x) = \dfrac{4x^4}{x^4-1}$

14 $f(x) = \dfrac{2x^2-2x}{12x^2-7x-10}$

15 $f(x) = \dfrac{x^2}{x^2+4x+1}$

16 $f(x) = \dfrac{x+1}{2x^2+7x+3}$

17 $f(x) = \dfrac{2x^2}{x^2+3x+2}$

18 $f(x) = \dfrac{x^2-2x+1}{3x^2+7x+2}$

19 $f(x) = \dfrac{2x}{(x-2)^2}$

20 $f(x) = \dfrac{4x^2+x-3}{x^2-1}$

21 $y = \dfrac{x^2-1}{x^2+1}$

22 $y = \dfrac{x^3}{x^3+8}$

23 $y = \dfrac{2x+1}{x^2}$

24 $y = \dfrac{x(x+4)}{(x+1)^2}$

25 Find a rational function whose graph possesses the graphical characteristics in the diagram.

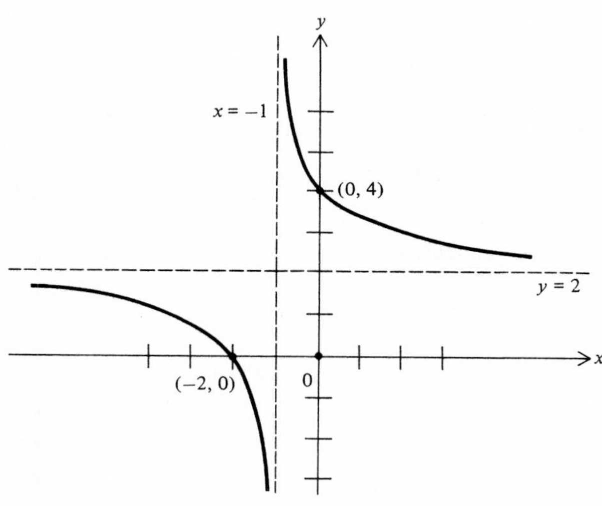

Compare the graphs of each pair of functions in Exercises 26 to 29.

26 $f(x) = \dfrac{1}{1+x}$ and $f(x) = \dfrac{x}{|x|+x^2}$

27 $f(x) = \dfrac{2x - 3}{x + 1}$ and $f(x) = \dfrac{2x + 3}{x - 1}$

28 $f(x) = \dfrac{(2x + 1)(x - 2)}{(x - 1)(x + 2)}$ and $f(x) = \dfrac{(2x - 1)(x + 2)}{(x + 1)(x - 2)}$

29 $f(x) = \dfrac{1}{x^2}$ and $f(x) = \dfrac{1}{x^2 + 1}$

30 Analyze $f(x) = 2 + \dfrac{x}{(x + 3)(x - 1)}$.

31 Analyze $f(x) = \dfrac{8 - 2x^2}{x^2 + 2x - 3}$.

32 Analyze $f(x) = \dfrac{(x - 2)^2}{(x + 3)^2}$.

33 Analyze $x^2y + ax + by = 0$ for:
(a) $a > 0$ and $b > 0$ **(b)** $a < 0$ and $b < 0$

34 How different are the graphs of

$$y = \frac{x^2}{x^2 - 4} \quad \text{and} \quad y = \frac{x^2}{x^2 + 4}$$

35 If $a = 2$ and $b = 4$, graph the function $f(x) = (x - a)(b - x)/x^2$ and examine the point on the graph corresponding to $x = 2ab/(a + b)$. What is the significance of this point?

36 Sketch the system

$$f(x) = \frac{2x}{x - 1} \qquad g(x) = 2x + 1$$

(a) At $x = -1$ and $x = 3$ on $f(x)$, find the equation of the lines passing through these points and parallel to $g(x)$.
(b) Find the equations of the lines perpendicular to the lines in part **(a)** at $x = -1$ and $x = 3$.
(c) Show that the lines in part **(b)** intersect $f(x)$ in *exactly* one point. (This implies that the lines are tangent.)
(d) Determine the solution of the inequality $f(x) \leq g(x)$.

6.3 OBLIQUE LINEAR AND NONLINEAR ASYMPTOTES

In the final example of Section 6.2 we encountered a difficulty when a horizontal asymptote was absent. This is easily remedied by reasoning in the same manner as we did for the horizontal asymptote. Take

$$h(x) = \frac{x^2 + 1}{x} = x + \frac{1}{x}$$

We are interested in the behavior of $h(x)$ as $|x| \to \infty$.

$$\frac{1}{x} = \frac{1}{100} = 0.01 \qquad \text{at } x = 100$$

$$\frac{1}{x} = \frac{1}{1000} = 0.001 \qquad \text{at } x = 1000$$

$$\frac{1}{x} = -\frac{1}{100} = -0.01 \qquad \text{at } x = -100$$

$$\frac{1}{x} = -\frac{1}{1000} = -0.001 \qquad \text{at } x = -1000$$

Obviously, $1/x$ has little effect on $h(x)$ as $|x| \to \infty$. The function thus behaves like the linear function $y = x$. Hence $y = x$ is an *oblique linear asymptote.* Diagram 11 shows the graph of this function.

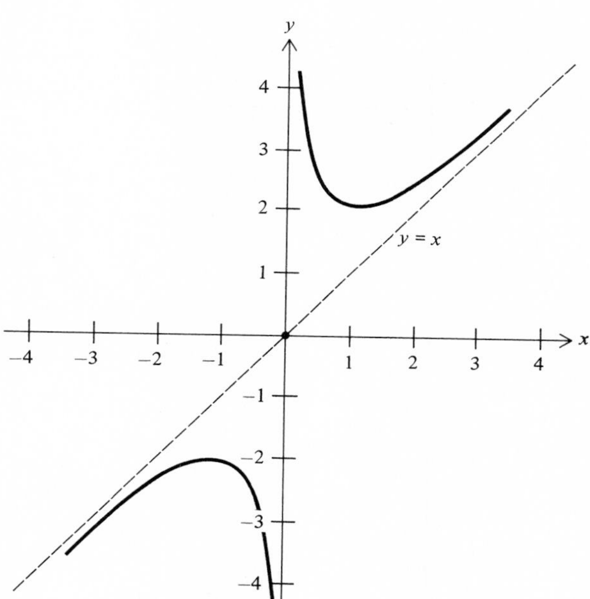

Diagram 11

COMMENT In general, when the degree of the numerator exceeds that of the denominator, the function can be written as the sum of a polynomial function and some *proper rational function.*

DEFINITION 1 A *proper rational function* is a function in the form $f(x) = r(x)/g(x)$, where the degree of $r(x)$ is less than the degree of $g(x)$.

If $$f(x) = a_m x^m + a_{m-1} x^{m-1} + a_{m-2} x^{m-2} + \cdots + a_0$$

and $$g(x) = b_n x^n + b_{n-1} x^{n-1} + b_{n-2} x^{n-2} + \cdots + b_0$$

where $m > n$ and $m > 1$, we can write

$$h(x) = \frac{f(x)}{g(x)} = q(x) + \frac{r(x)}{g(x)}$$

where $r(x)/q(x)$ is a proper rational function.

The following theorem is based on this information.

Theorem 1 **(a)** If the degree of $f(x)$ is *exactly one degree greater* than $g(x)$, that is,

$$m - n = 1$$

then $y = q(x)$ is an *oblique linear asymptote* of $h(x)$.
(b) If the degree of $f(x)$ is at *least two degrees greater* than $g(x)$, that is,

$$m - n \geq 2$$

then $y = q(x)$ is a *nonlinear asymptote* of $h(x)$.

Let us illustrate these theorems with some examples.

Example 1 Analyze $f(x) = \dfrac{x^3 + 2x}{x^2 - 1}$.
We start by factoring the right-hand side:

$$f(x) = \frac{x(x^2 + 2)}{(x + 1)(x - 1)}$$

STEP 1 The domain is all real numbers x, $x \neq \pm 1$.

STEP 2 The vertical asymptotes are $x = 1$ and $x = -1$.

STEP 3 There are no horizontal asymptotes. (Note that the degree of the numerator exceeds the degree of the denominator.)

STEP 4 We find the oblique linear asymptote as follows (we specify linear since the degree of the numerator is *exactly one more* than that of the denominator).

$$\begin{array}{r} x \phantom{{}+2x} \\ x^2 - 1 \overline{)\, x^3 + 2x} \\ \underline{x^3 - x} \\ 3x \end{array} \qquad f(x) = x + \frac{3x}{x^2 - 1}$$

Therefore $y = x$ is an oblique linear asymptote.

A good habit to get into is to check two things when an oblique linear asymptote occurs: (1) possible intersections with the graph of $f(x)$ and (2) possible intersections with other asymptotes. Checking (1) gives

$$\frac{x^3 + 2x}{x^2 - 1} = x \qquad x^3 + 2x = x^3 - x$$

$$x = 0$$

Therefore at (0, 0), $y = x$ intersects the graph of $f(x)$. Checking (2), we see that the two vertical asymptotes are $x = 1$ and $x = -1$. They intersect the oblique asymptote $y = x$ at (1, 1) and (−1, −1), respectively.

STEP 5 The x intercept as well as the y intercept are zero.

STEP 6 Two crucial values and one zero generate four intervals: $(-\infty, -1)$, $(-1, 0)$, $(0, 1)$, $(1, +\infty)$ (see Diagram 12).

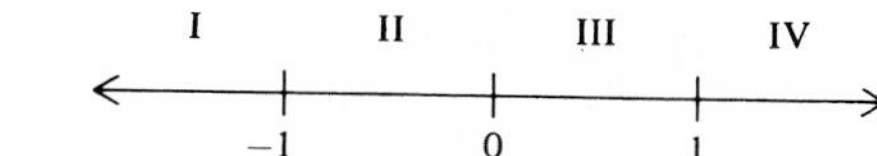

Diagram 12

STEP 7 Check each of the intervals for consistency of sign and compare your results with the graph (see Diagram 13).

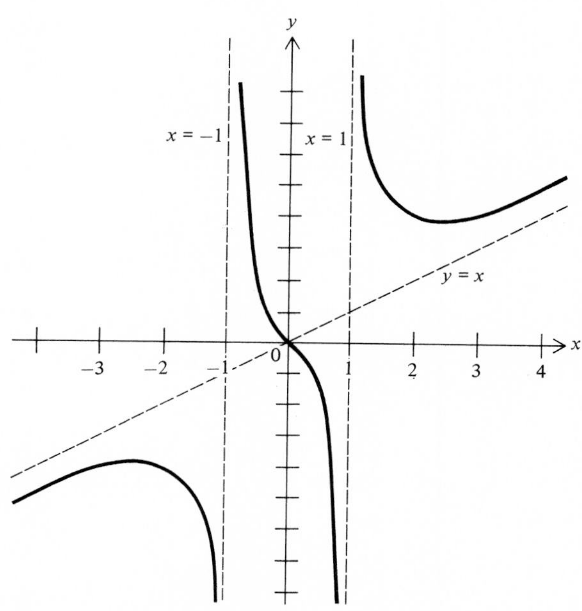

Diagram 13

COMMENT In Example 1 we expressed

$$f(x) = \frac{x^3 + 2x}{x^2 - 1} \qquad \text{as} \qquad f(x) = x + \frac{3x}{x^2 - 1}$$

using the process of long division. The same result can be obtained using an alternate procedure (give-and-take method).

$$\frac{x^3 + 2x}{x^2 - 1} = \frac{x^3 - x + 3x}{x^2 - 1} \qquad \text{Note that } 2x = -x + 3x$$

$$= \frac{x(x^2 - 1) + 3x}{x^2 - 1} \qquad \text{Factoring } x$$

Dividing each term in the numerator by $x^2 - 1$, we obtain

$$\frac{x^3 + 2x}{x^2 - 1} = x + \frac{3x}{x^2 - 1}$$

Example 2 Analyze $f(x) = \dfrac{x^3 + 1}{x}$.

First we factor

$$f(x) = \frac{(x + 1)(x^2 - x + 1)}{x}$$

STEP 1 The domain is all real numbers x, $x \neq 0$.

STEP 2 Since $f(x) = 0$ when $x = -1$, -1 is a zero. $f(0)$ is undefined; hence there is no y intercept.

STEP 3 The vertical asymptote is $x = 0$.

STEP 4 There is no horizontal asymptote since the degree of the numerator exceeds that of the denominator.

STEP 5 Find the nonlinear asymptote. We specify nonlinear since the degree of the numerator is 2 more than that of the denominator.

$$f(x) = \frac{x^3 + 1}{x} = \frac{x^3}{x} + \frac{1}{x} = x^2 + \frac{1}{x}$$

$y = x^2$ is the nonlinear asymptote.

STEP 6 One crucial value and one zero generate three intervals: $(-\infty, -1)$, $(-1, 0)$, $(0, +\infty)$. Be sure to check the signs in each interval (see Diagram 14).

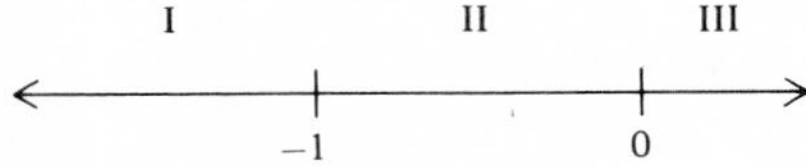

Diagram 14

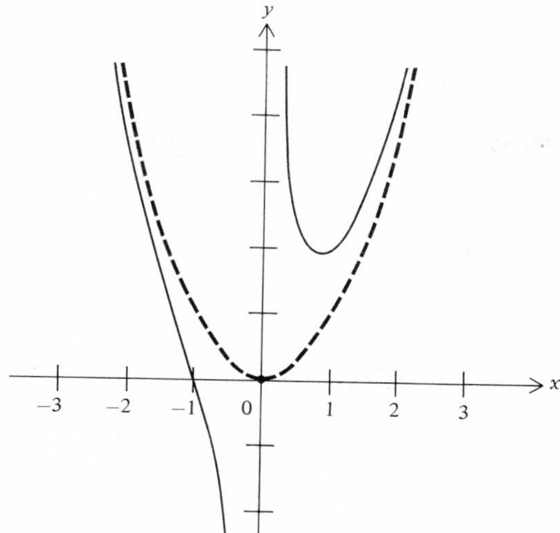

Diagram 15

STEP 7 Does the graph of $f(x)$ intersect the nonlinear asymptote $y = x^2$?

$$\frac{x^3 + 1}{x} = x^2 \implies x^3 + 1 = x^3 \implies 1 = 0 \qquad \text{impossible}$$

Therefore there is no intersection between the graph of the function and its nonlinear asymptote. Where does the vertical asymptote, $x = 0$, intersect the nonlinear asymptote $y = x^2$? Since $y = (0)^2 = 0$, at $(0, 0)$ the vertical asymptote and the nonlinear asymptote intersect. With the amassed information we now obtain the graph of $f(x)$ (see Diagram 15).

As a final exercise, let us examine a rather awesome looking rational function whose graph would be extremely difficult to sketch without the methods presented in this chapter.

Example 3 Analyze $f(x) = \dfrac{(x + 2)^2(x - 4)}{(x - 1)^2}$.

Solution

$$f(x) = \frac{(x - 4)(x^2 + 4x + 4)}{x^2 - 2x + 1} = \frac{x^3 - 12x - 16}{x^2 - 2x + 1}$$

Multiplying out numerator and denominator

STEP 1 The domain is all real numbers x, $x \neq 1$.

STEP 2 We find the x intercept by setting $f(x) = 0$. Hence 4 and -2 are the x intercepts. Recall from Chapter 5 that this root $x = -2$ is a double zero and implies that the graph of $f(x)$ will have a turning point on the x axis at $(-2, 0)$. Since $f(0) = -16$, -16 is the y intercept.

STEP 3 The vertical asymptote is $x = 1$.

STEP 4 There are no horizontal asymptotes.

STEP 5 We find the linear asymptote as follows:

$$\begin{array}{r|l} & x + 2 \\ \hline x^2 - 2x + 1 & x^3 \qquad\qquad - 12x - 16 \\ & x^3 - 2x^2 + \quad x \\ \hline & \quad 2x^2 - 13x - 16 \\ & \quad 2x^2 - \ 4x + \ 2 \\ \hline & \qquad\quad - \ 9x - 18 \end{array}$$

$$f(x) = (x + 2) + \frac{-9x - 18}{x^2 - 2x + 1} \Longrightarrow y = x - 2 \qquad \text{oblique asymptote}$$

STEP 6 One crucial value and two zeros generate four intervals: $(-\infty, -2)$, $(-2, 1)$, $(1, 4)$, and $(4, +\infty)$ (see Diagram 16).

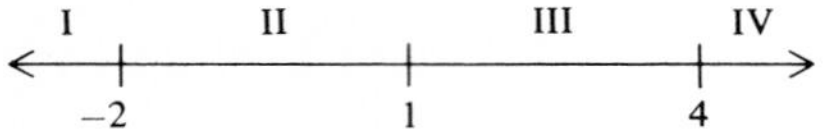

I II III IV

−2 1 4

Diagram 16

STEP 7 Does the graph of $f(x)$ intersect the linear asymptote $y = x + 2$?

$$\frac{x^3 - 12x - 16}{x^2 - 2x + 1} = x + 2$$

$$x^3 - 12x - 16 = x^3 - 3x + 2 \qquad \text{Multiplying both sides by } x^2 - 2x + 1$$

$$-9x = 18 \qquad \text{Simplifying}$$

$$x = -2$$

At $(-2, 0)$ the function $f(x)$ intersects its linear asymptote $y = x + 2$. (This is an interesting point, since this is also the location of the double zero determined in step 2.)

Where does the vertical asymptote $x = 1$ intersect the linear asymptote $y = x + 2$?

$$y = 1 + 2 = 3$$

Hence at $(1, 3)$ the vertical asymptote and the linear asymptote intersect. Diagram 17 summarizes all the information found above.

COMMENT In the beginning of this problem, we expressed $f(x)$ in three different yet equivalent forms. It should be noted that in problem solving it is often necessary to rewrite the original function. There is no hard-and-fast rule about which form is to be used. By solving many problems, you will develop a natural ability to choose the proper form at the proper time.

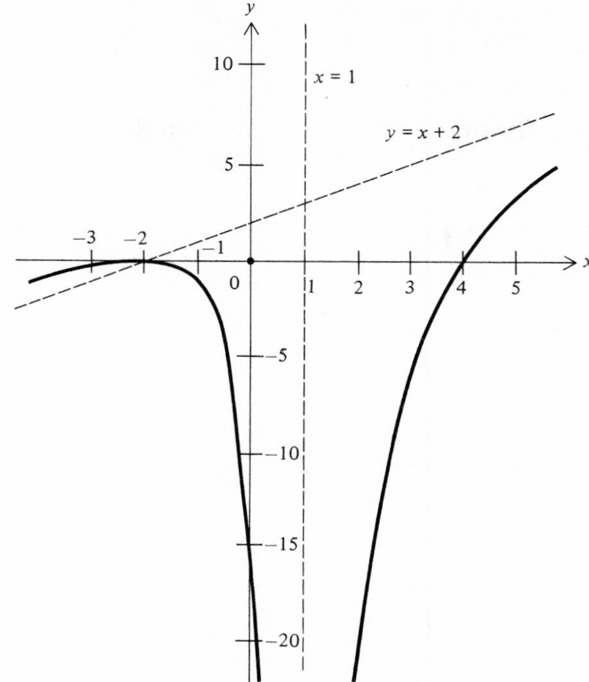

Diagram 17

EXERCISES

Sketch and analyze the rational functions in Exercises 1 to 14.

1 $f(x) = \dfrac{x^2 - 2x + 1}{x + 1}$

2 $f(x) = \dfrac{x^2 - 1}{x + 2}$

3 $f(x) = \dfrac{(x + 4)(x - 2)}{x + 1}$

4 $f(x) = \dfrac{x^2 - 1}{x}$

5 $f(x) = \dfrac{x^3 - 1}{x}$

6 $f(x) = \dfrac{x^4 - 1}{x}$

7 $f(x) = \dfrac{x^3 - 4x^2 + x + 6}{x - 4}$

8 $f(x) = \dfrac{(x + 2)(x - 1)}{2 - x}$

9 $f(x) = \dfrac{(x - 2)(x + 1)}{1 - x}$

10 $f(x) = \dfrac{x^2}{x - a}$

11 $f(x) = \dfrac{x(x - 1)}{x + 1}$

12 $f(x) = \dfrac{x^3 + 2x^2 - 4x - 8}{x^2 - 1}$

13 $f(x) = \dfrac{(x + 3)(x - 1)(x + 1)}{(x + 2)(x - 3)}$

14 $x^2 + xy + 2 = 0$

15 Analyze $f(x) = x^2 + a/x$ for **(a)** $a = -27$, **(b)** $a = 8$, and **(c)** $a = -1$.

16 Analyze $f(x) = \dfrac{a}{x} + bx$ when **(a)** $a = 1$ and $b = \frac{1}{4}$ and **(b)** $a = \frac{1}{4}$ and $b = -1$.

17 Compare the graph of

$$f(x) = \frac{x}{2} + 2 + \frac{2}{x} \text{ with that of } g(x) = \frac{x}{2} + 2$$

18 Analyze $f(x) = \dfrac{x^4 + 2x^3 - 19x^2 - 8x + 60}{x^3 + 3x^2 - x - 3}$

6.4 PARTIAL FRACTIONS

When you first took a course in elementary algebra, you probably came across a problem similar to the following. Collect and write as a single term

$$\frac{1}{x+2} + \frac{1}{x+3}$$

After the required algebra, your result should have been

$$\frac{2x+5}{(x+2)(x+3)} \quad \text{or} \quad \frac{2x+5}{x^2+5x+6}$$

In higher algebra, and especially in the calculus, it becomes necessary to do the exact opposite, that is, to take the expression $(2x + 5)/(x^2 + 5x + 6)$ and decompose it into the *sum of two fractions*. These fractions are named *partial fractions*.

It can be shown that a polynomial with real coefficients can be written as the product of linear and quadratic factors with real coefficients. For example,

$$x^3 + 1 = (x + 1)(x^2 - x + 1)$$

The factor $x^2 - x + 1$ is *irreducible* with respect to the real numbers. This means that $x^2 - x + 1$ cannot be factored into the form $(x + a)(x + b)$, where a and b are real numbers. It should be noted, however, that *all* real polynomials can be written as the product of quadratic and/or linear factors.

We now examine four cases that may arise. Each case is governed by Theorem 1 which will give the correct procedure for the decomposition.

Theorem 1 For every *proper fraction* $f(x)/g(x)$ there exists an equivalent algebraic sum of partial fractions which has one of the following four forms.

FORM 1 If a linear factor $ax + b$ occurs *once* as a factor of $g(x)$, there is a partial fraction of the form $A/(ax + b)$, where A is some constant.

FORM 2 If a linear factor $ax + b$ occurs N times as a factor of $g(x)$, there are N partial fractions of the form

$$\frac{A_1}{ax + b} + \frac{A_2}{(ax + b)^2} + \cdots + \frac{A_N}{(ax + b)^N}$$

FORM 3 If an irreducible quadratic factor $ax^2 + bx + c$ occurs *once* as a factor of $g(x)$, there is a partial fraction of the form

$$\frac{Ax + B}{ax^2 + bx + c} \qquad \text{where } A \text{ and } B \text{ are constants}$$

FORM 4 If an irreducible quadratic factor $ax^2 + bx + c$ occurs N times as a factor of $g(x)$, there are N partial fractions of the form

$$\frac{A_1x + B_1}{ax^2 + bx + c} + \frac{A_2x + B_2}{(ax^2 + bx + c)^2} + \cdots + \frac{A_Nx + B_N}{(ax^2 + bx + c)^N}$$

This theorem says quite a lot. Only by taking each case separately can we demonstrate how to find the constants. First we need a definition.

DEFINITION 1 Consider two polynomials

$$p(x) = a_mx^m + a_{m-1}x^{m-1} + a_{m-2}x^{m-2} + \cdots + a_0$$

and

$$q(x) = b_nx^n + b_{n-1}x^{n-1} + b_{n-2}x^{n-2} + \cdots + b_0$$

$$p(x) = q(x) \qquad \text{if and only if}$$

$$m = n \qquad \text{and} \qquad a_0 = b_0,\ a_1 = b_1,\ a_2 = b_2,\ \ldots,\ a_m = b_n$$

Example 1 $4x^3 - 2x^2 + 3x - 1 = ax^3 + bx^2 + cx + d$.
In order for the equality to be valid, we must have

$$a = 4 \qquad b = -2 \qquad c = 3 \qquad d = -1$$

The following examples illustrate Theorem 1.

FORM 1 $f(x)/g(x)$ is a proper fraction, and $g(x)$ has only distinct linear factors.

Example 2 Decompose into partial fractions

$$\frac{2x + 5}{x^2 + 5x + 6}$$

METHOD 1

Factor the denominator:

$$\frac{2x + 5}{(x + 2)(x + 3)} = \frac{A}{x + 2} + \frac{B}{x + 3} \tag{1}$$

Multiply both sides of Equation (1) by $(x + 2)(x + 3)$:

$$2x + 5 = A(x + 3) + B(x + 2) \tag{2}$$

Expand and collect like terms on the right side of Equation (2):

$$2x + 5 = Ax + 3A + Bx + 2B = Ax + Bx + 3A + 2B = (A + B)x + (3A + 2B)$$

According to the definition of the equality of polynomials, the corresponding terms on each side of the equation must have identical coefficients

$$2x + 5 = (A + B)x + (3A + 2B) \Longrightarrow \begin{matrix} A + B = 2 \\ 3A + 2B = 5 \end{matrix} \Longrightarrow A = 1, B = 1$$

$$\frac{A}{x+2} + \frac{B}{x+3} = \frac{1}{x+2} + \frac{1}{x+3} \qquad \text{Substituting } A = B = 1 \text{ in (1)}$$

METHOD 2

Work the problem out exactly as we did up to Equation (2):

$$2x + 5 = A(x + 3) + B(x + 2)$$

If you glance at the original fraction

$$\frac{2x + 5}{(x + 2)(x + 3)}$$

you will note two restrictions on x, namely, $x \neq -2$ or -3. If you use these values of x in Equation (2), you can solve directly for A and B.

From Equation (2), when $x = -2$,

$$2(-2) + 5 = A(-2 + 3) + B(-2 + 2)$$
$$1 = A$$

and when $x = -3$,

$$2(-3) + 5 = A(-3 + 3) + B(-3 + 2)$$
$$-1 = -B$$
$$1 = B$$

Example 3 Decompose into partial fractions

$$\frac{4x^2 + 25x - 3}{(x + 1)(x - 2)(x + 5)}$$

METHOD 1

$$\frac{4x^2 + 25x - 3}{(x + 1)(x - 2)(x + 5)} = \frac{A}{x+1} + \frac{B}{x-2} + \frac{C}{x+5}$$

Clearing fractions, we obtain

$$4x^2 + 25x - 3 = A(x - 2)(x + 5) + B(x + 1)(x + 5) + C(x + 1)(x - 2)$$

Expanding and collecting like terms, we obtain

$$4x^2 + 25x - 3 = (A + B + C)x^2 + (3A + 6B - C)x + (5B - 2C - 10A)$$

Equating coefficients produces the system

$$\begin{aligned} A + B + C &= 4 \\ 3A + 6B - C &= 25 \\ -10A + 5B - 2C &= -3 \end{aligned}$$

Solving for A, B, C gives $A = 2, B = 3, C = -1$. Therefore,

$$\frac{4x^2 + 25x - 3}{(x + 1)(x - 2)(x + 5)} = \frac{2}{x + 1} + \frac{3}{x - 2} - \frac{1}{x + 5}$$

METHOD 2

$$4x^2 + 25x - 3 = A(x - 2)(x + 5) + B(x + 1)(x + 5) + C(x + 1)(x - 2) \qquad (3)$$

Let $x = -1, 2$, and -5. When $x = -1$, Equation (3) becomes

$$-24 = -12A \Longrightarrow A = 2$$

$$63 = 21B \qquad \text{Substituting } x = 2 \text{ in (3)}$$

$$B = 3$$

$$-28 = 28C \qquad \text{Substituting } x = -5 \text{ in (3)}$$

$$C = -1$$

Using these values, you can now write the same solution as obtained by Method 1.

FORM 2 $f(x)/g(x)$ is a proper fraction, and $g(x)$ has linear factors, some or all of which repeat.

Example 4 Decompose into partial fractions

$$\frac{x}{(x + 1)(x + 2)^2}$$

We have two different linear factors in the denominator, and one factor repeats. We can make use of both forms 1 and 2 to obtain

$$\frac{x}{(x + 1)(x + 2)^2} = \frac{A}{x + 1} + \frac{B}{x + 2} + \frac{C}{(x + 2)^2} \qquad (4)$$

Clearing fractions, we find

$$x = A(x + 2)^2 + B(x + 1)(x + 2) + C(x + 1) \qquad (5)$$

Let us use Method 2 to solve for A, B, and C. Note that there are only two restricted values of x, namely, $x = -1$ and $x = -2$, to use Equation (5)

$$-1 = A \qquad \text{Substituting } x = -1 \text{ in (5)}$$

$$-2 = -C \quad \text{or} \quad C = 2 \qquad \text{Substituting } x = -2 \text{ in (5)}$$

What about B? Simply choose any other value for x, say $x = 0$. When $x = 0$, Equation (5) becomes

$$0 = 4A + 2B + C$$

but $A = -1$ and $C = 2$ and so

$$0 = -4 + 2B + 2 \Longrightarrow B = 1$$

Therefore we conclude that

$$\frac{x}{(x+1)(x+2)^2} = \frac{1}{x+2} + \frac{2}{(x+2)^2} - \frac{1}{x+1} \qquad \text{Substituting } A = -1,\ B = 1, \text{ and } C = 2 \text{ in (4)}$$

FORM 3 $f(x)/g(x)$ is a proper fraction, and $g(x)$ has an *irreducible* quadratic factor.

Example 5 Decompose into partial fractions

$$\frac{3x^2 + 2x - 9}{x(x^2+1)}$$

$$\frac{3x^2 + 2x - 9}{x(x^2+1)} = \frac{Ax + B}{x^2 + 1} + \frac{C}{x} \qquad \text{Using forms 1 and 3} \qquad (6)$$

$$3x^2 + 2x - 9 = (Ax + B)x + C(x^2 + 1) \qquad \text{Clearing fractions} \qquad (7)$$

The only restricted value is $x = 0$. However we will now choose two other values of x which seem *easy to work with,* $x = 1$ and $x = 2$. When $x = 0$, Equation (7) becomes

$$-9 = C$$

$$-4 = A + B + 2C \qquad \text{Substituting } x = 1 \text{ in (7)}$$

$$14 = A + B \qquad \text{But } C = -9 \qquad (8)$$

$$7 = 4A + 2B + 5C \qquad \text{Substituting } x = 2 \text{ in (7)}$$

$$26 = 2A + B \qquad \text{Again } C = -9 \qquad (9)$$

$$A = 12 \qquad B = 2 \qquad \text{Solving (8) and (9)}$$

Therefore,

$$\frac{3x^2 + 2x - 9}{x(x^2+1)} = \frac{12x + 2}{x^2 + 1} - \frac{9}{x} \qquad \text{Substituting } A = 12,\ B = 2,\ C = -9 \text{ in (6)}$$

FORM 4 $f(x)/g(x)$ is a proper fraction, and $g(x)$ has repeated irreducible quadratic factors.

Example 6 Decompose into partial fractions

$$\frac{x^2 + x + 1}{(x^2 + 1)^2}$$

$$\frac{x^2 + x + 1}{(x^2 + 1)^2} = \frac{Ax + B}{x^2 + 1} + \frac{Cx + D}{(x^2 + 1)^2} \qquad \text{Using form 4} \qquad (10)$$

$$x^2 + x + 1 = (Ax + B)(x^2 + 1) + (Cx + D) \qquad \text{Clearing fractions} \qquad (11)$$

For $x = 0$ Equation (13) becomes

$$1 = B + D \qquad \text{Substituting } x = 0 \text{ in (11)}$$

$$3 = 2A + 2B + C + D \qquad \text{Substituting } x = 1 \text{ in (11)}$$

$$7 = 10A + 5B + 2C + D \qquad \text{Substituting } x = 2 \text{ in (11)}$$

$$1 = -2A + 2B - C + D \qquad \text{Substituting } x = -1 \text{ in (11)}$$

We now have four equations and four unknowns. If we substitute $D = 1 - B$ in the last three equations, the system will reduce to

$$\begin{aligned} 2 &= 2A + B + C \\ 3 &= 5A + 2B + C \\ 0 &= -2A + B - C \end{aligned}$$

Solving this system, we arrive at the solution

$$A = D = 0 \qquad \text{and} \qquad B = C = 1$$

Therefore,

$$\frac{x^2 + x + 1}{(x^2 + 1)^2} = \frac{1}{x^2 + 1} + \frac{x}{(x^2 + 1)^2}$$

COMMENT In the examples shown, the coefficients always worked out to be integers. This is not always the case. Rational as well as irrational values can occur. The examples displayed two methods of attack but favored the second. Neither is superior to the other, and both methods should be practiced.

The form in which a particular problem resolved itself was due to the theorem. Depending on the factors of the denominator, any or all parts of the theorem can be used.

Example 7 $\dfrac{x^4 + 3x^2 + 6x + 1}{(x - 1)(x - 2)^2(x^2 + 1)(x^2 + x + 1)^2}.$

This rational fraction decomposes into

$$\underbrace{\frac{A}{x+1}}_{\text{Form 1}} + \underbrace{\frac{B}{(x-2)} + \frac{C}{(x-2)^2}}_{\text{Form 2}} + \underbrace{\frac{Dx+E}{x^2+1}}_{\text{Form 3}} + \underbrace{\frac{Fx+G}{x^2+x+1} + \frac{Hx+I}{(x^2+x+1)^2}}_{\text{Form 4}}$$

Example 8 Suppose $f(x)/g(x)$ is not a proper fraction.

Then all we need do is divide $g(x)$ into $f(x)$ and obtain a quotient containing a rational term which is proper.

$$f(x)/g(x) = \frac{x^3 + 6x^2 + 2x - 1}{x^2 + 3x + 2}$$

$$\begin{array}{r|l} & x + 3 \\ \hline x^2 + 3x + 2 & x^3 + 6x^2 + 2x - 1 \\ & \underline{x^3 + 3x^2 + 2x} \\ & 3x^2 \qquad\quad - 1 \\ & \underline{3x^2 + 9x + 6} \\ & - 9x - 7 \end{array}$$

$$f(x)/g(x) = x + 3 - \frac{9x + 7}{x^2 + 3x + 2}$$

It then follows that

$$\frac{x^3 + 6x^2 + 2x - 1}{x^2 + 3x + 2} = x + 3 + \frac{A}{x+2} + \frac{B}{x+1}$$

You can finish this problem as a warm-up to the following exercises.

EXERCISES

Decompose the rational fractions in Exercises 1 to 21 into partial fractions.

1 $\dfrac{2x + 1}{x(x + 2)(x - 1)}$

2 $\dfrac{x - 2}{x^2(x - 1)^2}$

3 $\dfrac{4 - 3x}{(2x - 1)(x^2 + 1)}$

4 $\dfrac{x^2 - 2x + 1}{(x^2 + 1)^2}$

5 $\dfrac{x^3 + 3x - 2}{x^2 - x - 2}$

6 $\dfrac{x^2 + 2x + 3}{x^3 - x^2 + x - 1}$

7 $\dfrac{x^2 - 2}{x(x - 1)^2}$

8 $\dfrac{3x^2 + 11x + 4}{x^3 + 4x^2 + x - 6}$

9 $\dfrac{(x + 2)(x - 4)}{x^3 - x^2 + 2}$

10 $\dfrac{2x^3 - x + 1}{(x^2 + 1)^2}$

11 $\dfrac{x^3 + 1}{x(x - 1)^3}$

12 $\dfrac{-x - 2}{(x + 1)(x^2 + 1)}$

13 $\dfrac{4x + 23}{(x - 3)(x + 2)(x + 4)}$

14 $\dfrac{2x^2 + 6x - 4}{x(x + 2)^2}$

15 $\dfrac{7x^2 - 25x + 6}{(3x - 2)(x^2 - 2x - 1)}$

16 $\dfrac{17x - 45}{x^3 - 2x^2 - 15x}$

17 $\dfrac{3x^2 - 8x + 9}{(x - 2)^3}$

18 $\dfrac{2x^2 + 7x + 23}{(x - 1)(x + 3)^2}$

19 $\dfrac{5x^2 - 8}{x(x^2 + 2x - 4)}$

20 $\dfrac{5x^2 + 8x + 21}{(x + 1)(x^2 + x + 6)}$

21 $\dfrac{2x^2 + 10x - 3}{x^3 + x^2 - 9x - 9}$

22 Examine the following problem, and find the error.

$$\frac{x^2 + 1}{x^2 - 1} = \frac{a}{x - 1} + \frac{b}{x + 1}$$

$$x^2 + 1 = a(x + 1) + b(x - 1) = (a + b)x + (a - b)$$

$$a + b = 1 \qquad \text{and} \qquad a - b = 1$$

Therefore $a = 1$ and $b = 0$, so

$$\frac{x^2 + 1}{x^2 - 1} = \frac{1}{x - 1}$$

Constructing Functions

7.1 ALGEBRAIC OPERATIONS

Can we add, subtract, multiply, or divide two or more known functions and obtain a new function? In order to answer this question let us first take a look at a practical problem.

A man fills his pool on the day his neighbor leaves for a summer vacation, when he can make use of his neighbor's garden hose. This outlet fills the pool so that the water level (in centimeters) at any time t (in seconds) is

$$f(t) = \frac{t}{60}$$

His own outlet, alone, can fill the pool according to the function

$$g(t) = \frac{t}{50}$$

In order to fill the pool quickly he uses both outlets, and so the water level at any time t is given by

$$h(t) = \frac{t}{60} + \frac{t}{50} = \frac{11t}{300}$$

If our friend in his excitement forgets to close the drain, this is known to decrease the water level according to the function $d(t) = t/40$. Then the water level increases according to the function

$$k(t) = \frac{11t}{300} - \frac{t}{40} = \frac{7t}{600}$$

In this example we actually formed a new function $h(t)$ by adding two functions $f(t)$ and $g(t)$. Likewise we constructed another function $k(t)$ by subtracting $d(t)$ from $h(t)$. Of course these functions do not make any practical sense unless $t > 0$; that is, $d(t)$ has no meaning unless there is water in the pool to be drained. Hence we see the need for a *common domain.*

It seems reasonable that functions can also be constructed by using products and quotients of other functions. In general, this is true, and we state the following definition.

DEFINITION 1 Consider two functions $f(x)$ and $g(x)$ that have a common domain D; then:
(a) $(f + g)(x) = f(x) + g(x)$.
(b) $(f - g)(x) = f(x) - g(x)$.
(c) $(fg)(x) = f(x)g(x)$.
(d) $(f/g)(x) = f(x)/g(x)$, excluding any x for which $g(x) = 0$.

COMMENT $(f + g)(x)$ is read "f plus g of x," and $(f - g)(x)$ is read "f minus g of x," and so on.

Example 1 If $f(x) = 2x$ and $g(x) = x - 1$, find $(f + g)(x)$.

Solution

$$(f + g)(x) = f(x) + g(x) = 2x + x - 1 = 3x - 1$$

Example 2 For $f(x)$ and $g(x)$ defined as in Example 1, find $(fg)(x)$.

Solution

$$(fg)(x) = f(x)g(x) = 2x(x - 1) = 2x^2 - 2x$$

Example 3 For $f(x)$ and $g(x)$ defined as in Example 1, find $(f/g)(x)$.

Solution

$$\frac{f}{g}(x) = \frac{f(x)}{g(x)} = \frac{2x}{x - 1}$$

The domain of $(f + g)(x)$ and $(fg)(x)$ is the same, namely, all real numbers. The domain of $(f/g)(x)$ is all real numbers *excluding* $x = 1$.

Example 4 If $f(x) = \sqrt{x-1}$ and $g(x) = 1 - x$, find $(f-g)(x)$ and $(f/g)(x)$.

Unlike the situation in Examples 1 to 3, the domains of these functions differ, and we must first obtain a *common domain.* The domain of $f(x)$ is $x \geq 1$, and the domain of $g(x)$ is all real numbers. Hence, the collection of numbers that satisfy *both* functions is $x \geq 1$. Now

$$\begin{aligned}(f-g)(x) &= f(x) - g(x) = \sqrt{x-1} - (1-x) \\ &= \sqrt{x-1} + x - 1\end{aligned}$$

The domain of $(f \cdot g)(x)$ is $x \geq 1$, and

$$\frac{f}{g}(x) = \frac{f(x)}{g(x)} = \frac{\sqrt{x-1}}{1-x}$$

The domain of $(f/g)(x)$ is $x \geq 1$ and $x \neq 1$, or simply $x > 1$.

7.2 COMPOSITE FUNCTIONS

Let us define $g(x)$ as

$$g(x) = \frac{1}{x+1} \qquad \text{for } x = 0, 1, 2$$

Then the range of $g(x)$ is the collection of three elements $\frac{1}{3}$, $\frac{1}{2}$, and 1. Suppose now we have a second function $f(x)$ defined as

$$f(x) = \frac{1}{x} \qquad \text{for } = \tfrac{1}{3}, \tfrac{1}{2}, 1$$

Then the range of $f(x)$ is the collection of three elements, namely, 1, 2, and 3. Note that we have used the range of $g(x)$ as the domain of $f(x)$ and in two separate steps we have transformed the elements in the domain of $g(x)$ to correspond to elements in the range of $f(x)$ (see Diagram 1). You can trace 0 to 1, 1 to 2, and 2 to 3.

We can cut out the middleman, so to speak by replacing the independent variable x of $f(x)$ by $g(x)$ and forming what is called the *composite function* of $f(x)$ and $g(x)$, that is,

$$f[g(x)] = \frac{1}{g(x)} = \frac{1}{1/(x+1)} = x + 1$$

For convenience, since $f[g(x)]$ is a function, we can use the symbol $h(x) = f[g(x)]$. Hence,

$$h(x) = x + 1$$

Note that $h(0) = 1$, $h(1) = 2$, and $h(2) = 3$.

One important point that cannot be overlooked is the domain of $f[g(x)]$.

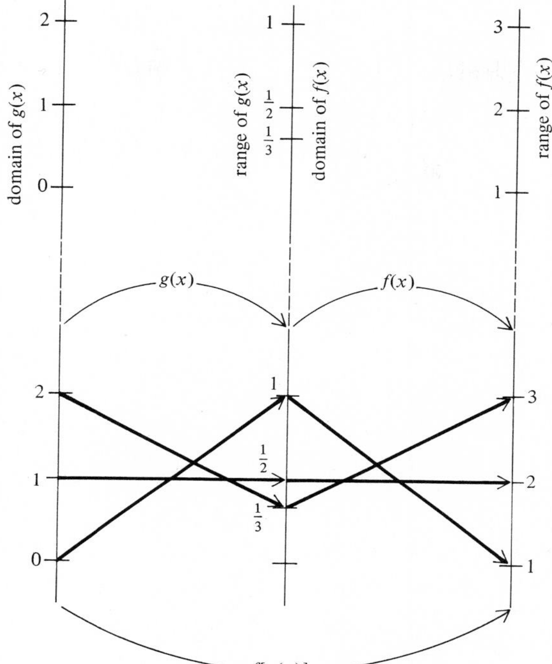

Diagram 1

If you look back over the example we just completed, you will see that the range values of $g(x)$, obtained via the domain values of this function, were in turn elements of the domain of $f(x)$. Therefore the domain of the composite function was equal to the domain of $g(x)$. It should be noted, however, that this will not always be the case. Let us summarize this discussion with the following definition.

DEFINITION 1 The *composite function* of two functions $f(x)$ and $g(x)$ is $f[g(x)]$. The domain of $f[g(x)]$ will be all those values a in the domain of $g(x)$ such that $g(a)$ is a value of the domain of $f(x)$.

Example 1 If $f(x) = x + 1$ and $g(x) = 3x - 2$, determine $f[g(x)]$.

Solution $f[g(x)] = g(x) + 1 = (3x - 2) + 1 = 3x - 1$

The domain is all real numbers since $g(x)$ has a domain of all real numbers and for each real number in the domain of $g(x)$, $3x - 2$ is a real number and contained in the domain of $f(x)$.

Example 2 Use $f(x)$ and $g(x)$ as defined in Example 1 to determine $g[f(x)]$.

Solution

$$g[f(x)] = 3f(x) - 2 = 3(x + 1) - 2 = 3x + 1$$

Note carefully that

$$f[g(x)] = 3x - 1 \neq 3x + 1 = g[f(x)].$$

In general, $f[g(x)] \neq g[f(x)]$.

Example 3 If $f(x) = \sqrt{x + 1}$ and $g(x) = x^2 - 2x$, determine $f[g(x)]$.

The domain of $g(x)$ is all real numbers. Its range is all real numbers greater than or equal to -1. The domain of $f(x)$ is all real numbers x, $x \geq -1$. Hence, the entire range of $g(x)$ is contained in the domain of $f(x)$. Therefore the domain of $f[g(x)]$ is equal to the domain of $g(x)$, which is all real numbers, and

$$f[g(x)] = \sqrt{g(x) + 1} = \sqrt{x^2 - 2x + 1} = \sqrt{(x - 1)^2} = |x - 1|$$

Example 4 Use $f(x)$ and $g(x)$ in Example 3 to determine $g[f(x)]$.

For the domain of this composite function we must first consider the domain of $f(x)$ and find all those real values x such that $f(x)$ belongs to the domain of $g(x)$.

For each value x, $x \geq 1$, $f(x) \geq 0$. Hence the domain of $g[f(x)]$ equals the domain of $f(x)$, which is all values x, $x \geq 1$. Now

$$g[f(x)] = [f(x)]^2 - 2f(x) = (\sqrt{x + 1})^2 - 2\sqrt{x + 1} = x + 1 - 2\sqrt{x + 1}$$

COMMENT Again we see that $f[g(x)] \neq g[f(x)]$.

After working through the last few examples, you may think that finding the domain of a composite function can easily be accomplished by observing the final formula obtained. For example, $f[g(x)] = |x - 1|$ would automatically have a domain of all real numbers and $g[f(x)] = x + 1 - 2\sqrt{x + 1}$ contains a square root; hence $x + 1 \geq 0$ or $x \geq -1$. For these two examples it works, but does it always work? Let us see in the next example.

Example 5 Determine $f[g(x)]$ if $f(x) = 1/(x + 1)$ and $g(x) = 1/x$.

Solution

$$f[g(x)] = \frac{1}{g(x) + 1} = \frac{1}{1/x + 1} = \frac{x}{x + 1}$$

From this formula you would simply claim that the domain of $f[g(x)]$ is all real numbers excluding $x = -1$. However, this is *incorrect*. Let us now follow the definition and determine the correct domain.

STEP 1 The domain of $g(x)$ is all real numbers x, $x \neq 0$. Hence $x = 0$ is immediately excluded from the domain of $f[g(x)]$.

STEP 2 For any real number in the domain of $g(x)$, does $1/x = -1$? Note that -1 is not a member of the domain of $f(x)$. Solving $1/x = -1$, we obtain

$$x = -1$$

STEP 3 $x = -1$ must be excluded from the domain of $g(x)$. Hence the domain of $f[g(x)]$ is all real numbers x, $x \neq 0$, $x \neq -1$.

EXERCISES

In Exercises 1 to 10 for each pair of functions find
(a) $(f + g)(x)$ **(b)** $(f - g)(x)$ **(c)** $(fg)(x)$ **(d)** $(f/g)(x)$
(e) The domain for parts (*a*) to (*d*)

1 $f(x) = 2x$; $g(x) = (x - 1)^2$

2 $f(x) = \sqrt{x}$; $g(x) = x$

3 $f(x) = \sqrt{x - 1}$; $g(x) = 5$

4 $f(x) = \dfrac{x + 1}{x}$; $g(x) = \dfrac{1}{x}$

5 $f(x) = \dfrac{x}{x - 1}$; $g(x) = \dfrac{x - 1}{x}$

6 $f(x) = \sqrt{x}$; $g(x) = x^2$

7 $f(x) = \sqrt{x}$; $g(x) = |x|$

8 $f(x) = x^2$; $g(x) = -\dfrac{1}{x^2}$

9 $f(x) = \sqrt{1 - x}$; $g(x) = \sqrt{2 + x}$

10 $f(x) = \sqrt{1 - x}$; $g(x) = \sqrt{x}$

11 Consider $f(x) = \dfrac{x + 1}{x - 1}$ and $g(x) = \dfrac{1}{x}$. Find
(a) $(f - g)(2)$ **(b)** $(f/g)(0)$

12 Consider $f(x) = x^2$ and $g(x) = \sqrt{2x - 1}$. Find
(a) $(f + g)(\frac{1}{2})$ **(b)** $(fg)(\frac{1}{2})$ **(c)** $(f/g)(\frac{1}{2})$ **(d)** $f[g(\frac{1}{2})]$

13 Consider $f(x) = x^2 - x$ and $g(x) = x - x^2$. Find
(a) $(f + g)(x)$ **(b)** The domain of $(f + g)(x)$

14 Using the information at the beginning of Section 7.1, solve the dishonest-neighbor pool problem for the following cases:
(a) He uses only his outlet.
(b) He uses both outlets.
(c) He uses both outlets and forgets to close the drain. His pool is:
(1) A rectangular pool 5 meters long, 3 meters wide, and 2 meters deep.
(2) A circular pool 4 meters deep with a 3-meter radius.

In Exercises 15 to 24 find **(a)** $f[g(x)]$ and **(b)** its domain.

15 $f(x) = 6x$; $g(x) = x - 4$

16 $f(x) = x^2 - 2$; $g(x) = x + 1$

17 $f(x) = 2$; $g(x) = \sqrt{x}$

18 $f(x) = \sqrt{2x}$; $g(x) = 2$

19 $f(x) = \sqrt{x}$; $g(x) = \dfrac{1}{x}$

20 $f(x) = \frac{1}{2}x$; $g(x) = \dfrac{1}{\sqrt{x}}$

21 $f(x) = \dfrac{x + 1}{x}$; $g(x) = \dfrac{x}{x + 1}$

22 $f(x) = \sqrt{4 - 2x}$; $g(x) = x^2 + 5$

23 $f(x) = \sqrt{\dfrac{x + 1}{x - 1}}$; $g(x) = \sqrt{x}$

24 $f(x) = \sqrt{4 - 2x}$; $g(x) = 2x - \dfrac{x^2}{2}$

25 Consider: $f(x) = mx + b$ and $g(x) = px + c$.
 (a) Find the slope of the graph of $f[g(x)]$.
 (b) Find the slope of the graph of $g[f(x)]$.
 (c) Are the slopes different?

26 Using the functions in Exercise 25:
 (a) Find the point where the graph of $f[g(x)]$ intersects the y axis.
 (b) Find the point where the graph of $g[f(x)]$ intersects the y axis.
 (c) Are these points different?

7.3 INVERSE FUNCTIONS

In Section 2.4 we discussed the cube-root function

$$f(x) = \sqrt{x}$$

If we choose a few specific values for x, say -8, -1, 0, 1, 8, we can calculate five ordered pairs of numbers that belong to this function. They are $(-8, -2)$, $(-1, -1)$, $(0, 0)$, $(1, 1)$, and $(8, 2)$. Similarly, we can choose the power function

$$g(x) = x^3$$

However, $(-2, -8)$, $(-1, -1)$, $(0, 0)$, $(1, 1)$, and $(2, 8)$ belong to this function. We could go on indefinitely generating as many ordered pairs as we care to; but for each ordered pair (a, b) of the cubing function we would find the ordered pair (b, a) for the cube-root function. Hence their domains and ranges are interchanged, and for $g(a) = b, f(b) = a$ for all a in the domain of $g(x)$. That is, $g(-2) = -8$ and $f(-8) = -2$.

We see here that the cubing function and the cube-root function undo each other. Therefore when we form the composite of $g(x)$ and $f(x)$, the result is

$$g[f(x)] = (\sqrt[3]{x})^3 = x \qquad \text{and} \qquad f[g(x)] = \sqrt[3]{x^3} = x$$

for all x in their domains.

When this occurs, we say that $g(x)$ is the *inverse* of $f(x)$ and $f(x)$ is the *inverse* of $g(x)$.

It is customary to denote the inverse of a function $f(x)$ by the symbol $f^{-1}(x)$, read "f inverse of x." The -1 is not to be taken as an exponent. For the functions we just discussed $g(x) = f^{-1}(x)$ and $f(x) = g^{-1}(x)$. We can now summarize this discussion with a definition.

DEFINITION 1 Consider two functions $f(x)$ and $g(x)$. If $f[g(x)] = x$ for each x in the domain of $g(x)$, and if $g[f(x)] = x$ for each x in the domain of $f(x)$, then $f(x)$ and $g(x)$ are said to be *inverses of each other,* and we write $f(x) = g^{-1}(x)$ and $g(x) = f^{-1}(x)$.

It is required that the domain of a function $f(x)$ be the range of its inverse function $f^{-1}(x)$ and the range of $f^{-1}(x)$ be the domain of $f(x)$. This sometimes creates a problem. We already know that in order to have a function the rule must determine a unique y in the range for each element x in its domain; however, if a function is to have an inverse, we must also have a unique x in its domain for each y in the range. If a function possesses this unique reversible correspondence, we say it is *one to one.*

DEFINITION 2 A function $f(x)$ is *one to one* if for each x and y belonging to its domain $f(x) = f(y)$ implies $x = y$.

To emphasize what we just discussed we state the following theorem.

Theorem 1 A function $f(x)$ has an inverse if and only if it is one to one.

We can determine whether or not a function $f(x)$ is one to one by making direct use of the definition. However there is another method often used when the graph of a function is easily accessible. That is to say, if this function is one to one, any horizontal line will intersect its graph in at most one point. This is known as the *horizontal-line test.*

Example 1 Show that $f(x) = 3x - 6$ is one to one.
For any real numbers x and y if $f(x) = f(y)$,

$$3x - 6 = 3y - 6 \Longrightarrow 3x = 3y \Longrightarrow x = y$$

Example 2 Show that $f(x) = \sqrt{x}$ is one to one.
The domain of $f(x)$ is $x \geq 0$. Hence if $f(x) = f(y)$,

$$\sqrt{x} = \sqrt{y} \quad \text{and} \quad x = y \qquad \text{Squaring both sides}$$

Example 3 Show that $g(x) = x^2$ is not one to one.
This can be done conveniently several different ways.

1 For any real number x and y if $g(x) = g(y)$

$$x^2 = y^2 \Longrightarrow x = y \text{ or } x = -y$$

but the definition does not allow $x = -y$. Hence $g(x)$ is *not* one to one.

2 By counter example

$$g(2) = g(-2) \qquad \text{but} \qquad 2 \neq -2$$

3 By the horizontal-line test (see Diagram 2). The line $y = c$ meets the graph at two points $(-c, c^2)$ and (c, c^2).

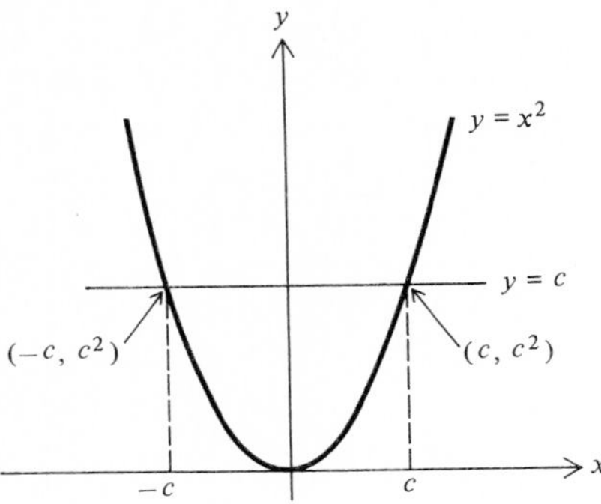

Diagram 2

The following examples obtain the inverse of a given function.

Example 4 Find $f^{-1}(x)$ given $f(x) = 2x - 1$.

Since $f(x)$ is one to one and has a range of all real numbers, $f^{-1}(x)$ now has a domain of all real numbers and by definition $f[f^{-1}(x)] = x = 2f^{-1}(x) - 1$. Hence, solving for $f^{-1}(x)$, we have

$$2f^{-1}(x) - 1 = x$$
$$2f^{-1}(x) = x + 1$$
$$f^{-1}(x) = \frac{x+1}{2}$$

Example 5 Find $f^{-1}(x)$ given $f(x) = x^2$, $x < 0$.

$$f[f^{-1}(x)] = x = [f^{-1}(x)]^2 \qquad \text{for each } x < 0$$
$$f^{-1}(x) = \sqrt{x} \text{ or } -\sqrt{x}$$

We reject $f^{-1}(x) = \sqrt{x}$ and write

$$f^{-1}(x) = -\sqrt{x}$$

Keep in mind that the domain of f becomes the range of $f^{-1}(x)$. Since the domain of $f(x)$ was all negative numbers and $\sqrt{x} \geq 0$, we had to reject this result and accept $-\sqrt{x}$ since $-\sqrt{x} \leq 0$.

Example 6 Find $f^{-1}(x)$ given that the range of $f(x)$ is all nonnegative real numbers excluding $y = 1$ and

$$f(x) = \sqrt{\frac{x+1}{x-1}}$$

Determine whether $f(x)$ is one to one. That is, $f(x) = f(y) \Longrightarrow x = y$.

$$\sqrt{\frac{x+1}{x-1}} = \sqrt{\frac{y+1}{y-1}}$$

$$(x + 1)(y - 1) = (x - 1)(y + 1) \qquad \text{Squaring both sides and cross multiplying}$$
$$x = y \qquad \text{Simplifying}$$

Therefore $f(x)$ is one to one and $f^{-1}(x)$ exists. Now

$$f[f^{-1}(x)] = x = \sqrt{\frac{f^{-1}(x) + 1}{f^{-1}(x) - 1}} \qquad \text{or} \qquad \sqrt{\frac{f^{-1}(x) + 1}{f^{-1}(x) - 1}} = x$$

$$\frac{f^{-1}(x) + 1}{f^{-1}(x) - 1} = x^2 \qquad \text{Squaring both sides}$$

$$f^{-1}(x) + 1 = x^2[f^{-1}(x) - 1] \qquad \text{Multiplying through by } f^{-1}(x) - 1$$

or

$$f^{-1}(x) + 1 = x^2 f^{-1}(x) - x^2 \qquad \text{Distributing } x^2 \text{ on the right side}$$

$$x^2 f^{-1}(x) - f^{-1}(x) = x^2 + 1 \qquad \text{Collecting the } f^{-1}(x) \text{ terms}$$

or

$$(x^2 - 1)f^{-1}(x) = x^2 + 1 \qquad \text{Factoring the left-hand side}$$

$$f^{-1}(x) = \frac{x^2 + 1}{x^2 - 1} \qquad \text{Dividing through by } x^2 - 1$$

For completeness we determine the domain of $f^{-1}(x)$. The range of $f(x)$ was stated in the problem. Since the domain of $f^{-1}(x)$ is the range of $f(x)$, we simply write

$$f^{-1}(x) = \frac{x^2 + 1}{x^2 - 1} \qquad x \geq 0, x \neq 1$$

We have one thing left to discuss, namely, the graph of a function and its inverse. For every point (a, b) belonging to the graph of a function, the point (b, a) belongs to the graph of its inverse. The line $y = x$ is the perpendicular bisector of the segment whose end points are (a, b) and (b, a). This means that each point (a, b) is the reflection of (b, a) in the line $y = x$. Hence the graph of $f^{-1}(x)$ can be obtained by reflecting the graph of $f(x)$ in this line.

We illustrate this graphing technique by graphing functions we have made use of in this chapter (see Diagram 3).

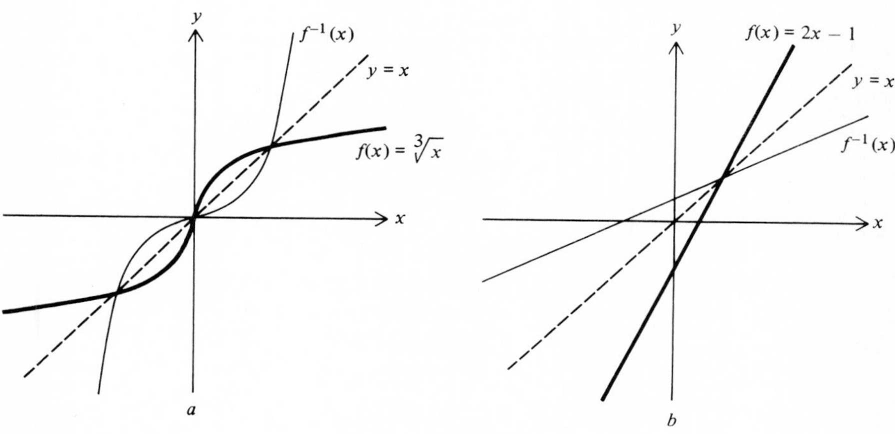

Diagram 3

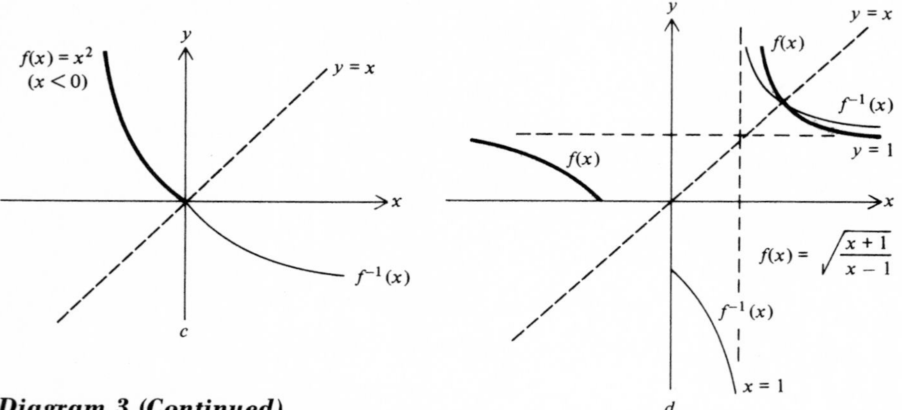

Diagram 3 (Continued)

EXERCISES

In Exercises 1 to 12, find $f^{-1}(x)$.

1 $f(x) = 2x$

2 $f(x) = 2 - x$

3 $f(x) = 4x - 3$

4 $f(x) = x^2 + 1, x \geq 0$

5 $f(x) = x^2 + x, x \leq -\frac{1}{2}$

6 $f(x) = x^2 - 2x + 1, x > 1$

7 $f(x) = \sqrt{x}$

8 $f(x) = \dfrac{1}{x}$

9 $f(x) = \dfrac{1}{x + 4}$

10 $f(x) = |x - 1|, x \geq 1$

11 $f(x) = \dfrac{2 - x}{1 + x}$

12 $f(x) = \sqrt{\dfrac{x - 2}{x + 2}}$

Show that the members of each pair of functions are inverses of each other:

13 $\sqrt{x}, x^2$

14 $\dfrac{1}{x}, \dfrac{1}{x}$

15 $\dfrac{1}{x^2}, \dfrac{\sqrt{x}}{x}$

16 $\sqrt[3]{x - 1}, x^3 + 1$

Using the definition for one-to-one functions, prove that each of the functions in Exercises 17 to 21 is one to one.

17 $f(x) = 2x - 3$ **18** $f(x) = \sqrt{x + 10}$ **19** $f(x) = x^3$

20 $f(x) = x^3 + 3x^2 + 3x + 1$ **21** $f(x) = 6 - \frac{3}{4}x$

22–31 Graph $f(x)$ and $f^{-1}(x)$ of Exercises 1 to 10. In each case $f(x)$ and $f^{-1}(x)$ should be on the same grid.

32 Prove that the line $y = x$ is the perpendicular bisector of any line segment having end points (a, b) and (b, a).

Use the horizontal line test to show the functions in Exercises 33 to 36 are not one to one.

33 $f(x) = x^2 + 2x$

34 $f(x) = x^3 - x$

35 $f(x) = \dfrac{1}{x^2}$

36 $f(x) = \dfrac{x - 2}{x^2 + 5x + 6}$

37 Consider $f(x) = x + 4$ and $g(x) = 3x$. Show that

$$(f[g(x)])^{-1} = g^{-1}[f^{-1}(x)]$$

38–41 Restrict the domains of the functions in Exercises 33 to 36 such that they are one to one.

7.4 PARAMETRIC EQUATIONS

Suppose a particle moves along a curve, so that at any time $t \geq 0$ its position is given by the pair of equations

$$x = t + 2 \qquad \text{and} \qquad y = t^2$$

Each value of t will determine a point (x, y), and the totality of all such points generated by letting t take on all its values will determine the graph of the curve. We make up a table to obtain some of the points.

t	x	y
0	2	0
0.5	2.5	0.25
1	3	1
2	4	4
3	5	9

Using these points, we can graph this curve (see Diagram 4).

Note that we do not plot t, only the ordered pair (x, y). t is an extra variable called a *parameter*, and the pair of equations involving t as an independent variable are called *parametric equations*.

DEFINITION 1 Let the coordinates (x, y) of a point on a curve be defined in terms of a third variable t such that

$$x = f(t) \qquad \text{and} \qquad y = g(t)$$

Then t is called a *parameter*, and the functions $x = f(t)$ and $y = g(t)$ are called *parametric equations* of the curve.

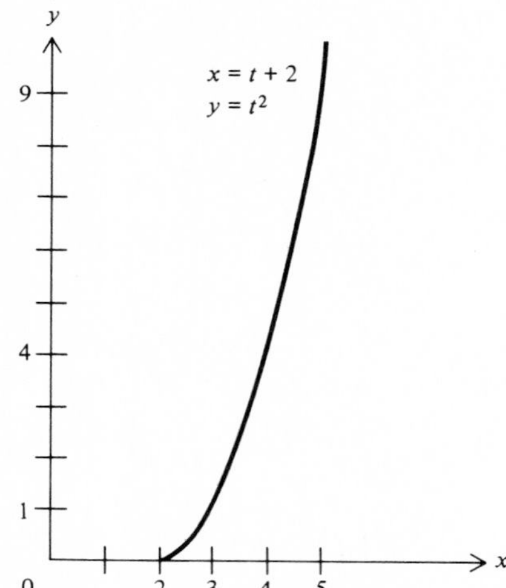

Diagram 4

Example 1 Define the curve discussed above in terms of $y = f(x)$.

We must eliminate the parameter for $x = t + 2$ and $y = t^2$. This can be done by solving for t in one equation and substituting the result in the other equation:

$$x = t + 2 \Longrightarrow t = x - 2$$

Hence $\quad y = t^2 = (x - 2)^2 \quad$ or $\quad y = x^2 - 4x + 4$

Now since $t \geq 0$, $x \geq 2$.

COMMENT Graph $y = (x - 2)^2$ for $x \geq 2$ and check your result with Diagram 4.

Example 2 Write $y = 3x - 2$ in parametric form.
Let $x = t$. Then $y = 3t - 2$.

Example 3 Find two parameterizations of the line $y = x$.
First, let $x = t$; then $y = t$. Second, let $x = t^3$, then $y = t^3$.

COMMENT As you can see, parametric equations for a given curve are not unique. We did not choose $x = t^2$ and $y = t^2$ as parametric equations for $y = x$ because for any real number t, $t^2 \geq 0$ and negative values of x, y would be impossible to obtain.

EXERCISES

In Exercises 1 to 9 eliminate the parameter and obtain an equation relating x to y.

1 $x = t + 1$; $y = t - 1$

2 $x = 4t$; $y = -4t$

3 $x = \sqrt{t}$; $y = t^2$

4 $x = t^2$; $y = \sqrt{t}$

5 $x = t^2 - t;\ y = \dfrac{1}{t}$

6 $x = \dfrac{t}{t+1};\ y = t^2$

7 $x = t^2 - 1;\ y = t^2 + 1$

8 $x = \dfrac{5t}{t+1};\ y = \dfrac{t+1}{t-1}$

9 $x = t^3 - 1;\ y = \dfrac{t^3}{3t^3 - 1}$

10 Graph the portion of the curve represented by $x = 1/t$ and $y = t + 1$ for $0 < t \leq 4$.

11 Eliminate the parameter in:
(a) $x = t$ and $y = t^2$ **(b)** $x = t^2$ and $y = t^4$
Do the two sets of parametric equations represent the same curve?

12 A projectile fired from ground level travels on a path given by $x = 40t$ and $y = 64t - 16t^2$.
(a) Plot the projectile's path for $0 \leq t \leq 4$.
(b) At $t = 3.5$ seconds name its position on the path.
(c) What happens at $t = 4$ seconds?

7.5 ALGEBRAIC FUNCTIONS

Recall that a constant function has the form $g(x) = c$, where c is a real number and the identity function is $f(x) = x$. If we use $f(x) = x$ as a factor 3 times, add this product to $f(x) = x$ used as a factor twice, and then subtract from this sum the constant function $g(x) = 2$, we obtain the polynomial function

$$y = x^3 + x^2 - 2 \tag{1}$$

Suppose we now divide the rule part of Equation (1) by the sum obtained when we add $f(x) = x$ to $g(x) = 4$. This yields a rational function

$$y = \frac{x^3 + x^2 - 2}{x + 4} \tag{2}$$

Further, if we now take the square root of the rule part of both (1) and (2), this yields

$$y = \sqrt{x^3 + x^2 - 2} \tag{3}$$

$$y = \sqrt{\frac{x^3 + x^2 - 2}{x + 4}} \tag{4}$$

We have just generated four different functions all belonging to a particular type, or class. We say these functions are *algebraic in x*.

DEFINITION 1 $f(x)$ is said to be an *algebraic function in x* if its rule is constructed by combining real numbers and x in addition, subtraction, multiplication, division, or raising to a rational power in a finite number of steps.

Functions like those in Equations (1) and (2) have been thoroughly examined in Chapters 5 and 6; however, those like Equations (3) and (4) have not been dealt with in detail. These functions require special care and an orderly procedure. With this in mind, we now analyze Equations (3) and (4).

Example 1 Discuss and graph $y = \sqrt{x^3 + x^2 - 2}$.

STEP 1 Since this function involves a square root, its domain will be limited to those values of x such that $x^3 + x^2 - 2 \geq 0$. If you observe that $x = 1$ is a root of $x^3 + x^2 - 2 = 0$, then it follows that $(x - 1)(x^2 + 2x + 2) \geq 0$. Now $x^2 + 2x + 2 \geq 1$, for all x and the inequality reduces to $x - 1 \geq 0$ or $x \geq 1$.

STEP 2 Zeros: $x = 1$.

STEP 3 There is no y intercept since $x = 0$ is not in the domain.

STEP 4 There are no vertical asymptotes (no crucial values involved).

STEP 5 To find the horizontal asymptotes

$$\sqrt{x^3 + x^2 - 2} = \sqrt{x^3\left(1 + \frac{1}{x} - \frac{2}{x^3}\right)}$$

As $x \to +\infty$, $$\sqrt{x^3\left(1 + \frac{1}{x} - \frac{2}{x^3}\right)} \longrightarrow \sqrt{x^3} \neq \text{constant}$$

Hence no horizontal asymptotes are present.

STEP 6 Since $x \geq 1$ and we are dealing with a square root function, the entire graph of this function is in the first quadrant, but since there are no asymptotes to guide us, we must obtain some points. A table of approximate values follows:

x	1	1.4	1.8	2.2	2.6	3
y	0	1.64	2.66	3.67	4.72	5.83

These points give a fair indication of the behavior of this function. If we now extend a smooth unbroken curve through these points, we obtain the graph of this function (see Diagram 5).

Example 2 Discuss and graph $y = \sqrt{\dfrac{x^3 + x^2 - 2}{x + 4}}$.

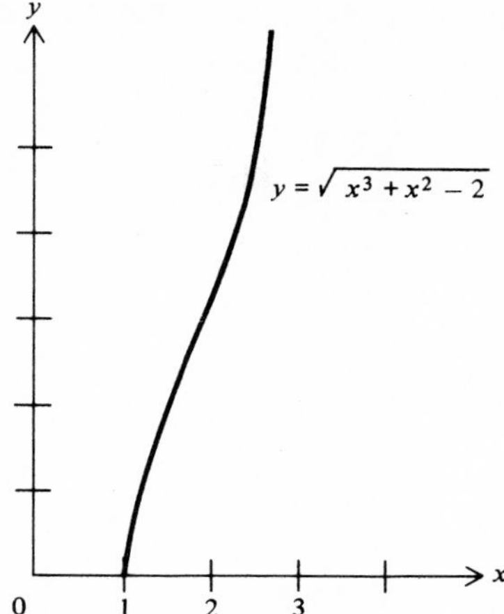

Diagram 5

STEP 1 To find the domain

$$\frac{x^3 + x^2 - 2}{x + 4} \geq 0$$

$$\frac{(x - 1)(x^2 + 2x + 2)}{x + 4} \geq 0 \qquad \text{Factoring}$$

$x < -4$ or $x \geq 1$ Obtaining the solution from the sign pattern (see Diagram 6)

−	−	+	$x - 1$
+	+	+	$x^2 + 2x + 2$
−	+	+	$1/(x + 4)$
+	−	+	$(x - 1)(x^2 + 2x + 2)[1/(x + 4)]$

−4 0 1

Diagram 6

STEP 2 This function has a zero at $x = 1$.

STEP 3 There is no y intercept since $x = 0$ is not in the domain.

STEP 4 We can only approach -4 from the left, and we write this as $x \to -4^-$. Hence $x \to -4^-$ as $y \to +\infty$; therefore $x = 4$ is a vertical asymptote.

STEP 5 To find oblique asymptotes, rewrite

$$\sqrt{\frac{x^3 + x^2 - 2}{x + 4}} \quad \text{as} \quad \sqrt{x^2 - 3x + 12 - \frac{50}{x + 4}} \qquad \text{Dividing } x + 4 \text{ into } x^3 + x^2 - 2$$

Factoring the radicand (the expression under the radical sign) gives

$$y = \sqrt{x^2\left[1 - \frac{3}{x} + \frac{12}{x^2} - \frac{50}{x^2(x+4)}\right]}$$

$$y \longrightarrow \begin{cases} \sqrt{x^2} = |x| = x & \text{as } x \longrightarrow +\infty \\ \sqrt{x^2} = |x| = -x & \text{as } x \longrightarrow -\infty \end{cases}$$

Hence we have *two* oblique asymptotes $y = x$ and $y = -x$.

STEP 6 $y = x$ and $y = -x$ do not intersect the graph of this function. To check:

$$\sqrt{\frac{x^3 + x^2 - 2}{x + 4}} = |x|$$

$$\frac{x^3 + x^2 - 2}{x + 4} = x^2 \qquad \text{Squaring both sides and recalling } |x|^2 = x^2$$

$$x^3 + x^2 - 2 = x^3 + 4x^2 \qquad \text{Multiplying through by } x + 4$$

$$-2 = 3x^2 \qquad \text{Simplifying}$$

$$-\tfrac{2}{3} = x^2$$

impossible since $x^2 \geq 0$ for all real numbers.

Recalling all this information, we can now obtain the graph of this function (see Diagram 7).

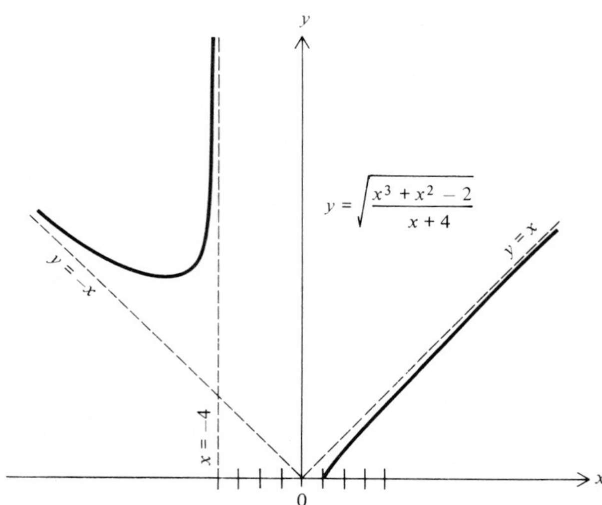

Diagram 7

Example 3 $y = \dfrac{1 - \sqrt{x - 2}}{x + 1}$.

STEP 1 Domain: $x - 2 \geq 0$ and $x \neq -1$; this is equivalent to $x \geq 2$.

STEP 2 Zeros: when $x = 3$.

STEP 3 There is no y intercept since $x = 0$ is not in the domain.

STEP 4 There are no vertical asymptotes. ($x = -1$ is *not* a vertical asymptote since the graph of this function is on and to the right of $x = 2$.)

STEP 5 Horizontal asymptotes: Rationalize the numerator

$$y = \frac{1 - \sqrt{x-2}}{x+1}\,\frac{1 + \sqrt{x-2}}{1 + \sqrt{x-2}} = \frac{3-x}{x+1}\,\frac{1}{1 + \sqrt{x-2}}$$

As $x \to +\infty$,

$$\frac{3-x}{x+1} = \frac{-x+3}{x+1} \longrightarrow -1 \qquad \text{and} \qquad \frac{1}{1+\sqrt{x-2}} \longrightarrow 0$$

Hence as $x \to +\infty, y \to (-1)(0) \to 0$. Therefore $y = 0$ is a horizontal asymptote.

Carefully paying attention to all these found facts, we can now graph the function (see Diagram 8).

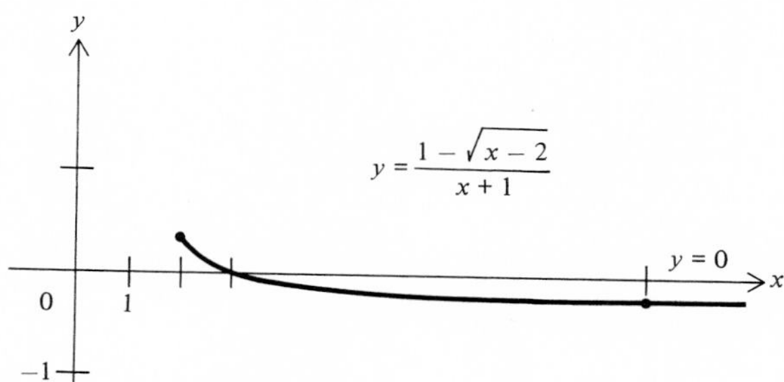

Diagram 8

EXERCISES

Follow the procedure used in the examples of this section to graph the functions in Exercises 1 to 17.

1 $y = \sqrt{x^2 - 2x + 1}$

2 $y = \dfrac{\sqrt{x-2} - 1}{x - 1}$

3 $y = \sqrt{\dfrac{x+1}{x-1}}$

4 $y = \sqrt{\dfrac{x-1}{x+1}}$

5 $y = \sqrt{x} + \sqrt[3]{x}$

6 $y = \dfrac{\sqrt{x-2} - 2}{x - 4}$

7 $y = \dfrac{x+3}{\sqrt{x+2}}$

8 $y = \sqrt{\dfrac{2x}{x-1}}$

9 $y = \dfrac{2x - 1}{\sqrt{x^2 - 1}}$

10 $y = x\sqrt{x^2 - 1}$

11 $y = x^2\sqrt{x + 1}$

12 $y = \dfrac{x}{\sqrt{x - 1}}$

13 $y = x + \sqrt{x^2 - 1}$

14 $y = x - \sqrt{x^2 - 1}$

15 $y = x\sqrt{\dfrac{x}{4 - x}}$

16 $y = \dfrac{x\sqrt{1 - x}}{1 + x}$

17 $y = \dfrac{x - 1}{\sqrt{x(x + 1)}}$

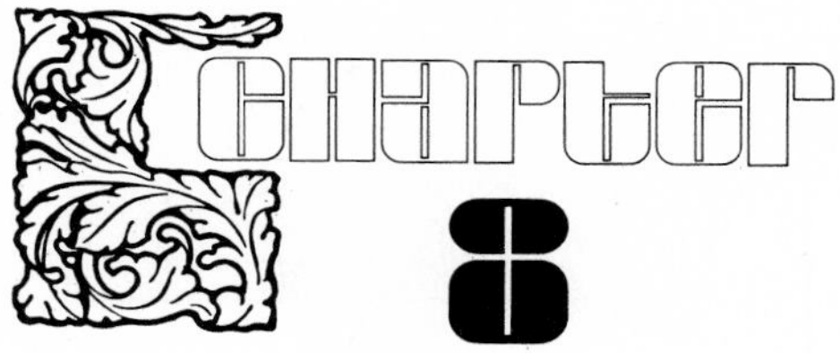

Trigonometry

8.1 INTRODUCTION

Trigonometry is the branch of mathematics that deals with the solution of triangles. This is, if we know one side and any other two parts of a triangle, we can find the remaining three parts. In order to solve triangles, certain relationships must be developed. This will be the object of the first section of this chapter. The remainder will focus on using trigonometry to solve applied problems.

8.2 THE RIGHT TRIANGLE

A triangle composed of two acute angles and one right angle is called a *right triangle.* In Diagram 1 the acute angles are denoted by the capital letters A and B, and the right angle is denoted by the capital letter C. The sides *opposite* these angles are denoted by the small letters a, b, and c, respectively. This notation of angles labeled with capital letters and the sides opposite these

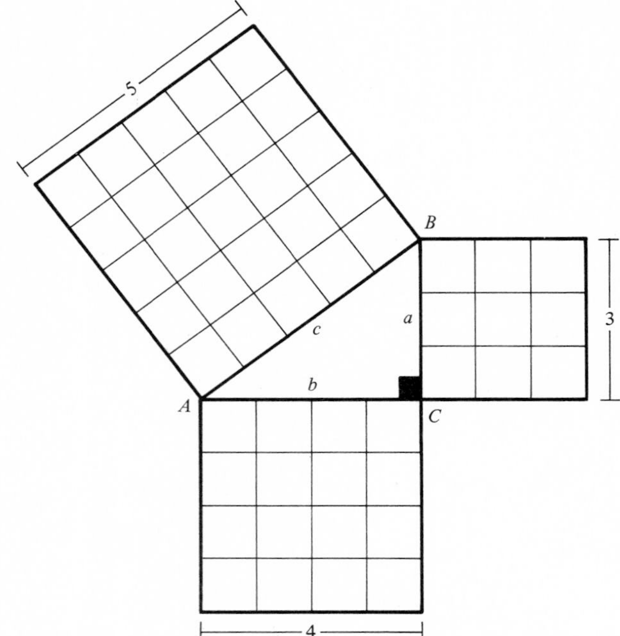

Diagram 1

angles labeled with corresponding small letters will be used throughout our development.

From plane geometry we know that the side opposite the largest angle is the longest side of the triangle. This side is called the *hypotenuse* and is side c in Diagram 1; it is always opposite the right angle. Diagram 1 also can be used to illustrate the first important relationship we need to know, the *pythagorean theorem.*

Pythagorean Theorem In any right triangle the square of the hypotenuse is equal to the sum of the squares of the other two sides. Hence, $c^2 = a^2 + b^2$.

It follows from this formula that $c = \sqrt{a^2 + b^2}$ or $a = \sqrt{c^2 - b^2}$ or $b = \sqrt{c^2 - a^2}$.

Example 1 In each of the triangles drawn in Diagram 2, find the length of the unknown side.

In part (a)

$$c = \sqrt{a^2 + b^2} = \sqrt{3^2 + 4^2} = \sqrt{9 + 16} = \sqrt{25} = 5$$

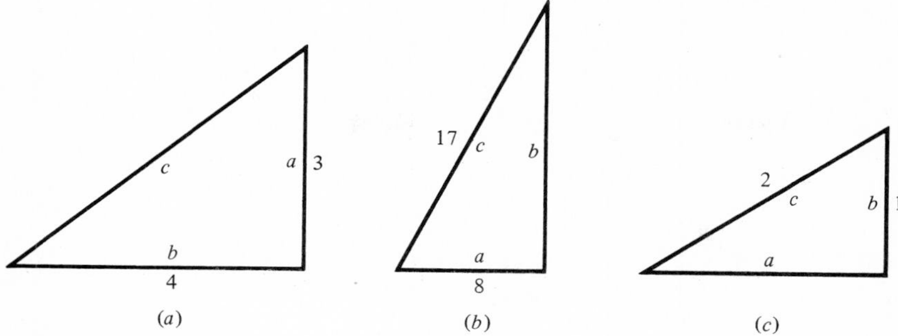

Diagram 2

In part (b)

$$b = \sqrt{c^2 - a^2} = \sqrt{17^2 - 8^2} = \sqrt{289 - 64} = \sqrt{225} = 15$$

In part (c)

$$a = \sqrt{c^2 - b^2} = \sqrt{2^2 - 1^2} = \sqrt{4 - 1} = \sqrt{3} \approx 1.732$$

It should be noted that the acute angles of a right triangle are complementary. In Diagram 1 this means that $A + B = 90°$ (or that $A = 90° - B$ or $B = 90° - A$).

Angles One method of measuring angles is by using *degrees.* It is generally understood that *one degree* (1°) is the measure of an angle formed by two radii of a circle which intercept an arc equal in length to $\frac{1}{360}$ times the circumference of the circle. The degree can be further subdivided into 60 equal parts, called *minutes.* For example, $24\frac{1}{2}$ degrees is the same as 24 degrees, 30 minutes, written 24°30′, where the symbol ′ denotes minutes. (Finer separations can be made in measuring angles, as each minute can again be divided into 60 equal parts called *seconds,* symbolized by ″.)

Example 2 Translate 64°20′.
This reads 64 degrees, 20 minutes, which is the same as 64 degrees plus $\frac{20}{60}$ degree.

Example 3 Subtract 39°40′ from 90°.
Let 90° be written 89°60′ (remember, that 1° = 60′). Then

$$\begin{array}{r} 89°60' \\ -\,39°40' \\ \hline 50°20' \end{array}$$

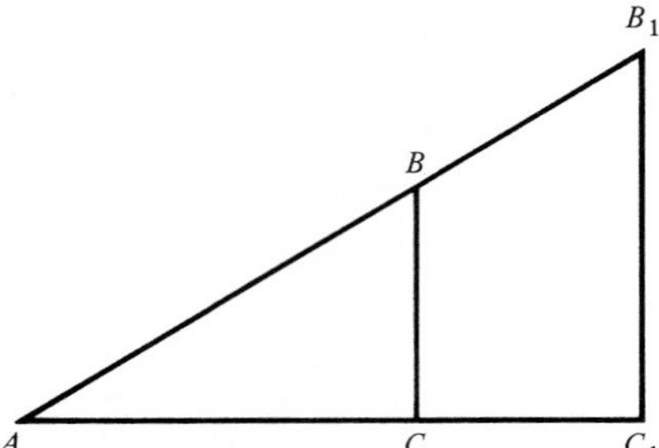

Diagram 3

The Trigonometric Ratios Diagram 3 represents a right triangle *ABC*, where *AB* has been extended to B_1 and *AC* has been extended to C_1 so that *BC* is parallel to B_1C_1. In geometry it can be demonstrated that triangle *ABC* and triangle AB_1C_1 are similar to each other. This implies that their corresponding sides are *proportional*. These proportions can be expressed as follows:

$$\frac{BC}{AB} = \frac{B_1C_1}{AB_1} \tag{1}$$

$$\frac{AC}{AB} = \frac{AC_1}{AB_1} \tag{2}$$

$$\frac{BC}{AC} = \frac{B_1C_1}{AC_1} \tag{3}$$

Regardless of how large we construct a right triangle by extending *AB* and *AC*, the ratios of Equations (1) to (3) will remain *constant* provided that the acute angle *A* remains fixed. These ratios are very important and are known as the *trigonometric ratios*.

Using the right triangle *ABC* in Diagram 4, we can now *define* the trigonometric ratios of acute angle *A*. The left-hand column gives the abbreviations of these ratios.

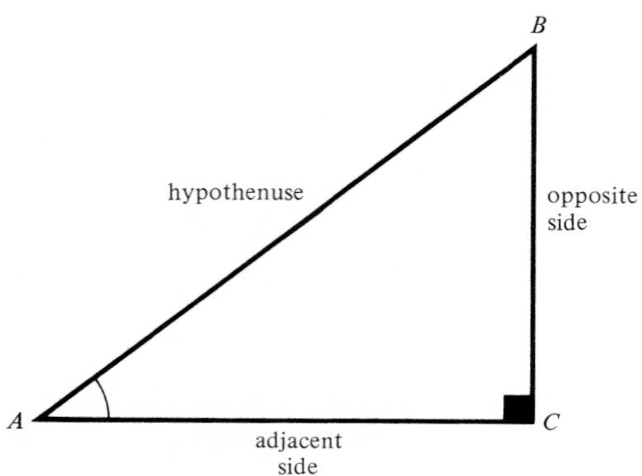

Diagram 4

DEFINITION

$$\sin A = \text{sine of angle } A = \frac{\text{opposite side}}{\text{hypotenuse}}$$

$$\cos A = \text{cosine of angle } A = \frac{\text{adjacent side}}{\text{hypotenuse}}$$

$$\tan A = \text{tangent of angle } A = \frac{\text{opposite side}}{\text{adjacent side}}$$

$$\cot A = \text{cotangent of angle } A = \frac{\text{adjacent side}}{\text{opposite side}}$$

$$\sec A = \text{secant of angle } A = \frac{\text{hypotenuse}}{\text{adjacent side}}$$

$$\csc A = \text{cosecant of angle } A = \frac{\text{hypotenuse}}{\text{opposite side}}$$

Example 4 Given triangle ABC, where $a = 12$ and $b = 5$. If angle $C = 90°$, find the values of *all* the trigonometric ratios of angle A and angle B.
Refer to Diagram 5. To find c, use the pythagorean theorem.

$$c = \sqrt{a^2 + b^2} = \sqrt{(12)^2 + (5)^2} = \sqrt{144 + 25} = \sqrt{169} = 13$$

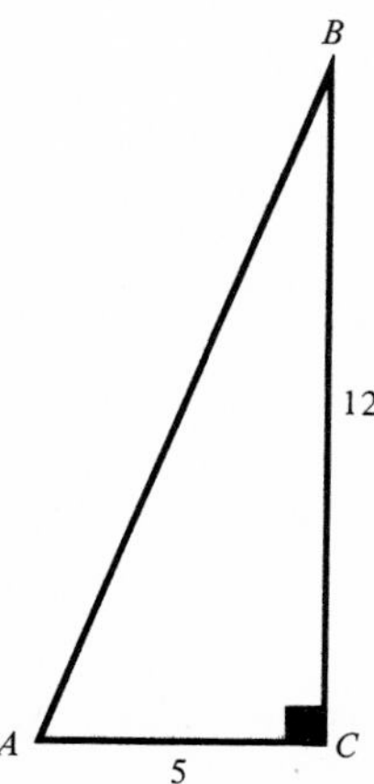

Diagram 5

From the definitions of the trigonometric ratios,

$\sin A = \frac{12}{13}$	$\cos A = \frac{5}{13}$	$\tan A = \frac{12}{5}$
$\cot A = \frac{5}{12}$	$\sec A = \frac{13}{5}$	$\csc A = \frac{13}{12}$
$\sin B = \frac{5}{13}$	$\cos B = \frac{12}{13}$	$\tan B = \frac{5}{12}$
$\cot B = \frac{12}{5}$	$\sec B = \frac{13}{12}$	$\csc B = \frac{13}{5}$

Trigonometric Ratios of 30°, 45°, and 60° Diagram 6 shows an equilateral triangle of side 2 and a square of side 1. Suppose that in the equilateral triangle we construct an altitude to the base of the triangle and in the square we draw a diagonal. In each case we will have formed two triangles (see Diagram 6). In the equilateral triangle, the altitude bisects angle B *and* side AC.

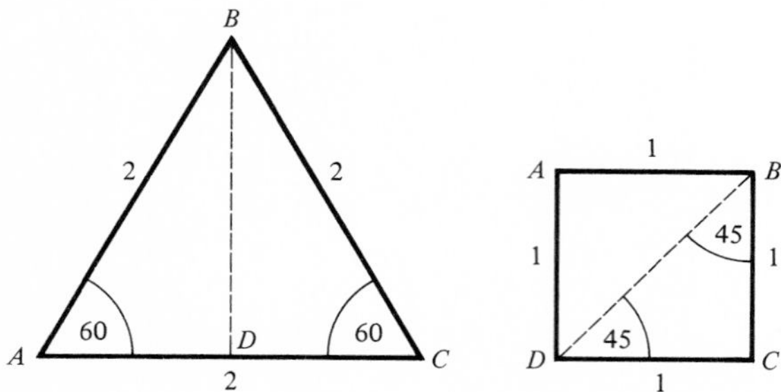

Diagram 6

Hence angle $ABD = 30°$ and $AD = 1$. In the square, the diagonal bisects angle D, and angle $BDC = 45°$. Two *reference triangles* result (Diagram 7).

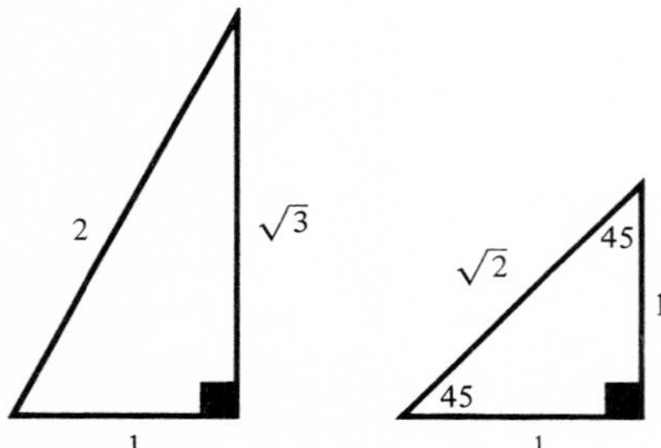

Diagram 7

One is called a 30 −60 −90 right triangle; the other is called a 45 −45 −90 right triangle. Note that the side opposite the 60° angle can easily be found by using the pythagorean theorem:

$$c^2 = a^2 + b^2 \qquad a^2 = c^2 - b^2$$

$$a = \sqrt{c^2 - b^2} = \sqrt{4 - 1} = \sqrt{3}$$

Likewise the hypotenuse of the 45° right triangle can be determined using $c = \sqrt{a^2 + b^2} = \sqrt{1 + 1} = \sqrt{2}$.

These triangles should be *memorized*. It will then be easy to find the values of any trigonometric ratio involving these special angles of 30, 45, and 60°.

Example 5 Using the reference triangles in Diagram 7, find **(a)** cos 60°, **(b)** tan 45°, and **(c)** sec 30°.

PART (*a*) The cosine of an angle is defined to be the ratio of the adjacent side to the hypotenuse. Therefore cos 60° = $\frac{1}{2}$.

PART (*b*)

$$\tan 45° = \frac{1}{1} = 1$$

PART (*c*)

$$\sec 30° = \frac{2}{\sqrt{3}}$$

The table below summarizes all possible values for the trigonometric ratios involving 30, 45, or 60°. The reader is urged to verify the results.

θ	30°	60°	45°
sin θ	$\frac{1}{2}$	$\frac{\sqrt{3}}{2}$	$\frac{1}{\sqrt{2}}$
cos θ	$\frac{\sqrt{3}}{2}$	$\frac{1}{2}$	$\frac{1}{\sqrt{2}}$
tan θ	$\frac{1}{\sqrt{3}}$	$\sqrt{3}$	1
cot θ	$\sqrt{3}$	$\frac{1}{\sqrt{3}}$	1
sec θ	$\frac{2}{\sqrt{3}}$	2	$\sqrt{2}$
csc θ	2	$\frac{2}{\sqrt{3}}$	$\sqrt{2}$

Using Trigonometric Tables Involving Degrees A question which immediately arises is: How are the ratios of acute angles *not equal to* 30, 45, or 60° generally found? The answer is that we use a table.

A four-place table that graduates angles in increments of 10 minutes will be sufficient for the material in this chapter. When the angle in question is *less than 45°*, we refer to the row headings at the *top* of the page and the angle will be found in the *left*-hand column (see Appendix Table 6). Note that as you read *down* the page the angles *increase* in size. If the angle in question is *greater than 45°*, we refer to the row headings at the *bottom* of the page and the angle will be found in the *right*-hand column. As you read *up* the page, the angles *increase* in size.

Example 6 Find **(a)** tan 20°40′ and **(b)** sin 69°10′.

PART (*a*) 20°40′ is found in the left-hand column. Using the Tan heading on the *top* of the page (since the angle is less than 45°), we find tan 20°40′ = 0.3772.

PART (*b*) Since 69°10′ is greater than 45°, we must use the right-hand column. It is absolutely necessary to make use of the heading at the *bottom* of the page and remember to read *up*. Doing this gives sin 69°10′ = 0.9346.

COMMENT If you thought that 0.9283 was the correct answer, you forgot to read *up*.

Suppose we now consider the above example is reverse. If cos $A = 0.3827$, how do we find the measure of angle A? This will require us to search the columns labeled Cos until we locate the given value. In this case 0.3827 is found in the column as you read *up*. This means that we must use the right-hand column to find the required angle. Hence, cos $A = 0.3827$ and $A = 67°30'$.

COMMENT If you thought $A = 68°30'$, you forgot to read *up*. If you thought $A = 22°30'$ you forgot to read the *right*-hand column.

A final explanation will now be given for using the table in instances when the *exact value* is not in the table.

Example 7 sin $A = 0.3533$. Find to the nearest 10 minutes the measure of angle A.

This particular value is not in the table, but we do find the entries 0.3529 and 0.3557. These two numbers bound the given number 0.3533. We will then choose the number *closest* to 0.3533. In this case it is 0.3529, and the angle which corresponds to this value is 20°40′. Hence if sin $A = 0.3533$, then $A = 20°40'$ to the *nearest 10 minutes.*

Solving a Right Triangle Let us now consider the concepts just discussed and solve three problems involving right triangles.

Example 8 Given triangle ABC, where $A = 20°$, $a = 8$, and $C = 90°$. Find side b to the nearest integer (refer to Diagram 8).

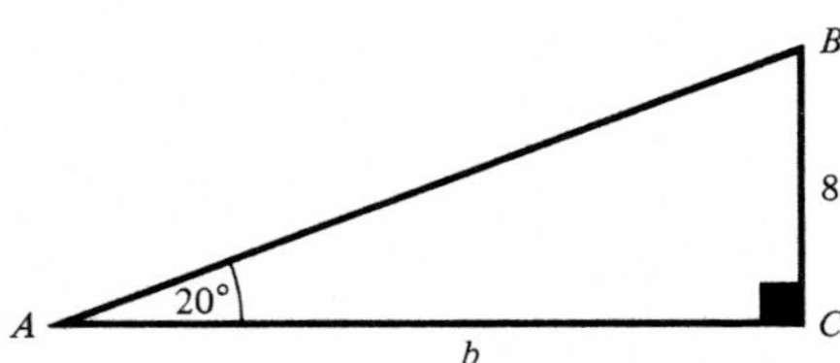

Diagram 8

Since the given side is *opposite* the given angle and the required side is *adjacent* to the given angle, we can use either the cotangent or the tangent. In this problem, let us use cot A.

$\cot A = \dfrac{b}{a}$	Definition
$a \cot A = b$	Multiplying both sides by a
$8 \cot 20° = b$	Substituting the given information
$8(2.7475) = b$	From trigonometric tables
$21.9800 = b$	
$22 = b$	Rounding off to the nearest integer

COMMENT The tangent ratio could have also been used in this problem: $\tan A = a/b$. Solving for b gives

$$b = \frac{a}{\tan A} = \frac{8}{0.3640} = 22$$

We have avoided using the tan ratio to bypass the operation of division. (This precaution is unnecessary when using a calculator.)

Example 9 Given triangle *ABC,* where $B = 62°10'$, $b = 10$, and $C = 90°$. Find side c to the nearest tenth (refer to Diagram 9).

Since the given side is *opposite* the given angle and the required side is the *hypotenuse,* we can use the sine ratio.

$\sin B = \dfrac{b}{c}$	Definition
$c = \dfrac{b}{\sin B}$	Solving for c
$c = \dfrac{10}{\sin 62°10'}$	Substituting the given information
$c = \dfrac{10}{0.8843} = 11.308$	
$= 11.3$	Rounding off to the nearest tenth

COMMENT As in Example 8, we could have used an alternate method for finding c. Since $\csc B = c/b$, $c = b \csc B$ or $c = 10(1.131) = 11.3$.

Example 10 Given a right triangle *ABC,* where $a = 3$, $b = 4$, and $c = 5$. Find A and B to the nearest 10 minutes (refer to Diagram 10).

We may use *any* of the six trigonometric ratios to find A. Two different ways are as follows.

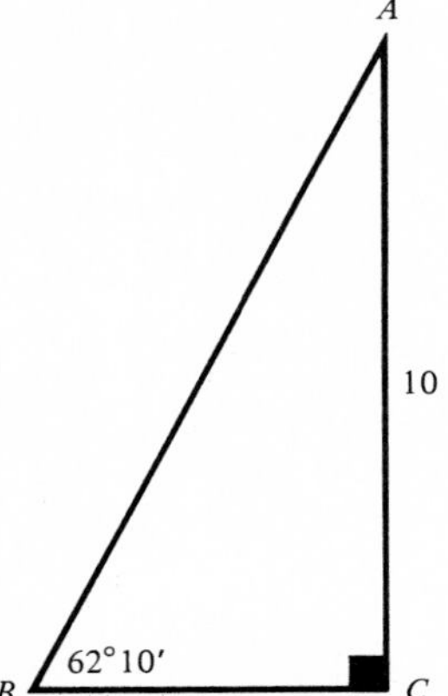

Diagram 9

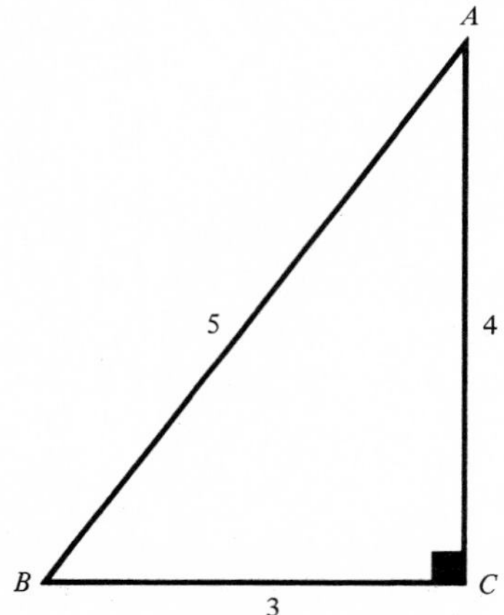

Diagram 10

METHOD 1 If we decide to use the opposite side and the hypotenuse, then

$$\sin A = \frac{a}{c} = \frac{3}{5} = 0.6000$$

Looking for 0.6000 in the table yields $A = 36°50'$.

METHOD 2 If we decide to use the opposite side and the adjacent side, then

$$\tan A = \frac{a}{b} = \frac{3}{4} = 0.7500 \qquad \text{and} \qquad A = 36°50'$$

Since $A + B = 90°$,

$$B = 90° - A = 90 - 36°50' = 53°10'$$

EXERCISES

In Exercises 1 to 7, the right triangle ABC is given, and $C = 90°$. Use the pythagorean theorem to find the missing side.

1 $a = 5, b = 12$
2 $a = 24, c = 25$
3 $a = \sqrt{3}, b = 1$
4 $a = \sqrt{3}, b = \sqrt{6}$
5 $a = 10, c = 20$
6 $b = 4, c = 6$
7 $a = \frac{3}{4}, b = \frac{5}{4}$

In Exercises 8 to 16 use Appendix Table 6 to find the required values.

8 sin 29°
9 tan 47°
10 cot 18°50′
11 cos 81°
12 sec 44°
13 sin 67°20′
14 csc 20.5°
15 cos $77\frac{1}{3}°$
16 tan $69\frac{5}{6}°$

In Exercises 17 to 26 find A to the nearest 10 minutes.

17 $\sin A = 0.5348$
18 $\cos A = 0.5592$
19 $\tan A = 2.2460$
20 $\sec A = 1.8360$
21 $\sin A = 0.8812$
22 $\tan A = 0.4351$
23 $\csc A = 1.1120$
24 $\cos A = 0.4450$
25 $\cot A = 2.0400$
26 $\sin A = 1.5000$

In Exercises 27 to 37, solve for x. If x represents a length, round off to the nearest tenth; if x represents an angle, express x to the nearest 10 minutes.

27

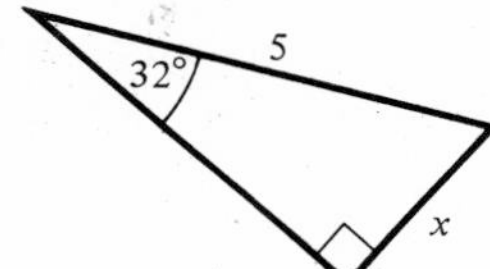

28

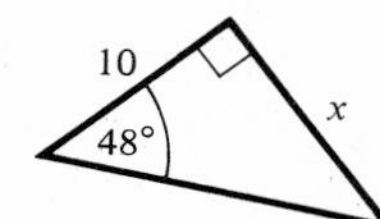

29

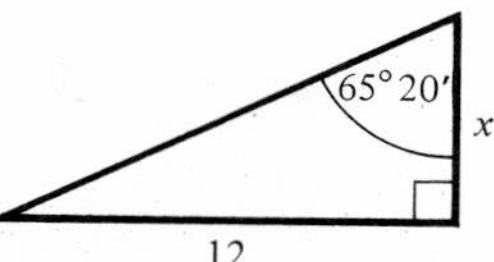

30

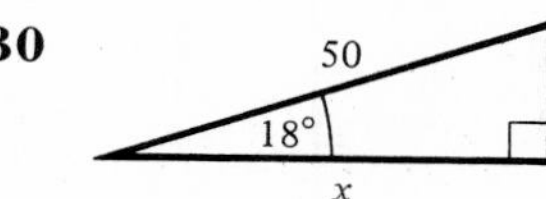

31

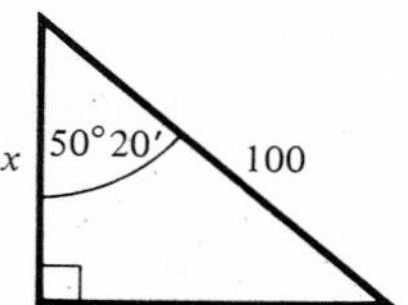

32

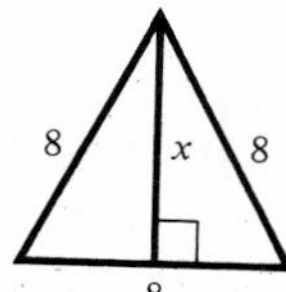

33

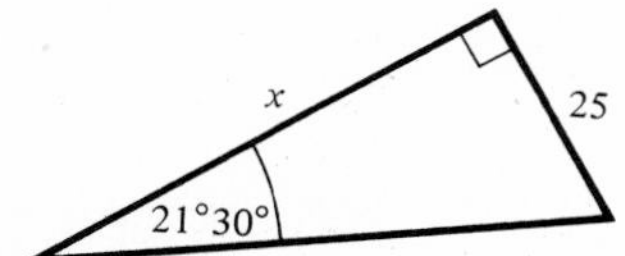

34

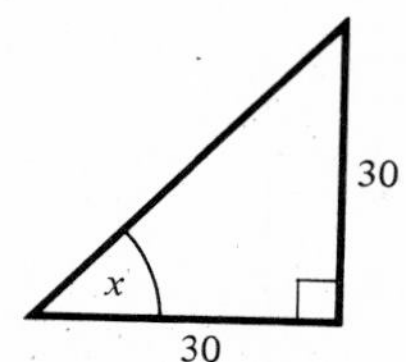

35

36

37

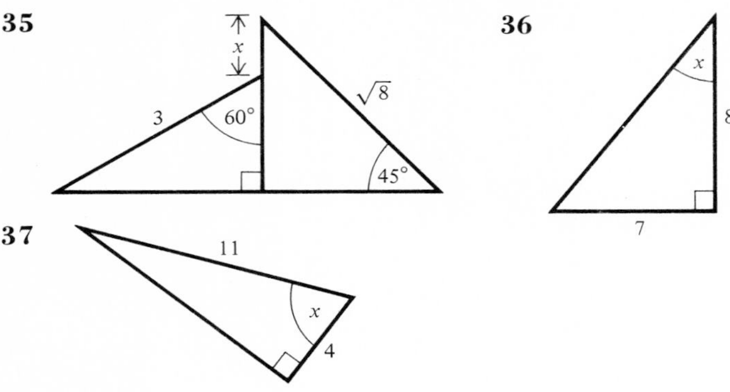

8.3 APPLYING TRIGONOMETRY IN RIGHT TRIANGLES

In this section we will be to apply our knowledge of the trigonometric ratios in solving problems containing right triangles. To *solve* a right triangle will mean to use the given numerical information and find as many of the remaining sides and/or angles as required. Determining the height of a flagpole, the width of river, or an angle of sight are typical problems which can be solved using trigonometry.

Procedure for Solving Right Triangles

1 Draw an accurate diagram of the triangle.

2 On the diagram label *all* the known parts.

3 Indicate on the diagram the parts which are to be found.

4 Determine the trigonometric function to be used.

5 Find the required parts of the triangle.

6 Check the computations (if possible).

Example 1 In triangle ABC, $C = 90°$, $b = 12$, $a = 8$. Find the measure of the other angles (to the nearest degree) and the length of side c (to the nearest tenth) (refer to Diagram 11).

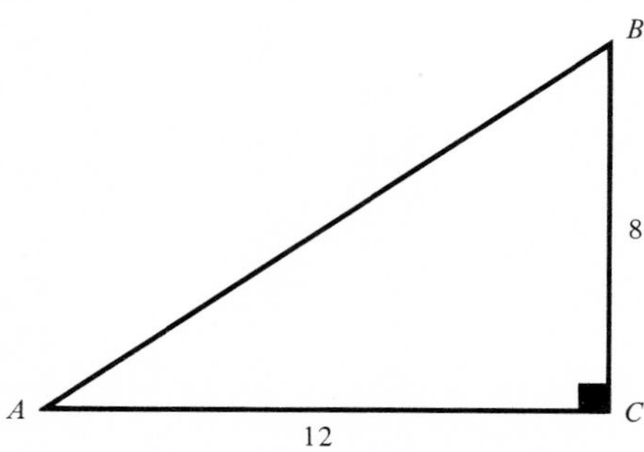

Diagram 11

Solution

$$\tan B = \frac{b}{a} = \frac{12}{8} = 1.5 \qquad \tan A = \frac{a}{b} = \frac{8}{12} = 0.6667$$

$$B = 56° \qquad A = 34°$$

(or simply, $A = 90° - B$, $A = 90 - 56° = 34°$).

$c = \sqrt{a^2 + b^2}$ Pythagorean theorem

$= \sqrt{(12)^2 + (8)^2}$ Substituting the given data

$= \sqrt{144 + 64} = \sqrt{208} \approx 14.4$

In many practical problems, an angle is described as an angle of elevation or an angle of depression (see Diagram 12). If a person is standing at A and looks to B, he must elevate his eyes from the horizontal line AC. AB, the line segment connecting the eye of the observer at A to the position B, is called the *line of sight.* The angle between the horizontal line AC and the line of sight is called the *angle of elevation* (angle BAC in Diagram 12).

Suppose, however, the person is at B looking *down* to point C. He must depress his eyes from the horizontal line BD. The angle between the horizontal line BD and the line of sight is called the *angle of depression* (angle CBD in Diagram 12).

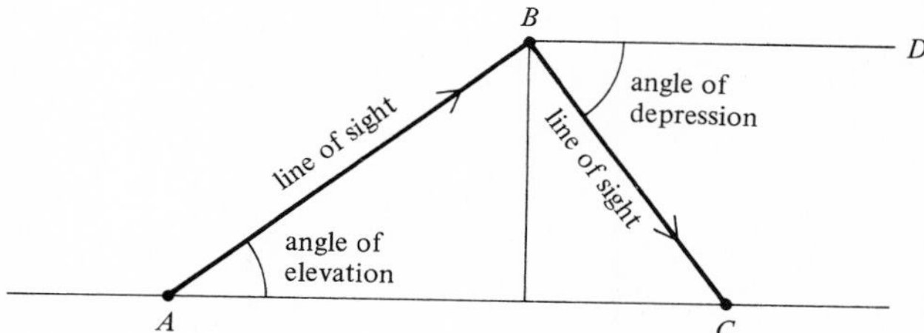

Diagram 12

COMMENT The angle of depression from B to C is the same as the angle of elevation from C to B.

Example 2 From the top of a lighthouse 120 feet above sea level the angle of depression of a sailboat is 40°. Find, to the nearest foot, the distance from the sailboat to the foot of the lighthouse (refer to Diagram 13).

Solution

$\angle DAB = \angle ABC$ Angle of elevation equals angle of depression

$\cot 40° = \dfrac{a}{120}$ Definition

$120 \cot 40° = a$ Multiplying both sides by 120

$120(1.1918) = a$ Using the tables and substituting

$143 = a$

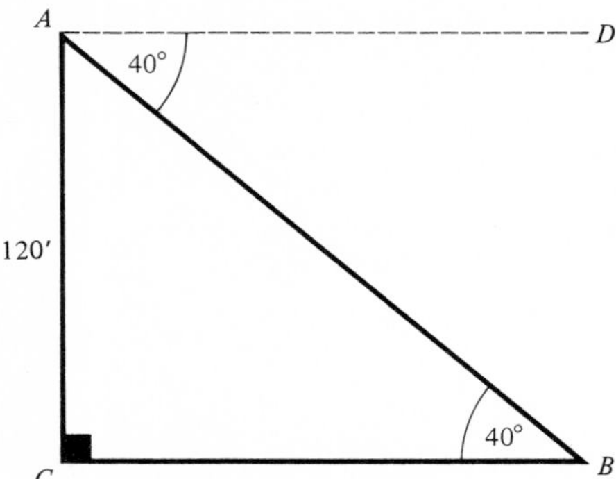

Diagram 13

Example 3 A boy's kite has a string 54 feet long and is flying 36 feet above his eyes. Find the angle of elevation of the kite to the nearest 10 minutes (refer to Diagram 14).

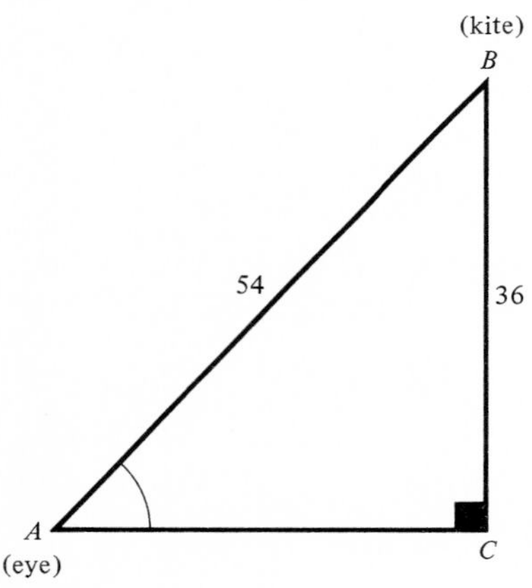

Diagram 14

Solution

$$\sin A = \frac{36}{54} = 0.6667$$

$$A = 41°50'$$

REMINDER In the trigonometric tables, $\sin 41°50' = 0.6670$ and $\sin 41°40' = 0.6648$. Since 0.6667 is closer to 0.6670, A is closer to $41°50'$.

The problems and solutions illustrated so far have been rather straightforward. Verbal problems may be much more difficult, involving more than one right triangle. Your ability to solve one of these problems is a much more challenging test of your understanding. Unless an accurate sketch is drawn, many of these problems will be impossible to solve. Let us illustrate this idea with two examples.

Example 4 A flagpole is on top of a building. From a point 200 feet from the base of the building, the angle of elevation of the top of the flagpole is 32°20′, and the angle of elevation of the bottom of the flagpole is 27°50′. How high is the flagpole (refer to Diagram 15)?

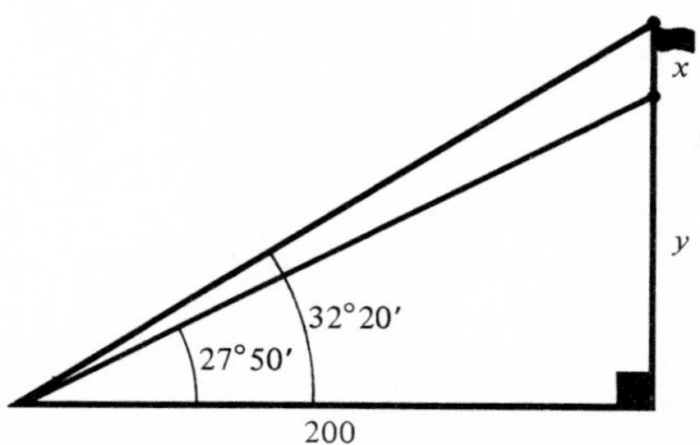

Diagram 15

Solution

$$\tan 32°20' = \frac{x + y}{200} \qquad \text{Diagram 15} \qquad (1)$$

$$\tan 27°50' = \frac{y}{200} \qquad (2)$$

$$(200) \tan 27°50' = y \qquad \text{Solving for } y \text{ in (2)}$$

$$(200)(0.5280) = y = 105.6$$

$$(200) \tan 32°20' = x + y$$

$$(200) \tan 32°20' - y = x \qquad \text{Solving for } x \text{ in (1)}$$

$$(200)(0.6330) - 105.6 = x$$

$$126.6 - 105.6 = x$$

$$21 = x \qquad \text{Height of flagpole}$$

Example 5 In order to determine the width of a river, an observer standing on one bank of the river sights the top of a tree on the opposite bank. The angle of elevation is 50°. The observer now walks 200 feet back from this point in a direct line with the tree. The angle of elevation of the top of tree is now 24°. Find the width of the river (to the nearest foot) (refer to Diagram 16).

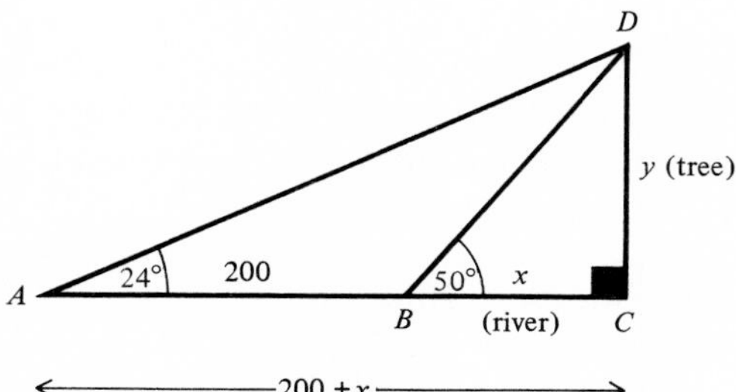

Diagram 16

In right triangle DBC: $\tan 50° = \frac{y}{x}$ (1)

In right triangle DAC: $\tan 24° = \frac{y}{x + 200}$ (2)

Since there are two equations with two unknowns, it is possible to solve them by traditional methods.

$$y = x \tan 50°$$

and $y = (x + 200) \tan 24°$ Solving for y in (1) and (2)

$x \tan 50° = (x + 200) \tan 24°$ Substituting

$x(1.1918) = (x + 200)(0.4452)$ Using the tables

$$1.1918x = 0.4452x + 89.04$$

$$0.7466x = 89.04$$

$x = \frac{89.04}{0.7466} = 119$ feet width of river

The height of the tree y can be determined by using Equation (1):

$\tan 50° = \frac{y}{119}$ and $119 \tan 50° = y$

$119(1.1918) = y = 142$ feet height of tree

In many applications, the direction from one location to another or the direction which an object must travel to reach a target come under consideration. The most common convention for this is a rotation from a north-south line.

The *bearing* of a point P from a point O is always expressed as an acute angle measured from the north-south line (refer to Diagram 17). In (*a*), to indicate the bearing of P from O, we first ask: Is P north or south of O? (It is north.) Then we ask whether P is east or west of O (it is west). The bearing of P from O is written N30°W. In (*b*), the bearing of P from O is S60°E. In (*c*) we say that P is due east of O.

A second method used to indicate direction is by measuring the angle determined by a *clockwise* rotation from the north line. This is referred to as a *course*. In Diagram 18, using the first method we would say that the bearing of P from O is S40°W. Using the second method, no directions are stated, merely an angle. Hence in Diagram 18, we simply say that the course followed by P is 220°.

Example 6 At 8:30 A.M. a ship sailing due east at 16 miles per hour, observes a lighthouse due south. (Ships' speeds are correctly given in knots; miles per hour are used here for simplicity.) At 10 A.M. the bearing of the

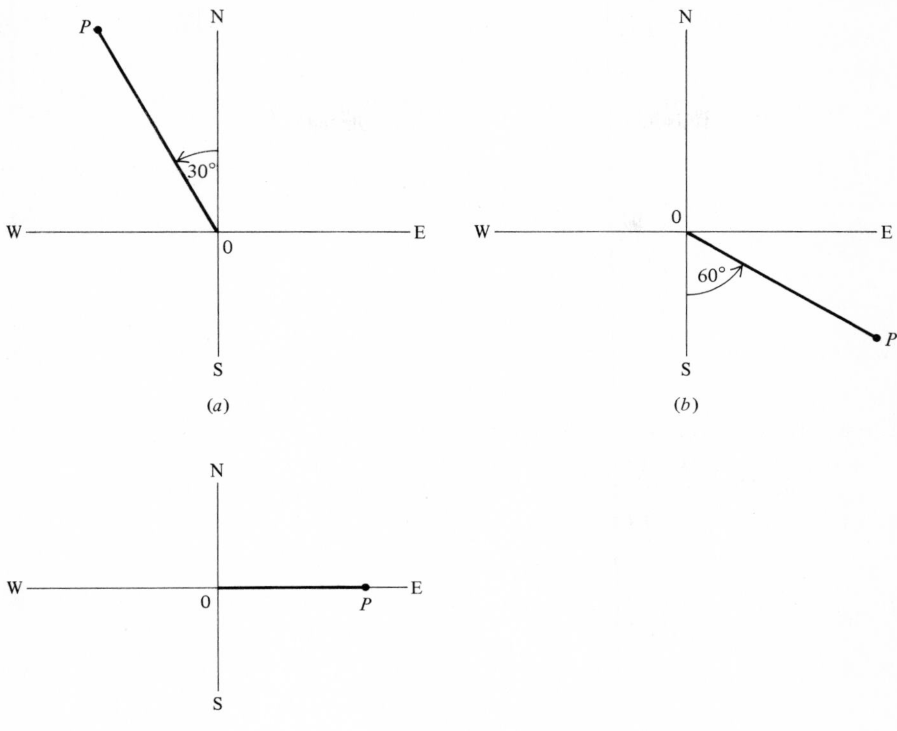

Diagram 17

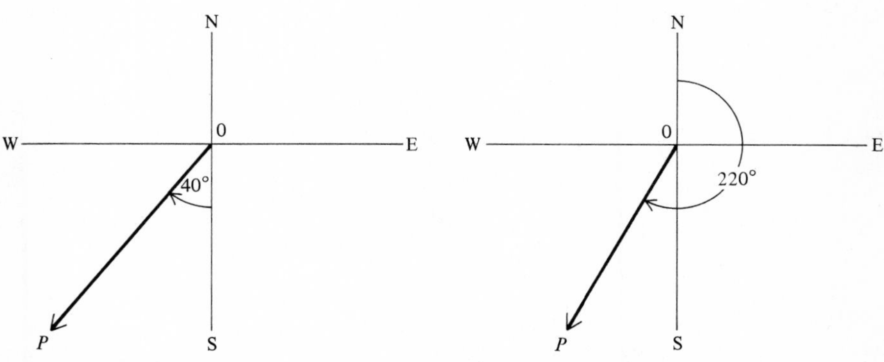

Diagram 18

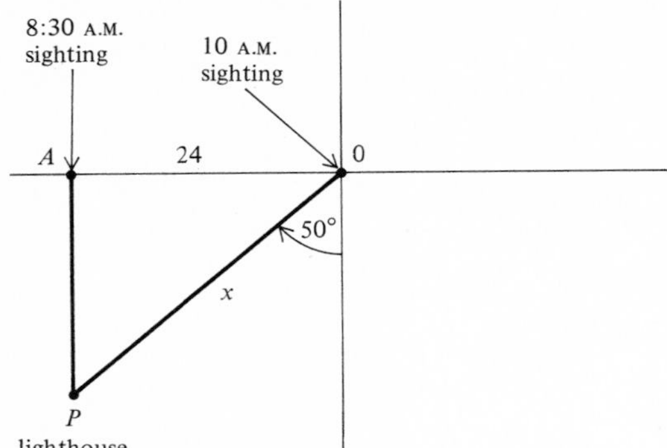

Diagram 19

same lighthouse is S50°W. Find the distance from the ship to the lighthouse at the time of the second observation (refer to Diagram 19).

Examine the labeled diagram, which represents the given data. Let x represent the required distance.

$$\angle AOP = 40°$$

$$\cos 40° = \frac{24}{x} \qquad \text{Definition}$$

$$x = \frac{24}{\cos 40°} \qquad \text{Solving for } x$$

$$= \frac{24}{0.7660} \qquad \text{Using the tables}$$

$$= 31 \text{ miles}$$

Example 7 A plane flies 80 miles due west and then changes direction and flies 60 miles due south. Find the direction of the plane from its starting point (to the nearest 10 minutes). Use the two methods for expressing direction (refer to Diagram 20).

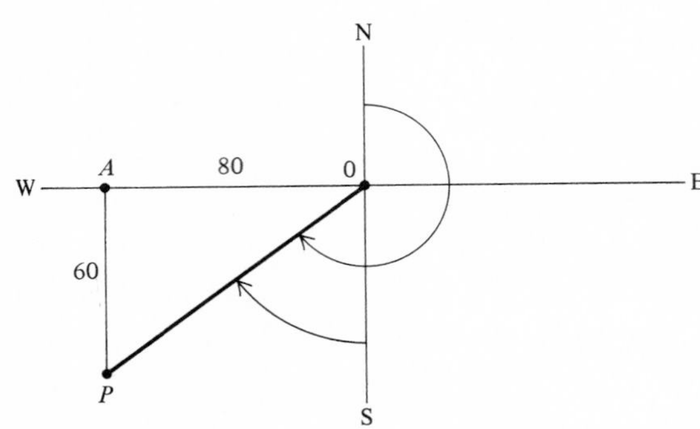

Diagram 20

Examine the labeled diagram, which represents the given data. By finding the measure of $\angle AOP$ we can answer the question.

$$\tan \angle AOP = \frac{60}{80} = 0.75 \qquad \text{Definition}$$

$$\angle AOP = 36°50' \qquad \text{Using the tables}$$

METHOD 1 Bearing

$$\angle POS = 90 - 36°50' = 53°10'$$

Therefore, the bearing of P from O is S53°10′W.

METHOD 2 Course

$$\angle NOP = 180 + 53°10' = 233°10'$$

Therefore the course traveled by the plane is 233°10′.

EXERCISES

In Exercises 1 to 5, find the unknown sides (to the nearest integer) and the unknown angles (to the nearest degree). In each triangle $C = 90°$.

1 $A = 56°, c = 40$
2 $A = 15°, a = 25$
3 $a = 8, b = 15$
4 $A = 84°, b = 16$
5 $a = 8, c = 20$

In Exercises 6 to 33, answers should be expressed to the same degree of accuracy as the given data unless otherwise specified; all angles should be expressed to the nearest 10 minutes.

6 From an observation post 54 feet directly above point A, the angle of elevation of point B is 6°10′. Find the distance from A to B.

7 Town A is 10 miles due north of town C. Town B is 5 miles east of town C. Find the bearing of B from A.

8 A is 100 miles N44°E of B. C is due north of B and due west of A. Find the distance from B to C.

9 A is 50 miles due north of B, and C is 40 miles due east of A. Find the bearing of C from B.

10 A ship at A is 150 miles due west of a ship at B. Both ships are heading toward the same port, with A on a bearing N72°E and B on a bearing N18°W. Find the distance from the port to the closer ship at this point.

11 A ship sails 80 miles on a course of 144°. How far east has it gone?

12 A plane flies 50 miles due west and then 30 miles due south. Find the direction of the plane from its starting point. Express the answer as a bearing *and* a course.

13 Three ships are situated as follows: A is 40 miles south of C, and B is 54 miles east of C.

(a) Find the bearing of B from A.

(b) What course would you have to follow in going from B to A?

14 A is 18 miles east of C, and B is south of C. The bearing of B from A is S20°40′W.

(a) How far is it from B to A?
(b) What course would you follow in going from A to B?

15 A is 20 miles north of C, and B is 30 miles east of C.
(a) What is the bearing of B from A?
(b) What course would you follow in going from B to A?

16 When the sun is 40° above the horizon, how long a shadow is cast (to the nearest inch) by a 6-foot man?

17 A person sights the top of a tree and determines the angle of elevation to be 24°. He then moves 60 feet closer to the tree and determines the angle of elevation of the top of the tree to be 48°. Find the height of the tree to the nearest tenth.

18 Two poles are 40 feet apart. From the top of the higher pole, 30 feet high, the angle of depression to the top of the lower pole is 32°. Find the height of the lower pole.

19 A machinist wants to put 15 equally spaced holes on the outside edge of a circular disk of radius 8 inches. Find the distance between the centers of the holes to the nearest tenth.

20 An 8-foot ladder in a room makes a 50° angle when it touches one wall. From the same point it makes a 32° angle when it touches the opposite wall. How wide is the room?

21 A surveyor wishes to determine the height of a building without going to the top of the building. He stands an unknown distance from the base of the building and measures the angle of elevation of the top of the building to be 23°20′. He walks 400 feet closer to the base and now determines the angle of elevation to be 38°40′. How high is the building?

22 A 60-foot flagpole stands atop a ramp which is inclined 15° to the horizontal. From the top of the flagpole to the foot of the ramp the angle of depression is 32°30′. How long is the ramp?

23 Two men 200 feet apart are on opposite sides of a tree. The men and the tree are in a straight line. Both men sight the top of the tree, and the angles of elevation are 46°50′ and 33°10′. Determine the height of the tree.

24 From the top of a 150-foot building the angle of depression of a car due north is 24°40′. If the car now travels due east for 4.5 minutes, the angle of depression is 16°20′. Compute the speed of the car.

25 In order to determine the width of a river, an observer spots a tree growing on the opposite bank. The angle of elevation of the top of the tree from a point on the opposite bank is 42°, and 100 feet back from this point and in line with the tree the angle of elevation is 25°. Find the width of the river and the height of the tree.

26 An airplane flying at a height of 1000 feet sights the top of the tower at an angle of depression of 56°20′ and the base of the tower at an angle of depression of 64°40′. Find the height of the tower.

27 A radio antenna is on top of a building. From a point 80 feet from the base of the building the angle of elevation of the top of the antenna is 65°, while the angle of elevation of the bottom of the antenna is 59°. How high is the antenna?

28 Two men and a flagpole are in the same horizontal plane. One man is due west while the other is due east of a 75-foot flagpole. If the angles of elevation to the top of the flagpole are 37° and 21°30′, find the distance between the two men.

29 A baseball diamond is a 90-foot square, and the pitcher's mound is 60.5 feet from home plate. How far is the pitcher's mound from first base to the nearest tenth?

30 Three houses are located along the same shoreline. From a boat due west of the first house an observation is made. The angle between the first and second house is 27°, and the angle between the second and third house is also 27°. The distance from the boat to the shoreline is exactly 1 mile. How far is the second house from the boat?

31 Two birdwatchers are due west of a high nest of a rare species of hawk. They are 250 feet from each other, and they respectively make angles of elevation of 48°20′ and 28°30′ from where they stand to the nest. How far would the closest birdwatcher have to walk to get directly beneath the nest?

32 A surveyor observes the angle of elevation of the top of a tower to be 39°40′ from a point 6 feet above the ground. At a point 40 feet closer to the tower the angle of elevation is 44°10°, also from a height of 6 feet. Find the height of the tower.

33 A tower is opposite a building. From the top of the building, the angle of elevation of the top of the tower is 10°50′, and the angle of depression of the base of the tower is 32°20′. If the building is 100 feet high, find the height of the tower.

8.4
THE GENERAL ANGLE

A ray is part of a line which extends in one direction from a point called an end point. Suppose a ray rotates about a point O (called the initial position OA) to a terminal position OB (refer to Diagram 21). If this rotation is counterclockwise, a positive angle is formed; if the rotation is clockwise, a negative angle is formed.

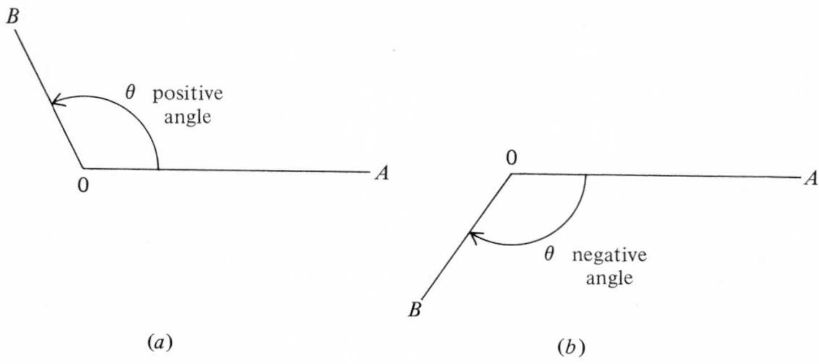

Diagram 21

Suppose the vertex of an angle is positioned at the origin of the rectangular plane and OA is the initial side of this angle along the positive side of the x axis (refer to Diagram 22). Then the angle AOB is said to be in *standard position.*

An angle of any measure can be generated by simply rotating OB to any position in the plane. The angle is said to lie in the quadrant where the terminal side is located (see Diagram 23). It should be noted here that an angle can be *greater* than 360° by rotating through more than 1 revolution. Again, the

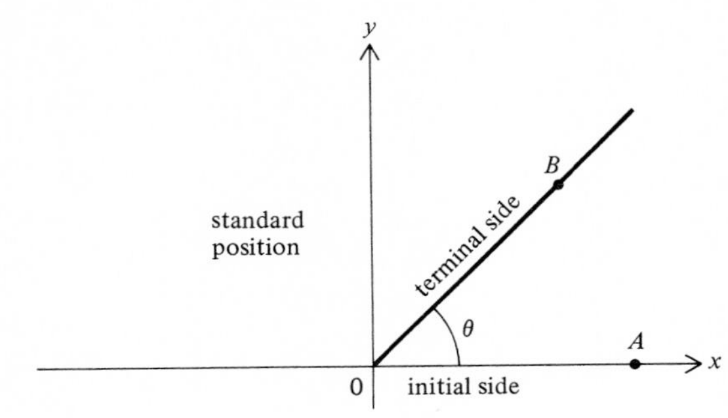

Diagram 22

θ in quadrant II

θ in quadrant IV

θ in quadrant III

Diagram 23

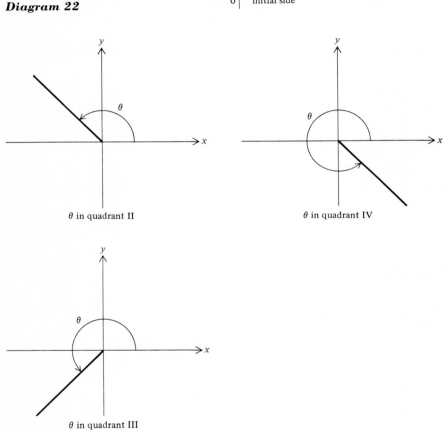

angle formed is said to lie in the quadrant of the terminal side. In Diagram 24, $\angle AOB = 450°$ and lies in quadrant I.

If θ is any angle in standard position, let P represent any point on the terminal side of θ (Diagram 25). The coordinates of P will be labeled (x, y). If we drop a perpendicular from P to the x axis, we construct a right triangle. Call this point of intersection M. In Diagram 25 this has been done in each of the four quadrants.

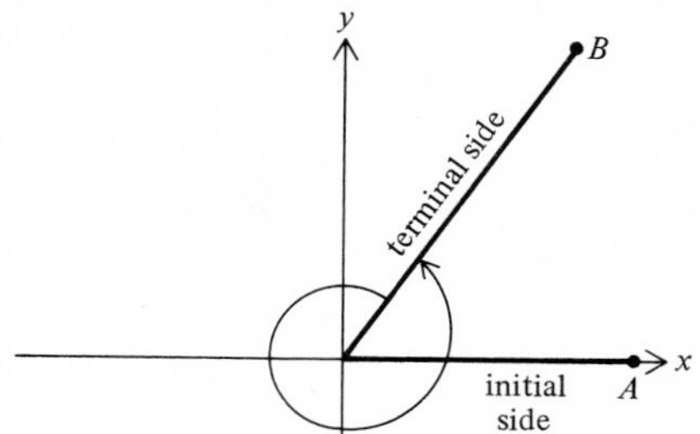

Diagram 24

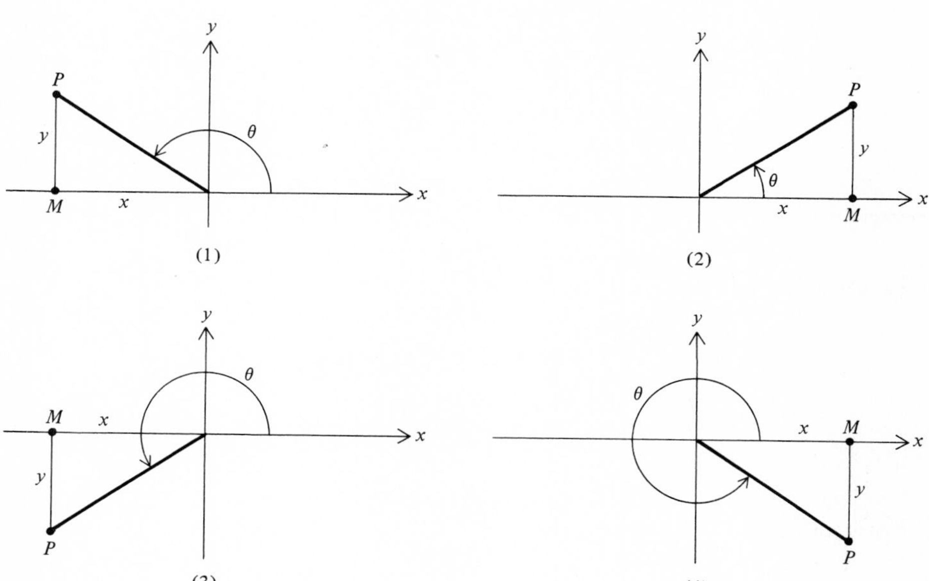

Diagram 25

The following observations can be made regarding Diagram 25.

1 Regardless of what quadrant θ is in, P is the same distance from the origin (call this distance r). Therefore,

$$r = \sqrt{(OM)^2 + (MP)^2} = \sqrt{x^2 + y^2} > 0$$

This implies that *r is always positive.*

2 The *signs* of the coordinates of P depend upon the quadrant in which θ lies (see Diagram 26).

The triangles in Diagram 26 are called *reference triangles.* We can now define the six trigonometric ratios in terms of r and the coordinates of P, that is, x and y.

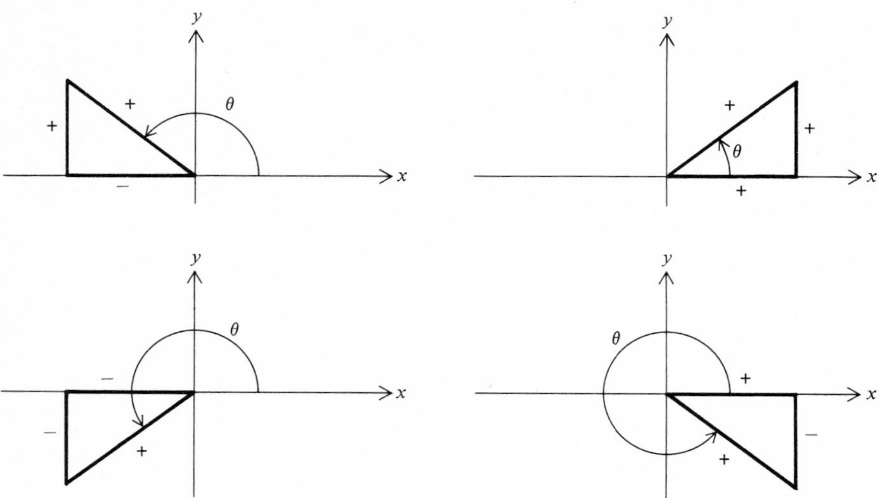

Diagram 26

DEFINITION

$$\sin\theta = \frac{y}{r} \qquad \csc\theta = \frac{r}{y}$$

$$\cos\theta = \frac{x}{r} \qquad \sec\theta = \frac{r}{x}$$

$$\tan\theta = \frac{y}{x} \qquad \cot\theta = \frac{x}{y}$$

Care must be taken when determining the ratio of an angle which *does not lie* in quadrant I. The values of these ratios depend on the *signs* of x and y (since r is *always* positive).

Example 1 If θ terminates in Quadrant II, determine the *signs* of each of the six trigonometric ratios (refer to Diagram 26 *and* the definitions of the trigometric ratios).

Solution

$$\sin\theta = \frac{y}{r} = \frac{+}{+} = + \qquad \csc\theta = \frac{r}{y} = \frac{+}{+} = +$$

$$\cos\theta = \frac{x}{r} = \frac{-}{+} = - \qquad \sec\theta = \frac{r}{x} = \frac{+}{-} = -$$

$$\tan\theta = \frac{y}{x} = \frac{+}{-} = - \qquad \cot\theta = \frac{x}{y} = \frac{-}{+} = -$$

COMMENT Knowing *both* the definition of the trigonometric ratios *and* the quadrant in which the angle lies, we can determine the *sign* of the given angle.

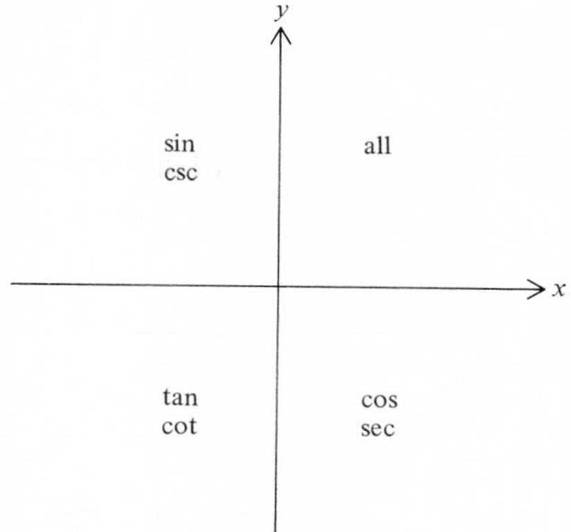

Diagram 27

To help us we summarize the results for *all* quadrants. Diagram 27 indicates which ratios are *positive* in each of the four quadrants (by reviewing the definitions, these can easily be verified).

Let us now examine the four angles (θ_1, θ_2, θ_3, θ_4) which are associated with the four points, $P_1(4, 3)$, $P_2(-4, 3)$, $P_3(-4, -3)$ and $P_4(4, -3)$ (refer to Diagram 28). In each of the four cases, $r = 5$ (by the pythagorean theorem). Also, each of the four triangles OMP can be formed by dropping a perpendicular line segment to the *x axis.* The four triangles formed are *congruent* (identical). Therefore, the acute angle R is the same in each of the four quadrants. This acute angle R is called the *reference angle* for θ.

COMMENT It should be carefully observed that reference angle R is a *positive acute angle* and is *always* formed by the terminal side of the angle in question and the *x axis.*

Example 2 Using the numerical data depicted in Diagram 28, compute the sine, cosine, and tangent values of θ_1, θ_2, θ_3, and θ_4.

Solution

$\sin\theta_1 = \frac{3}{5}$
$\sin\theta_2 = \frac{3}{5}$ → positive signs in quadrants I and II

$\sin\theta_3 = -\frac{3}{5}$
$\sin\theta_4 = -\frac{3}{5}$ → negative signs in quadrant III and IV

$\cos\theta_1 = \frac{4}{5}$
$\cos\theta_2 = -\frac{4}{5}$
$\cos\theta_3 = -\frac{4}{5}$ → negative signs in quadrants II and III
$\cos\theta_4 = \frac{4}{5}$ ($\cos\theta_1$, $\cos\theta_4$) → positive signs in quadrants I and IV

$\tan\theta_1 = \frac{3}{4}$
$\tan\theta_3 = \frac{3}{4}$ → positive signs in quadrants I and III

$\tan\theta_2 = -\frac{3}{4}$
$\tan\theta_4 = -\frac{3}{4}$ → negative signs in quadrants II and IV

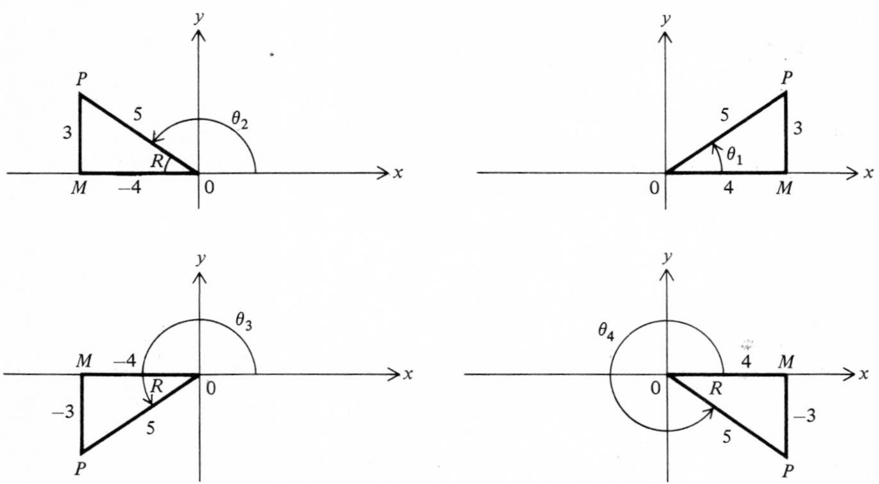

Diagram 28

Example 3 Find cos 100° (refer to Diagram 29).

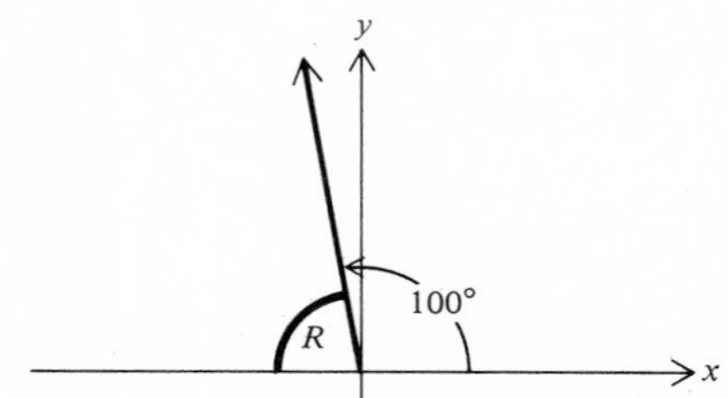

Diagram 29

Sketch $\theta = 100°$ and observe that it lies in quadrant II. Label the reference angle R.

$\cos 100° = -\cos R$	Cosine is *negative* in quadrant II
$R = 180 - 100° = 80°$	Determining R (Diagram 28)
$\cos 100° = -\cos 80°$	Substitution
$= -0.1736$	Using Appendix Table 6

Example 4 Find tan 244° (refer to Diagram 30).
In Diagram 30, 244° is sketched; it lies in quadrant III.

$\tan 244° = \tan R$	Tangent is positive in quadrant III
$R = 244° - 180° = 64°$	Determining R
$\tan 244° = \tan 64° = 2.0503$	Using Appendix Table 6

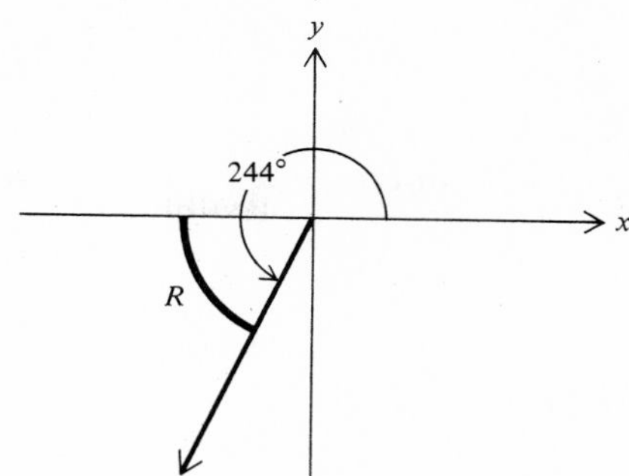

Diagram 30

Suppose now that the problem is reversed and we are given the value of the ratio and asked to find the *angle* which corresponds to this value.

Example 5 Find θ, $0° < \theta < 360°$, such that $\sin \theta = -0.7660$.

STEP 1 A negative value of θ indicates that θ lies in either quadrant III or IV since these are the two quadrants where the sine is negative. Therefore there will be *two* solutions for θ.

STEP 2 The reference angle R associated with θ can be found by saying $\sin R = 0.7660$ and referring to Appendix Table 6:

$$R = 50°$$

STEP 3 From step 1, we sketch θ and indicate the reference angle R (see Diagram 31).

STEP 4 θ_3 and θ_4 can now be calculated as Diagram 31 indicates:

$$\theta_3 = 180° + 50° = 230° \qquad \theta_4 = 360° - 50° = 310°$$

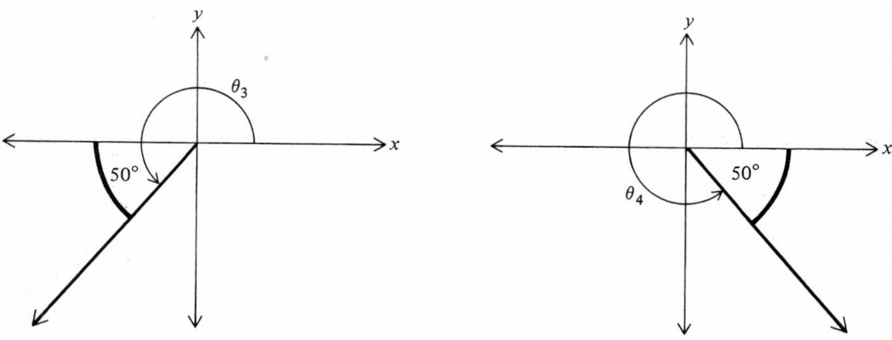

Diagram 31

Example 6 For $90° < \theta < 180°$, if $\tan \theta = -\frac{8}{15}$, find $\sin \theta$ and $\cos \theta$.

STEP 1 Draw a reference triangle from the given information (see Diagram 32). Since the tangent is defined to be y/x and θ is in quadrant II, $x = -15$ and $y = 8$.

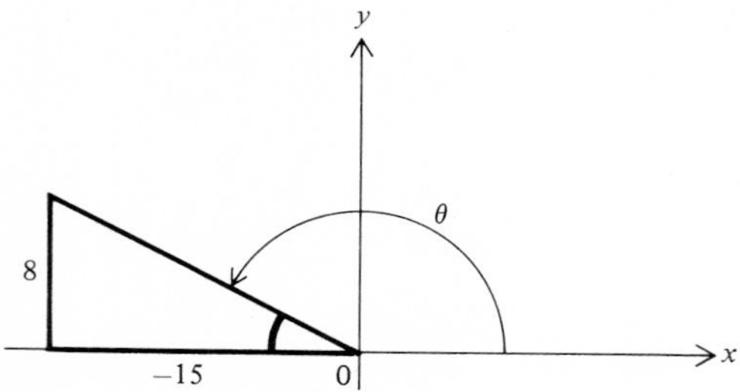

Diagram 32

STEP 2

$$r = \sqrt{x^2 + y^2} = \sqrt{(-15)^2 + (8)^2} \qquad \text{Pythagorean theorem}$$
$$= \sqrt{289} = 17$$

STEP 3

$$\sin \theta = \frac{8}{17} \quad \text{and} \quad \cos \theta = -\frac{15}{17} \qquad \text{Definition of sine and cosine}$$

Example 7 Prove $\sin^2 \theta + \cos^2 \theta = 1$.

For convenience select P on the terminal side of θ such that $OP = 1$. Again, for convenience, select θ in quadrant I, although it could be in *any* of the other quadrants (see Diagram 33). Since we do not know the coordinates

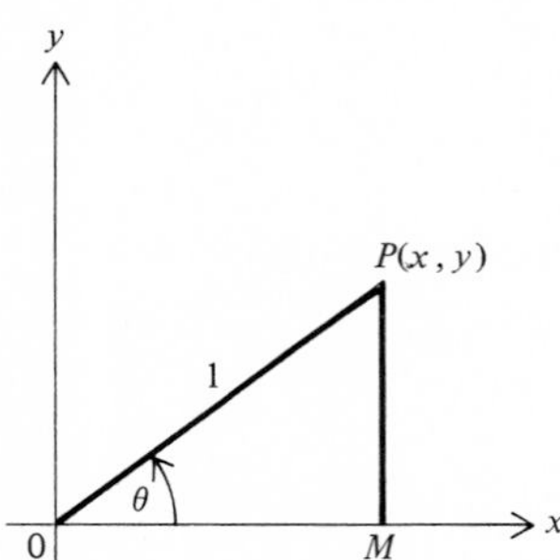

Diagram 33

of P, we drop a perpendicular line segment from P to the x axis and form a right triangle OMP. By the pythagorean theorem

$$(OM)^2 + (MP)^2 = (OP)^2 \tag{1}$$

also

$$\cos\theta = \frac{OM}{1} = OM \qquad \sin\theta = \frac{MP}{1} = MP \qquad \text{Definition of sine and cosine}$$

Recall, $OP = 1$

Substituting into Equation (1), we obtain

$$\cos^2\theta + \sin^2\theta = 1$$

COMMENT This result will be used in the next section in the proof of the law of cosines.

EXERCISES

In Exercises 1 to 20 find the value of each of the following, expressing the answer in radical form.

1 sin 210° **2** cos 240° **3** tan 135° **4** sec 510°
5 csc 120° **6** cot 480° **7** sin 315° **8** cos 225°
9 tan 300° **10** sec 150° **11** csc 495° **12** cot 330°
13 sin(−30°) **14** cos(−60°) **15** tan(−210°) **16** sec(−225°)
17 csc(−480°) **18** cot(−150°) **19** sin(−120°) **20** cos(−510°)

In Exercises 21 to 44, express each of the given ratios in terms of a positive acute angle, that is, the reference angle.

21 sin 230° **22** cos 152° **23** tan 212° **24** sec 440°
25 csc 324° **26** cot 95° **27** sin 174° **28** cos 248°40′
29 tan 340°50′ **30** sec 1000° **31** csc 272° **32** cot 190°20′
33 sin 342° **34** cos 296° **35** tan 162°10′ **36** sin(−170°)
37 cos(−212°) **38** tan(−54°) **39** sec(−315°30′) **40** csc(−204°40′)
41 cot(−152°) **42** sin(−800°) **43** cos(−181°) **44** tan(−449°)

In Exercises 45 to 56 find all the possible angles which correspond to the given values (to the nearest 10 minutes), $0° \leq x \leq 360°$.

45 $\sin x = 0.4384$ **46** $\cos x = -0.6018$ **47** $\tan x = 0.1495$
48 $\sec x = -1.5400$ **49** $\csc x = 57.3000$ **50** $\cot x = -0.4950$
51 $\sin x = -0.9740$ **52** $\cos x = 0.8134$ **53** $\tan x = -4$
54 $\sec x = 2.5$ **55** $\csc x = -1.2$ **56** $\cot x = 2.5$

Using the given information in Exercises 57 to 68, draw the angle and find the values of the remaining five trigonometric ratios.

57 $\sin A = \frac{5}{13}$, A in quadrant I **58** $\cos A = -\frac{4}{5}$, A in quadrant II
59 $\tan A = \frac{27}{4}$, A in quadrant III **60** $\sec A = \frac{17}{8}$, A in quadrant IV
61 $\cot A = \frac{4}{5}$, A in quadrant I **62** $\tan A = -2$, A in quadrant II
63 $\sin A = -\dfrac{4}{\sqrt{41}}$, A in quadrant III **64** $\csc A = -\dfrac{7}{\sqrt{13}}$, A in quadrant IV
65 $\cot A = 4$, A in quadrant I **66** $\cos A = -\dfrac{1}{5\sqrt{2}}$, A in quadrant III
67 $\csc A = \dfrac{1}{\sqrt{5}}$, A in quadrant II **68** $\sec A = 3$, A in quadrant IV

Using the techniques of Example 7, prove:

69 $\tan^2 A + 1 = \sec^2 A$ **70** $\cot^2 A + 1 = \csc^2 A$

8.5 OBLIQUE TRIANGLES; THE LAW OF SINES AND COSINES

A triangle which *does not contain* a right angle is called an *oblique triangle.* The procedures used in Section 8.3 applied only when right triangles were available. Even though it may be easy to separate any oblique triangle into separate right triangles, this may be time-consuming. Formulas can be derived to solve oblique triangles *without* using right triangles.

Given triangle ABC (refer to Diagram 34). Drop a perpendicular line segment from C to AB. We now have two right triangles, $\triangle CDB$ and $\triangle CDA$.

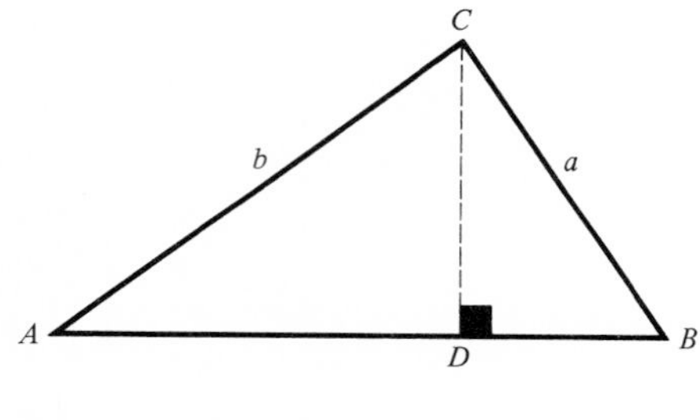

Diagram 34

Note that CD is common to both triangles.

In $\triangle CDA$: $\sin A = \dfrac{CD}{b}$ and $CD = b \sin A$ (1)

In $\triangle CDB$: $\sin B = \dfrac{CD}{a}$ and $CD = a \sin B$ (2)

Then $b \sin A = a \sin B$ Equating (1) and (2)

or $\dfrac{\sin A}{a} = \dfrac{\sin B}{b}$ Dividing both sides by ab

Using similar reasoning, it can be shown that

$$\frac{\sin A}{a} = \frac{\sin C}{c}$$

LAW OF SINES In any triangle ABC, the following relationship is true:

$$\frac{\sin A}{a} = \frac{\sin B}{b} = \frac{\sin C}{c}$$

COMMENT Since three of the entries must be known in order to determine the fourth, we must know either:

1 Two angles and one side opposite one of the given angles or
2 Two sides and one angle opposite one of the given sides

Example 1 In $\triangle ABC$, $A = 80°$, $B = 56°$, $b = 16$. Determine the remaining parts of the triangle (refer to Diagram 35).

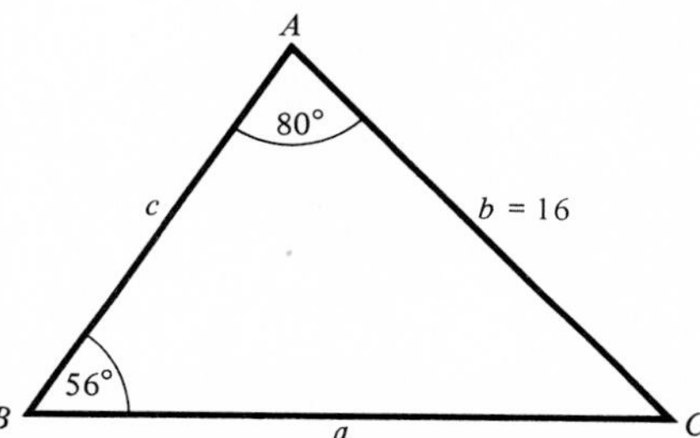

Diagram 35

Solution

$$C = 180° - (80° + 56°) = 180° - 136° = 44°$$

Applying the law of sines gives

$$\frac{\sin 80°}{a} = \frac{\sin 56°}{16} \qquad \text{and} \qquad 16 \sin 80° = a \sin 56°$$

$$a = \frac{16(\sin 80°)}{\sin 56°} = \frac{16(0.9848)}{0.8290} \approx 19$$

To find the length of side c, we use the law of sines again:

$$\frac{\sin 44°}{c} = \frac{\sin 56°}{16}$$

$$c = \frac{16 \sin 44°}{\sin 56°} = \frac{16(0.6947)}{0.8290} \approx 13.4$$

COMMENT Note that we use side $b = 16$ rather than side $a = 19$. Since $b = 16$ is given as an exact value and a is only a rounded-off value.

When two angles and one side are given, the data *always* determine a unique triangle. However, when two sides and the angle opposite one of them are given, the data may determine:

1 No triangle (Example 2)
2 One triangle (Example 4)
3 Two triangles (Example 3)

Since different possibilities may exist, this case is usually referred to as the *ambiguous case.*

Example 2 In $\triangle ABC$ $A = 64°$, $a = 20$, $c = 32$ (refer to Diagram 36). Find the remaining parts of ABC.

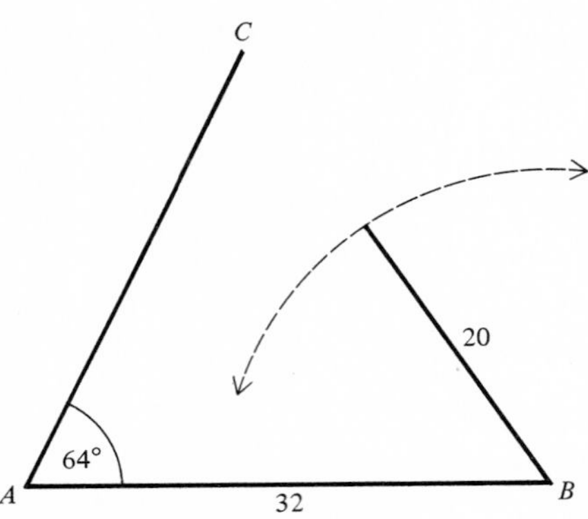

Diagram 36

Using the law of sines,

$$\frac{\sin 64°}{20} = \frac{\sin C}{32}$$

$$\sin C = \frac{32 \sin 64°}{20} \qquad \text{Solving for } \sin C$$

$$= \frac{32(0.8988)}{20} = 1.4381\ ?$$

Since $\sin C > 1$, there is no triangle with this given information (see Diagram 36, which represents the actual data).

Example 3 $\triangle ABC$, $A = 40°$, $a = 12$, $b = 16$. Find the remaining parts of the triangle (refer to Diagram 37).

Using the law of sines,

$$\frac{\sin 40°}{12} = \frac{\sin B}{16}$$

$$\sin B = \frac{16 \sin 40°}{12} \qquad \text{Solving for } \sin B$$

$$= \frac{16(0.6428)}{12} = 0.8571$$

There are *two* possible solutions to the equation $\sin B = 0.8571$ in the interval $0° < B < 180°$ (refer to Section 8.3).

SOLUTION 1 If $0° < B < 90°$, $B = 59°$.

SOLUTION 2 If $90° < B < 180°$, we must first determine the reference angle R:

$$R = 180° - 59° = 121°$$

SOLUTION 1 If $B = 59°$, then $C = 180° - (40° + 59°) = 81°$.

SOLUTION 2 If $B = 121°$, then $C = 180° - (40° + 121°) = 19°$.

COMMENT Therefore there are two possible solutions to this problem (see Diagram 37):

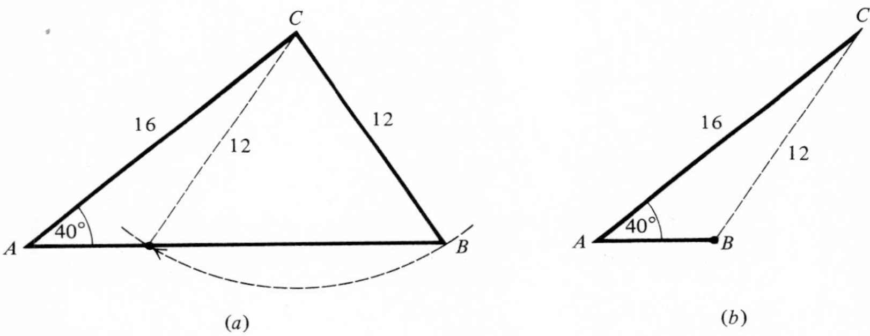

Diagram 37

Triangle 1: $A = 40°$ $B = 59°$ $a = 12$ $b = 16$ $C = 81°$
Triangle 2: $A = 40°$ $B = 121°$ $a = 12$ $b = 16$ $C = 19°$

Solving triangle 1 gives

$$\frac{\sin 81°}{c} = \frac{\sin 40°}{12}$$

$$c = \frac{12 \sin 81°}{\sin 40°} \qquad \text{Solving for } c$$

$$= \frac{12(0.9877)}{0.6428} \approx 18.4$$

Solving triangle 2 gives

$$\frac{\sin 19°}{c} = \frac{\sin 40°}{12}$$

$$c = \frac{12 \sin 19°}{\sin 40°} \qquad \text{Solving for } c$$

$$= \frac{12(0.3256)}{0.6428} \approx 6.1$$

COMMENT There is no need to devise special tests for the number of solutions in the various situations which arise. When

$$\sin A = \frac{a \sin B}{b}$$

find *all possible values of A* that can be angles of a triangle. Then determine the number of solutions. The problem which follows is a practical situation in which only *one* triangle is possible. Let us mechanically examine the data and discover *why* there is a unique solution.

Example 4 City C is N64°W of city B. City A is 4.7 miles south of city B. If it is 9 miles from city C to city A, find the direction of city C from city A (refer to Diagram 38).

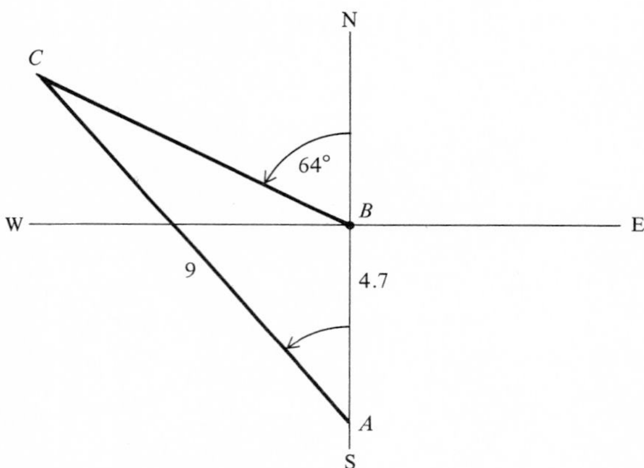

Diagram 38

STEP 1 Diagram 38 represents the given information:

$$\angle ABC = 180° - 64° = 116°$$

STEP 2 The objective is to determine angle A. Since we do not know the side opposite A, we must first determine C. Then $A = 180° - (C + 116°)$.

STEP 3 $\dfrac{\sin 116°}{9} = \dfrac{\sin C}{4.7}$ Law of sines

$$\sin C = \frac{4.7(0.8988)}{9} = 0.4694 \qquad \text{Solving for } \sin C$$

NOTE Just a reminder: $\sin 116° = +\sin 64°$, using the concepts from Section 8.4.

STEP 4 As in the previous example, there are two possible values for C in the interval $0° < C < 180°$. Since $\sin C = 0.4694$,

$C = 28°$ and $A = 180° - (28° + 116°) = 36°$ From step 2

FINAL ANSWER City C is N36°W of city A.

STEP 5 How do we know this solution is *unique*?

If $C = 180° - 28° = 152°$ reference angle in quadrant II

$A = 180° - (152° + 116)$ From step 2

$= 180° - (268°) = -88°?$

CONCLUSION There is only *one* triangle which satisfies the given data.

When two sides and the included angle are given *or* when three sides are given, the law of sines is not *directly* applicable in solving the triangle. We can derive a second law useful in solving triangles when these data are known. It is called the *law of cosines.*

Consider $\triangle ABC$ in Diagram 39. If we drop a perpendicular line segment from C to AB, we form two right triangles, $\triangle BDC$ and $\triangle ADC$. From $\triangle BDC$,

$$a^2 = (BD)^2 + (CD)^2 \qquad \text{Pythagorean theorem} \qquad (1)$$

The objective of this proof will be to express BD and CD in terms of the parts of the given $\triangle ABC$.

In $\triangle ADC$,

$\sin A = \dfrac{CD}{b}$ and $\cos A = \dfrac{AD}{b}$ — Definition of sine and cosine

$CD = b \sin A$ and $AD = b \cos A$ — Solving for CD and AD

$c = AD + BD$ — Diagram 39

$BD = c - AD = c - b \cos A$ — Solving for BD; Substituting the value of AD

$a^2 = (c - b \cos A)^2 + (b \sin A)^2$ — Substituting the values of BD and CD into (1)

$= c^2 - 2bc \cos A + b^2 \cos^2 A + b^2 \sin^2 A$ — Expanding terms

$= b^2(\sin^2 A + \cos^2 A) + c^2 - 2bc \cos A$ — Factoring b^2

$\sin^2 A + \cos^2 A = 1$ — From Example 7

$a^2 = b^2 + c^2 - 2bc \cos A$ — Law of cosines (2)

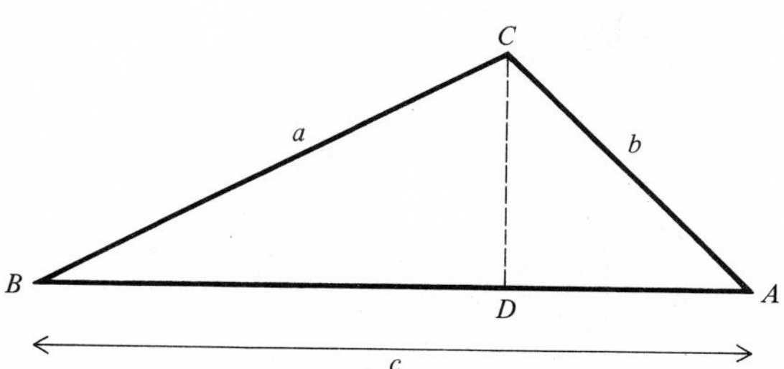

Diagram 39

Equation (2) is called the law of cosines and can be rewritten in two alternate forms by interchanging a with b *or* a with c:

FORM 1 $b^2 = a^2 + c^2 - 2ac \cos B$ Interchanging a with b

FORM 2 $c^2 = a^2 + b^2 - 2ab \cos C$ Interchanging a with c

In general, the law of cosines can be stated verbally as follows:

LAW OF COSINES The square of a side of a triangle is equal to the sum of the squares of the other two sides *minus* twice the product of these two sides with the cosine of the angle between them.

The law of cosines is used under two conditions:

1. When two sides and an included angle are given and you want to find the side opposite the given angle
2. When three sides are given and you want to find an angle opposite any one of the sides

Example 5 In $\triangle ABC$, $a = 4$, $b = 6$, $c = 68°$. Find the length of side c (refer to Diagram 40).

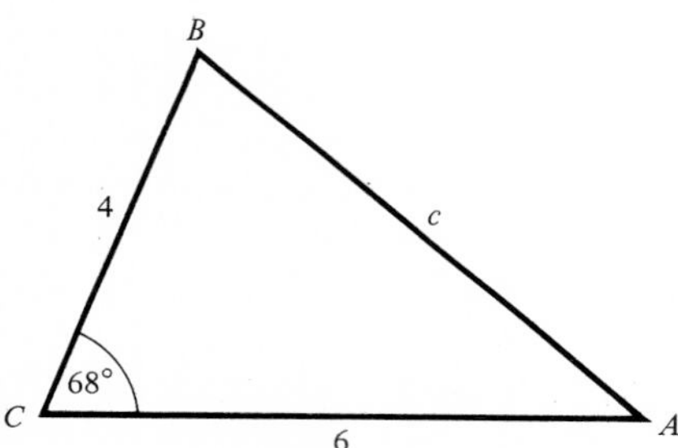

Diagram 40

Using the law of cosines,

$$c^2 = 4^2 + 6^2 - 2(4)(6) \cos 68° = 16 + 36 - 48(0.3746)$$
$$= 52 - 48(0.3746) = 52 - 18 = 34 \qquad c \approx 5.8$$

Example 6 A ship sails from port A S18°W for 12 miles to port B. It then changes direction to S71°W for 8 miles to port C. Find, to the nearest mile, the distance from port A to port C (refer to Diagram 41).

STEP 1 Referring to Diagram 41, $\angle NBA = 18°$ (if two parallel lines are cut by a transversal, the alternate interior angles are equal).

STEP 2 $\angle SBC = 71°$; therefore $\angle WBC = 19°$ Complementary angles

STEP 3 $B = 18° + B = 18° + 90° + 19° = 127°$ Diagram 41

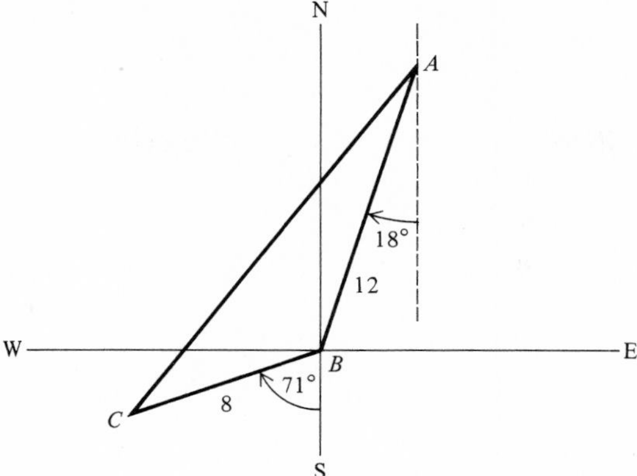

Diagram 41

STEP 4 $b^2 = 8^2 + 12^2 - 2(8)(12) \cos 127°$ Law of cosines
$= 64 + 144 - 192 \cos 127°$
$= 208 - 192 \cos 127°$

Remember that in finding cos 127° we first determine the reference angle R to be $180° - 127°$ or 53°. Since 127° is in quadrant II, $\cos 127° = -\cos 53° = -0.6018$.

STEP 5 Now

$$b^2 = 208 - 192(-0.6018) = 208 + 116 = 324$$

$$b = 18 \text{ miles}$$

The next example will consider the case when three sides of a triangle are known.

Example 7 Islands A and B are 20 miles apart. The bearing of B from A is S40°E. Two ships leave A and B at the same time at speeds of 8 and 9 miles per hour, respectively. They both take exactly 4 hours to reach port C, which is northeast of both A and B. Find, to the nearest 10 minutes, the direction followed by the ship leaving from A (refer to Diagram 42).

Referring to Diagram 42, our objective will be to find $\angle NAC$. Since this angle is *not* in $\triangle ABC$, we first determine $\angle BAC$ using the law of cosines (we refer to $\angle BAC$ as angle A).

$a^2 = b^2 + c^2 - 2bc \cos A$ Law of cosines

$a^2 - b^2 - c^2 = -2bc \cos A$ Transposing $b^2 + c^2$

$b^2 + c^2 - a^2 = 2bc \cos A$ Multiplying both sides by (-1)

$\dfrac{b^2 + c^2 - a^2}{2bc} = \cos A$ Dividing both sides by $2bc$ (1)

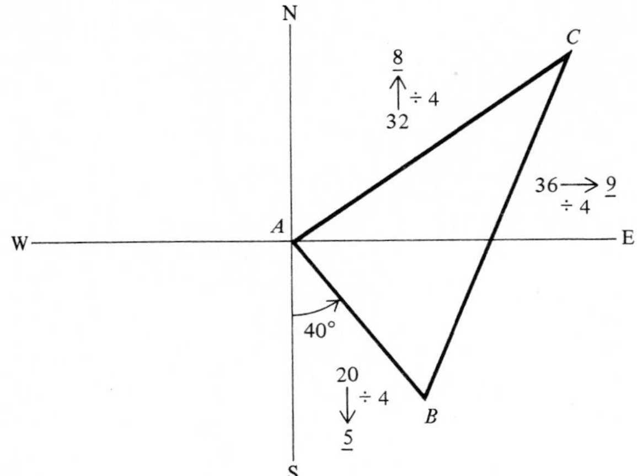

Diagram 42

COMMENT Since the sides of $\triangle ABC$ are 20-32-36, they can be divided by a common factor *without* altering the angles (this is due to the geometric property of similarity). Therefore 20-32-36 becomes 5-8-9 upon division by 4. This will simplify the computations and is only an arithmetic convenience (it is *not* necessary).

STEP 1 Now $a = 9$, $b = 8$, and $c = 5$.

$$\cos A = \frac{8^2 + 5^2 - 9^2}{2(8)(5)}$$ Substituting the given data into (1)

$$= \frac{64 + 25 - 81}{80} = \frac{8}{80} = 0.1$$

$$A = 84°20'$$ Using the tables

STEP 2 $\angle NAC = 180° - (40° + 84°20')$ Definition of supplementary angles

$$= 180° - 124°20'$$

$$= 55°40'$$

STEP 3 The bearing of C from A is N55°40′E.

EXERCISES

In Exercises 1 to 9 find the required parts of triangle ABC using the law of sines. (If there are two solutions, find all possibilities.) Express all sides to the nearest tenth and all angles to the nearest 10 minutes.

1 $A = 32°$, $B = 64°$, $c = 20$; find a and b
2 $A = 36°20'$, $B = 44°40'$, $a = 12$; find b and c
3 $a = 10$, $b = 16$, $B = 150°$; find c and A

4 $a = 3, b = 5, A = 32°40'$; find c, B, and C
5 $c = 6, b = 4.8, C = 54°20'$; find a and A
6 $a = 6, b = 8, C = 102°$; find c, A, and B
7 $a = 4, b = 6, A = 44°$; find c, B, and C
8 $a = 50, c = 40, A = 128°$; find b, B, and C
9 $a = 32, b = 36, A = 54°$; find c, B, and C

In Exercises 10 to 17 find the required parts using the law of cosines. Express all sides to the nearest tenth and all angles to the nearest 10 minutes.
10 $a = 12, b = 20, C = 24°20'$; find c
11 $b = 5, c = 12, A = 116°50'$; find a
12 $a = 3, b = 4, c = 6$; find the smallest angle of the triangle
13 $a = 4.0, b = 3.2, c = 5.6$; find the largest angle of the triangle
14 $a = 5, c = 8, B = 48°40'$; find b
15 $a = 6, b = 9, c = 10$; find A, B, and C
16 $a = 6, b = 3, c = 8$; find A, B, and C
17 $a = 4, b = 8, C = 36°$; find c, A, and B

In Exercises 18 to 32, answers should be expressed to the same degree of accuracy as the given data unless otherwise specified; all angles should be expressed to the nearest 10 minutes

18 Port B is 4.7 miles due north of port A. Port C is N64°W of B and N36°W of A. A ship sails from A to C at 6 miles per hour. Find the time required to make the trip.

19 An airplane flies 80 miles in the direction N36°W and then turns back *supposedly* on the same line of flight. Through an error the plane now flies S48°E. How far from his original starting point will he be to the nearest tenth?

20 A submarine leaves its home port and travels due east at 4 miles per hour. After 2 hours it changes course to N36°E and travels 3 hours at the same rate before reaching an island. How far is the island from the home port?

21 Port B is 70.6 miles due south of port A. An aircraft carrier is located at C, N75°E from A and N32°E from B. A plane flies in a straight line from B to C at 150 miles per hour. Find (to the nearest minute) the time required for the plane to make the flight.

22 A ship sails from port A in the direction S28°W for 9 miles to port B. It then changes course to S47°E for 14 miles to port C. Find the distance from port A to port C.

23 Atop a ski slope of 22° is a lone pine tree standing vertically. The tree casts a 170-foot shadow down the entire slope when the angle of elevation of the sun is 35°. Find the height of the tree.

24 A plane flies from an airport on a course of 160° at a speed of 250 miles per hour for 2 hours. It then changes course to 230° and continues at the same rate for another 3 hours. At the end of this 5-hour flight, how far is the plane from the airport, *and* in what direction from airport is the plane?

25 In $\triangle ABC$, $A = 52°10'$, $a = 54$, and $b = 63$. If B is obtuse, find C.

26 A ship sailing due east observes a light bearing N62°E. After the ship travels 30 miles, the bearing of the light is N48°E. If the ship continues its course, what is the closest the ship will come to the light?

27 A captain sights a lighthouse bearing N38°E. After sailing N22°30′W for 6.4 miles, he finds the bearing of the lighthouse to be N72°E. Find the distance of the ship from the lighthouse at the time of the second sighting.

28 Two ships leave a harbor at the same time, one sailing 116° at 12 miles per hour and the other 264° at 15 miles per hour. Find the distance between them after 2 hours.

29 A radar operator in an airport observes two planes flying on a collision course. One plane is N48°E of the tower, flying 500 miles per hour, and 40 miles from the tower. The other plane is N54°E of the tower, flying 400 miles per hour, and 50 miles from the tower.

(a) How far apart are the planes at this instant?

(b) If the planes continue on this collision course, how long do they have to take remedial action to avoid a crash?

30 A ship travels due south for 36 miles. It then changes direction to S35°20′E for 24 miles, where it anchors at a a home port.

(a) Find the bearing of the home port from the starting point.

(b) How many miles would the ship have saved had it traveled in a straight line from the starting point to the home port?

31 If the sides of an oblique triangle are a, b, c, the perimeter is equal to $a + b + c$. If s represents half the perimeter, $s = \frac{1}{2}(a + b + c)$. In any triangle, it can be shown that

$$\tan \frac{A}{2} = \sqrt{\frac{(s - b)(s - c)}{s(s - a)}}$$

(a) If three sides of a triangle are 36, 24, and 50, find the largest angle of the triangle.

(b) If three sides of a triangle are 4.8, 6.2, and 7.4, find the smallest angle of the triangle.

32 The formula

$$\frac{a + b}{c} = \frac{\cos[(A - B)/2]}{\sin(C/2)}$$

is known as *Mollweide's equation.* It is an extremely useful tool for checking solutions of triangles since it includes *all six parts* of a triangle.

(a) Verify the above equations for a triangle whose parts are $A = 30°$, $B = 60°$, $C = 90°$, $a = 1$, $b = \sqrt{3}$, and $c = 2$.

(b) In $\triangle ABC$, $B = 21°$, $C = 30°20'$, and $a = 34$. Find b and c to the nearest integer and check the problem using Mollweide's equation.

(c) In $\triangle ABC$, $b = 15$, $c = 25$, and $A = 68°$. Find b to the nearest integer and A and B to the nearest degree. Check the solution with Mollweide's equation.

8.6 VECTORS

Many of the ideas of algebra, geometry, and trigonometry can be represented by *vectors.* A quantity that is described by its direction as well as its magnitude is known as a vector quantity, for example a wind velocity of 40 miles per hour south or the displacement of an object 50 feet to the left. To represent these physical quantities we use directed line segments (or arrows).

DEFINITION A directed line segment is a line segment to which a positive direction has been assigned, denoted by an arrowhead at the end of the segment (refer to Diagram 43).

Symbolically a vector is denoted by two letters, the first letter indicating the initial point and the second letter indicating the terminal point (Diagram 43). The vector from A to B is written $\overrightarrow{AB}$. The *length* of the vector $\overrightarrow{AB}$ represents the magnitude and is denoted by $|\overrightarrow{AB}|$.

Diagram 43

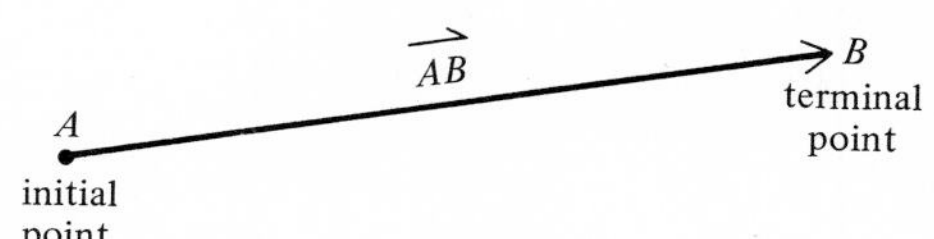

Diagram 44 shows a ship which has sailed 50 miles due east from A ($\overrightarrow{AB}$) then sailed 30 miles in the direction N30°E to C ($\overrightarrow{BC}$). The vector $\overrightarrow{AC}$ is called the sum (or *resultant*) of $\overrightarrow{AB}$ and $\overrightarrow{BC}$ and is written $\overrightarrow{AC} = \overrightarrow{AB} + \overrightarrow{BC}$.

Measuring $\overrightarrow{AC}$ to the scale of the drawing, we can *estimate* the magnitude of the resultant $|\overrightarrow{AC}|$ to be 70 miles and the angle between the resultant and the horizon to be 20°.

Diagram 44

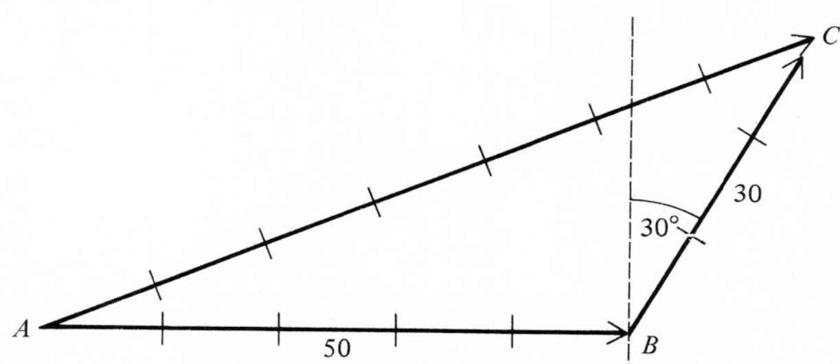

In the last example, the initial point of the second vector started at the terminal point of the first vector. However, when two forces are applied at the same point, there exists a single force which has the same effect. This single force is known as the *resultant* of the forces and can be determined using the *parallelogram law*. Let us illustrate this concept with the following example.

Example 1 A horizontal force of 130 pounds and a vertical force of 75 pounds act on an object (refer to Diagram 45). Determine the magnitude of the resultant force and the angle between the resultant and the horizontal force.

STEP 1 Draw a scale diagram representing the two forces $\overrightarrow{AB}$ and $\overrightarrow{AC}$.

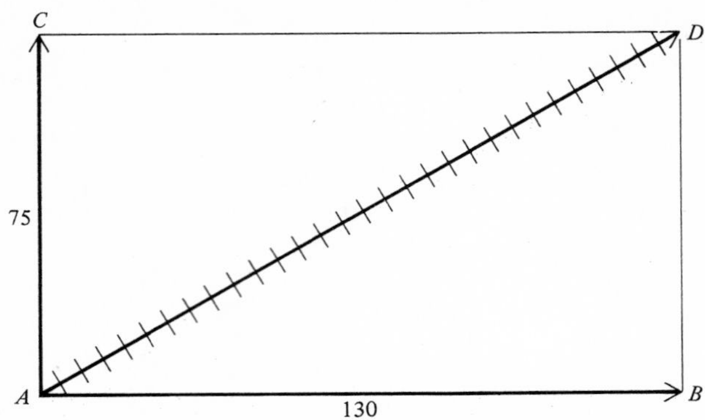

Diagram 45

STEP 2 Construct a parallelogram having $\overrightarrow{AB}$ and $\overrightarrow{AC}$ as two adjacent sides.

STEP 3 Diagonal $\overrightarrow{AD}$ represents the resultant force. *Note:* Due to the geometric properties of the parallelogram, $\overrightarrow{BD} = \overrightarrow{AC}$; therefore

$$\overrightarrow{AD} = \overrightarrow{AB} + \overrightarrow{AC} \qquad \text{or} \qquad \overrightarrow{AD} = \overrightarrow{AB} + \overrightarrow{BD}$$

STEP 4 To find AD use the pythagorean theorem:

$$AD = \sqrt{(130)^2 + (75)^2} = \sqrt{16{,}900 + 5625} = \sqrt{22{,}525} \approx 150 \text{ pounds}$$

STEP 5 $\angle DAB$ is the angle between the resultant and the horizontal.

$$\tan \angle DAB = \frac{75}{130} = 0.5769 \Longrightarrow \angle DAB = 30°$$

COMMENT Two vectors $\overrightarrow{AC}$ and $\overrightarrow{AB}$ whose sum is $\overrightarrow{AD}$ are called the *components* of $\overrightarrow{AD}$. Expressing a vector as the sum of two components is called *resolving a vector.* In Example 1, $\overrightarrow{AB}$ (130 pounds) is the horizontal component of $\overrightarrow{AD}$, and $\overrightarrow{AC}$ (75 pounds) is the vertical component of $\overrightarrow{AD}$.

SUMMARY

A vector in *standard position* has its initial point at the origin and the positive x axis as a reference line. θ is the *direction* of the vector (Diagram 46). If a perpendicular line segment is drawn from B to C on the x axis:

1 $\overrightarrow{AC}$ is called the horizontal component of $\overrightarrow{AB}$.
2 $\overrightarrow{CB}$ is called the vertical component of $\overrightarrow{AB}$.
3 $\overrightarrow{AB}$ is called the resultant of $\overrightarrow{AC}$ and $\overrightarrow{CB}$.

4 $$\tan \theta = \frac{|\overrightarrow{CB}|}{|\overrightarrow{AC}|}$$

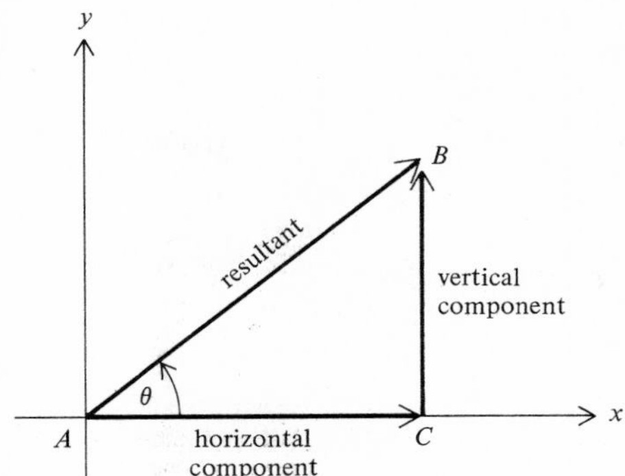

Diagram 46

5 $|\overrightarrow{CB}| = |\overrightarrow{AB}| \sin\theta$ Since $\sin\theta = \dfrac{|\overrightarrow{CB}|}{|\overrightarrow{AB}|}$

$|\overrightarrow{AC}| = |\overrightarrow{AB}| \cos\theta$ Since $\cos\theta = \dfrac{|\overrightarrow{AC}|}{|\overrightarrow{AB}|}$

6 $|\overrightarrow{AB}| = \sqrt{|\overrightarrow{AC}|^2 + |\overrightarrow{CB}|^2}$ Pythagorean theorem

Example 2 A vector has a magnitude of 12 pounds and a direction of 60°. Find the horizontal and vertical components of the vector (refer to Diagram 47).

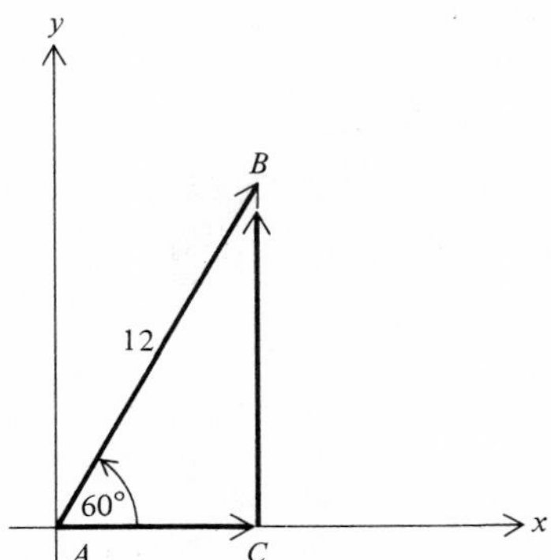

Diagram 47

Solution

$$|\overrightarrow{AC}| = |\overrightarrow{AB}|(\cos 60°) = 12(0.5) = 6 \text{ pounds}$$

$$|\overrightarrow{CB}| = |\overrightarrow{AB}|(\sin 60°) = 12(0.8660) = 10.4 \text{ pounds}$$

Therefore, $\overrightarrow{AC}$ is a 6-pound force acting to the right, and $\overrightarrow{CB}$ is a 10.4-pound force acting upward.

Example 3 The horizontal and vertical components of a vector are 24 and 8 pounds, respectively. Determine the magnitude and direction of the vector (refer to Diagram 48).

$$\tan\theta = \frac{7}{24} = 0.2917 \qquad \theta = 16° \qquad \text{direction}$$

$$|\overrightarrow{AB}| = \sqrt{(24)^2 + (7)^2} = \sqrt{576 + 49} = \sqrt{625}$$

$$= 25 \text{ pounds} \qquad \text{magnitude}$$

Various different types of practical applications can now be illustrated using the vector concept and the techniques of trigonometry.

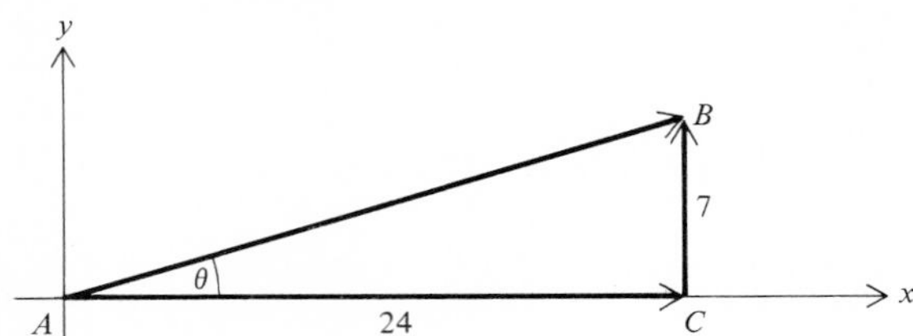

Diagram 48

Example 4 A 4000-pound automobile is parked on an inclined road that makes an angle of 22°20′ with the horizontal. Find the components of the weight (refer to Diagram 49).

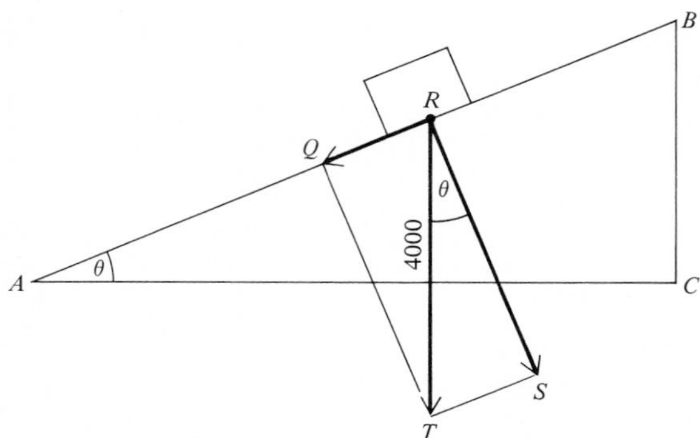

Diagram 49

$\overrightarrow{RT}$ represents the force of gravity pulling the object toward the center of the earth (perpendicular to the horizontal AC) with a force equivalent to the weight of the automobile (4000 pounds). The components of $\overrightarrow{RT}$ are $\overrightarrow{RQ}$ (parallel to the inclined plane) and $\overrightarrow{RS}$ (perpendicular to the inclined plane). $\overrightarrow{RQ}$ represents the force pulling the automobile *down* the plane (or the minimum force required to prevent the car from rolling down the incline).

Why does $\angle A = \angle TRS$? By examining Diagram 49, we can give the following justification:

1 $\overrightarrow{RT}$ is parallel to BC and forms a right triangle such that

$$\angle A + \angle QRT = 90° \quad \text{or} \quad \angle A = 90° - \angle QRT \tag{1}$$

2 Since $\angle QRS = 90°$,

$$\angle QRT + \angle TRS = 90° \quad \text{and} \quad \angle QRT = 90° - \angle TRS \tag{2}$$

3 Substituting (2) into (1), we obtain

$$\angle A = 90 - (90 - \angle TRS) = \angle TRS$$

To determine the magnitude of $\overrightarrow{ST}$,

$$\sin\theta = \frac{|\overrightarrow{ST}|}{4000} \quad \text{or} \quad |\overrightarrow{ST}| = 4000 \sin 22°30' = 4000(0.3800)$$

$$= 1520 \text{ pounds} \qquad \text{minimum braking force}$$

To determine the magnitude of $\overrightarrow{RS}$,

$$\cos\theta = \frac{|\overrightarrow{RS}|}{4000} \quad \text{or} \quad |\overrightarrow{RS}| = 4000 \cos 22°30' = 4000(0.9250)$$

$$= 3700 \text{ pounds}$$

Therefore, the parallel and perpendicular components of the 4000-pound weight are 1520 and 3700 pounds, respectively.

Example 5 Two forces of 24 and 16 pounds act on an object at an angle of 44° (Diagram 50). Find to the nearest pound the magnitude of the resultant. Find to the nearest 10 minutes the angle that the resultant makes with the smaller force.

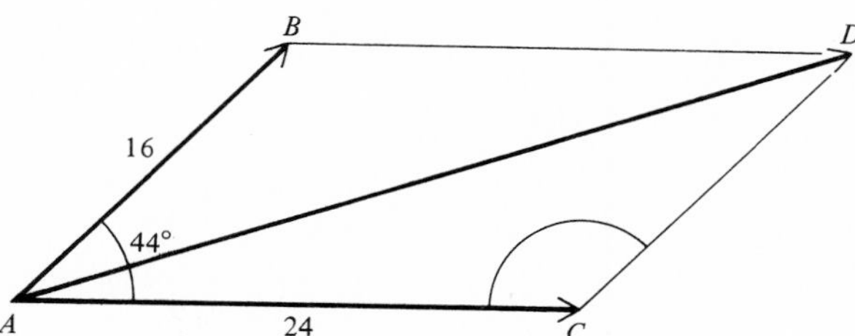

Diagram 50

Use the geometric properties of a parallelogram, $\angle ACD = 136°$ (consecutive angles of a parallelogram are supplementary) and $|\overrightarrow{CD}| = 16$ (opposite sides of a parallelogram are equal).

STEP 1 To find $|\overrightarrow{AD}|$, use the law of cosines:

$$|\overrightarrow{AD}|^2 = (16)^2 + (24)^2 - 2(16)(24)\cos 136° = 256 + 576 - 768(-0.7193)$$
$$= 832 + 552 = 1384$$

$$|\overrightarrow{AD}| \approx 37 \text{ pounds}$$

STEP 2 To find $\angle ADC$, use the law of sines:

$$\frac{\sin \angle ADC}{24} = \frac{\sin 136°}{37}$$

$$\sin \angle ADC = \frac{24 \sin 136°}{37} = \frac{24(0.6947)}{37} = 0.4506$$

$$\angle ADC = 26°50'$$

COMMENT In determining cos 136°, the reference angle R is 44°. Since the cosine is *negative* in quadrant II, $\cos 136° = -\cos 44°$.

Example 6 A river flows from north to south. To swim from the west bank directly across to the east bank, a mathematical swimmer computed that he must maintain a direction of N64°10′E. If the trip took 30 minutes and the swimmer travels at a constant velocity of 2 miles per hour in this direction, find **(a)** the width of the river and **(b)** the rate of the current (assuming that the swimmer made a perfect prediction). Refer to Diagram 51.

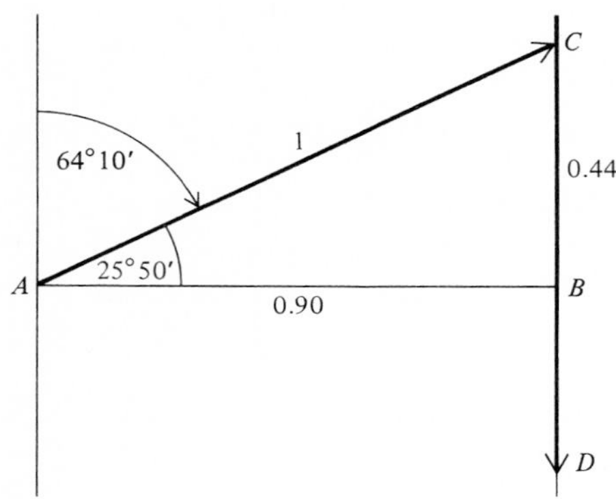

Diagram 51

STEP 1 $|\overrightarrow{AC}|$ can be determined easily by using the formula, distance = rate × time.

Hence
$$|\overrightarrow{AC}| = 2 \text{ miles/hour} \times \tfrac{1}{2} \text{ hour} = 1 \text{ mile}$$

STEP 2 To find AB (the width of the river) note that

$$\angle CAB = 25°50' \qquad \text{and} \qquad \cos 25°50' = \frac{AB}{1}$$

Now

$$AB = \cos 25°50' = 0.9001 \approx 0.90 \text{ miles}$$

STEP 3 To find CB

$$\sin 25°50' = \frac{CB}{1}$$

and $$CB = \sin 25°50' = 0.4358 \approx 0.44 \text{ mile}$$

STEP 4 To find the rate of the current, use

$$\text{Rate} = \frac{\text{distance}}{\text{time}}$$

$$\text{Rate of current} = \frac{0.44 \text{ mile}}{\frac{1}{2} \text{ hour}} = 0.88 \text{ miles per hour}$$

COMMENT In Diagram 51, $\overrightarrow{CD}$ represents 0.88 miles per hour south.

When several forces act on the same point of an object, their vector sum must equal zero in order for *balance* to occur. An application of this principle can be illustrated in the following example.

Example 7 Two cables support a 400-pound weight. One cable is at 20° to the horizontal, and the other is at 40° (Diagram 52). Find the magnitude of the force that each cable exerts in holding the weight stationary.

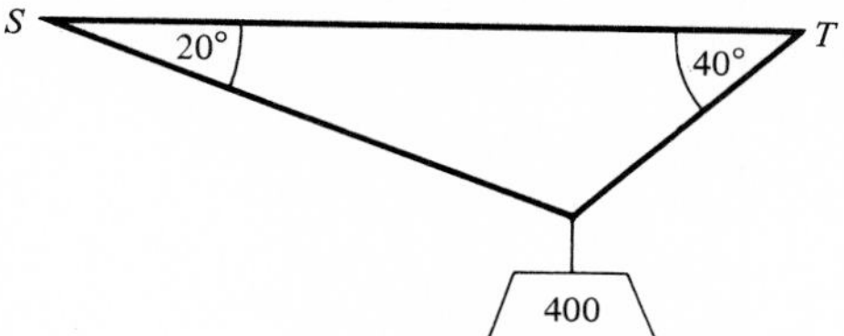

Diagram 52

COMMENT The 400-pound weight acts vertically downward, while the two cables at S and T exert forces in such a way that the weight remains motionless. Diagram 53 is called a *force diagram* for this system. In the solution to this problem we let $\mathbf{T}$, $\mathbf{T}_x$, $\mathbf{T}_y$, $\mathbf{S}$, $\mathbf{S}_x$, and $\mathbf{S}_y$ represent the vectors in the system.

The forces acting vertically upward ($\mathbf{S}_y$ and $\mathbf{T}_y$) must equal the force acting vertically downward (400 pounds):

$$\mathbf{S}_y + \mathbf{T}_y = 400 \qquad (1)$$

The force acting to the left $\mathbf{S}_x$ must equal the force acting to the right $\mathbf{T}_x$ in order for the system to be motionless.

$$\mathbf{S}_x = \mathbf{T}_x \qquad \text{or} \qquad \mathbf{S}_x - \mathbf{T}_x = 0 \qquad (2)$$

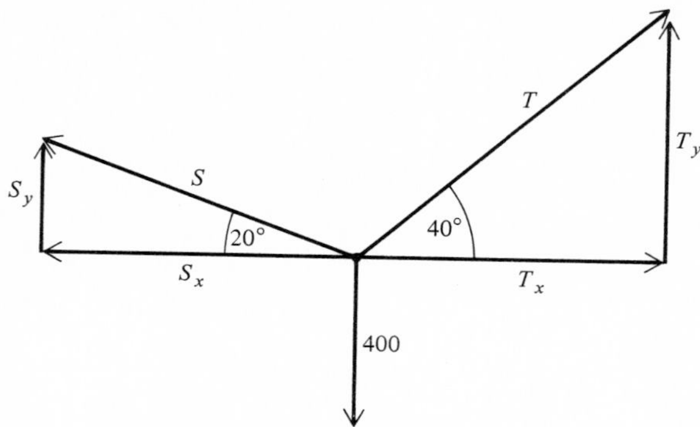

Diagram 53

$\sin 20° = \dfrac{\mathbf{S}_y}{\mathbf{S}}$ $\cos 20° = \dfrac{\mathbf{S}_x}{\mathbf{S}}$	Left triangle in Diagram 53	
$\mathbf{S}_y = \mathbf{S} \sin 20°$ $\mathbf{S}_x = \mathbf{S} \cos 20°$	Solving for vertical component $\mathbf{S}_y$ and horizontal component $\mathbf{S}_x$ of $\mathbf{S}$	(3) (4)
$\sin 40° = \dfrac{\mathbf{T}_y}{\mathbf{T}}$ $\cos 40° = \dfrac{\mathbf{T}_x}{\mathbf{T}}$	Right triangle in Diagram 53	
$\mathbf{T}_y = \mathbf{T} \sin 40°$ $\mathbf{T}_x = \mathbf{T} \cos 40°$	Solving for vertical component $\mathbf{T}_y$ and horizontal component $\mathbf{T}_x$ of $\mathbf{T}$	(5) (6)

Substituting (3) and (5) into (1) and (4) and (6) into (2), we obtain

$$\mathbf{S} \sin 20° + \mathbf{T} \sin 40° = 400 \tag{7}$$
$$\mathbf{S} \cos 20° - \mathbf{T} \cos 40° = 0 \tag{8}$$

Equations (7) and (8) form a 2 × 2 linear system which can be solved using the method of determinants as described in Section 1.8.

$$\mathbf{S} = \frac{\begin{vmatrix} 400 & \sin 40° \\ 0 & -\cos 40° \end{vmatrix}}{\begin{vmatrix} \sin 20° & \sin 40° \\ \cos 20° & -\cos 40° \end{vmatrix}} = \frac{-400 \cos 40°}{-\sin 20° \cos 40° - \sin 40° \cos 20°}$$

$$= \frac{-400(0.7660)}{-(0.3420)(0.7660) - (0.6428)(0.9397)}$$

$$= \frac{-306.4}{-0.8660} \approx 354 \text{ pounds} = \text{force exerted by cable } S$$

$$\mathbf{T} = \frac{\begin{vmatrix} \sin 20^\circ & 400 \\ \cos 20^\circ & 0 \end{vmatrix}}{-0.8660} = \frac{0 - 400 \cos 20^\circ}{-0.8660}$$

$$= \frac{-400(0.9397)}{-0.8660}$$

$$= \frac{-375.9}{-0.8660} \approx 434 \text{ pounds} = \text{force exerted by cable } T$$

EXERCISES

In Exercises 1 to 5 the first entry is the horizontal force acting on an object, and the second entry is the vertical force. Determine the magnitude of the resultant force and the angle between the resultant and the horizontal force (the direction of the vector).

1 $10\sqrt{3}$, 10 **2** 90, 48 **3** 120, 35 **4** 72, 30 **5** 50, 25

In Exercises 6 to 9 the first entry is the magnitude of a vector, and the second entry is the direction of the vector. Determine the horizontal and vertical components of the vector.

6 50, 53° **7** 65, 22°40′ **8** 85, 62° **9** 30, 30°

10 A pole is supported in a vertical position by a wire which makes an 18° angle with the pole and exerts a *pull* of 200 pounds. Find the horizontal and vertical components of the force exerted by the wire.

11 A car is moving horizontally at 100 feet per second when an object is thrown from the car at 40 feet per second at right angles to the path of the car. Find the magnitude and direction of the object.

12 A block weighing 200 pounds rests on a ramp inclined 32° with the horizontal.
(a) Find the force tending to move the block down the ramp.
(b) Find the force of the block on the ramp (this is the force perpendicular to the ramp).

13 Two forces act on a point at an angle of 102°. The first is a force of 80 pounds. If the resultant makes an angle of 38°20′ with this force, find the magnitude of the resultant (to the nearest pound).

14 Two forces, one of 72 pounds and the other 48 pounds, act on an object at an angle of 67°20′. Find to the nearest 10 minutes the angle between the resultant and the smaller force *and* the magnitude of the resultant.

15 Two forces of 22 and 38 pounds act at an angle of 43°20′. Find the angle between the resultant and the larger force *and* the magnitude of the resultant.

16 A river flows due south at the rate of 2 feet per second. A boat traveling 8 feet per second in still water heads due west across the river.
(a) Find the direction the boat will be moving.
(b) Find the speed at which the boat will be traveling.
(c) How far down river, from the point directly west of the starting point, will the boat land if the river is 400 feet wide?

17 In what direction should an airplane traveling 200 miles per hour head in order to reach a point due north if the wind is blowing 80 miles per hour from the east?

18 Suppose an airplane flying 400 miles per hour wishes to go N50°E from city A to city B. If a 60 miles per hour wind is coming *from* S36°E, what course should the plane plan to follow in order to reach city B?

19 A jet has an airspeed of 600 miles per hour and is traveling N45°E. A wind starts moving due east at 80 miles per hour. If no course corrections are made, how far will the jet be blown off course on a 1000-mile trip? How fast is the jet now traveling?

20 Given three concurrent forces acting on an object (see diagram). If the resultant of all of these forces is given by

$$\mathbf{R} = \sqrt{(\mathbf{R}_x)^2 + (\mathbf{R}_y)^2}$$

where $\mathbf{R}_x$ is the sum of horizontal components of A, B, and C and $\mathbf{R}_y$ is the sum of vertical components of A, B, and C. Determine $\mathbf{R}$ if $C = 30$ pounds, $B = 20$ pounds, $A = 40$ pounds; find the direction of $\mathbf{R}$.

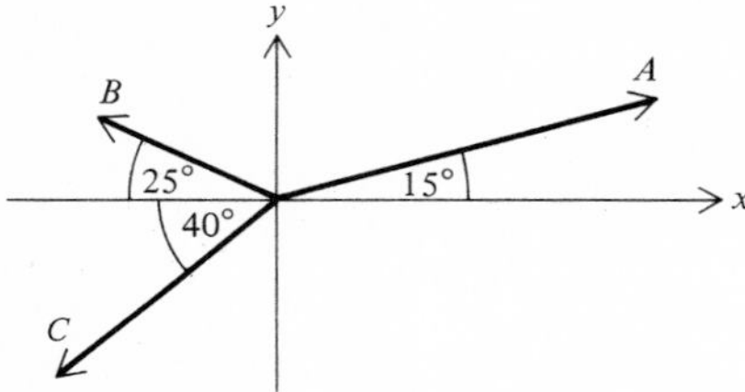

21 Find the magnitude and direction of the resultant of the forces shown in the diagram below.

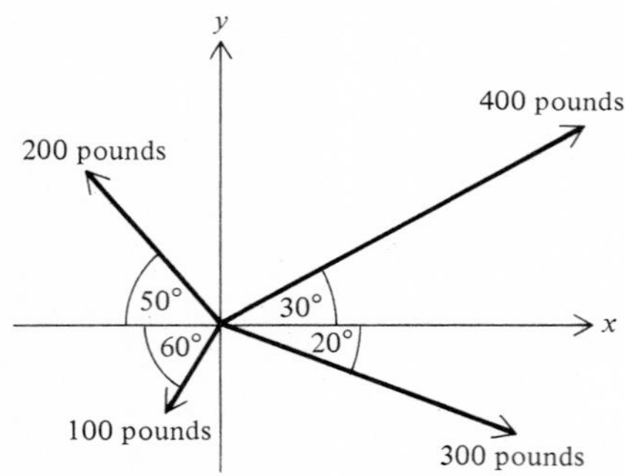

22 A block weighing 500 pounds rests on a ramp inclined 22°40′ with the horizontal.

(a) Find the force tending to move the block down the ramp.

(b) If the angle of the ramp were halved, how would that affect the answer to part **(a)**?

23 A cable is attached to the top of a pole and a 16-pound weight at the other end of the cable swings out parallel to the ground. When the cable makes an angle of 24° with the pole, find the centrifugal force on the ball $\overrightarrow{DB}$ and the force exerted by the cable $\overrightarrow{DA}$. *Note:* When an object is in motion with a constant velocity, the sum of the forces acting on the object equals zero; that is, $\overrightarrow{DA} + \overrightarrow{DB} + \overrightarrow{DC} = 0$. Refer to the diagram below.

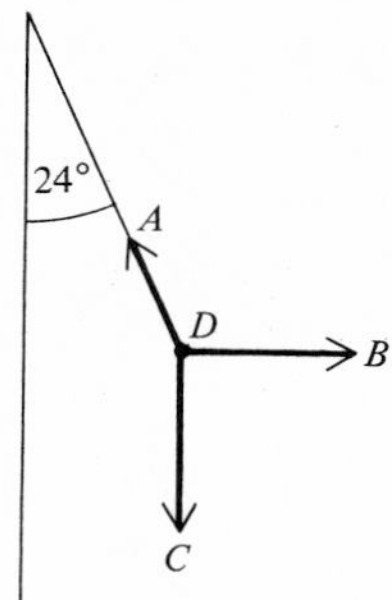

In Exercises 24 to 27 find the tensions on cables S and T such that the system is in equilibrium. Examine the diagrams for each problem.

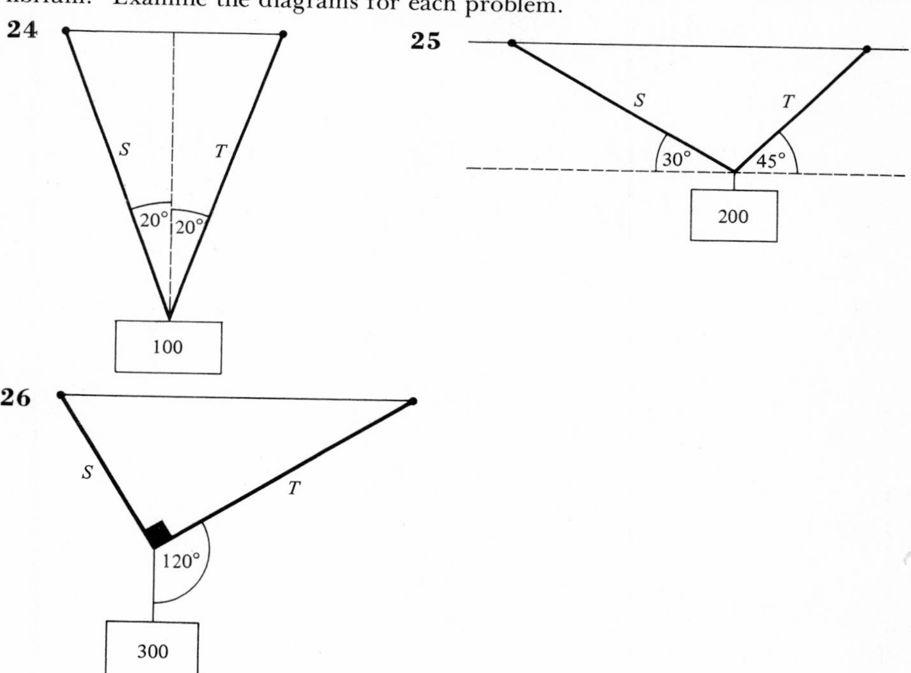

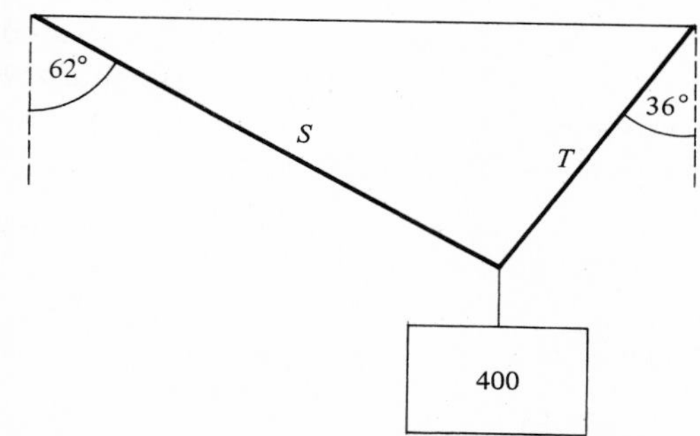

Trigonometric Functions of Real Numbers

9.1 INTRODUCTION

The functions we study in this chapter are not only interesting mathematically but also very important tools to the engineer and physicist. In the study of electricity and vibrating motion these functions appear in a natural way.

The common name of trigonometric functions defined over the real numbers is *circular functions.* We avoid this name because many students, feeling that it is a disguise to hide some dark secrets, never grasp that these are simply trigonometric functions defined over the real numbers. We stress that we are dealing with the sine or cosine of real numbers *not* the sine or cosine of so many degrees (sin 3 *not* sin 3°).

9.2 THE WINDING FUNCTION

In order to define the trigonometric functions over the real numbers, we must start by constructing and defining a new type of function, *new* in the sense that its domain is all real numbers and its range is the collection of all ordered pairs that are coordinates on the unit circle.

To start, let us consider a unit circle C (Diagram 1) (a circle of radius 1 unit), whose center is at the origin of a rectangular grid. Next, construct a

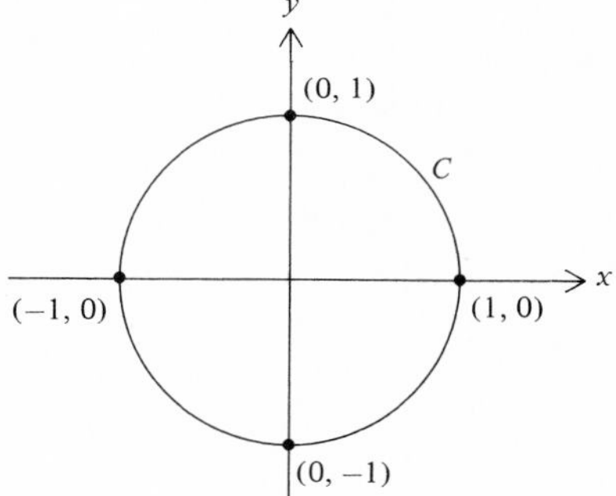

Diagram 1

real-number line L, having the same units as the x and y axes, such that L is parallel to the y axis and its zero point coincides with the point (1, 0) of the circle C (Diagram 2). Now think of L as an endless flexible cord and tightly

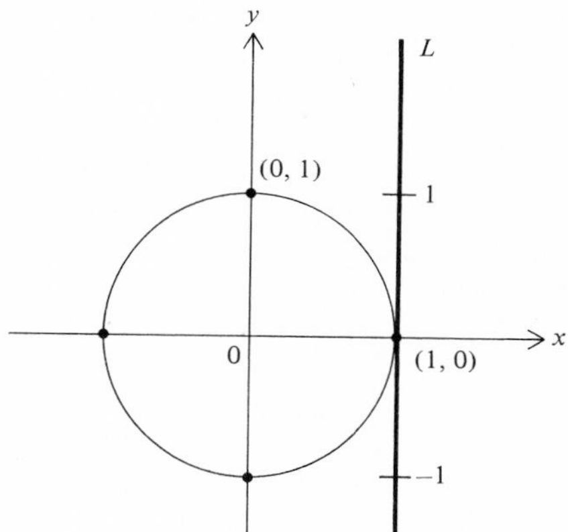

Diagram 2

wind L about C in a counterclockwise direction. This causes each point on L, associated with a positive real number, to coincide with a unique point on C. Likewise if we wind the lower end of the cord in a clockwise fashion, each point of L associated with a negative real number coincides again with a unique point on C (Diagram 3).

Easy points to detect would be points on L associated with $\pm\pi/2$, $\pm\pi$, $\pm 3\pi/2$, $+2\pi$, etc., which coincide with the *quadrantal* points of C. (A point determined by the intersection of the circle and the coordinate axes is referred to as a quadrantal point of C.) When the cord is wound tightly, you can visualize $\pi/2$ of L coinciding with (0, 1) of C and $-\pi/2$ coinciding with (0, -1) of C (Diagram 4).

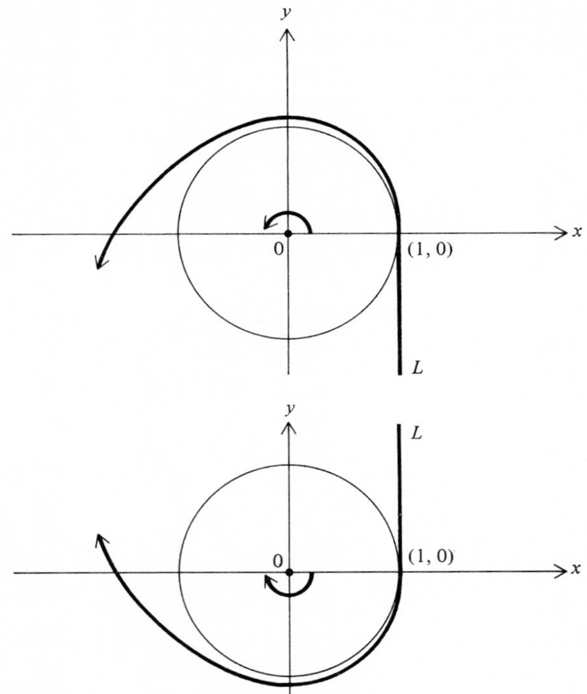

Diagram 3

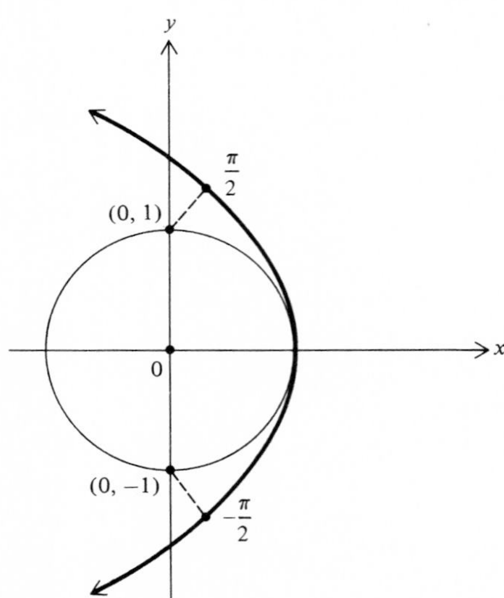

Diagram 4

You can see that since the unit circle has a circumference of length 2π, each time we wind the cord around C we use up a 2π length of L. Since we can wind as many revolutions as we want (either clockwise or counterclockwise), any real number, no matter how large in absolute value, can be made to correspond to some point on the circle C. Suppose a point on L does not coincide with any of the quadrantal points on C. How can the coordinate of C be determined? There are many points on C whose coordinates can be determined by elementary knowledge of the plane geometry of a circle. For convenience let Z represent a number on L.

Example 1 Let $Z = \pi/4$. Find $P_0(x_0, y_0)$ on circle C which is associated with Z (see Diagram 5).

The length of the arc AB is $\frac{1}{4}$, the circumference of the unit circle 2π. Therefore, arc $AB = \pi/2$.

Since $Z = \pi/4$ and L is tightly wound onto C, arc $AP_0 = \pi/4$. Therefore, arc AP_0 = arc P_0B. From geometry, since these arcs are equal, the *chords* that subtend them will also be equal; that is, chord AP_0 = chord P_0B. Using the distance formula, we obtain

$$\sqrt{(x_0 - 1)^2 + y_0^2} = \sqrt{x_0^2 + (y_0 - 1)^2}$$

$$(x_0 - 1)^2 + y_0^2 = x_0^2 + (y_0 - 1)^2 \qquad \text{Squaring both sides}$$

$$x_0^2 - 2x_0 + 1 + y_0^2 = x_0^2 + y_0^2 - 2y_0 + 1 \qquad \text{Squaring out the binomial in each side}$$

$$-2x_0 + 1 = -2y_0 + 1 \qquad \text{Simplifying}$$

$$x_0 = y_0$$

(See Diagram 5.)

Now all points on C must satisfy

$$x^2 + y^2 = 1 \tag{1}$$

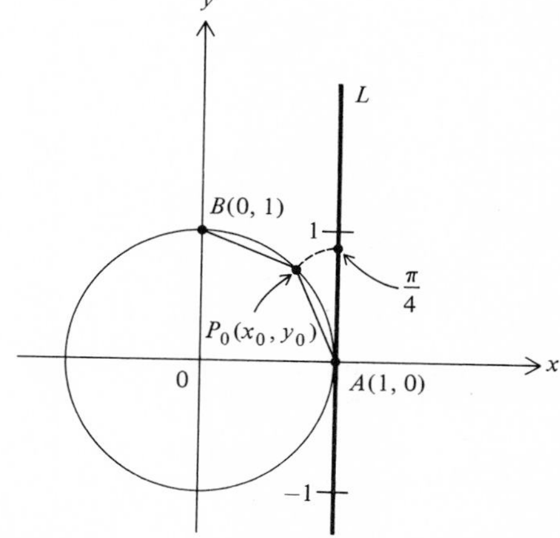

Diagram 5

(For a complete review of the equation of a circle, refer to Section 11.1.) Therefore, substituting x_0 for y_0 in Equation (1) gives

$$x_0^2 + x_0^2 = 1$$
$$2x_0^2 = 1$$
$$x_0 = \pm\sqrt{\tfrac{1}{2}} \qquad \text{Taking the square root of each side} \qquad (2)$$

Since P_0 is in the first quadrant, $x_0 > 0$ and $y_0 > 0$; hence, taking the positive square root of Equation (2), we get

$$x_0 = \sqrt{\frac{1}{2}} \quad \text{or} \quad \frac{1}{\sqrt{2}}$$

Therefore, $$x_0 = y_0 = \frac{1}{\sqrt{2}} \Longrightarrow P_0(x_0, y_0) = \left(\frac{1}{\sqrt{2}}, \frac{1}{\sqrt{2}}\right)$$

Example 2 Let $Z = \pi/6$. Find $P_0(x_0, y_0)$ on circle C which is associated with Z (see Diagram 6).

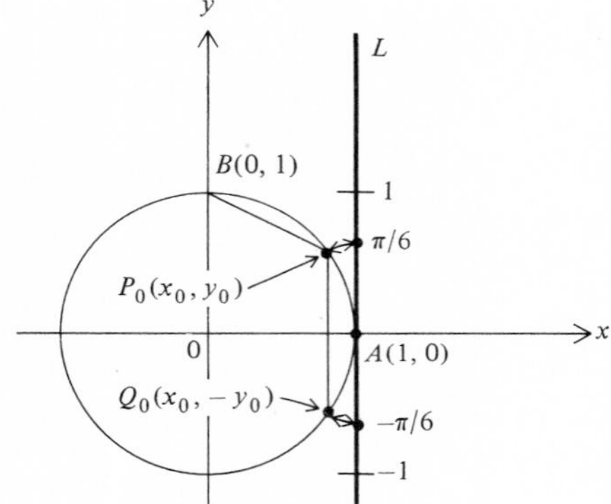

Diagram 6

Let $P_0(x_0, y_0)$ be the point on C at a distance $\pi/6$ from $A(1, 0)$. Let $Q_0(x_0, -y_0)$ also be a point on C at a distance $\pi/6$ from $A(1, 0)$ in the *opposite direction.*

$$\text{Arc } P_0Q_0 = \text{arc } AP_0 + \text{arc } AQ_0 = \frac{\pi}{6} + \frac{\pi}{6} = \frac{\pi}{3}$$

The length of arc P_0B is also $\pi/3$ since

$$\text{Arc } P_0B = \text{arc } AB - \text{arc } AP_0 = \frac{\pi}{2} - \frac{\pi}{6} = \frac{\pi}{3}$$

Since arcs of equal length on a circle are subtended by chords of equal length, chord P_0B = chord P_0Q_0. Hence, using the distance formula gives

$$\sqrt{(x_0 - 0)^2 + (y_0 - 1)^2} = 2y_0$$
$$x_0^2 + y_0^2 - 2y_0 + 1 = 4y_0^2 \qquad \text{Squaring both sides} \qquad (3)$$

Since $P_0(x_0, y_0)$ is a point on C, $x_0^2 + y_0^2 = 1$. Substituting $x_0^2 = 1 - y_0^2$ into Equation (3) leads to

$$1 - 2y_0 + 1 = 4y_0^2$$

or

$$4y_0^2 + 2y_0 - 2 = 0 \qquad \text{Subtracting } 4y_0^2 \text{ from each side}$$

$$2y_0^2 + y_0 - 1 = 0 \qquad \text{Dividing through by 2}$$

$$(2y_0 - 1)(y_0 + 1) = 0 \qquad \text{Factoring}$$

$$y_0 = \tfrac{1}{2} \quad \text{or} \quad y_0 = -1 \qquad \text{Setting each factor equal to zero and solving}$$

Since $y_0 > 0$, we reject $y_0 = -1$ and use $y_0 = \frac{1}{2}$. Again $x_0^2 + y_0^2 = 1$ and $x_0 > 0$

$$x_0^2 + \tfrac{1}{4} = 1 \Longrightarrow x_0^2 = \tfrac{3}{4} \Longrightarrow x_0 = \frac{\sqrt{3}}{2}$$

Therefore,

$$P_0(x_0, y_0) = \left(\frac{\sqrt{3}}{2}, \frac{1}{2}\right)$$

Also, we now have the point on C associated with $z = -\pi/6$, namely,

$$Q_0(x_0, -y_0) = \left(\frac{\sqrt{3}}{2}, -\frac{1}{2}\right)$$

Example 3 Let $Z = \pi/3$. Find $R_0(x_0, y_0)$ on circle C which is associated with Z (see Diagram 7).

In Example 2 we found $P(\sqrt{3}/2, \frac{1}{2})$ and $Q(\sqrt{3}/2, -\frac{1}{2})$. And the arc $P_0Q_0 = \pi/3$. Hence arc AR_0 = arc P_0Q_0; therefore, chord AR_0 = chord P_0Q_0. Since chord $P_0Q_0 = 1$, chord $AR_0 = 1$. And the distance from A to R_0 is

$$\sqrt{(x_0 - 1)^2 + y_0^2} = 1 \qquad (1)$$

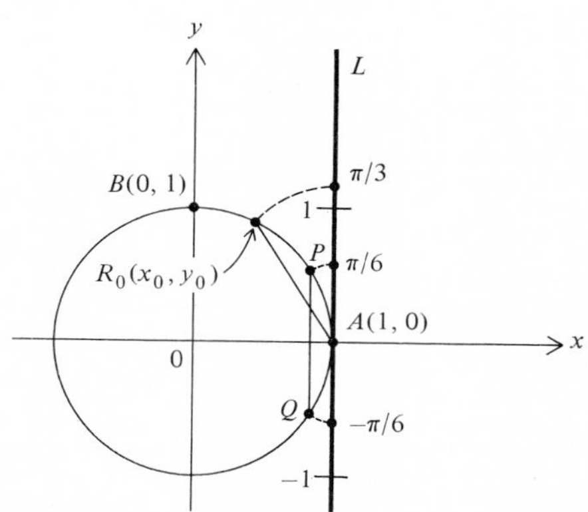

Diagram 7

$$x_0^2 - 2x_0 + 1 + y_0^2 = 1 \qquad \text{Squaring both sides of (1)} \qquad (2)$$

$$x_0^2 + y_0^2 = 1 \Longrightarrow y_0^2 = 1 - x_0^2$$

$$-2x_0 + 2 = 1 \qquad \text{Substituting } 1 - x_0^2 \text{ for } y_0^2 \text{ in (2)}$$

$$2x_0 = 1 \Longrightarrow x_0 = \tfrac{1}{2}$$

Again, since $x_0^2 + y_0^2 = 1$, we have

$$\tfrac{1}{4} + y_0^2 = 1 \qquad \text{Substituting } x_0 = \tfrac{1}{2}$$

$$y_0^2 = \tfrac{3}{4} \Longrightarrow y_0 = \frac{\sqrt{3}}{2} \qquad \text{Since } y_0 > 0$$

Therefore,
$$P_0(x_0, y_0) = \left(\frac{1}{2}, \frac{\sqrt{3}}{2}\right)$$

These three examples illustrate that for any real number Z on the line L we can associate a point P on the circle C. In fact, for any real number, $Z + 2\pi n$, where n is an integer, the line L wraps itself about C in such a way that P is associated with Z *and* $Z + 2\pi n$.

Since π on L associates itself with $(-1, 0)$ on the unit circle, we know that $\pi + 2\pi = 3\pi$ also associates itself with $(-1, 0)$.

Recalling the definition of a function (Chapter 2), we can see that the *winding process* which associates real numbers Z on L with points (x, y) on C establishes a function.

DEFINITION 1 The *winding function* is $W(Z) = (x, y)$, where Z belongs to L and (x, y) belongs to C. The correspondence between z and (x, y) is established by the winding process.

Using functional notation, we can now write

$$W(0) = (1, 0) \qquad W\left(\frac{\pi}{3}\right) = \left(\frac{1}{2}, \frac{\sqrt{3}}{2}\right)$$

$$W\left(\frac{\pi}{4}\right) = \left(\frac{1}{\sqrt{2}}, \frac{1}{\sqrt{2}}\right) \qquad W\left(\frac{\pi}{2}\right) = (0, 1)$$

$$W\left(\frac{\pi}{6}\right) = \left(\frac{\sqrt{3}}{2}, \frac{1}{2}\right) \qquad W(\pi) = (-1, 0)$$

We have observed that for all real Z

$$W(Z + 2n\pi) = W(Z) \qquad n \text{ an integer}$$

This gives birth to an important property of the winding function: it is *periodic* (repeating at regular intervals).

Example 4

$$W(5\pi) = W(\pi + 2[2]\pi) = W(\pi) = (-1, 0)$$
$$W(-\pi) = W(\pi + 2[-1]\pi) = W(\pi) = (-1, 0)$$
$$W(314) \approx W(0 + 2[50]\pi) \approx W(0) \approx (1, 0)$$

Observe that $314 \approx 100\pi$. Therefore, since the circumference of C is 2π,

$$\frac{100\pi}{2\pi} = 50$$

which means that the number 50 represents the number of counterclockwise windings of L about C. Therefore, $Z = 314$ on L associates itself *approximately* with $(1, 0)$ on C.

DEFINITION 2 Consider a function $f(x)$ with a domain D such that:
(a) Whenever x belongs to D, $x + p$ belongs to D (p is a positive number).
(b) $f(x + p) = f(x)$ for each x in D.
Then $f(x)$ is a *periodic function* of period p.

The winding function is an example of a periodic function since

$$W(Z + 2n\pi) = W(Z)$$

$2n\pi$ are the periods of W. However, $p = 2\pi$ is the least positive period and is known as the *fundamental period of W*.

EXERCISES

Find (x, y) in Exercises 1 to 4.

1 $W\left(\frac{3\pi}{2}\right)$ **2** $W\left(\frac{-3\pi}{2}\right)$ **3** $W(18\pi)$ **4** $W(-3\pi)$

Using the symmetry of the circle and the results of the examples of Section 9.2

$$W\left(\frac{\pi}{4}\right) = \left(\frac{1}{\sqrt{2}}, \frac{1}{\sqrt{2}}\right) \qquad W\left(\frac{\pi}{3}\right) = \left(\frac{1}{2}, \frac{\sqrt{3}}{2}\right) \qquad W\left(\frac{\pi}{6}\right) = \left(\frac{\sqrt{3}}{2}, \frac{1}{2}\right)$$

determine (x, y) in Exercises 5 to 13.

5 $W\left(\frac{5\pi}{6}\right)$ **6** $W\left(\frac{7\pi}{6}\right)$ **7** $W\left(\frac{11\pi}{6}\right)$ **8** $W\left(\frac{3\pi}{4}\right)$

9 $W\left(\frac{5\pi}{4}\right)$ **10** $W\left(\frac{7\pi}{4}\right)$ **11** $W\left(\frac{2\pi}{3}\right)$ **12** $W\left(\frac{4\pi}{3}\right)$

13 $W\left(\frac{5\pi}{3}\right)$

In Exercises 14 to 22, determine (x, y) and verify the results by using the relationship $W(Z) = W(Z + 2\pi)$.

14 $W\left(\frac{-5\pi}{6}\right)$ **15** $W\left(\frac{-7\pi}{6}\right)$ **16** $W\left(\frac{-11\pi}{6}\right)$ **17** $W\left(\frac{-3\pi}{4}\right)$

18 $W\left(\frac{-5\pi}{4}\right)$ **19** $W\left(\frac{-7\pi}{4}\right)$ **20** $W\left(\frac{-2\pi}{3}\right)$ **21** $W\left(\frac{-4\pi}{3}\right)$

22 $W\left(\frac{-5\pi}{3}\right)$

If $0 < Z < \pi/2$ and $W(Z) = (x, y)$, simplify expressions 23 to 26.

23 $W(\pi + Z)$ **24** $W(\pi - Z)$ **25** $W(2\pi - Z)$ **26** $W(4\pi + Z)$

27 Show that the winding function is *not* one to one.

28 Suppose that $\pi/2 < Z < \pi$. Show that no value of k exists such that $W(Z) = k[W(-Z)]$.

9.3
THE TRIGONOMETRIC FUNCTIONS COSINE AND SINE

In Section 9.2 we developed the winding function. This was specifically done to supply a means to an end, the end being the definition of the cosine and sine functions (commonly abbreviated cos and sin).

DEFINITION 1 Consider $W(Z) = (x, y)$; then

$$x = \cos Z \qquad \text{and} \qquad y = \sin Z$$

This definition compactly says that the domain for the cosine and sine functions is all real numbers. For any real value Z the ranges for $\cos Z$ and $\sin Z$ are simply the first and second coordinates of W, respectively, where W represents the winding function.

Specifically every ordered pair (x, y) of W must lie on the unit circle, $x^2 + y^2 = 1$, and it follows that $|x| \leq 1$ and $|y| \leq 1$, which means

$$|\cos Z| \leq 1 \qquad \text{and} \qquad |\sin Z| \leq 1 \tag{1}$$

Accepting these functions as continuous and restating expressions (1), we see that the ranges of these functions are confined to the interval $[-1, 1]$.

By definition,

$$W(Z) = (x, y) = (\cos Z, \sin Z) \tag{2}$$

Hence, Equation (2) gives us a method of determining $\cos Z$ and $\sin Z$ for specific values of Z.

Example 1 Find $\sin(\pi/4)$ and $\cos(\pi/4)$.
From Section 9.2 we found $W(\pi/4) = (1/\sqrt{2}, 1/\sqrt{2})$; hence,

$$\sin\frac{\pi}{4} = \frac{1}{\sqrt{2}} \qquad \text{and} \qquad \cos\frac{\pi}{4} = \frac{1}{\sqrt{2}}$$

Example 2 Find $\cos 36\pi$.
Since $36\pi = 2(18)\pi$ and W has a fundamental period of 2π, we know that

$$W(0 + 2[18]\pi) = W(0) = (1, 0) \qquad \text{and} \qquad \cos 36\pi = \cos 0 = 1$$

The winding function is periodic, and its fundamental period is 2π. It follows that the cosine and sine functions are also periodic and have the same fundamental period 2π.

Hence for $n = 0, 1, 2, 3, \ldots$

$$\cos(Z \pm 2n\pi) = \cos Z \qquad \sin(Z \pm 2n\pi) = \sin Z$$

Example 3 Find $\sin(\frac{19}{3}\pi)$.
First, $\frac{19}{3}\pi = 6\frac{1}{3}\pi = 6\pi + \pi/3$. Now since

$$\sin(Z + 2n\pi) = \sin Z \qquad n = 0, 1, 2, 3, \ldots$$

if $n = 3$,

$$\sin\left[\frac{\pi}{3} + 2(3)\pi\right] = \sin\frac{\pi}{3} = \frac{\sqrt{3}}{2}$$

Example 4 Find $\sin 1.20$.
Appendix Table 2 can be used to find the approximate answer

$$\sin 1.20 = 0.93204$$

Example 5 Find $\cos 1.00$ using Appendix Table 2.

Solution $\cos 1.00 = 0.5403$

EXERCISES

1 Complete the table on pages 262 and 263.

In Exercises 2 to 5 $W(Z) = (x, y)$.

2 If $\cos Z > 0$, and $\sin Z < 0$, in what quadrant is (x, y)?
3 If $\cos Z < 0$, and $\sin Z > 0$, in what quadrant is (x, y)?
4 If $\cos Z < 0$, and $\sin Z = -\frac{5}{13}$, find $\cos Z$.
5 If $\cos Z = \frac{3}{5}$, and $\sin Z < 0$, find $\sin Z$.

Using Appendix Table 2, find to the nearest hundredth the values of the expressions in Exercises 6 to 11.

6 $\sin 0.5$ **7** $\sin 1.00$ **8** $\cos 1.45$ **9** $\cos 0.75$
10 $\sin 0.82$ **11** $\cos 1.67$

Find the value of each expression in Exercises 12 to 15.

12 $\sin(\pi + 0.5)$ **13** $\sin(3\pi - 0.5)$ **14** $\cos(2\pi + 1.00)$
15 $\cos(2\pi - 1.00)$

Z	$W(Z)$	$\cos Z$	$\sin Z$
0			
$\frac{\pi}{3}$			
$\frac{2\pi}{3}$			
$\frac{4\pi}{3}$			
$\frac{5\pi}{3}$			
$\frac{\pi}{6}$			
$\frac{5\pi}{6}$			
$\frac{7\pi}{6}$			
$\frac{11\pi}{6}$			
$\frac{\pi}{4}$			
$\frac{3\pi}{4}$			
$\frac{5\pi}{4}$			
$\frac{7\pi}{4}$			
$\frac{\pi}{2}$			
π			
$\frac{3\pi}{2}$			

Z	$W(Z)$	$\cos Z$	$\sin Z$
2π			
100π			
$\frac{22\pi}{3}$			
$\frac{43\pi}{4}$			
$\frac{-9\pi}{4}$			
$\frac{37\pi}{6}$			
$\frac{-13\pi}{6}$			

9.4 SOME IMPORTANT TRIGONOMETRIC FORMULAS

Theorem 1 For all real numbers Z,

$$\cos^2 Z + \sin^2 Z = 1 \tag{1}$$

Proof For any point (x, y) on C, $x^2 + y^2 = 1$. Now $x = \cos Z$, and $y = \sin Z$. Substituting into $x^2 + y^2 = 1$, we obtain

$$(\cos Z)^2 + (\sin Z)^2 = 1$$

which is usually written $$\cos^2 Z + \sin^2 Z = 1$$

Example 1 Verify Theorem 1 for $Z = \pi/3$.

Solution

$$\sin \frac{\pi}{3} = \frac{\sqrt{3}}{2} \qquad \cos \frac{\pi}{3} = \frac{1}{2} \Longrightarrow \sin^2 \frac{\pi}{3} = \frac{3}{4} \text{ and } \cos^2 \frac{\pi}{3} = \frac{1}{4}$$

$$\Longrightarrow \sin^2 \frac{\pi}{3} + \cos^2 \frac{\pi}{3} = \frac{3}{4} + \frac{1}{4} = 1$$

Example 2 Verify Theorem 1 for $Z = 1$ (refer to Appendix Table 2).

Solution

$$\sin 1 \approx 0.84147 \qquad \cos 1 \approx 0.54030$$
$$\sin^2 1 \approx 0.70807 \qquad \cos^2 1 \approx 0.29192$$
$$\sin^2 1 + \cos^2 1 \approx 0.70807 + 0.29192 \approx 0.99999$$

COMMENT We stress the fact that sin 1 is *not* sin 1° but rather the sine of the *real number* 1.

To derive other useful formulas, it will be convenient to start with two real numbers, Z_1 and Z_2. For this example, $Z_2 > Z_1$, and since $Z_2 - Z_1$ is a positive real number, using Theorem 1 yields

$$\sin^2(Z_2 - Z_1) + \cos^2(Z_2 - Z_1) = 1$$

Diagram 8 shows what we are doing. Our objective in the following development is to express $\cos(Z_2 - Z_1)$ in terms of sine and cosine.

On L we show Z_1 and Z_2. The length of the segment between Z_1 and Z_2 is d. We will construct d from $A(1, 0)$ to a point $Z_2 - Z_1$ on L. When we wind L onto C, arc AB has the same length as arc DE, and therefore chord AB has the same length as chord DE.

Examine Diagram 8 carefully and verify (x, y) for B, D, and E.

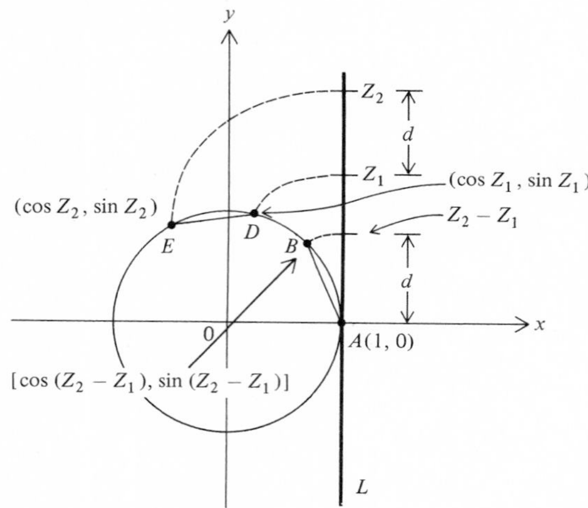

Diagram 8

Now, using the distance formula,

$$AB = \sqrt{[\cos(Z_2 - Z_1) - 1]^2 + \sin^2(Z_2 - Z_1)}$$

and

$$DE = \sqrt{(\cos Z_2 - \cos Z_1)^2 + (\sin Z_2 - \sin Z_1)^2}$$

Since $AB = DE$, squaring both radical expressions and simplifying yields

$$\underline{\cos^2(Z_2 - Z_1)} - 2\cos(Z_2 - Z_1) + 1 + \underline{\sin^2(Z_2 - Z_1)} =$$
$$\underline{\cos^2 Z_2} - 2\cos Z_1 \cos Z_2 + \underline{\cos^2 Z_1} + \underline{\sin^2 Z_2} - 2\sin Z_1 \sin Z_2 + \underline{\sin^2 Z_1}$$

Using Theorem 1, $\cos^2 Z + \sin^2 Z = 1$, and applying the theorem to the underlined terms, we have

$$2 - 2\cos(Z_2 - Z_1) = 2 - 2\cos Z_1 \cos Z_2 - 2\sin Z_1 \sin Z_2$$

Simplifying the above equation leads to

$$\cos(Z_2 - Z_1) = \cos Z_1 \cos Z_2 + \sin Z_1 \sin Z_2 \tag{2}$$

Example 3 Verify Equation (2) for $Z_1 = \pi/6$ and $Z_2 = \pi/3$,

Solution

$$Z_2 - Z_1 = \frac{\pi}{3} - \frac{\pi}{6} = \frac{\pi}{6}$$

and we know

$$\cos\frac{\pi}{6} = \frac{\sqrt{3}}{2}$$

Now

$$\cos\left(\frac{\pi}{3} - \frac{\pi}{6}\right) = \cos\frac{\pi}{3}\cos\frac{\pi}{6} + \sin\frac{\pi}{3}\sin\frac{\pi}{6}$$

$$= \frac{1}{2}\frac{\sqrt{3}}{2} + \frac{\sqrt{3}}{2}\frac{1}{2} = \frac{\sqrt{3}}{4} + \frac{\sqrt{3}}{4} = \frac{\sqrt{3}}{2}$$

Hence,

$$\cos\frac{\pi}{6} = \cos\left(\frac{\pi}{3} - \frac{\pi}{6}\right)$$

Equation (2) can now be stated as a theorem.

Theorem 2 For all real numbers Z_1 and Z_2

$$\cos(Z_2 - Z_1) = \cos Z_2 \cos Z_1 + \sin Z_2 \sin Z_1$$

Several theorems now follow.

Theorem 3 For any real number Z,

$$\cos(-Z) = \cos Z$$

Proof Write $\cos(-Z) = \cos(0 - Z)$. Using Theorem 2,

$$\cos(0 - Z) = \cos 0 \cos Z + \sin 0 \sin Z = 1 \cos Z + 0 \sin Z = \cos Z$$

Theorem 4 For any real number Z,

$$\cos\left(\frac{\pi}{2} - Z\right) = \sin Z$$

Proof Using Theorem 2

$$\cos\left(\frac{\pi}{2} - Z\right) = \cos\frac{\pi}{2}\cos Z + \sin\frac{\pi}{2}\sin Z = 0\cos Z + 1\sin Z = \sin Z$$

Theorem 5 For any real number Z

$$\sin\left(\frac{\pi}{2} - Z\right) = \cos Z$$

Proof $\cos\left(\frac{\pi}{2} - Z\right) = \sin Z$ Applying Theorem 4

Let $T = \pi/2 - Z$; then $Z = \pi/2 - T$, and we have

$$\cos T = \sin\left(\frac{\pi}{2} - T\right) \Longrightarrow \sin\left(\frac{\pi}{2} - Z\right) = \cos Z$$ Replacing T with Z

Theorem 6 For any real number Z,

$$\sin(-Z) = -\sin Z$$

Proof Using Theorem 4,

$$\sin(-Z) = \cos\left[\frac{\pi}{2} - (-Z)\right]$$ Applying Theorem 4

$$= \cos\left(\frac{\pi}{2} + Z\right)$$ Simplifying

Rewriting the right-hand member of the equation to fit the form of Theorem 2, we obtain

$$\sin(-Z) = \cos\left[Z - \left(-\frac{\pi}{2}\right)\right]$$

$$\sin(-Z) = \cos Z\cos\left(-\frac{\pi}{2}\right) + \sin Z\sin\left(-\frac{\pi}{2}\right)$$ Using Theorem 2

$$= (\cos Z)(0) + (\sin Z)(-1) = -\sin Z$$

The proofs of Theorems 7 to 9 are left as exercises.

Theorem 7 For all real numbers Z_1 and Z_2

$$\cos(Z_1 + Z_2) = \cos Z_1\cos Z_2 - \sin Z\sin Z_2$$

Theorem 8 For all real numbers Z_1 and Z_2

$$\sin(Z_1 + Z_2) = \sin Z_1 \cos Z_2 + \cos Z_1 \sin Z_2$$

Theorem 9 For all real numbers Z_1 and Z_2

$$\sin(Z_1 - Z_2) = \sin Z_1 \cos Z_2 - \cos Z_1 \sin Z_2$$

Example 4 Find $\cos(\pi/12)$.
Since $\pi/12$ can be written $\pi/3 - \pi/4$, we can use Theorem 2 and get

$$\cos\frac{\pi}{12} = \cos\left(\frac{\pi}{3} - \frac{\pi}{4}\right) = \cos\frac{\pi}{3}\cos\frac{\pi}{4} + \sin\frac{\pi}{3}\sin\frac{\pi}{4} = \frac{1}{2}\frac{1}{\sqrt{2}} + \frac{\sqrt{3}}{2}\frac{1}{\sqrt{2}}$$

$$= \frac{1 + \sqrt{3}}{2\sqrt{2}}$$

Example 5 Find $\cos(-\pi/12)$.
By Theorem 3 and Example 4,

$$\cos\left(-\frac{\pi}{12}\right) = \cos\frac{\pi}{12} = \frac{1 + \sqrt{3}}{2\sqrt{2}}$$

Example 6 Find $\cos(5\pi/12)$.

Since $\dfrac{5\pi}{12} = \dfrac{\pi}{4} + \dfrac{\pi}{6}$

$$\cos\frac{5\pi}{12} = \cos\left(\frac{\pi}{4} + \frac{\pi}{6}\right) = \cos\frac{\pi}{4}\cos\frac{\pi}{6} - \sin\frac{\pi}{4}\sin\frac{\pi}{6} \qquad \text{Using Theorem 7}$$

$$= \frac{1}{\sqrt{2}}\frac{\sqrt{3}}{2} - \frac{1}{\sqrt{2}}\frac{1}{2} = \frac{\sqrt{3} - 1}{2\sqrt{2}}$$

Example 7 Evaluate $\sin\dfrac{\pi}{3}\cos\dfrac{\pi}{6} + \cos\dfrac{\pi}{3}\sin\dfrac{\pi}{6}$.

There are two basic methods.

METHOD 1

$$\sin\frac{\pi}{3}\cos\frac{\pi}{6} + \cos\frac{\pi}{3}\sin\frac{\pi}{6} = \frac{\sqrt{3}}{2}\frac{\sqrt{3}}{2} + \frac{1}{2}\frac{1}{2} = \frac{3}{4} + \frac{1}{4} = 1$$

METHOD 2

Recognizing the problem to be *in the form* of the right-hand member of Theorem 8, we can conclude that

$$\sin \frac{\pi}{3} \cos \frac{\pi}{6} + \cos \frac{\pi}{3} \sin \frac{\pi}{6} = \sin \left(\frac{\pi}{3} + \frac{\pi}{6} \right) = \sin \frac{\pi}{2} = 1$$

We have just stated nine formulas. There are many others, and they are often valuable. You should strive to learn them. The best way to master these formulas and become familiar with the trigonometric functions is by working out many problems yourself. The exercises for this chapter will give you this opportunity.

EXERCISES

Use the methods of this section to find the value of each expression in Exercises 1 to 16.

1 $\sin \frac{\pi}{12}$ **2** $\sin \left(-\frac{\pi}{12}\right)$ **3** $\sin \frac{5\pi}{12}$ **4** $\cos \frac{7\pi}{12}$

5 $\cos \frac{13\pi}{12}$ **6** $\sin \left(-\frac{5\pi}{12}\right)$ **7** $\cos \left(-\frac{5\pi}{12}\right)$ **8** $\sin \frac{49\pi}{12}$

9 $\cos \left(-\frac{49\pi}{12}\right)$ **10** $\sin(-\pi)$ **11** $\cos(-\pi)$ **12** $\cos \frac{5\pi}{6}$

13 $\sin \frac{5\pi}{6}$ **14** $\cos \left(-\frac{5\pi}{12}\right)$ **15** $\sin \frac{17\pi}{6}$ **16** $\cos \left(-\frac{17\pi}{6}\right)$

17 Show that $\cos \left(\frac{3\pi}{2} - Z\right) = -\sin Z$.

18 Show that $\sin \left(\frac{3\pi}{2} - Z\right) = -\cos Z$.

19 Prove Theorem 7. **20** Prove Theorem 8. **21** Prove Theorem 9.
22 Prove that $\cos(\pi - Z) = -\cos Z$.
23 Prove that $\cos(\pi + Z) = -\cos Z$.
24 Prove that $\sin(\pi - Z) = \sin Z$.
25 Prove that $\sin(\pi + Z) = -\sin Z$.
26 Prove that $\cos 2Z = \cos^2 Z - \sin^2 Z$.
27 Prove that $\sin 2Z = 2 \sin Z \cos Z$.

28 Prove that $\cos \frac{Z}{2} = \pm \sqrt{\frac{1 + \cos Z}{2}}$, and explain when to use + or −.

29 Prove that $\sin \frac{Z}{2} = \pm \sqrt{\frac{1 - \cos Z}{2}}$, and explain when to use + or −.

30 Prove that $2 \sin Z_1 \cos Z_2 = \sin(Z_1 + Z_2) + \sin(Z_1 - Z_2)$.
31 Prove that $2 \cos Z_1 \cos Z_2 = \cos(Z_1 + Z_2) + \cos(Z_1 - Z_2)$.
32 Prove that $2 \sin Z_1 \sin Z_2 = \cos(Z_1 - Z_2) - \cos(Z_1 + Z_2)$.
33 Derive a formula for $\cos 3Z$ in terms of $\cos Z$.
34 Derive a formula for $\sin 3Z$ in terms of $\sin Z$.
35 Is the following statement true or false? Analyze carefully.

$$\frac{\sin Z_1}{\sin Z_2} = \frac{Z_1}{Z_2}$$

36 Using two different methods, evaluate

$$\cos \frac{3\pi}{2} \cos \frac{\pi}{6} - \sin \frac{3\pi}{2} \sin \frac{\pi}{6}$$

37 Using two different methods, evaluate

$$\cos \frac{3\pi}{2} \sin \frac{\pi}{4} - \cos \frac{\pi}{4} \sin \frac{3\pi}{2}$$

38 Using two different methods, evaluate

$$\cos \frac{\pi}{2} \sin \frac{\pi}{3} + \cos \frac{\pi}{3} \sin \frac{\pi}{2}$$

9.5 NOTATION

We defined the cosine and sine functions as $x = \cos Z$ and $y = \sin Z$, where x and y are the coordinates of any point (x, y) belonging to the unit circle C. A notation problem arises. We have two different functions each having the independent variable Z; but in the case of the cosine function we have a dependent variable x, and in the case of the sine function we have a dependent variable y. If we try to graph both functions on the same grid, the confusion of plotting ordered pairs (x, Z) and (y, Z) becomes evident. We can simplify the situation by renaming the independent and dependent variables as we please.

Example 1

$$y = \sin \theta \qquad y = \cos \theta$$
$$v = \sin u \qquad v = \cos u$$
$$y = \sin x \qquad y = \cos x$$

In functional notation we can now write

$$f(x) = \cos x \qquad \text{and} \qquad g(x) = \sin x$$

9.6 GRAPHING $f(x) = \cos x$ AND $g(x) = \sin x$

Let us recall some earlier findings about $f(x)$ and $g(x)$:

1 Both functions are periodic with a fundamental period 2π.

2 $-1 \le \cos x \le 1$, or $|\cos x| \le 1$, and $-1 \le \sin x \le 1$, or $|\sin x| \le 1$.

3 The domain of both functions is all real numbers.

What do these facts tell us about $f(x)$ and $g(x)$?

1 Once we have a picture of $f(x)$ and $g(x)$ for any interval of length 2π, this part of the graph repeats indefinitely in either direction.

2 The graphs of $f(x)$ and $g(x)$ lie within the lines $y = 1$ and $y = -1$.

3 A complete graph of either function will traverse the entire horizontal axis.

All we need now is enough points to give us confidence to plot a fairly accurate graph of each function. First we set up a convenient set of points for $f(x) = \cos x$ and then graph them (Diagram 9).

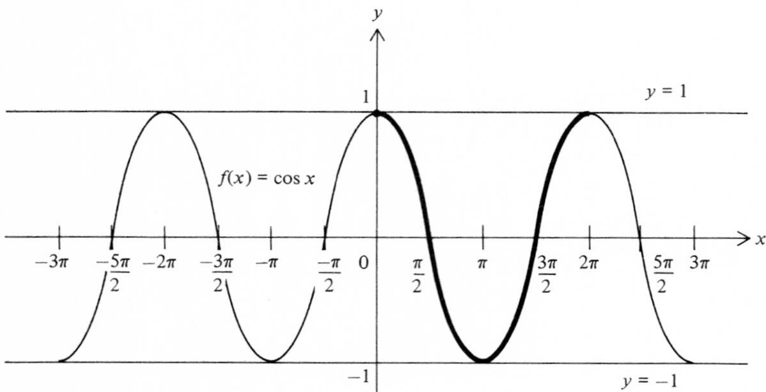

Diagram 9

x	$f(x) = \cos x$	x	$f(x) = \cos x$
0	1	$\frac{5\pi}{4}$	~ -0.71
$\frac{\pi}{4}$	$\frac{1}{\sqrt{2}} \approx 0.71$	$\frac{4\pi}{3}$	-0.50
$\frac{\pi}{3}$	0.50	$\frac{3\pi}{2}$	0
$\frac{\pi}{2}$	0	$\frac{5\pi}{3}$	0.50
$\frac{2\pi}{3}$	-0.50	$\frac{7\pi}{4}$	~ 0.71
$\frac{3\pi}{4}$	$-\frac{1}{\sqrt{2}} \approx -0.71$	2π	1
π	-1		

COMMENT We refer to the graph of $f(x)$ in the interval $[0, 2\pi]$ as the *characteristic curve* of the cosine function. The graph repeats itself in a pattern, extending to the right and to the left of the $[0, 2\pi]$ interval. At $x = \pi/2 + n\pi$ (n is an integer), $f(x) = 0$ (*zeros of the function*). At $x = 2n\pi$ (n is an integer), $f(x) = 1$ (*maximum value of the function*). At $x = \pi + 2n\pi$ (n is an integer), $f(x) = -1$ (*minimum value of the function*). The graph of $f(x)$ is bounded by the lines $y = 1$ and $y = -1$. In a similar manner, let us graph

$$g(x) = \sin x$$

(See Diagram 10.)

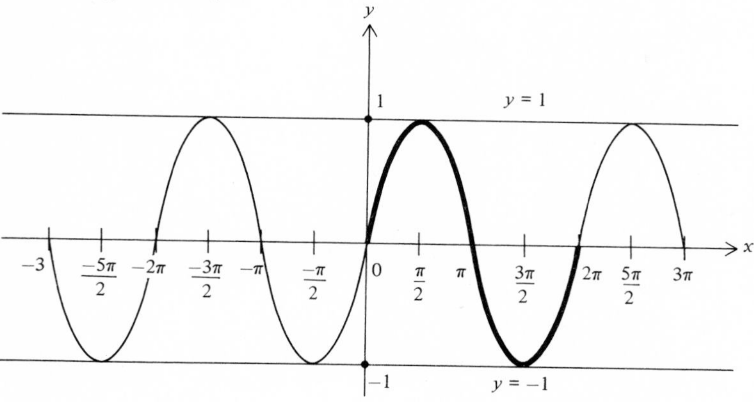

Diagram 10

x	$g(x) = \sin x$	x	$g(x) = \sin x$
0	0	$\frac{7\pi}{6}$	-0.50
$\frac{\pi}{6}$	0.50	$\frac{5\pi}{4}$	~ -0.71
$\frac{\pi}{4}$	~ 0.71	$\frac{3\pi}{2}$	-1
$\frac{\pi}{2}$	1	$\frac{7\pi}{4}$	~ -0.71
$\frac{3\pi}{4}$	~ 0.71	$\frac{11\pi}{6}$	-0.50
$\frac{5\pi}{6}$	0.50	2π	0
π	0		

Similar comments can be made for the sine function. The graph of $g(x)$ in the interval $[0, 2\pi]$ is the *characteristic curve* of the sine function. The graph repeats itself in a pattern, extending to the right and to the left of the $[0, 2\pi]$ interval. At $x = n\pi$ (n is an integer), $g(x) = 0$ (zeros of the function). At

$x = \pi/2 + 2n\pi$ (n is an integer), $g(x) = 1$ (maximum value of the function). At $x = 3\pi/2 + 2n\pi$ (n is an integer), $g(x) = -1$ (minimum value of the function).

Example 1 Analyze and sketch $f(x) = \cos x$, $-\pi/2 \leq x < \pi/2$.

A convenient way to graph $f(x)$ is to refer to Diagram 9. The portion of the graph we need will be between $x = -\pi/2$ and $x = \pi/2$ including the point $(\pi/2, f(-\pi/2))$ and excluding the point $(\pi/2, f(\pi/2))$ (see Diagram 11). An analysis of the graph immediately supplies us with the following information:

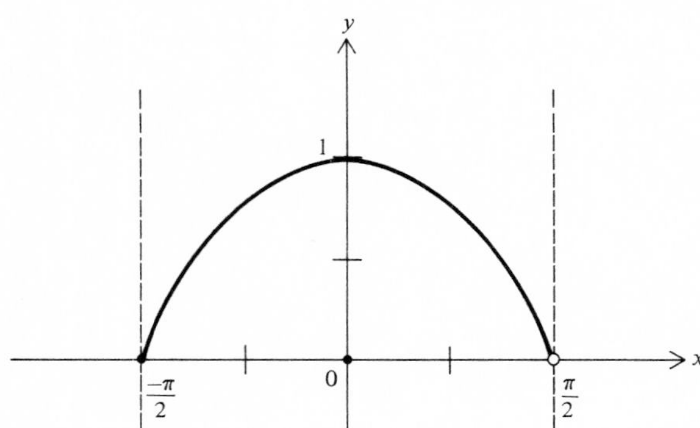

Diagram 11

1 The range of $f(x)$ is $0 \leq y \leq 1$.

2 $f(-\pi/2) = 0$ is the minimum; $x = -\pi/2$, is the *zero* of the function.

3 $f(0) = 1$ is the maximum of $f(x)$.

Example 2 Analyze and graph $g(x) = \sin x$, $-\pi/2 \leq x \leq 3$.

We must approximate $x = 3$ on the horizontal axis. This is easy, since we know that $3 < \pi$ and $\pi \approx 3.14$ (Diagram 12).

An analysis of the graph supplies us with the following information.

1 The range of $g(x)$ is $-1 \leq y \leq 1$.

2 Since $g(0) = 0$, $x = 0$ is a zero of the function.

3 $g(-\pi/2) = -1$ is the minimum of $g(x)$.

4 $g(\pi/2) = 1$ is the maximum of $g(x)$.

EXERCISES

In Exercises 1 to 10 analyze and graph each function in the interval indicated.

1 $f(x) = \cos x;\ \left[-2\pi, -\frac{\pi}{2}\right]$

2 $g(x) = \sin x;\ [1, 2]$

3 $f(x) = \cos x;\ [0, 4]$

4 $g(x) = \sin x;\ [\pi, 3\pi]$

5 $g(x) = \sin x;\ (\pi, 2\pi)$

6 $f(x) = \cos x;\ (-\pi, 1]$

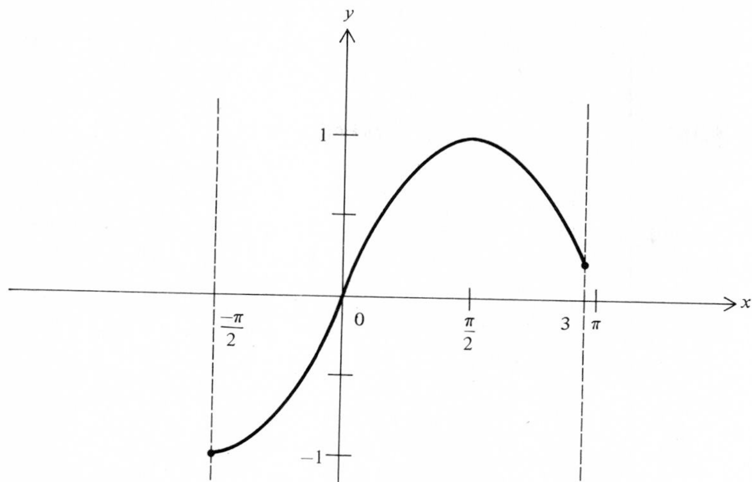

Diagram 12

7 $g(x) = \sin x; (-\sqrt{2}, \sqrt{3})$

8 $g(x) = \sin x; [4\pi, 15)$

9 $f(x) = \cos x; [-3\pi, -2\pi]$

10 $f(x) = \cos x; x = 0, 1, \frac{\pi}{2}$

11 Graph $f(x) = \cos x$ and $g(x) = \sin x$ on the same grid. For what values of x, $0 \le x \le 2\pi$, does $f(x) = g(x)$?

12 The horizontal lines $y = 1$ and $y = -1$ bound $f(x) = \cos x$ and $g(x) = \sin x$. Are $y = 1$ and $y = -1$ horizontal asymptotes of $f(x)$ and $g(x)$? Explain.

9.7 FURTHER GRAPHING TECHNIQUES

In this section let A and B be any real number other than zero. Let us analyze

$$f(x) = A \cos Bx \quad \text{and} \quad g(x) = A \sin Bx \tag{1}$$

If $A = B = 1$, Equations (1) simplify to

$$f(x) = \cos x \quad \text{and} \quad g(x) = \sin x$$

We have already studied these basic functions. Suppose $A \ne 1$ and $B = 1$; then Equations (1) reduce to

$$f(x) = A \cos x \quad \text{and} \quad g(x) = A \sin x \tag{2}$$

What effect will A have on the behavior of $f(x) = \cos x$ and $g(x) = \sin x$? We answer this question by examining the graphs of $f(x) = 3 \cos x$ and $g(x) = \frac{1}{2} \sin x$.

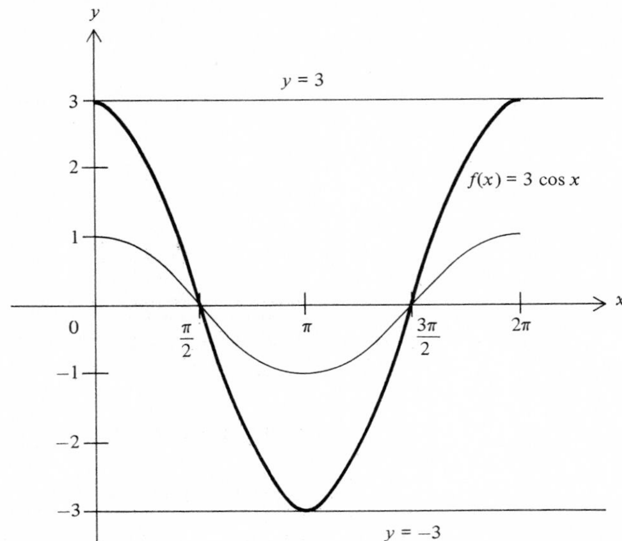

Diagram 13

Plotting the points for the ordered pairs listed in the Table 9.1 leads to the graph of $f(x) = 3 \cos x$ (see Diagram 13).

Plotting points for the ordered pairs listed in Table 9.2 leads to the graph of $g(x) = \frac{1}{2} \sin x$ (see Diagram 14).

We can now see the effects coefficients other than one have on these trigonometric functions. Comparing them with $y = \cos x$ and $y = \sin x$ shows that:

1 The *period* of each function is *unchanged*.

2 The *zeros* of each function are *unchanged*.

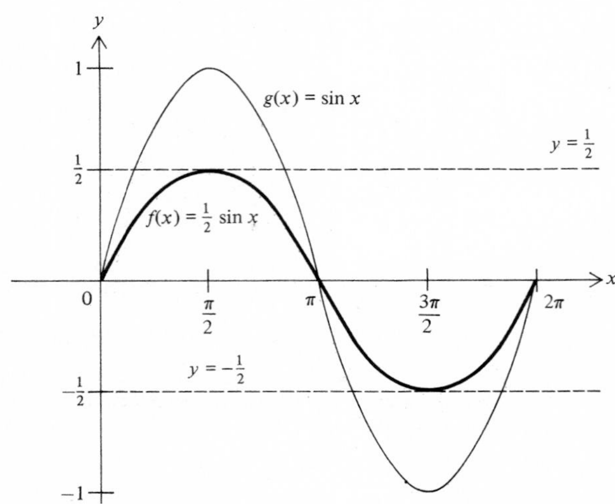

Diagram 14

TABLE 9.1

x	$\cos x$	$3 \cos x$
0	1	3
$\frac{\pi}{6}$	$\frac{\sqrt{3}}{2}$	$\frac{3\sqrt{3}}{2}$
$\frac{\pi}{3}$	$\frac{1}{2}$	$\frac{3}{2}$
$\frac{\pi}{2}$	0	0
$\frac{2\pi}{3}$	$-\frac{1}{2}$	$-\frac{3}{2}$
$\frac{5\pi}{6}$	$-\frac{\sqrt{3}}{2}$	$-\frac{3\sqrt{3}}{2}$
π	-1	-3
$\frac{7\pi}{6}$	$-\frac{\sqrt{3}}{2}$	$-\frac{3\sqrt{3}}{2}$
$\frac{4\pi}{3}$	$-\frac{1}{2}$	$-\frac{3}{2}$
$\frac{3\pi}{2}$	0	0
$\frac{5\pi}{3}$	$\frac{1}{2}$	$\frac{3}{2}$
$\frac{11\pi}{6}$	$\frac{\sqrt{3}}{2}$	$\frac{3\sqrt{3}}{2}$
2π	1	3

TABLE 9.2

x	$\sin x$	$\frac{1}{2} \sin x$
0	0	0
$\frac{\pi}{6}$	$\frac{1}{2}$	$\frac{1}{4}$
$\frac{\pi}{3}$	$\frac{\sqrt{3}}{2}$	$\frac{\sqrt{3}}{4}$
$\frac{\pi}{2}$	1	$\frac{1}{2}$
$\frac{2\pi}{3}$	$\frac{\sqrt{3}}{2}$	$\frac{\sqrt{3}}{4}$
$\frac{5\pi}{6}$	$\frac{1}{2}$	$\frac{1}{4}$
π	0	0
$\frac{7\pi}{6}$	$-\frac{1}{2}$	$-\frac{1}{4}$
$\frac{4\pi}{3}$	$-\frac{\sqrt{3}}{2}$	$-\frac{\sqrt{3}}{4}$
$\frac{3\pi}{2}$	-1	$-\frac{1}{2}$
$\frac{5\pi}{3}$	$-\frac{\sqrt{3}}{2}$	$-\frac{\sqrt{3}}{4}$
$\frac{11\pi}{6}$	$-\frac{1}{2}$	$-\frac{1}{4}$
2π	0	0

3 The maximum and minimum values of $f(x) = 3 \cos x$ are 3 times the maximum and minimum values of $f(x) = \cos x$.

4 The maximum and minimum values of $g(x) = \frac{1}{2} \sin x$ are $\frac{1}{2}$ times the maximum and minimum values of $g(x) = \sin x$.

5 Since

$$\begin{aligned} |\cos x| &\leq 1 \\ 3|\cos x| &\leq 3 \\ |3 \cos x| &\leq 3 \\ \underset{m}{\underset{\uparrow}{-3}} \leq 3 \cos x &\leq \underset{M}{\underset{\uparrow}{3}} \end{aligned}$$

Therefore, the maximum value M of $f(x)$ is 3, and the minimum value m of $f(x)$ is -3.

Similarly,

$$|\sin x| \le 1$$
$$\tfrac{1}{2}|\sin x| \le \tfrac{1}{2}$$
$$|\tfrac{1}{2}\sin x| \le \tfrac{1}{2}$$
$$\underset{\substack{\uparrow \\ m}}{-\tfrac{1}{2}} \le \tfrac{1}{2}\sin x \le \underset{\substack{\uparrow \\ M}}{\tfrac{1}{2}}$$

The maximum value M of $g(x)$ is $\frac{1}{2}$, and the minimum value m of $g(x)$ is $-\frac{1}{2}$. This leads us to the following definition.

DEFINITION 1 If M is the maximum value of a periodic function and m is the minimum value, then the positive number $(M - m)/2$ is called the *amplitude* of the function.

The amplitude of $f(x) = 3\cos x$ is

$$\frac{3 - (-3)}{2} = 3$$

The amplitude of $g(x) = \frac{1}{2}\sin x$ is

$$\frac{\frac{1}{2} - (-\frac{1}{2})}{2} = \frac{1}{2}$$

Theorem 1 If $f(x) = A\cos x$ and $g(x) = A\sin x$ (A is any nonzero real number), then $|A|$ is the amplitude of these functions.

Proof For $A > 0$

$$|\cos x| \le 1$$
$$-1 \le \cos x \le 1 \qquad \text{Property 9a of Chapter 4}$$
$$\underset{\substack{\uparrow \\ m}}{-A} \le A\cos x \le \underset{\substack{\uparrow \\ M}}{A} \qquad \text{Property 3 of Chapter 4}$$
$$\tfrac{1}{2}[A - (-A)] = A \qquad \text{By Definition 1}$$

For $A < 0$

$$|\cos x| \le 1$$
$$-1 \le \cos x \le 1 \qquad \text{Property 9a of Chapter 4}$$
$$-A \ge A\cos x \ge A \qquad \text{Property 4 of Chapter 4}$$
$$\underset{\substack{\uparrow \\ m}}{A} \le A\cos x \le \underset{\substack{\uparrow \\ M}}{-A} \qquad \text{By Definition 1}$$
$$\tfrac{1}{2}[-A - (A)] = -A$$

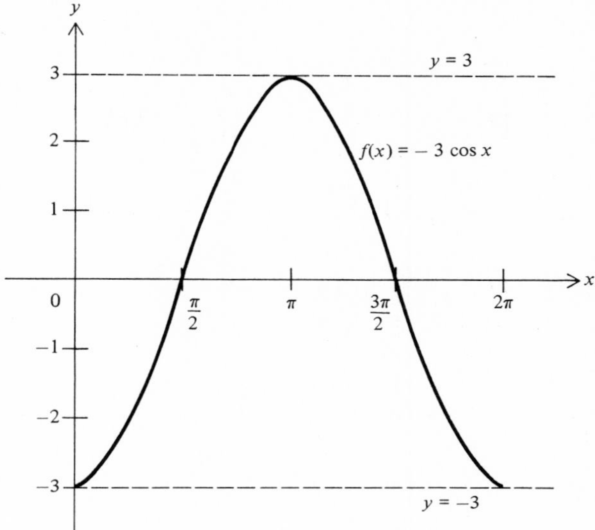

Diagram 15

Since the amplitude of $f(x)$ is A when $A > 0$ and $-A$ when $A < 0$, we can say that the amplitude of $f(x)$ is $|A|$.

Example 1 Graph $f(x) = -3 \cos x$, $0 \leq x \leq 2\pi$.

The amplitude is $|-3| = 3$. The graph of $f(x)$ will be contained within the lines $y = 3$ and $y = -3$. The negative sign, however, *reverses* the sign of each value of $f(x) = 3 \cos x$. Therefore, this graph will be the same as $f(x) = 3 \cos x$ if we flip it about the x axis (see Diagram 15 and compare it with Diagram 13).

In general, the graph of any function of the form $y = A \cos x$ or $y = A \sin x$, $A \neq 0$, is periodic, with fundamental period 2π and amplitude $|A|$.

Let us now consider the graph of function $y = A \sin Bx$, where $A = 1$ and $B \neq 1$. Suppose we analyze $f(x) = \sin 2x$ and compare it with $g(x) = \sin x$.

To draw the graph of $f(x) = \sin 2x$, note that as x increases from 0 to π, $2x$ increases from 0 to 2π. Therefore, $\sin 2x$ completes one cycle of sine values as x goes from 0 to π. This can be verified by examining Table 9.3 (see Diagram 16).

COMMENTS The fundamental period of $f(x) = \sin 2x$ is one-half the fundamental period of $g(x) = \sin x$. Hence, the constant B affects the *period* of the sine function. We can summarize these findings in the following theorem.

Theorem 2 Let $f(x) = \cos Bx$ and $g(x) = \sin Bx$, $B \neq 0$. Then $f(x)$ and $g(x)$ have the period

$$p = \frac{2\pi}{|B|}$$

TABLE 9.3

x	$2x$	$\sin 2x$	x	$2x$	$\sin 2x$
0	0	0	$\frac{7\pi}{12}$	$\frac{7\pi}{6}$	$-\frac{1}{2}$
$\frac{\pi}{12}$	$\frac{\pi}{6}$	$\frac{1}{2}$	$\frac{2\pi}{3}$	$\frac{4\pi}{3}$	$-\frac{\sqrt{3}}{2}$
$\frac{\pi}{6}$	$\frac{\pi}{3}$	$\frac{\sqrt{3}}{2}$	$\frac{3\pi}{4}$	$\frac{3\pi}{2}$	-1
$\frac{\pi}{4}$	$\frac{\pi}{2}$	1	$\frac{5\pi}{6}$	$\frac{5\pi}{3}$	$-\frac{\sqrt{3}}{2}$
$\frac{\pi}{3}$	$\frac{2\pi}{3}$	$\frac{\sqrt{3}}{2}$	$\frac{11\pi}{12}$	$\frac{11\pi}{6}$	$-\frac{1}{2}$
$\frac{5\pi}{12}$	$\frac{5\pi}{6}$	$\frac{1}{2}$	π	2π	0
$\frac{\pi}{2}$	π	0			

Proof We know that $\cos(Z \pm 2\pi) = \cos Z$. If $Z = Bx$, then $\cos(Bx \pm 2\pi) = \cos Bx$ or, regrouping factors,

$$\cos\left[B\left(x \pm \frac{2\pi}{B}\right)\right] = \cos Bx$$

Now we see that the above equation is in the form

$$f\left(x \pm \frac{2\pi}{B}\right) = f(x)$$

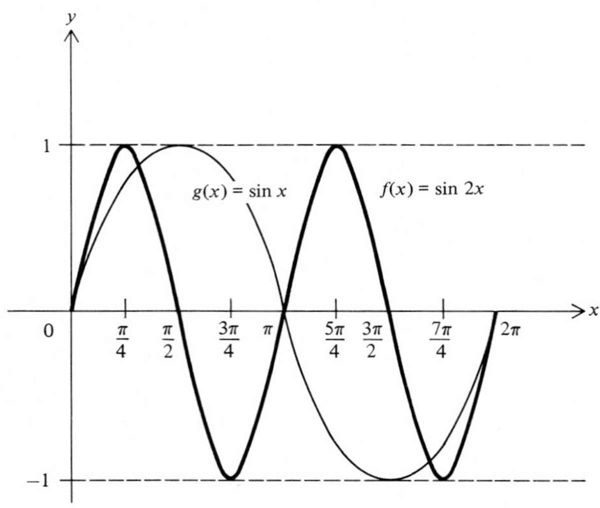

Diagram 16

This implies that $f(x) = \cos Bx$, $B \neq 0$, has a fundamental period of $p = 2\pi/|B|$, which is what we wanted to prove. [We leave the other part of this theorem as an exercise, that is, proving that $g(x) = \sin Bx$, $B \neq 0$, has a fundamental period of $2\pi/|B|$.]

It should be emphasized at this point that you can quickly sketch graphs of the form $y = A \cos Bx$ *or* $y = A \sin Bx$ by determining

1 The characteristic curve

2 The effects of A and B

3 The maximum and minimum points

4 The points where the graph crosses the x axis (zeros of the function)

Example 2 Graph $f(x) = 2 \cos 4x$, $0 \leq x \leq 2\pi$.
The amplitude is $|A| = 2$. The period is

$$p = \frac{2\pi}{|4|} = \frac{\pi}{2}$$

From Section 9.5, for $f(x) = \cos x$, $f(x)$ was a maximum at $x = 2n\pi$. For $f(x) = 2 \cos 4x$, we let

$$4x = 2n\pi \qquad \text{or} \qquad x = \frac{\pi}{2} \qquad n \text{ an integer}$$

In the interval $[0, 2\pi]$ the maximum values occur at $0, \pi/2, \pi, 3\pi/2$, and 2π. $f(x)$ was a minimum at $x = \pi + 2n\pi$. For $f(x) = 2 \cos 4x$, we let

$$4x = \pi + 2n\pi \qquad \text{or} \qquad x = \frac{\pi}{4} + \frac{n\pi}{2} \qquad n \text{ an integer}$$

In the interval $[0, 2\pi]$ the minimum values occur at $\pi/4$, $3\pi/4$, $5\pi/4$, and $7\pi/4$.

From Section 9.5, for $f(x) = \cos x$, the zeros of $f(x)$ occur at $x = \pi/2 + n\pi$. For $f(x) = 2 \cos 4x$, we let

$$4x = \frac{\pi}{2} + n\pi \qquad \text{or} \qquad x = \frac{\pi}{8} + \frac{n\pi}{4} \qquad n \text{ an integer}$$

In the interval $[0, 2\pi]$ the zeros occur at $\pi/8$, $3\pi/8$, $5\pi/8$, $7\pi/8$, $9\pi/8$, $11\pi/8$, $13\pi/8$, and $15\pi/8$. Taking these important observations into account, and knowing the shape of the characteristic cosine curve, we obtain the graph in Diagram 17.

Example 3 Graph $g(x) = \sqrt{2} \sin \frac{1}{2}x$, $-2\pi \leq x \leq 2\pi$.
The amplitude is $|A| = \sqrt{2}$. The period is

$$p = \frac{2\pi}{|\frac{1}{2}|} = 4\pi$$

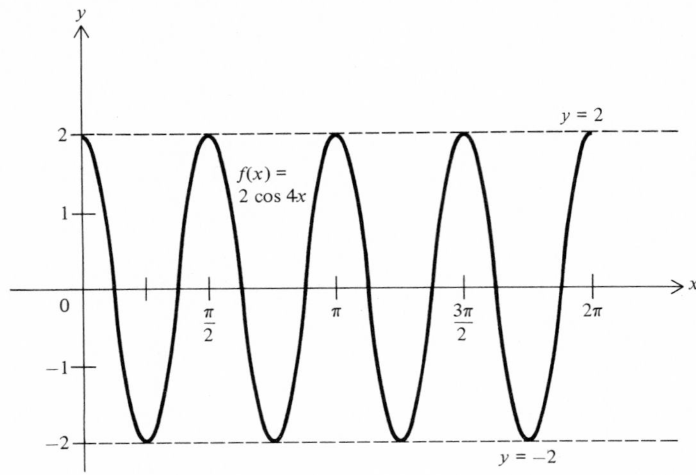

Diagram 17

For $g(x) = \sin x$, $g(x)$ was a maximum at

$$x = \frac{\pi}{2} + 2n\pi$$

For $g(x) = 2 \sin \frac{1}{2}x$,

$$\frac{x}{2} = \frac{\pi}{2} + 2n\pi \qquad \text{or} \qquad x = \pi + 4n\pi \qquad n \text{ an integer}$$

In the interval $[-2\pi, 2\pi]$, the only maximum value occurs at π. For $g(x) = \sin x$, $g(x)$ was a minimum at $x = 3\pi/2 + 2n\pi$. For $g(x) = 2 \sin \frac{1}{2}x$,

$$\frac{x}{2} = \frac{3\pi}{2} + 2n\pi \qquad \text{or} \qquad x = 3\pi + 4n\pi \qquad n \text{ an integer}$$

In the interval $[-2\pi, 2\pi]$ the only minimum value occurs at $-\pi$.

For $g(x) = \sin x$, the zeros of $g(x)$ occur at $x = n\pi$. For $g(x) = 2 \sin \frac{1}{2}x$,

$$\frac{x}{2} = n\pi \qquad \text{or} \qquad x = 2n\pi \qquad n \text{ an integer}$$

In the interval $[-2\pi, 2\pi]$ the zeros occur at -2π, 0, and 2π.

Knowing the shape of the characteristic sine curve, and using the above information, we obtain the graph in Diagram 18.

Example 4 Graph $f(x) = 2.2 \cos(-\pi x)$, $-3 \le x \le 3$.

Since $\cos(-Z) = \cos Z$, we can then rewrite the given function as $f(x) = 2.2 \cos \pi x$, and analyze its graph in this form. The amplitude is $|A| = 2.2$. The period is $p = 2\pi/|\pi| = 2$. To find the maximum points take

$$\pi x = 2n\pi$$
$$x = 2n$$

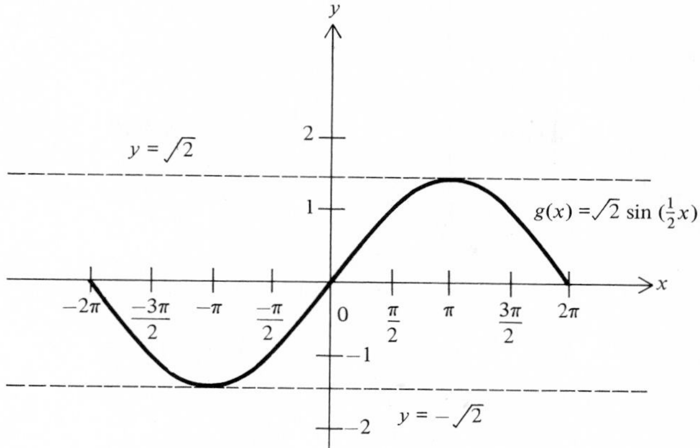

Diagram 18

In the interval $[-3, 3]$ the maximum values occur at $-2, 0, 2$. To find the minimum points take

$$\pi x = \pi + 2n\pi$$
$$x = 1 + 2n$$

In the interval $[-3, 3]$ the minimum values occur at $-3, -1, 1$, and 3. Zeros are found as follows:

$$\pi x = \frac{\pi}{2} + n\pi$$

$$x = \tfrac{1}{2} + n$$

In the interval $[-3, 3]$ the zeros occur at $-\frac{5}{2}, -\frac{3}{2}, -\frac{1}{2}, \frac{1}{2}, \frac{3}{2}$, and $\frac{5}{2}$.

We can now graph $f(x)$ in two steps.

STEP 1 Locate the special points of interest.

STEP 2 Graph $f(x)$ through the points in Step 1, recalling the characteristic cosine curve (see Diagram 19).

In Example 4, because the B term in $y = A \cos Bx$ contained a factor of π, it was advantageous to use integer values on the x axis instead of the customary $\pi/4, \pi/2, \ldots$. Flexibility of this sort is important for rapid, successful graphing.

We conclude this section with two more examples involving sine and cosine functions.

Example 5 Graph $g(x) = \sin\left(-\dfrac{\pi x}{2}\right)$, $-p \le x \le p$, where p is the period of $g(x)$.

The amplitude is $|A| = 1$. The period is

$$p = \frac{2\pi}{|-\pi/2|} = 4$$

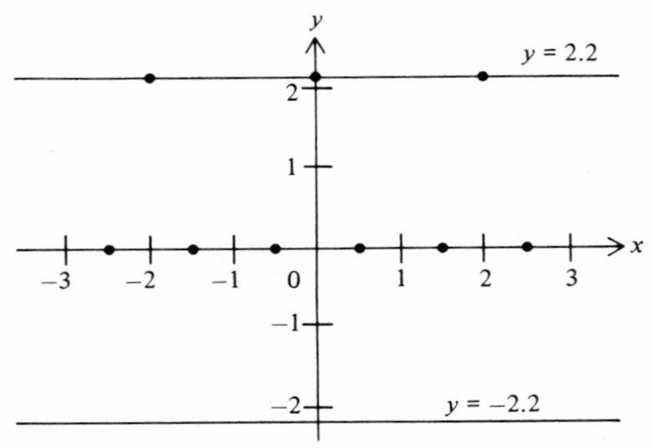

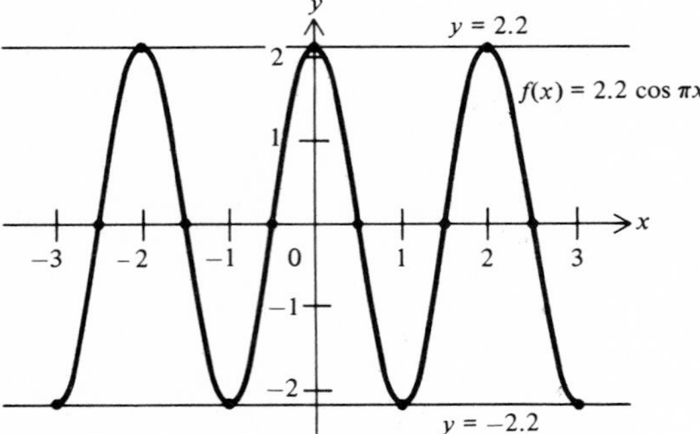

Diagram 19

Therefore, the domain of $g(x)$ is

$$-4 \leq x \leq 4$$

To find the maximum points, take

$$-\frac{\pi x}{2} = \frac{\pi}{2} + 2n\pi$$

$$x = -1 - 4n$$

In the interval $[-4, 4]$ the maximum values occur at $-1, 3$. To find the minimum points, take

$$-\frac{\pi x}{2} = \frac{3\pi}{2} + 2n\pi$$

$$x = -3 - 4n$$

In the interval $[-4, 4]$ the minimum values occur at $-3, 1$. To find the zeros take

$$-\frac{\pi x}{2} = n\pi$$

$$x = -2n$$

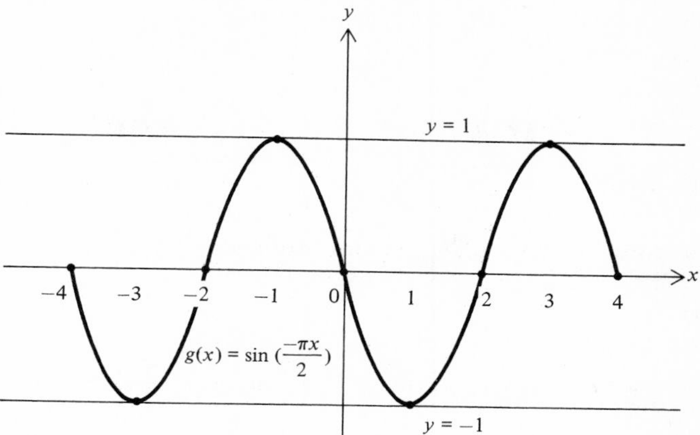

Diagram 20

In the interval $[-4, 4]$ the zeros occur at $-4, -2, 0, 2$, and 4.

This is adequate information to graph $g(x) = \sin(-\pi x/2)$ (see Diagram 20).

COMMENT Note that the x values for the maximum and minimum points occur *midway* between the zeros. This is an important fact to remember.

Example 6 Consider $f(x) = \cos x$ and $g(x) = \sin x$. Graph $(f + g)(x)$, $0 \leq x \leq 2\pi$.

We recall the addition of functions from Chapter 7. The technique to use on problems of this nature is to sketch both functions lightly on the same grid for the domain of the function $(f + g)(x)$. Then adding corresponding ordinates of $f(x)$ and $g(x)$ (range values) and tracing a smooth curve through these new points, we obtain the graph of $h(x) = (f + g)(x)$. [*Note:* For convenience we let $h(x)$ stand for $(f + g)(x)$.] (See Diagram 21.)

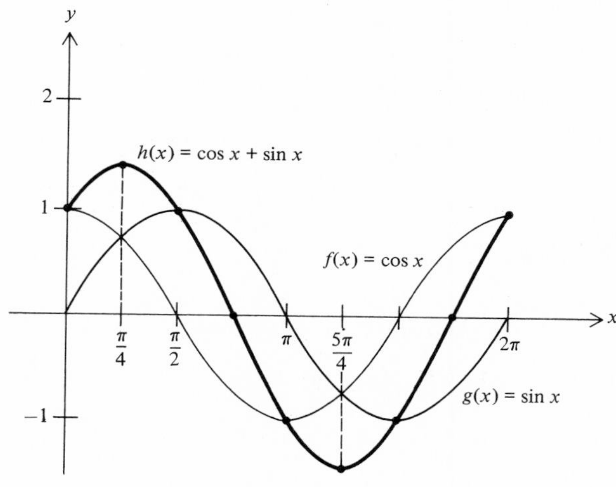

Diagram 21

OBSERVATIONS $h(x)$ takes its maximum value when $f(x) = g(x)$ in the interval $[0, \pi/2]$. $h(\pi/4) = \sqrt{2} \approx 1.414$. $h(x)$ takes its minimum value when $f(x) = g(x)$ in the interval $[\pi, 3\pi/2]$. $h(5\pi/4) = -\sqrt{2} \approx -1.414$.

EXERCISES

For Exercises 1 to 6, graph each pair of functions on the same grid. Use the interval $0 \le x \le 2\pi$.

1 $f(x) = 4 \sin x;\ g(x) = \frac{1}{4} \sin x$

2 $f(x) = \frac{3}{2} \cos x;\ g(x) = \frac{2}{3} \cos x$

3 $f(x) = 2 \sin x;\ g(x) = -2 \sin x$

4 $f(x) = \frac{1}{2} \cos x;\ g(x) = -\frac{1}{2} \cos x$

5 $f(x) = \pi \cos x;\ g(x) = 2\pi \cos x$

6 $f(x) = \frac{\pi}{2} \sin x;\ g(x) = -\frac{\pi}{2} \cos x$

In each of Exercises 7 to 9 write a cosine function and sine function with the given characteristics.

7 $p = \pi;\ A = \frac{1}{2}$ **8** $p = 2;\ A = 2$ **9** $p = \frac{\pi}{2};\ A = 2\pi$

Graph each of the functions 10 to 26 over the interval $-p \le x \le p$, where p represents the period.

10 $f(x) = \cos \frac{x}{4}$

11 $g(x) = \sin \frac{x}{4}$

12 $f(x) = -2 \cos 2x$

13 $g(x) = -\frac{1}{2} \sin 2x$

14 $f(x) = \cos 6x$

15 $g(x) = \sin 3x$

16 $f(x) = -\cos(-2x)$

17 $g(x) = -\sin(-2x)$

18 $f(x) = -2 \cos \frac{x}{2}$

19 $g(x) = \frac{1}{2} \sin \frac{x}{3}$

20 $f(x) = \cos \pi x$

21 $g(x) = \sin 2\pi x$

22 $f(x) = \cos \frac{\pi x}{2}$

23 $g(x) = \sin \frac{\pi x}{6}$

24 $f(x) = \sqrt{3} \cos \frac{2\pi x}{3}$

25 $g(x) = \sin \frac{x}{\pi}$

26 $f(x) = \cos \frac{x}{2\pi}$

Graph functions 27 to 34 over the $0 \le x \le 2\pi$ interval.

27 $f(x) = \sin 2x + \cos x$

28 $f(x) = 2 \cos x + \cos 2x$

29 $f(x) = \sin \frac{x}{2} + \sin x$

30 $f(x) = \cos x - \sin x$

31 $f(x) = x + \sin x$

32 $f(x) = x - \cos x$

33 $f(x) = 2 + \sin x$

34 $f(x) = \cos x - 2$

35 In general, what is the effect of k on the graphs of $f(x)$ and $g(x)$ if

$$f(x) = k + A \cos Bx \qquad \text{and} \qquad g(x) = k + A \sin Bx$$

9.8
PHASE SHIFT: $f(x) = A\cos(Bx + C)$ AND $g(x) = A\sin(Bx + C)$

Consider $f(x) = \cos(x - \pi/4)$. For each x in the domain, $\pi/4$ is subtracted from this value and the cosine of this difference is computed to give a corresponding range value. The basic character of the cosine function is preserved; that is,

$$\text{Amplitude} = |A| = 1 \qquad \text{and} \qquad p = \frac{2\pi}{|B|} = 2\pi$$

Let us then graph $\cos x$ and $\cos(x - \pi/4)$ on the same grid and make some further observations (Diagram 22).

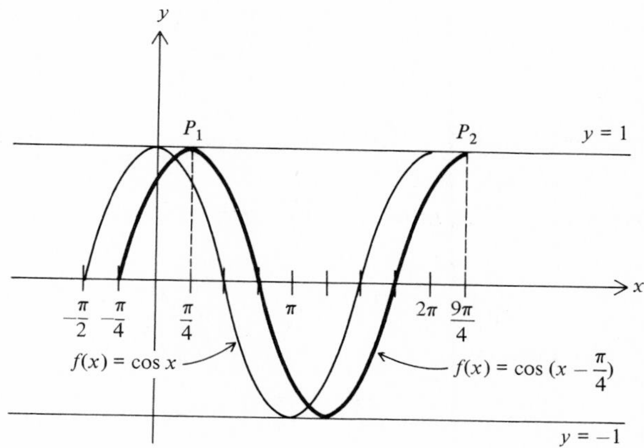

Diagram 22

Note that the graph of $f(x) = \cos(x - \pi/4)$ in the interval $[\pi/4, 9\pi/4]$ is indeed the wave of the graph of $f(x) = \cos x$ in the interval $[0, 2\pi]$. This is easily explained. As x increases from $\pi/4$ to $9\pi/4$, $x - \pi/4$ increases from 0 to 2π. What has happened is that $\pi/4$ in $f(x) = \cos(x - \pi/4)$ represents a *shift to the right* of $\pi/4$ unit of the function $f(x) = \cos x$.

Similarly, if we graph $f(x) = \cos(x + \pi/4)$, we note a *shift of $\pi/4$ unit to the left* of $f(x) = \cos x$ (Diagram 23).

This leads us to the following definition.

DEFINITION 1 Consider the functions

$$f(x) = \cos(x + C) \qquad \text{and} \qquad g(x) = \sin(x + C)$$

The constant C is called the *phase shift* of the graphs of these functions. If $C > 0$, the graphs of $f(x)$ and $g(x)$ will translate C units to the *left* of the graphs of $\cos x$ and $\sin x$, respectively. If $C < 0$, the graphs of $f(x)$ and $g(x)$ will translate $|C|$ units to the *right* of the graphs of $\cos x$ and $\sin x$, respectively.

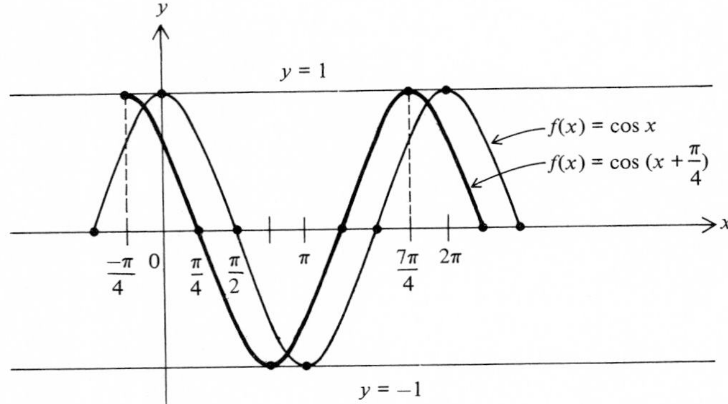

Diagram 23

Using this definition, we see that in the first case $f(x) = \cos(x - \pi/4)$ can be rewritten as $f(x) = \cos[x + (-\pi/4)]$, $C = -\pi/4 < 0$. Hence we conclude that $f(x)$ has a *phase shift* of $\pi/4$ unit to the *right* of the basic cosine function. In the case of $f(x) = \cos(x + \pi/4)$, $C = \pi/4 > 0$, and the phase shift is $\pi/4$ unit to the *left* of $f(x) = \cos x$. Note that in the general form of the definition the coefficient of x is 1.

We now graph a function involving a phase shift of a more general nature.

Example 1 Graph $g(x) = 2\sin(3x - \pi)$.
First rewrite $g(x)$ as

$$g(x) = 2\sin 3\left(x - \frac{\pi}{3}\right) \qquad \text{Factoring out 3}$$

The amplitude is $|A| = 2$. The period is

$$p = \frac{2\pi}{|B|} = \frac{2\pi}{3}$$

Now

$$g(x) = 2\sin 3\left[x + \left(-\frac{\pi}{3}\right)\right]$$

Since $B \neq 1$, we factor out the 3 from $3x - \pi$ and hence the phase shift is not $-\pi$ but $-\pi/3 < 0$ and the shifting is to the right. (The coefficient of x must be 1 before we can determine the proper phase shift.) To determine one cycle of $g(x)$ we note that

$$3x - \pi = \begin{cases} 0 & \text{when } x = \dfrac{\pi}{3} \\ 2\pi & \text{when } x = \pi \end{cases}$$

Hence we can graph one cycle of the wave for the interval $[\pi/3, \pi]$ then repeat this pattern to the left and right of this interval for a complete graph of $g(x)$ (Diagram 24).

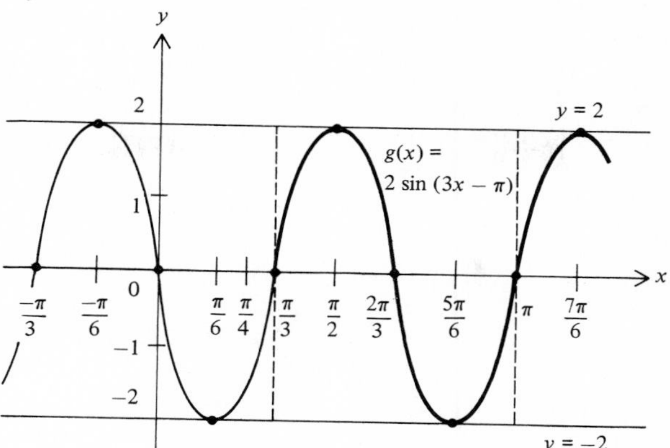

Diagram 24

Some students may have difficulty fitting the curve in a given interval. The following should be helpful. For $g(x)$ we found the interval $[\pi/3, \pi]$. The x values of particular importance (zeros, maximums, and minimums) will occur at the interval end points and one-fourth, two-fourths, and three-fourths of the way through the interval. This can easily be seen by dividing the given interval into four equal parts (half and then half again) obtaining $x_1 = \pi/2$, $x_2 = 2\pi/3$, and $x_3 = 5\pi/6$ (Diagram 25). Evaluating $g(x)$ at the endpoints of this interval and at points x_1, x_2, and x_3 should be sufficient to get you started with such a graph.

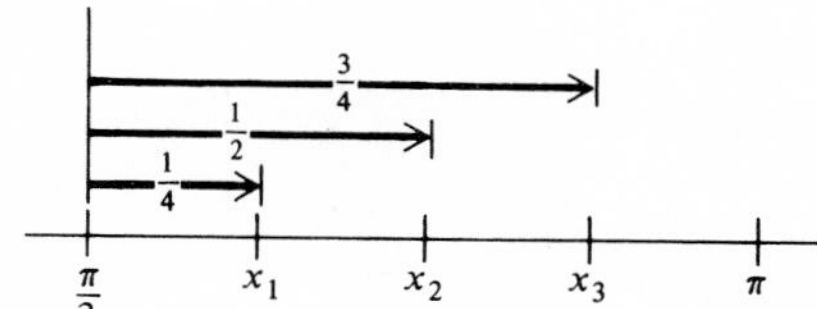

Diagram 25

It is instructive to analyze $g(x) = 2\sin(3x - \pi)$ another way. Recalling Theorem 9, Section 9.4, $\sin(Z_1 - Z_2) = \sin Z_1 \cos Z_2 - \cos Z_1 \sin Z_2$, let $Z_1 = 3x$ and $Z_2 = \pi$; we see

$$\begin{aligned} g(x) = 2\sin(3x - \pi) &= 2(\sin 3x \cos \pi - \cos 3x \sin \pi) \\ &= 2[(\sin 3x)(-1) - (\cos 3x)(0)] = -2\sin 3x \end{aligned}$$

Hence the graph of $g(x) = -2\sin 3x$ is Diagram 24. We have eliminated the analysis of the phase shift for $g(x)$. This, of course, was possible because $z_2 = \pi$.

We conclude this section with one more example demonstrating a pointwise method.

Example 2 Graph $f(x) = \frac{1}{2}\cos(2x + \pi)$.
The *pointwise* method requires the location of special points of inter-

est (maximums, minimums, and zeros) as explained briefly at the end of Example 1.

Let us also locate one complete cycle of $f(x)$ and apply the method referred to in the last comment. Rewriting $f(x)$ in proper form, we obtain

$$f(x) = \tfrac{1}{2}\cos 2\left(x + \frac{\pi}{2}\right)$$

The amplitude is $|A| = \tfrac{1}{2}$. The period is

$$p = \frac{2\pi}{|2|} = \pi$$

Since $f(x) = \tfrac{1}{2}\cos 2(x + \pi/2)$, $C = \pi/2$. Since $C > 0$, the graph of $f(x)$ is translated $\pi/2$ units to the left of $\cos 2x$.

To find one complete cycle of $f(x)$, we compute

$$2x + \pi = \begin{cases} 0 & \text{when } x = -\dfrac{\pi}{2} \\[2ex] 2\pi & \text{when } x = \dfrac{\pi}{2} \end{cases}$$

Since one cycle appears in the interval $[-\pi/2, \pi/2]$, refer to Diagram 26 and divide the interval into four equal partitions. Therefore, at $-\pi/2$, $-\pi/4$, 0, $\pi/4$, $\pi/2$, the zeros and maximum and minimum values of $f(x)$ will occur.

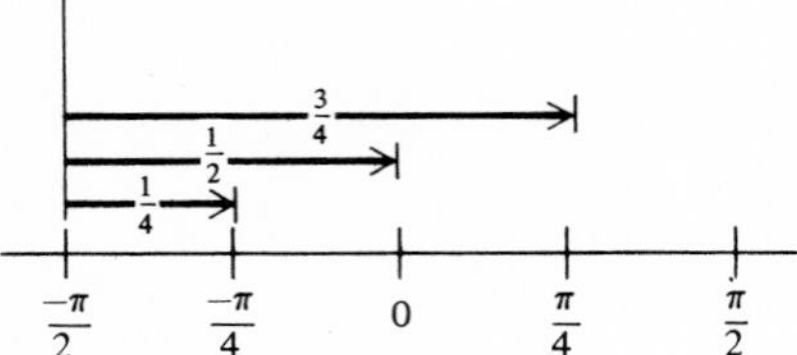

Diagram 26

Verification

To find the zeros, in the interval $[-\pi/2, \pi/2]$,

$$2x + \pi = \frac{\pi}{2} + n\pi$$

Recall Section 9.6 ($\cos z = 0$ when $z = \pi/2 + n\pi$, where n is an integer).

$$x = -\frac{\pi}{4} + \frac{n\pi}{2}$$

$$x = \begin{cases} -\dfrac{\pi}{4} & \text{when } n = 0 \\[2ex] \dfrac{\pi}{4} & \text{when } n = 1 \end{cases}$$

Find the maximum points: in the interval $[-\pi/2, \pi/2]$

$$2x + \pi = 2n\pi$$

Recall Section 9.6 ($\cos z = 1$ when $z = 2n\pi$, where n is an integer).

$$x = -\frac{\pi}{2} + n\pi$$

$$x = \begin{cases} -\dfrac{\pi}{2} & \text{when } n = 0 \\ \dfrac{\pi}{2} & \text{when } n = 1 \end{cases}$$

Find the minimum points: in the interval $[-\pi/2, \pi/2]$,

$$2x + \pi = \pi + 2n\pi$$

Recall Section 9.6 ($\cos z = -1$ when $z = \pi + 2n\pi$, where n is an integer).

$$x = n\pi$$
$$x = 0 \qquad \text{when } n = 0$$

We can now plot the five points and graph one cycle of $f(x)$ for the interval $[-\pi/2, \pi/2]$. Although this method is time-consuming, it is one method of systematically analyzing the graph of a function of the form

$$y = A\cos(Bx + C)$$

(see Diagram 27).

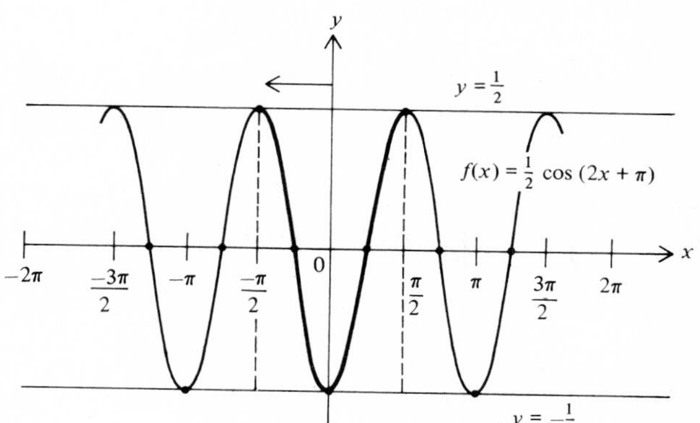

Diagram 27

COMMENT Using Theorem 7, Section 9.4, $f(x) = \frac{1}{2}\cos(2x + \pi)$ could have been written

$$\begin{aligned} f(x) &= \tfrac{1}{2}[\cos(2x)\cos\pi - \sin(2x)\sin\pi] \\ &= \tfrac{1}{2}[(\cos 2x)(-1) - (\sin 2x)(0)] = -\tfrac{1}{2}\cos 2x \end{aligned}$$

Hence, the graph of $f(x) = -\frac{1}{2}\cos 2x$ is also represented by Diagram 27, eliminating the phase-shift analysis.

EXERCISES

Graph each of the functions in Exercises 1 to 15.

1 $f(x) = \cos(x + \pi)$

2 $g(x) = \sin 2(x - \pi)$

3 $f(x) = \cos\left(2x - \frac{\pi}{4}\right)$

4 $g(x) = \sin(2x + \pi)$

5 $f(x) = -2\cos\left(x + \frac{\pi}{3}\right)$

6 $g(x) = -\frac{1}{2}\sin\left(x - \frac{\pi}{6}\right)$

7 $f(x) = \cos\left(\frac{\pi}{2} + \pi\right)$

8 $g(x) = \sin\left(2x + \frac{\pi}{4}\right)$

9 $f(x) = \frac{1}{2}\cos\left(\frac{\pi}{2} - x\right)$

10 $g(x) = 2\sin\left(3x - \frac{\pi}{4}\right)$

11 $f(x) = \cos(2x + 4)$

12 $g(x) = 2\sin(3x - 6)$

13 $f(x) = -\cos(x - 1)$

14 $g(x) = 1 + \sin\left(2x + \frac{\pi}{2}\right)$

15 $f(x) = -1 + \cos\left(\frac{\pi}{2} - x\right)$

16 Graph $y = \sin(x - \pi/2)$ and $y = \sin(x + \pi/2)$ on the same grid for the interval $0 \le x \le 2\pi$. What is obtained by graphing

$$y = \sin\left(x - \frac{\pi}{2}\right) + \sin\left(x + \frac{\pi}{2}\right)$$

17 Graph $f(x) = \cos\dfrac{4x}{2 - \pi}$.

9.9 INVERSE COSINE AND INVERSE SINE

In Chapter 7 we introduced the concept of the inverse function. Recall two facts:

1 In order to find an inverse function $f^{-1}(x)$, $f(x)$ must be one to one.

2 The inverse function $f^{-1}(x)$ has for its domain the range of $f(x)$ and has for its range the domain of $f(x)$.

Let us sketch $f(x) = \cos x$ (Diagram 28).

Observe line L parallel to the x axis in Diagram 28. In the interval $[-2\pi, 2\pi]$, L cuts the graph of $f(x)$ at x_0, x_1, x_2, and x_3. This shows that $f(x_0) = f(x_1) = f(x_2) = f(x_3)$, indicating that $f(x)$ *is not one to one*. We are now faced with the

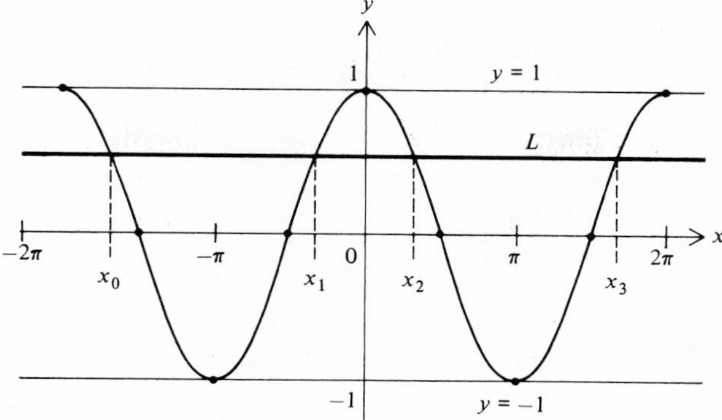

Diagram 28

problem of *restricting the domain of* $f(x)$ so as to create a cosine function which is one to one. We would like to choose a domain so that this new function is not only one to one but also has a range that is the same as the basic cosine function, namely, $[-1, 1]$.

Looking again at Diagram 28, and realizing the cosine function is periodic, we see that there are infinitely many intervals for which this is possible. In defining the cosine function which is one to one, it is customary to choose the interval $[0, \pi]$.

DEFINITION 1 Let $y = f(x)$; then $y = \cos x$ denotes the function $f(x) = \cos x,\ 0 \le x \le \pi$.

To solve for x in $y = \cos x$, we introduce a new notation to denote the *inverse cosine,* namely,

$$x = \cos^{-1} y$$

read "x is the inverse cosine function of y." Since we are more familiar with x as the independent variable and y as the dependent variable, we can write the inverse cosine function as

$$y = \cos^{-1} x$$

DEFINITION 2 $y = \cos^{-1} x$ if and only if $\cos y = x$ whenever $-1 \le x \le 1$.

A word of caution here about the new notation, $\cos^{-1} x$: Do not make the common mistake of thinking that $\cos^{-1} x$ is the same as $(\cos x)^{-1}$. It is *not.*

$$\cos^{-1} x \ne (\cos x)^{-1} = \frac{1}{\cos x}$$

In order to graph $y = \cos^{-1} x$, let us generate a table of values relating x and y.

For $y = \cos x$, as x increases from 0 to π, y decreases from 1 to -1. For $y = \cos^{-1} x$, as x increases from -1 to 1, y decreases from π to 0. The domain and range are interchanged (examine the tables below).

Function	Domain	Range
$y = \cos x$	$0 \leq x \leq \pi$	$-1 \leq y \leq 1$
$y = \cos^{-1} x$	$-1 \leq x \leq 1$	$0 \leq y \leq \pi$

x	$\cos^{-1} x$
-1	π
$-\frac{1}{2}$	$\frac{2\pi}{3}$
0	$\frac{\pi}{2}$
$\frac{1}{2}$	$\frac{\pi}{3}$
1	0

(See Diagrams 29 and 30.)

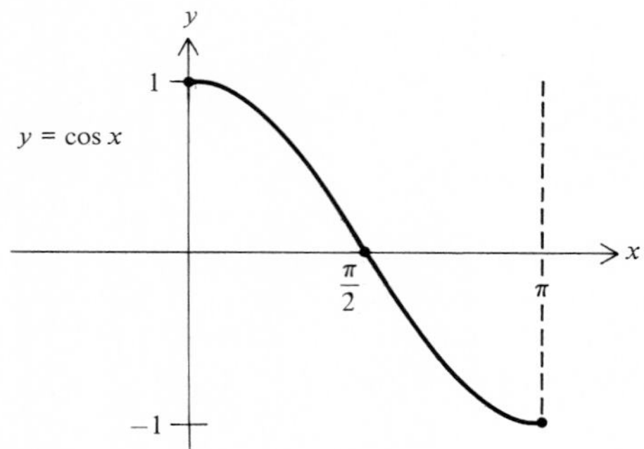

Diagram 29

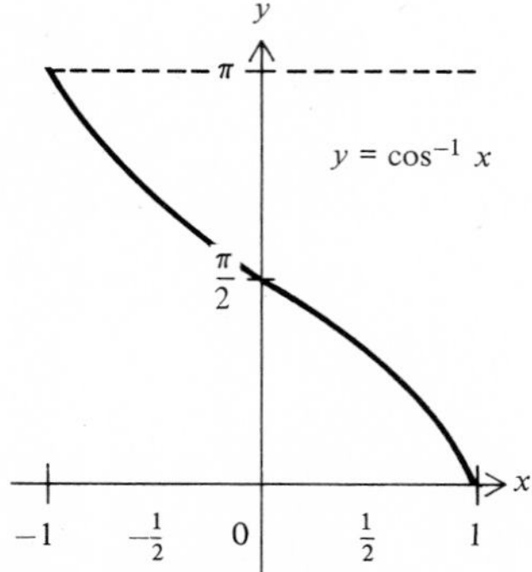

Diagram 30

Another notation commonly used to express the inverse cosine is $y = \arccos x$, read "y is the arc whose cosine is x." We use both notations for the inverse function.

For the sine function $g(x) = \sin x$ we define its inverse in a similar manner. First we define a sine function which is one to one.

DEFINITION 3 Let $y = g(x)$; then $y = \sin x$ denotes the function $g(x) = \sin x$, $-\pi/2 \le x \le \pi/2$.

We define the inverse sine function in a similar way.

DEFINITION 4 $y = \sin^{-1} x$ if and only if $\sin y = x$ whenever $-1 \le x \le 1$.

Observe (Diagram 31) that as the sine function increases through values from $-\pi/2$ to $\pi/2$, the value of y also increases from $y = -1$ to $y = 1$.

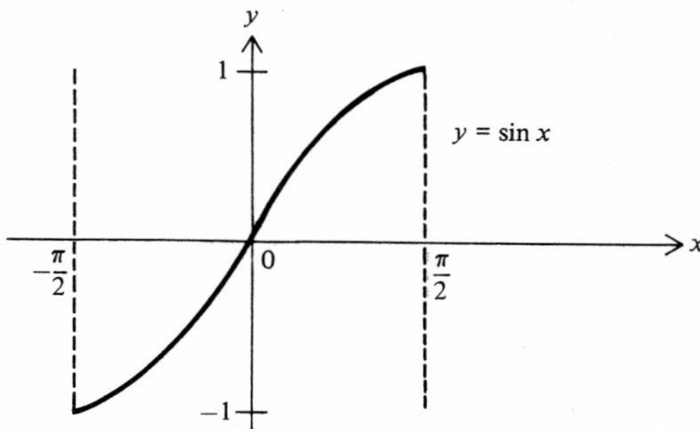

Diagram 31

Therefore for $y = \sin^{-1} x$, as x increases from -1 to 1, y increases from $-\pi/2$ to $\pi/2$. See the tables below and refer to Diagram 32.

Function	Domain	Range
$y = \sin x$	$-\frac{\pi}{2} \le x \le \frac{\pi}{2}$	$-1 \le y \le 1$
$y = \sin^{-1} x$	$-1 \le x \le 1$	$-\frac{\pi}{2} \le y \le \frac{\pi}{2}$

x	$\sin^{-1} x$
-1	$-\frac{\pi}{2}$
$-\frac{1}{2}$	$-\frac{\pi}{6}$
0	0
$\frac{1}{2}$	$\frac{\pi}{6}$
1	$\frac{\pi}{2}$

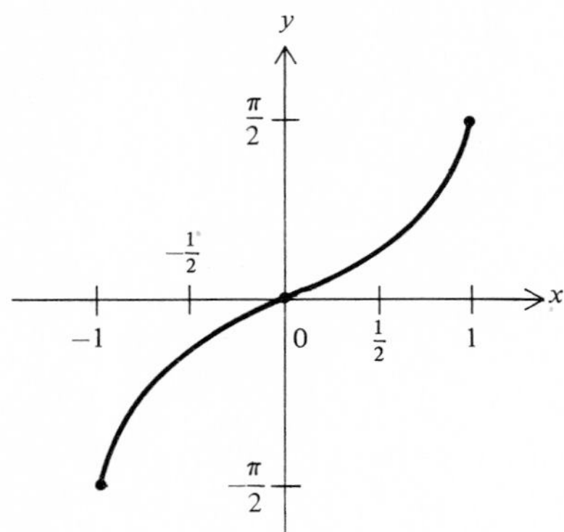

Diagram 32

Also, $y = \arcsin x = \sin^{-1} x$. To provide a working knowledge of $\cos^{-1} x$ and $\sin^{-1} x$ some examples are in order.

Example 1 Find $\cos^{-1} 0$, $\cos^{-1}(1/\sqrt{2})$, and $\cos^{-1} \frac{1}{2}$.

Solution

$$\cos^{-1} 0 = \frac{\pi}{2} \quad \text{because} \quad \cos \frac{\pi}{2} = 0$$

$$\cos^{-1} \frac{1}{\sqrt{2}} = \frac{\pi}{4} \quad \text{because} \quad \cos \frac{\pi}{4} = \frac{1}{\sqrt{2}}$$

$$\cos^{-1} \frac{1}{2} = \frac{\pi}{3} \quad \text{because} \quad \cos \frac{\pi}{3} = \frac{1}{2}$$

Example 2 Find $\sin^{-1} 0$, $\sin^{-1} \frac{1}{2}$, and $\sin^{-1}(\sqrt{3}/2)$.

Solution

$$\sin^{-1} 0 = 0 \quad \text{because} \quad \sin 0 = 0$$

$$\sin^{-1} \frac{1}{2} = \frac{\pi}{6} \quad \text{because} \quad \sin \frac{\pi}{6} = \frac{1}{2}$$

$$\sin^{-1} \frac{\sqrt{3}}{2} = \frac{\pi}{3} \quad \text{because} \quad \sin \frac{\pi}{3} = \frac{\sqrt{3}}{2}$$

Example 3 Find $\cos^{-1} 2$.

There is no real number y such that $\cos y = 2$; hence no solution is possible.

Example 4 Find $\sin^{-1} 0.48$.
$y = \sin^{-1} 0.48$ if and only if $\sin y = 0.48$. Using Appendix Table 2, we approximate $y \approx 0.50$.

Example 5 Find $\cos(\arcsin \frac{1}{2})$.
Let $\cos(\arcsin \frac{1}{2}) = \cos t$. Then $\arcsin \frac{1}{2} = t$, which means $t = \pi/6$. Hence,

$$\cos\left(\arcsin \frac{1}{2}\right) = \cos \frac{\pi}{6} = \frac{\sqrt{3}}{2}$$

Example 6 Find $\sin^{-1} [\sin(\pi/6)]$.

Solution Let $\sin^{-1}[\sin(\pi/6)] = \sin^{-1} t$. Then $\sin(\pi/6) = t = \frac{1}{2}$. Hence,

$$\sin^{-1}\left(\sin \frac{\pi}{6}\right) = \sin^{-1} \tfrac{1}{2} = \pi/6$$

This example shows that $\sin^{-1}(\sin x) = x$. This is generally true for x in the interval $[-\pi/2, \pi/2]$ since $\sin^{-1} x$ and $\sin x$ are inverses of each other.

Example 7 Find $\cos^{-1} (\cos 0)$.
Since this is similar to the last example, we can immediately conclude that $\cos^{-1} (\cos 0) = 0$.

Example 8 Find $\cos(\sin^{-1} \frac{3}{5})$.
Let $y = \sin^{-1} \frac{3}{5}$; then $\sin y = \frac{3}{5}$. We know that $\cos^2 y + \sin^2 y = 1$ or

$$\cos^2 y = 1 - \sin^2 y = 1 - (\tfrac{3}{5})^2 = \tfrac{16}{25} \qquad \cos y = \pm\sqrt{\tfrac{16}{25}} = \pm\tfrac{4}{5}$$

We further notice that if $y = \sin^{-1} \frac{3}{5}$, $0 < y < 1$ (see Diagram 31). This implies that $0 < \cos y < 1$; therefore

$$\cos y = \cos(\sin^{-1} \tfrac{3}{5}) = \tfrac{4}{5}$$

EXERCISES

Evaluate each expression in Exercises 1 to 20.

1 $\sin^{-1} \frac{\pi}{2}$

2 $\sin^{-1} 2$

3 $\sin^{-1} \frac{1}{2}$

4 $\sin^{-1} (-\frac{1}{2})$

5 $\sin^{-1} 0.2$

6 $\cos^{-1} 3$

7 $\cos^{-1}\left(-\frac{\sqrt{3}}{2}\right)$

8 $\cos^{-1}\left(-\frac{1}{\sqrt{2}}\right)$

9 $\cos^{-1} \frac{4}{5}$

10 $\cos^{-1}(-1)$

11 $\sin\left(\sin^{-1}\dfrac{1}{\sqrt{2}}\right)$

12 $\cos\left[\sin^{-1}\left(-\dfrac{1}{\sqrt{2}}\right)\right]$

13 $\sin^{-1}\left(\sin\dfrac{\pi}{3}\right)$

14 $\cos^{-1}\left(\cos\dfrac{\pi}{6}\right)$

15 sin(arccos 0.25)

16 cos(arcsin 0.25)

17 $\cos\left[\sin^{-1}\left(\sin\dfrac{\pi}{2}\right)\right]$

18 $\sin^{-1}[\cos(\sin^{-1} 1)]$

19 arcsin(sin 2)

20 arccos(cos 4)

21 Prove that $\arccos(1/\sqrt{2}) + \arcsin(1/\sqrt{2}) = \arcsin 1$.

22 Prove that $\cos^{-1}\frac{1}{2} + \sin^{-1}\frac{1}{2} = \arccos 0$.

23 Evaluate $\cos[\arcsin\frac{1}{2} - \arcsin(1/\sqrt{2})]$.

24 Evaluate $\sin(\arccos\frac{3}{5} + \arcsin\frac{3}{5})$.

25 Prove that $\cos^{-1} x + \sin^{-1} x = \pi/2$.

26 Prove that $\sin(\cos^{-1} Z) = \sqrt{1 - Z^2}$.

9.10 OTHER TRIGONOMETRIC FUNCTIONS

In the previous sections we defined and analyzed the functions $f(x) = \cos x$ and $g(x) = \sin x$.

$f(x)$ and $g(x)$ might very well be called the *parent functions* of the trigonometric family because by combining these functions as quotients of each other and by finding their reciprocals we create *four* other standard trigonometric functions. The four new functions are called tangent, cotangent, secant, and cosecant, abbreviated tan, cot, sec, and csc, respectively. We now define these functions.

DEFINITION 1 $t(x) = \tan x = \dfrac{\sin x}{\cos x}$ for $x \neq \dfrac{\pi}{2} + n\pi$.

DEFINITION 2 $c(x) = \cot x = \dfrac{\cos x}{\sin x}$ for $x \neq n\pi$.

DEFINITION 3 $s(x) = \sec x = \dfrac{1}{\cos x}$ for $x \neq \dfrac{\pi}{2} + n\pi$.

DEFINITION 4 $h(x) = \csc x = \dfrac{1}{\sin x}$ for $x \neq n\pi$.

Let us take a closer look at $t(x) = \tan x$.

$$\tan(x + \pi) = \frac{\sin(x + \pi)}{\cos(x + \pi)} = \frac{-\sin x}{-\cos x} = \tan x$$

Therefore, we know that $t(x) = \tan x$ is periodic.

Let us examine the behavior of $t(x)$ as x approaches values for which the function is not defined. Since $\tan x = (\sin x)/(\cos x)$, $x = \pm\pi/2$ gives two such values for which $\tan x$ is undefined. As x approaches $\pi/2$ from left to right ($x \to \pi^-/2$), $\sin x \to 1$ and $\cos x \to 0^+$. Hence $t(x)$ increases without bound and $x = \pi/2$ is a *vertical asymptote* of $t(x)$. Likewise, as x approaches $-\pi/2$ from the right ($x \to -\pi^+/2$), $\sin x \to -1$ and $\cos x \to 0^+$. Hence $\tan x$ decreases without bound, and $x = -\pi/2$ is *another vertical asymptote* of $t(x)$.

Diagram 33 is the graph of $y = \tan x$, $-\pi/2 < x < \pi/2$.

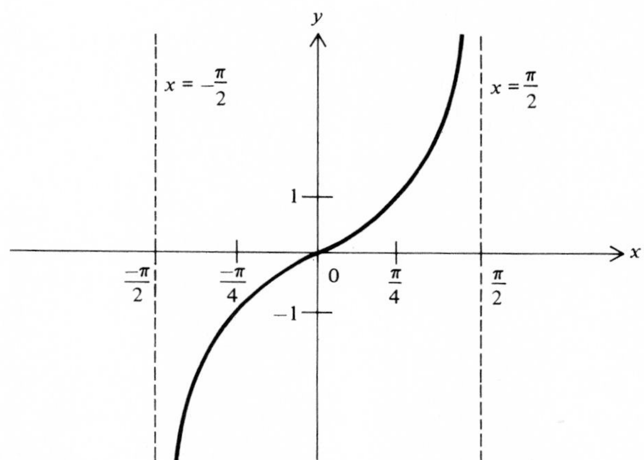

Diagram 33

The same behavior exists for $t(x)$ for every consecutive interval of $n\pi$. Diagram 34 shows $t(x) = \tan x$ in its repetitive pattern.

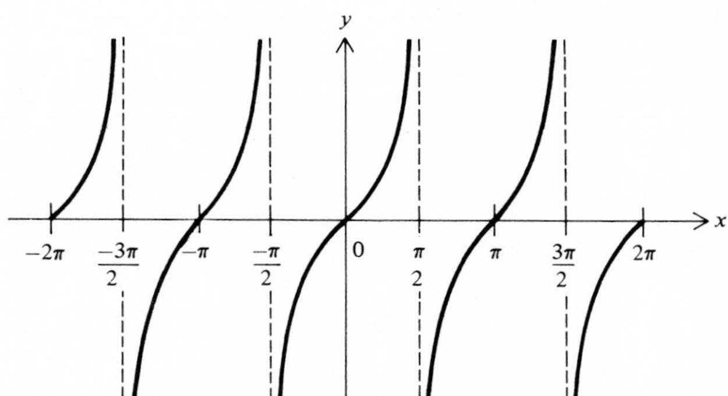

Diagram 34

From Diagram 34 we can surmise that the fundamental period of the tangent function is π and

$$\tan(-x) = -\tan x$$

Theorem 1 For all x in the domain of the tangent function $\tan(-x) = -\tan x$.

Proof
$$\tan(-x) = \frac{\sin(-x)}{\cos(-x)} = -\frac{\sin x}{\cos x} = -\tan x$$

For the cosine and the sine we have addition formulas. Let us determine an addition formula for the tangent.

Theorem 2 For all x_1 and x_2 in the domain of $t(x)$

$$\tan(x_1 + x_2) = \frac{\tan x_1 + \tan x_2}{1 - \tan x_1 \tan x_2}$$

Proof

$$\tan(x_1 + x_2) = \frac{\sin(x_1 + x_2)}{\cos(x_1 + x_2)}$$

$$= \frac{\sin x_1 \cos x_2 + \cos x_1 \sin x_2}{\cos x_1 \cos x_2 - \sin x_1 \sin x_2}$$

$$= \frac{\dfrac{\sin x_1 \cos x_2}{\cos x_1 \cos x_2} + \dfrac{\cos x_1 \sin x_2}{\cos x_1 \cos x_2}}{\dfrac{\cos x_1 \cos x_2}{\cos x_1 \cos x_2} - \dfrac{\sin x_1 \sin x_2}{\cos x_1 \cos x_2}}$$

Dividing each term by $\cos x_1 \cos x_2$

$$= \frac{\tan x_1 + \tan x_2}{1 - \tan x_1 \tan x_2}$$

Simplifying

Using our knowledge of the parent trigonometric functions and the following definitions of the other trigonometric functions, we could go on indefinitely deducing facts about them. We will work out several general examples on these functions and leave the remainder of the exploration for the exercises.

Example 1 Evaluate $\tan(\pi/4)$

Solution
$$\tan\frac{\pi}{4} = \frac{\sin(\pi/4)}{\cos(\pi/4)} = \frac{1/\sqrt{2}}{1/\sqrt{2}} = 1$$

Example 2 Graph $h(x) = \csc x$, $0 < x < \pi$.

Solution $\csc x = 1/(\sin x)$ but $\sin x = 0$ when $x = 0$ and $x = \pi$. This means $\csc x$ is undefined for these values and $x = 0$ and $x = \pi$ are asymptotes for the graph of $h(x) = \csc x$. $\sin(\pi/2) = 1$; therefore $\csc(\pi/2) = 1/[\sin(\pi/2)] = 1$. $(\pi/2, 1)$ is a point of intersection of the graphs of

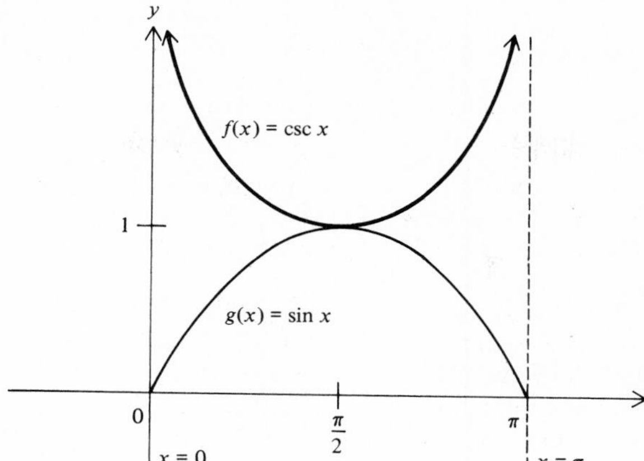

Diagram 35

$h(x) = \csc x$ and $g(x) = \sin x$. It follows that as $y = \sin x$ increases its reciprocal $y = \csc x$ decreases and vice versa. The graph of $h(x)$ is as shown for $(0, \pi)$ in Diagram 35.

Example 3 Evaluate $\cot(\pi/3)$.

Solution

$$\cot\frac{\pi}{3} = \frac{\cos(\pi/3)}{\sin(\pi/3)} = \frac{1/2}{\sqrt{3}/2} = \frac{1}{\sqrt{3}}$$

Example 4 Evaluate $\sec(\pi/3)$.

Solution

$$\sec\frac{\pi}{3} = \frac{1}{\cos\pi/3} = \frac{1}{\frac{1}{2}} = 2$$

EXERCISES

Evaluate the expression in Exercises 1 to 15.

1 $\tan 3\pi$ **2** $\cot\frac{\pi}{2}$ **3** $\csc\frac{\pi}{3}$ **4** $\sec\frac{\pi}{4}$

5 $\sec 0$ **6** $\tan 5\pi$ **7** $\tan\frac{3\pi}{4}$ **8** $\tan\frac{7\pi}{4}$

9 $\cot 2\pi$ **10** $\csc 2\pi$ **11** $\sec 2\pi$ **12** $\tan\frac{\pi}{12}$

13 $\tan\frac{7\pi}{12}$ **14** $\csc\frac{\pi}{6}$ **15** $\tan\frac{5\pi}{12}$

16 Show that for all x

$$1 + \tan^2 x = \sec^2 x \qquad \text{where} \qquad \cos x \neq 0$$

17 Show that for all x

$$1 + \cot^2 x = \csc^2 x \qquad \text{where} \qquad \sin x \neq 0$$

18 Show that

$$\tan(x_1 - x_2) = \frac{\tan x_1 - \tan x_2}{1 + \tan x_1 \tan x_2}$$

and state any restrictions.

19 Graph $f(x) = \cos x$ and $S(x) = \sec x$ on the same grid for the interval $0 \leq x \leq 2\pi$. Then answer the following questions:

(a) For what values of x is $S(x) = \sec x$ undefined?

(b) What is the domain of the secant?

(c) What is the range of the secant?

(d) What are the equations of the asymptotes of the secants?

20 Graph $g(x) = \sin x$ and $C(x) = \csc x$ on the same grid for the interval $0 \leq x \leq 2\pi$. Then answer the same questions as in Exercise 19 for the cosecant function.

21 Graph $T(x) = \tan x$ and $C(x) = \cot x$ on the same grid for the interval $0 \leq x \leq 2\pi$. Answer the same questions as in Exercise 19 for the cotangent function.

22 Determine whether $C(x) = \cot x$ is periodic or not. If so, what is its fundamental period?

23 Derive formulas for

(a) $\cot(x + y) = \dfrac{\cot x \cot y - 1}{\cot x + \cot y}$

(b) $\cot(x - y) = \dfrac{\cot x \cot y + 1}{\cot y - \cot x}$

State any restrictions on x.

24 In terms of $\tan x$, derive a formula for $\tan 2x$. State any restrictions.

25 In terms of $\cot x$, derive a formula for $\cot 2x$. State any restrictions.

26 Show that $\cot(-x) = -\cot x$.

27 Using the formulas derived in Exercise 23, evaluate:

(a) $\cot \dfrac{7\pi}{12}$ **(b)** $\cot \dfrac{\pi}{12}$ **(c)** $\cot \dfrac{5\pi}{12}$

28 Graph $T(x) = \tan 2x$ and on another grid graph $T(x) = \tan 3x$.

29 What is the period of each of the functions in Exercise 28?

30 If $T(x) = \tan bx$, $b \neq 0$, what would be the general formula for the period of t?

31 For $T(x) = \tan x$, amplitude is not be considered; why?

32 Find the fundamental period for the secant and cosecant functions.

In Exercises 33 to 40, graph each given function.

33 $f(x) = 2 \tan 3x$ **34** $h(x) = 3 \cot 2x$

35 $f(x) = \frac{1}{2} \csc 2x$ **36** $g(x) = 4 \sec 4x$

37 $t(x) = -2 \tan\left(-\dfrac{\pi}{4}\right)$ **38** $t(x) = 2 \cot\left(2x + \dfrac{\pi}{2}\right)$

39 $c(x) = \csc\left(x + \dfrac{\pi}{3}\right)$ **40** $t(x) = \pi \tan \pi x$

41 Prove that $\csc(\pi - x) = \csc x$.

42 Prove that $\csc(\pi + x) = -\csc x$.

Exponential Functions

10.1 REVIEW

Before defining the first of our new functions let us review the basic rules for exponents. Recall from elementary algebra the definition for x^n, where x is any real number and n is a positive integer.

DEFINITION 1 x^n represents a product where x is the unique factor n times. Hence, $x^1 = x$, $x^2 = x \cdot x$, $x^3 = x \cdot x \cdot x$, ...

In the above definition, x is called the *base,* and n is called the *exponent.* For example, in the expression 2^5, 2 is the base and 5 is the exponent. In the expression $(\frac{1}{2})^3$, $\frac{1}{2}$ is the base and 3 is the exponent.

For the sake of consistency and the need to include *all* integers (positive, negative, and zero), two more definitions follow.

DEFINITION 2 $x^0 = 1, x \neq 0$.

DEFINITION 3 $x^{-n} = \dfrac{1}{x^n}$, $x \neq 0$.

From these definitions a summary of the basic rules for exponents follows. Consider x and y as nonzero real numbers and m and n as integers.

RULE 1 $x^m x^n = x^{m+n}$.

RULE 2 $(x^m)^n = x^{mn}$.

RULE 3 $(xy)^m = x^m y^m$.

RULE 4 $\left(\dfrac{x}{y}\right)^m = \dfrac{x^m}{y^m}$.

RULE 5 $\dfrac{x^m}{x^n} = x^{m-n}$.

Example 1

$$2^2(2^3) = (2 \times 2)(2 \times 2 \times 2) = 2^5 \qquad \text{or} \qquad 2^2(2^3) = 2^{2+3} = 2^5$$

Example 2

$$(2^2)^3 = (2^2)(2^2)(2^2) = (2 \times 2)(2 \times 2)(2 \times 2) = 2^6 \qquad \text{or} \qquad (2^2)^3 = 2^{2(3)} = 2^6$$

Example 3

$$(3ax)^2 = (3ax)(3ax) = 3 \cdot 3 \cdot a \cdot a \cdot x \cdot x = 3^2 \cdot a^2 \cdot x^2 = 9a^2x^2$$

Example 4 $\left(\dfrac{3}{4}\right)^3 = \dfrac{3 \cdot 3 \cdot 3}{4 \cdot 4 \cdot 4} = \dfrac{27}{64}$

Example 5

$$\frac{5^6}{5^2} = \frac{5 \cdot 5 \cdot 5 \cdot 5 \cdot 5 \cdot 5}{5 \cdot 5} = 5 \cdot 5 \cdot 5 \cdot 5 = 5^4 \qquad \text{or} \qquad \frac{5^6}{5^2} = 5^{6-2} = 5^4$$

Example 6

$$\frac{5^2}{5^6} = \frac{5 \cdot 5}{5 \cdot 5 \cdot 5 \cdot 5 \cdot 5 \cdot 5} = \frac{1}{5 \cdot 5 \cdot 5 \cdot 5} = \frac{1}{5^4} = 5^{-4} \qquad \text{or} \qquad \frac{5^2}{5^6} = 5^{2-6} = 5^{-4}$$

Example 7 $2^0 = 1$.

Example 8 $(2x)^0 = 2^0 \cdot x^0 = 1 \cdot 1 = 1.$

Example 9 $2x^0 = 2 \cdot x^0 = 2 \cdot 1 = 2.$

So far our discussion of exponents has included only integers. Surely, we would like to extend this notion beyond integers, say to fractions, such as $x^{1/2}$ or $y^{2/3}$. We would also like to be consistent with the five basic rules for exponents. A logical way to pursue this is to consider a positive integer n; for example, $n = 2$, and examine the expression $(x^{1/2})^n$ or $(x^{1/2})^2$. If we follow the basic rules, then considering $x^{1/2}$ as the base and 2 as the exponent, we have $(x^{1/2})^2 = x^{(1/2)(2)} = x^1 = x$. This says that $x^{1/2}$ is a number whose *square* equals x, or in other words $x^{1/2}$ is the *square root* of x. Therefore, $x^{1/2} = \sqrt{x}$. For example, $16^{1/2} = \sqrt{16} = 4$. We can generalize this discussion in the following definition.

DEFINITION 4 For $x > 0$ and r a positive integer

$$x^{1/r} = \sqrt[r]{x}$$

The number designated by the notation $\sqrt[r]{x}$ is called the *principal rth root of x.*

For example, $16^{1/2} = \sqrt{16}$. The symbol $\sqrt{16}$ designates only the positive number 4; that is $\sqrt{16} = 4$. The symbol $-\sqrt{16}$ designates the negative number -4; that is $-\sqrt{16} = -4$.

COMMENT It should be noted that $\sqrt[r]{x}$ is undefined in the real-number system if $x < 0$ *and* r is *even.* However, if r is *odd* and $x > 0$, then

$$\sqrt[r]{-x} = -\sqrt[r]{x}$$

For instance, $(-8)^{1/3} = \sqrt[3]{-8} = -\sqrt[3]{8} = -2$

Example 10 $x^{1/3} = \sqrt[3]{x}.$

Example 11 $(-1)^{1/5} = \sqrt[5]{-1} = -\sqrt[5]{1} = -1.$

Example 12 $9^{1/2} = \sqrt{9} = 3.$

Example 13 $27^{1/3} = \sqrt[3]{27} = 3.$

DEFINITION 5 If p is any integer and $x > 0$,

$$x^{p/r} = (x^{1/r})^p = (\sqrt[r]{x})^p$$

The notation $x^{p/r}$ may be more easily understood by associating the letter p with *power* and r with *root.*

In formulating definitions for the rational exponents, we have preserved the validity of the basic rules for exponents.

Example 14 $8^{2/3} = (8^{1/3})^2 = (\sqrt[3]{8})^2 = 2^2 = 4.$

Example 15 $81^{3/4} = (81^{1/4})^3 = (\sqrt[4]{81})^3 = 3^3 = 27.$

Example 16 $3^{1/2}(3^{1/3}) = 3^{1/2+1/3} = 3^{5/6}$ or $(\sqrt[6]{3})^5.$

Example 17 $(3^{1/3})^{1/2} = 3^{(1/3)(1/2)} = 3^{1/6}$ or $\sqrt[6]{3}.$

Example 18 $(4x)^{1/2} = 4^{1/2}x^{1/2} = \sqrt{4}\sqrt{x} = 2\sqrt{x}.$

Example 19 $(3/2)^{1/2} = \dfrac{3^{1/2}}{2^{1/2}} = \dfrac{\sqrt{3}}{\sqrt{2}}$ or $\dfrac{\sqrt{6}}{2}.$

Example 20 $\dfrac{2^{3/2}}{2^{1/2}} = 2^{3/2-1/2} = 2^1 = 2.$

Example 21 $(64)^{-2/3} = \dfrac{1}{(64)^{2/3}} = \dfrac{1}{(\sqrt[3]{64})^2} = \dfrac{1}{4^2} = \dfrac{1}{16}.$

Example 22 $\sqrt[3]{\sqrt{64}} = \sqrt[3]{8} = 2.$

Example 23 $\sqrt{\sqrt[3]{64}} = \sqrt{4} = 2.$

COMMENT Examples 22 and 23 illustrate the property $(x^{1/2})^{1/3} = (x^{1/3})^{1/2}$ or in general $(x^p)^q = (x^q)^p$.

EXERCISES

Simplify each of the expressions and express all answers with positive exponents.

1 $3^2 3^4$ **2** $\pi^3\pi^4$ **3** $(4^3)^2$ **4** $(\frac{1}{2})^3$

5 $(2by)^3$ **6** $(-v/2s)^3$ **7** $\dfrac{8^6}{8^3}$ **8** $\dfrac{4^4}{4^6}$

9 π^0 **10** $(\pi r^2)^0$ **11** $2\pi r^0$ **12** $3(4^0)$

13 $[3\ (4)]^0$ **14** $64^{1/3}$ **15** $16^{1/4}$ **16** $(-32)^{1/5}$

17 $576^{1/2}$ **18** $343^{1/3}$ **19** $16^{3/2}$ **20** $8^{-2/3}$

21 $243^{3/5}$ **22** $\dfrac{3^{-2}}{9^{-1}}$ **23** $2^{1/3}2^{1/4}$ **24** $(2^{1/3})^{1/4}$

25 $(8x)^{1/3}$ **26** $\left(\dfrac{361}{441}\right)^{1/2}$ **27** $(27^{4/3})(9^{-3/2})$ **28** $\sqrt[3]{512}$

29 $256^{1/4}$ **30** $\dfrac{4^{5/2}}{4^{3/2}}$ **31** $(125)^{-2/3}$ **32** $\sqrt[3]{\sqrt{729}}$

33 $\sqrt{\sqrt[4]{256}}$ **34** $[(108)^{1/2}]^{1/3}$ **35** $2^{2.25}2^{1.05}$

36 $[3^2(9^2)^{1/4} + (10 + 9^{3/2})]^{-1/3}$ **37** $\dfrac{b^{5m}}{b^{3m-2}}$ **38** $b^{-n}(b^{3n} - b^{2n})$

39 $b^{-n}(b^{3n} + b^{-n})$ **40** $\left(\dfrac{1}{b^{-1}}\right)^{-1}$ **41** $(\frac{1}{2})^{-1}$

42 $(\frac{8}{27})^{-2/3}$ **43** $\dfrac{b^k}{b^{2k}}$ **44** $(10^{-2})^{-2}$

In Exercises 45 to 49 find x.

45 $2^x = 32$ **46** $2^{1/x} = 32$ **47** $10^x = 0.0001$

48 $2^x = \frac{1}{16}$ **49** $8^x = 4$

50 If b is any real number, answer true or false and comment on your answer.
(a) $(b^2)^{1/2} = b$ **(b)** $(b^3)^{1/3} = b$

10.2 EXPONENTIAL FUNCTIONS

Earlier we examined $f(x) = x^2$, $g(x) = x^3$, $h(x) = x^4$, ... These functions were simple examples of polynomial functions. If we make an innocent-looking modification by interchanging the base and the exponent so that the base is now the constant and the exponent becomes the variable, we will have achieved a major change in the type of function. The result is that we now have a function which is exponential in character.

Polynomial	**Exponential**
$f(x) = x^2$	$T(x) = 2^x$
$g(x) = x^3$	$T(x) = 3^x$
$h(x) = x^4$	$T(x) = 4^x$

These T functions are examples of *exponential* functions.

Before defining the term exponential function, let us examine $T(x) = b^x$ and see what is desirable in terms of the base b and the domain and range for $T(x)$.

QUESTION 1 Can the base b be any real number or are there restrictions on b?

Suppose $b = 0$. If we define the domain to be $x > 0$, $T(x) = 0^x$ and this behaves like the constant function $f(x) = 0$. Hence it loses its exponential character and so $b \neq 0$.

Similarly, if $b = 1$, 1 raised to any real power is 1. Again we see that $T(x) = 1^x$ resembles the constant function $f(x) = 1$, and so $b \neq 1$.

Suppose $b < 0$. Say $b = -2$. Then $T(x) = (-2)^x$ creates a problem in which the graph possesses infinitely many holes. Whenever the domain value is a reduced rational value p/r and r is an even positive integer, the function will be undefined. For example, $T(\frac{1}{2}) = (-2)^{1/2} = \sqrt{-2}$, which is not a real number. We avoid this situation by saying $b \nless 0$.

This leaves all positive real numbers, excluding $b = 1$. We can therefore conclude that

$$b > 0 \qquad \text{and} \qquad b \neq 1$$

QUESTION 2 Considering any base allowable (from Question 1), what is the largest possible domain for the exponential function?

This is a profound question when we consider irrational exponents. We have already discussed positive and negative rational exponents; however, we have evaded the discussion of the irrational exponent, for example, $2^{\sqrt{2}}$. This was done for a good reason. To answer this question involves the knowledge of sequences and the concept of limit which are discussed in the calculus. Let us sacrifice mathematical accuracy for an intuitive argument at this point. The justification given for the use of irrational numbers is as follows.

We know that $\sqrt{2}$ can be approximated to any desired degree of accuracy. In succession we can obtain different rational *approximations*

$$\sqrt{2} \approx 1.4 \approx 1.41 \approx 1.414 \approx 1.4142 \approx 1.41421 \approx \cdots$$

Note that $1.4 < 1.41 < 1.414 < 1.4142 < 1.41421 < \cdots$

Since each successive approximation is greater than its predecessor, we create a squeeze, that is,

$$\begin{array}{lcl} 1.4 & < \sqrt{2} < & 1.5 \\ 1.41 & < \sqrt{2} < & 1.42 \\ 1.414 & < \sqrt{2} < & 1.415 \\ 1.4142 & < \sqrt{2} < & 1.4143 \\ 1.41421 & < \sqrt{2} < & 1.41422 \\ \cdots & \cdots & \cdots \end{array}$$

The squeeze indicates that although the successive approximations of $\sqrt{2}$ increase, they are *bounded* (controlled), and so the approximations get closer and closer to a particular value which we call $\sqrt{2}$. $2^{\sqrt{2}}$ can now be inspected in the same fashion.

Since the base, 2, is positive and greater than 1, the larger its exponent the greater its value; that is,

$$2^{1.4} < 2^{1.41} < 2^{1.414} < 2^{1.4142} < 2^{1.41421} < \cdots$$

but

$$\begin{array}{l} 2^{1.4} < 2^{\sqrt{2}} < 2^{1.5} \\ 2^{1.41} < 2^{\sqrt{2}} < 2^{1.42} < \cdots \end{array}$$

Hence we see that the successive approximations of $2^{\sqrt{2}}$ increase but are bounded. We call this boundary number $2^{\sqrt{2}}$.

$$2^{1.4} \approx 2^{1.41} \approx 2^{1.414} \approx 2^{1.4142} \approx \cdots \approx 2^{\sqrt{2}}$$

This can be done similarly for other irrational exponents, such as $\sqrt{3}$.

We therefore conclude that the domain of the exponential function $T(x) = b^x$ ($b > 0$ and $b \neq 1$) is *all the real numbers.*

QUESTION 3 What is the range of the exponential function?

Again, let us inspect $T(x) = b^x$ where $b = 2$ [$T(x) = 2^x$]. Consider the following table for some convenient nonnegative values of x.

x	0	1	2	3	$x \to +\infty$
2^x	1	2	4	8	$2^x \to +\infty$

In a second table we evaluate 2^x for convenient negative x values:

x	-1	-2	-3	-4	$x \to -\infty$
2^x	$\frac{1}{2}$	$\frac{1}{4}$	$\frac{1}{8}$	$\frac{1}{16}$	$2^x \to 0$

Plotting these points (from both tables) we obtain the points in Diagram 1. We could now make up a third table evaluating 2^x for both rational and irrational values to fill in the gaps. Since this would be tedious and use a procedure we have not developed yet, we will assume that the graph of $y = 2^x$ is a smooth, unbroken curve.

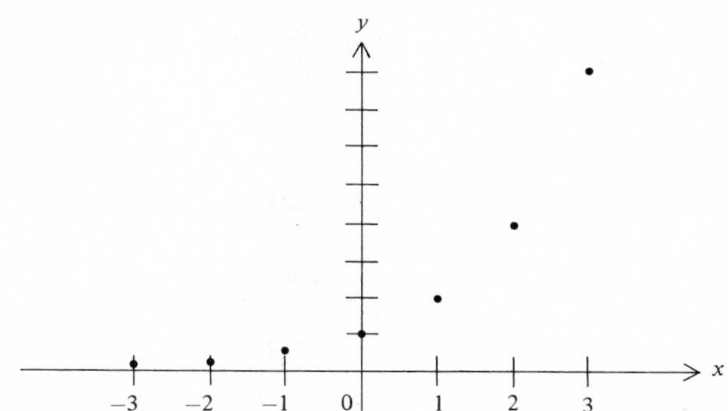

Diagram 1

Viewing Diagram 2 and studying the behavior of $T(x) = 2^x$, we see that the range of $y = T(x)$ consists of all real values of y in the interval $(0, +\infty)$ (all positive real numbers, zero not included) and that the negative portion of the x axis serves as a horizontal asymptote.

Before formally defining the exponential function, let us examine another example for which the base b now is $0 < b < 1$. Say $T(x) = (\frac{1}{2})^x$.

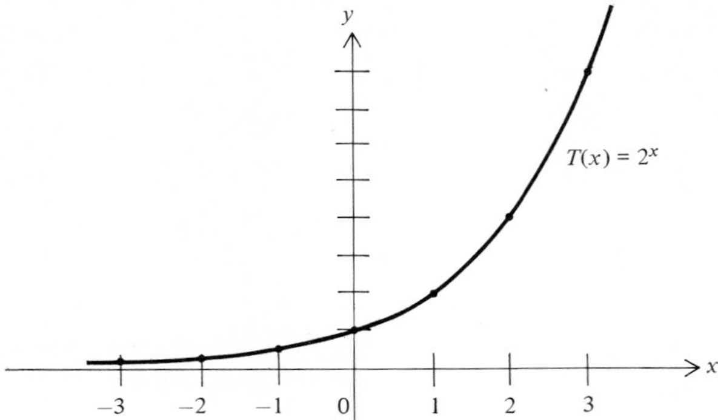

Diagram 2

Start with a table of convenient values.

x	-1	-2	-3	$x \to -\infty$
$T(x)$	2	4	8	$T(x) \to +\infty$

x	0	1	2	3	$x \to +\infty$
$T(x)$	1	$\frac{1}{2}$	$\frac{1}{4}$	$\frac{1}{8}$	$T(x) \to 0$

A definite pattern is established whereby the function behaves opposite to that of 2^x. Let us plot the points and graph $T(x) = (\frac{1}{2})^x$ (see Diagram 3).

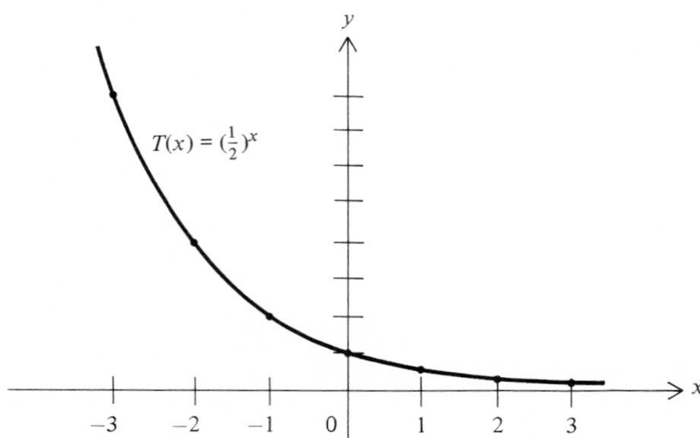

Diagram 3

We make the following observations:

1 The domain and range of $T(x) = (\frac{1}{2})^x$ are the same as for $T(x) = 2^x$. The domain is all real values of x in the interval $(-\infty, +\infty)$, and the range is all real values of y in the interval $(0, \infty)$.

2 The *positive* portion of the x axis acts as a horizontal asymptote.

3 The two graphs have the point (0, 1) in common. (This is their *only* common point.)

We now define an exponential function.

DEFINITION Consider b, a real number, such that $b > 0$ and $b \neq 1$. Then a function of the form

$$T(x) = b^x$$

is called an *exponential function.* If not restricted, the domain of T is all real numbers.

Theorem *All* exponential functions of the form $T(x) = b^x$ have the point $(0, 1)$ in common.

Proof By definition, $b^0 = 1$ and $T(0) = b^0 = 1$.

As an aid in graphing exponential functions, we conclude this section with two reference graphs (see Diagram 4).

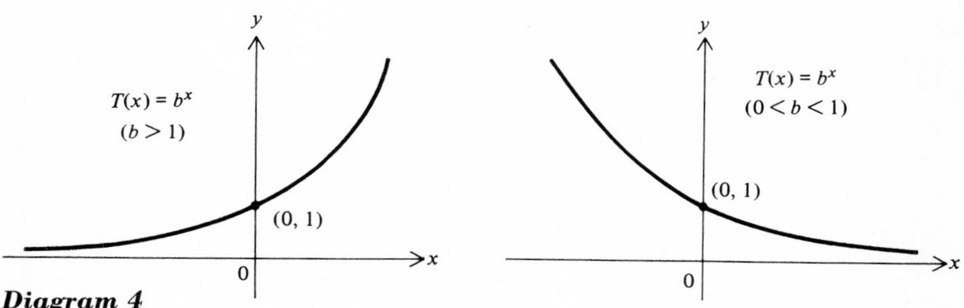

Diagram 4

EXERCISES

1 Graph $T(x) = 2^{-x}$. Check your results with Diagram 3. Comment.

2 **(a)** Make a careful sketch of $T(x) = 3^x$ using only $0, \pm 1, \pm 2$ and from the sketch approximate the value of $T(\sqrt{2})$. Check your result with Appendix Table 5.

(b) Approximate $T(\frac{1}{2})$. Check your result with Appendix Table 7.

3 On the same grid graph $f(x) = x^2$ and $g(x) = 2^x$.

(a) How many values of x are there such that $f(x) = g(x)$?

(b) What are the rational values of x for which part *(a)* is true?

4 Find the graphic approximations for **(a)** 2^π and **(b)** $2^{\sqrt{3}}$.

5 On the same grid graph $T(x) = 2^x$ and $J(x) = 2^{2x}$.

6 On the same grid graph $T(x) = 4(\frac{1}{2})^x$ and $S(x) = (\frac{1}{2})^x$.

7 Graph $f(x) = 2^{(x-1)}$. Why is the point $(0, 1)$ not on the graph of the function?

8 Graph $f(x) = 2^{(x^2)}$.

9 On the same grid graph $P(x) = (\frac{1}{2})^x$, $R(x) = (\frac{1}{3})^x$, and $S(x) = (\frac{1}{4})^x$.

(a) What behavioral change is caused by making the base smaller?

(b) Where would you place the graphs of $T(x) = (\frac{2}{3})^x$, $T(x) = (\frac{1}{8})^x$, and $T(x) = (\frac{3}{8})^x$ with respect to $P(x), R(x)$, and $S(x)$ of part *(a)*?

10 Sketch an *accurate* graph of $y = 10^x$, $-1 \leq x \leq 1$, and from the graph approximate:

(a) $10^{0.2}$ **(b)** $10^{-0.2}$ **(c)** $10^{0.8}$
(d) $10^{-0.8}$ **(e)** $10^{0.5}$ **(f)** $10^{-0.5}$

11 From the graph in Exercise 10, find x if:

(a) $10^x = 4$ **(b)** $10^x = 0.4$ **(c)** $10^x = 8$
(d) $10^x = 0.8$ **(e)** $10^x = 6.4$ **(f)** $10^x = -2$

12 Using a desk calculator or other computer, sketch a graph of $f(x) = (\sqrt{2})^x$ and approximate the value of $(\sqrt{2})^{\sqrt{2}}$.

10.3 THE CONSTANT *e*

The letter e represents one of the most fascinating real numbers. Like the constant π, it is irrational, and its uses in mathematics are varied. Euler (pronounced "oiler"), a great Swiss mathematician of the eighteenth century, was the first to publish e ("Introductio in analysin infinitorum," 1748). He used the constant in calculations of population-growth and bank-interest problems.

To express e as a real number, we use what is called an *infinite series.*

$$e = 1 + \frac{1}{1} + \frac{1}{1 \cdot 2} + \frac{1}{1 \cdot 2 \cdot 3} + \frac{1}{1 \cdot 2 \cdot 3 \cdot 4} + \cdots \qquad (1)$$

In expression (1) we cannot possibly add up *all* the terms. However, since the number of terms is endless, we can get a better and better approximation of e by adding more and more terms in the series. In fact, it is possible to compute e to any degree of accuracy.

Example 1 Find e to the nearest thousandth.
Since the eighth term of the series is

$$\frac{1}{1 \cdot 2 \cdot 3 \cdot 4 \cdot 5 \cdot 6 \cdot 7} = \frac{1}{5040} < 0.0005$$

e to the nearest thousandth can be found by adding the first seven terms.

$$e \approx 1 + \tfrac{1}{1} + \tfrac{1}{2} + \tfrac{1}{6} + \tfrac{1}{24} + \tfrac{1}{120} + \tfrac{1}{720} \approx 2.718$$

$f(x) = e^x$ is a special exponential function because of its common use in advanced mathematics. The base e is a real number, and the rules for exponents are still applicable. Hence

$$\begin{aligned} f(0) &= e^0 = 1 \\ f(1) &= e^1 = e \\ f(x + y) &= e^{x+y} = e^x e^y = f(x)f(y) \\ &\cdots\cdots\cdots\cdots \end{aligned}$$

Since $e > 1$, the graph of $f(x) = e^x$ resembles the reference graph in Section 10.2. Also $2 < e < 3$; therefore, the graph will be between the graphs of $T(x) = 2^x$ and $T(x) = 3^x$ and share the common point (0, 1) (see Diagram 5).

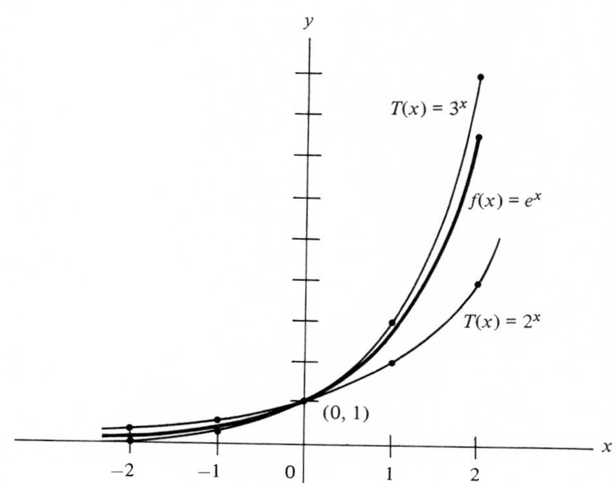

Diagram 5

COMMENT The graph of $f(x) = e^x$ is "closer" to the graph of $T(x) = 3^x$ than to the graph of $T(x) = 2^x$, since e is closer to 3 than it is to 2.

It can be shown in the calculus that the tangent line to any given point (x_0, y_0) of the graph of $f(x) = e^x$ is given by the equation

$$y - y_0 = e^{x_0}(x - x_0) \tag{2}$$

Example 2 Find the equation of the tangent line to the graph of $f(x) = e^x$ at the point (0, 1).

Solution $(x_0, y_0) = (0, 1)$ $\quad x_0 = 0 \quad y_0 = 1$

From Equation (2)

$$y - 1 = e^0(x - 0) \quad \text{and} \quad y = x + 1$$

Graphically we have Diagram 6.

COMMENT It should be noted that Equation (2) applies only to $f(x) = e^x$ and is not applicable to other exponential functions. We showed earlier that all functions of the form $T(x) = b^x$ have (0, 1) as a common point. That *does not mean* that $y = x + 1$ is the tangent line to all graphs of these functions at (0, 1). $f(x) = e^x$ is the *only* exponential function whose tangent line at (0, 1) has the slope $m = 1$.

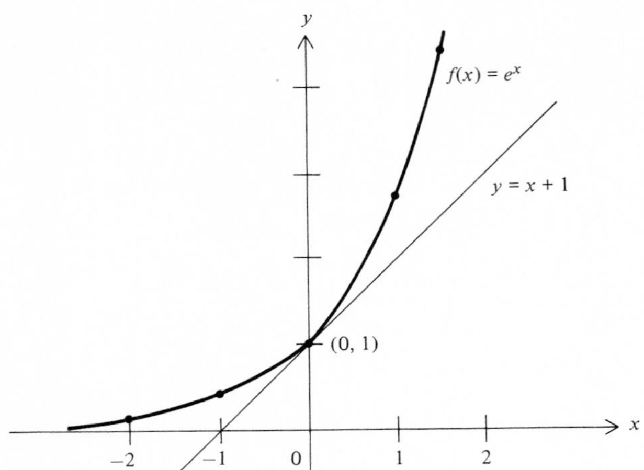

Diagram 6

EXERCISES

In the calculus it can be shown that

$$e^x = 1 + x + \frac{x^2}{2} + \frac{x^3}{2 \cdot 3} + \frac{x^4}{2 \cdot 3 \cdot 4} + \cdots$$

Using this infinite series, find the value of the expressions in Exercises 1 to 6 to the nearest thousandth.

1 $e^{0.5}$ **2** $e^{0.1}$ **3** $e^{0.3}$
4 $e^{-1/2}$ **5** $e^{-0.1}$ **6** $e^{0.75}$
7 The infinite series above defines e^x. If we use it for $x = 0$, we get $e^0 = 1$. Why is this value exact and not an approximation?

On four separate grids on a sheet of graph paper sketch the functions in Exercises 8 to 11.

8 $f(x) = e^x$ **9** $f(x) = e^{-x}$ **10** $f(x) = -e^{-x}$ **11** $f(x) = -e^x$
12 Describe a method for obtaining the graph of Exercise 11 from that of Exercise 8.
13 Describe a method for obtaining the graph of Exercise 10 from that of Exercise 9.
14 Graph $f(x) = e^{-x^2}$.

The exponential function e^x is used to define the hyperbolic function. Two such functions are

$$\sinh x = \frac{e^x - e^{-x}}{2} \qquad \text{and} \qquad \cosh x = \frac{e^x + e^{-x}}{2}$$

called the *hyperbolic sine* and the *hyperbolic cosine*, respectively. Using your knowledge of e^x and Table 1 in the Appendix, evaluate the functions in Exercises 15 to 19.

15 sinh 0 **16** cosh 0 **17** sinh 1
18 cosh 1 **19** sinh 0.5
20 Using the definition above, show that $\cosh x + \sinh x = e^x$.
21 Using the definition above, show that $\cosh^2 x - \sinh^2 x = 1$.
22 A line perpendicular to the tangent to a curve at a point is said to be the *normal* at

that point. Find the equation of the normal to the curve $y = e^x$ at the point (0, 1); use Equation (2).

23 If $f(x) = (1 + 1/x)^x$, using a calculator show that as $x \to +\infty$, $f(x) \to e$.

10.4 LOGARITHMIC FUNCTIONS

For convenience, we show the general reference graphs for exponential functions of the form $T(x) = b^x$ (Diagram 7).

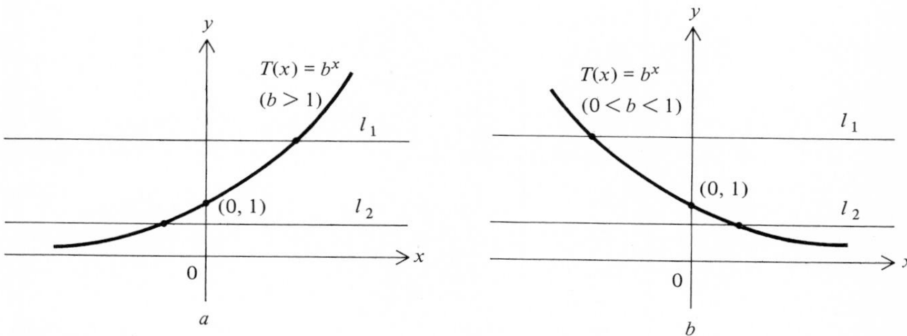

Diagram 7

Recalling the *horizontal-line test* (Chapter 7), we apply it here to these reference graphs. Note that l_1 and l_2 are arbitrarily drawn parallel to the x axis. In each case, the line intersects the graph at one and only one point. Hence, we see that $T(x) = b^x$ is one to one, and therefore $T(x)$ *possesses an inverse.* The inverse of an exponential function is called a *logarithmic function.*

The domain of an exponential function $T(x)$ is all the real numbers, and the range is all the *positive* real numbers. In keeping with the definition of an inverse function, the domain and the range must be interchanged. That is, the *domain* of the logarithmic function becomes the range of the exponential function (all positive real numbers), and the *range* of the logarithmic function becomes the domain of the exponential function (all real numbers).

DEFINITION If b is any positive real number, $b \neq 1$, then $y = \log_b x$ if and only if $x = b^y$.

$y = \log_b x$ is read "y is the logarithm of x to the base b"

COMMENT The graph of the exponential function and the graph of its inverse function, the logarithmic function, are shown in Diagram 8 with the base $b > 1$. This diagram again illustrates the fact from Chapter 7 that the graph of a function and its inverse are symmetric about the line $y = x$.

Function	Domain	Range	Comment
$y = b^x$	All real values of x in the interval $(-\infty, +\infty)$	All real values of y in the interval $(0, +\infty)$	For any base $b > 0$, and $b \neq 1$, $(0, 1)$ is contained in the graph of $y = b^x$
$y = \log_b x$	All real values of x in the interval $(0, +\infty)$	All real values of y in the interval $(-\infty, +\infty)$	The point $(1, 0)$ is contained in the graphs of all logarithmic functions

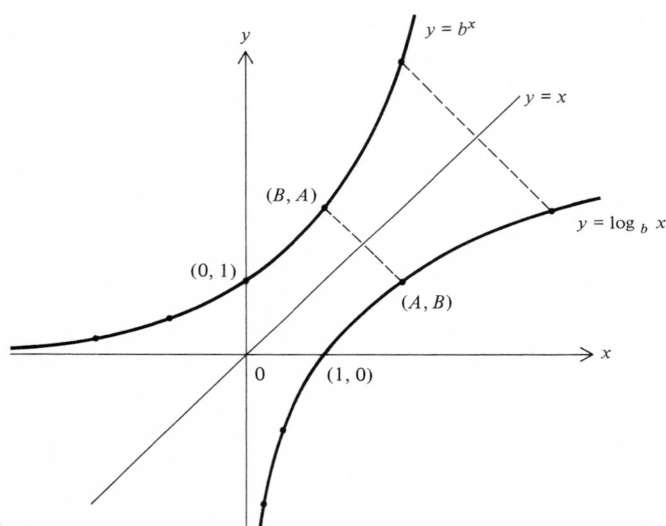

Diagram 8

Example 1 Since $x = b^y$ if and only if $y = \log_b x$,

$$8 = 2^3 \Leftrightarrow 3 = \log_2 8 \qquad \text{and} \qquad \tfrac{1}{8} = 2^{-3} \Leftrightarrow -3 = \log_2 \tfrac{1}{8}$$

or

$$\log_{16} 4 = \tfrac{1}{2} \Leftrightarrow 4 = 16^{1/2} \qquad \text{and} \qquad \log_4 \tfrac{1}{16} = -2 \Leftrightarrow \tfrac{1}{16} = 4^{-2}$$

Example 2 Using your knowledge of inverses, graph $y = \log_2 x$.

First examine $y = 2^x$. (Why?) Making use of the table in Section 10.2, we note the points $(-2, \frac{1}{4})$, $(-1, \frac{1}{2})$, $(0, 1)$, $(1, 2)$, and $(2, 4)$. Hence, the points $(\frac{1}{4}, -2)$, $(\frac{1}{2}, -1)$, $(1, 0)$, $(2, 1)$, and $(4, 2)$ are points of the graph of $y = \log_2 x$. Just for review, we make a separate graph showing the line $y = x$, which is the perpendicular bisector of the line segments shown in Diagram 9. We then make a graph of $y = \log_2 x$ (Diagram 9*b*).

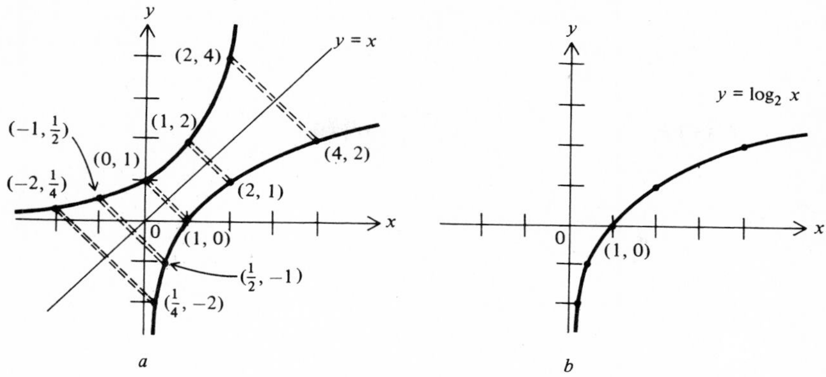

Diagram 9

COMMENT From Diagram 9*b* you can see where the logarithm is negative:

$$y < 0 \qquad \text{whenever } 0 < x < 1$$
$$\log_2 1 = 0$$
$$y > 0 \qquad \text{whenever } x > 1$$

Example 3 Graph $y = \log_{1/2} x$.

As in the previous example, generate a set of points by first investigating $y = (\frac{1}{2})^x$: $(-2, 4)$, $(-1, 2)$, $(0, 1)$, $(1, \frac{1}{2})$, and $(2, \frac{1}{4})$. Now we have a set of points for $y = \log_{1/2} x$: $(4, -2)$, $(2, -1)$, $(1, 0)$, $(\frac{1}{2}, 1)$, and $(\frac{1}{4}, 2)$. We plot the second set of points, tracing a smooth curve through them (Diagram 10).

COMMENT For $y = \log_{1/2} x$, the base, is *less than 1*.

$$y > 0 \qquad \text{whenever } 0 < x < 1$$
$$y < 0 \qquad \text{whenever } x > 1$$
$$\log_{1/2} 1 = 0$$

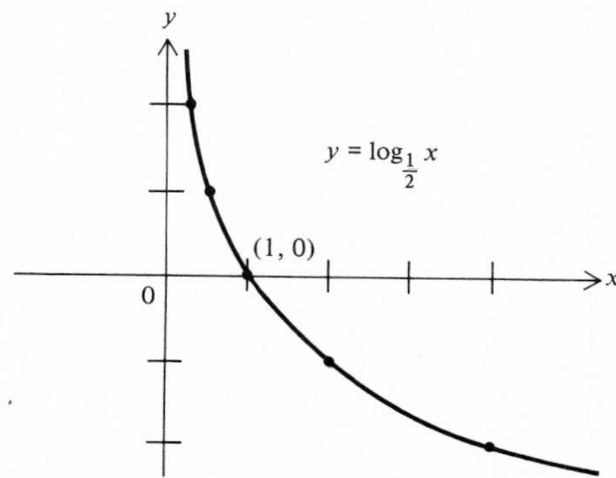

Diagram 10

Diagram 11 is a reference graph for the graphs of logarithmic functions.

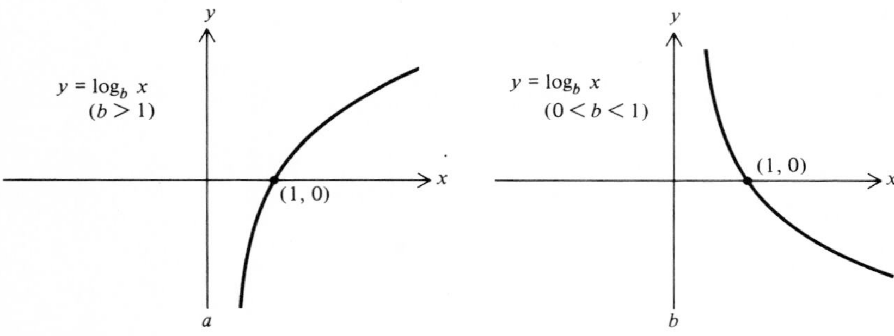

Diagram 11

EXERCISES

Write each statement in Exercises 1 to 5 in exponential form.

1 $\log_2 16 = 4$ **2** $\log_4 1 = 0$ **3** $\log_{10} 10 = 1$
4 $\log_{1/2} 8 = -3$ **5** $\log_{2/3} \frac{9}{4} = -2$

In Exercises 6 to 10, write each exponential statement in logarithmic form.

6 $64 = 2^6$ **7** $256 = 4^4$ **8** $\frac{1}{2} = 2^{-1}$
9 $\frac{1}{64} = 8^{-2}$ **10** $1 = e^0$

In Exercises 11 to 16, what value of x is necessary for a valid statement?

11 $\log_x 27 = 3$ **12** $\log_x \frac{1}{2} = \frac{1}{2}$ **13** $\log_4 x = -\frac{1}{2}$
14 $\log_2 256 = x$ **15** $\log_3 x = -1$ **16** $\log_2 x = 0$

17 Draw an accurate graph of $f(x) = 2^x$, $-3 \le x \le 3$. Make the graph as large as possible and use it to construct the inverse of the function, namely, $y = \log_2 x$.

18 From the graph in Exercise 17, find:
(a) $\log_2 3$ **(b)** $\log_2 5$ **(c)** $\log_2 1$ **(d)** $\log_2 0.6$

19 Using the graph in Exercise 17, find x if:
(a) $\log_2 x = 1.5$ **(b)** $\log_2 x = 0.5$
(c) $\log_2 x = 2.5$ **(d)** $\log_2 x = -0.5$

20 As in Exercise 17, construct a large, accurate graph of $f(x) = 10^x$, $-1 \le x \le 1$, and use it to construct its inverse, namely, $y = \log_{10} x$.

21 From the graph in Exercise 20, find:
(a) $\log_{10} 2$ **(b)** $\log_{10} 3$ **(c)** $\log_{10} 4$ **(d)** $\log_{10} 8$

22 From the graph in Exercise 20, find x if:
(a) $\log_{10} x = 0.25$ **(b)** $\log_{10} x = 0.50$
(c) $\log_{10} x = 0.75$ **(d)** $\log_{10} x = -0.5$

Graph the functions in Exercises 23 to 28.

23 $y = \log_e x$ **24** $y = \log_{1/3} x$ **25** $y = \log_{10} x$
26 $y = \log_{0.6} x$ **27** $y = \log_2 |x|$ **28** $y = |\log_2 x|$

29 Find the inverse function of Exercises 23 to 28.

30 Prove that $\log_b 1 = 0$ for any allowable base.

31 Does $y = \log_b x$ have any vertical or horizontal asymptotes? If so, state the equation or equations.

10.5 COMPUTATIONS

Because the real-number system is base 10 and because the base e has many practical applications and unique properties, logarithms to these particular bases have greater importance.

DEFINITION 1 $y = \log_{10} x$ is denoted simply by $y = \log x$ and is called the *common logarithmic function.*

DEFINITION 2 $y = \log_e x$ is denoted by $y = \ln x$ and is called the *natural logarithmic function.*

Before we perform any computations with logarithms we must introduce the basic properties of these functions.

Theorem 1 If $x > 0$ and $y > 0$, then $\log_b xy = \log_b x + \log_b y$.

Proof Let $Z = \log_b xy$, $U = \log_b x$, and $V = \log_b y$. Then

$$b^Z = xy \qquad b^U = x \qquad \text{and} \qquad b^V = y$$

Now
$$xy = b^Z = b^U b^V = b^{U+V}$$

Hence
$$b^Z = b^{U+V} \qquad \text{and} \qquad Z = U + V$$

Therefore
$$\log_b xy = \log_b x + \log_b y$$

Theorem 2 Let $x > 0$ and $y > 0$. Then $\log_b \dfrac{x}{y} = \log_b x - \log_b y$.

Theorem 3 Let $x > 0$ and y be real. Then $\log_b x^y = y \log_b x$.

We can now demonstrate some basic uses of the preceding theorems and definitions.

Example 1 $\log 6\pi = \log 6 + \log \pi$.

Example 2 $\log \dfrac{6}{\pi} = \log 6 - \log \pi$.

Example 3 $\log(\sqrt[4]{6})^3 = \log 6^{3/4} = \frac{3}{4} \log 6.$

Example 4 Simplify log 10 and ln e.

Solution

$$\log 10 = \log_{10} 10$$

$$\log_{10} 10 = y \text{ if and only if } 10^y = 10 \qquad y = 1$$

and

$$\ln e = \log_e e$$

$$\log_e e = y \text{ if and only if } e^y = e \qquad y = 1$$

Therefore $\quad \log 10 = 1 \qquad \ln e = 1$

This brings up an interesting fact, namely, $\log_b b = 1$.

In order to use logarithms as a computational aid, we introduce the following theorem.

Theorem 4 If c is an integer and x is a real number such that $1 \leq x < 10$, then

$$\log(x \times 10^c) = c + \log x$$

Proof

$$\begin{aligned} \log(x \times 10^c) &= \log x + \log 10^c && \text{Using Theorem 1} \\ &= \log x + c \log 10 && \text{Using Theorem 3} \\ &= \log x + c && \log 10 = 1 \\ &= c + \log x \end{aligned}$$

Example 5 Rewrite log 250 in the form suggested in Theorem 4. Since 250 can be written as 2.5×10^2,

$$\log 250 = \log(2.5 \times 10^2)$$

Here, $c = 2$ and $x = 2.5$. (Note that $1 < 2.5 < 10$.) Hence, $\log(2.5 \times 10^2) = 2 + \log 2.5$.

DEFINITION 3 In $\log(x \times 10^c) = c + \log x$, c is called the *characteristic* and $\log x$ is called the *mantissa*

COMMENT A logarithm is nothing more than an exponent. We know that $10^0 = 1$ and $10^1 = 10$. Therefore, 0 is the logarithm of 1 and 1 is the logarithm of 10. For any number x, $1 \leq x < 10$, $0 \leq \log x < 1$. From the previous definition, $\log x$ is the mantissa. Hence, the mantissa is a number between 0 and 1, including 0 but not 1. These mantissas can be found in Appendix Table 3. For instance, to complete the solution in Example 5,

$$\log 250 = 2 + \log 2.5$$
$$= 2 + 0.3979 \qquad \text{Appendix Table 3}$$

For computational purposes, we express the characteristic and mantissa separately, in the form $c + \log x$.

Example 6 Find log 4.86.

Solution

$$\log 4.86 = \log(4.86 \times 10^0) = 0 + \log 4.86 = 0 + 0.6866$$

Example 7 Find log 4860.

Solution

$$\log 4860 = \log(4.86 \times 10^3) = 3 + \log 4.86 = 3 + 0.6866$$

Example 8 Find log 0.486.

Solution

$$\log 0.486 = \log(4.86 \times 10^{-1}) = -1 + \log 4.86 = -1 + 0.6866$$

Example 9 Find log 0.00486.

Solution

$$\log 0.00486 = \log(4.86 \times 10^{-3}) = -3 + \log 4.86 = -3 + 0.6866$$

Suppose $\log N = 1 + 0.3160$. How do we determine N? In this problem, we are given the exponent of the base 10, that is, $10^{1+0.3160}$. Using Property 1, Section 10.1, we have

$$10^{1+0.3160} = 10^1 \times 10^{0.3160} = 10 \times 2.07 = 20.7$$

COMMENT Appendix Table 3 is also used to find a number whose logarithm is given. This number is called the *antilogarithm.* Hence, 20.7 is the antilogarithm of 1 + 0.3160.

Example 10 Find N if $\log N = 1 + 0.7672$.
From Appendix Table 3

$$\log 5.85 = 0.7672$$

$$\log N = 1 + \log 5.85 = \log(5.85 \times 10^1) = \log 58.5$$

If $\log n = \log x$, then $n = x$. Hence, $N = 58.5$.

Example 11 Find N if $\log N = -2 + 0.7672$.

Solution

$$\log N = -2 + \log 5.85 = \log(5.85 \times 10^{-2}) = \log 0.0585$$

Hence, $N = 0.0585$.

Example 12 Find N if $\log N = 0.7672$.
Expressing $\log N$ in proper form, we obtain

$$\log N = 0 + 0.7672 = 0 + \log 5.85 = \log(5.85 \times 10^{0}) = \log(5.85 \times 1).$$

Hence, $N = 5.85$.

The mantissa for numbers up to three digits can be found directly from Appendix Table 3. Suppose we want to find log 3.544. This can be accomplished by *interpolation,* which we now demonstrate.

	x			$\log x$	
0.010	3.550			0.5502	0.0012
	3.544	0.004	k	$0.5490 + k$	
	3.540			0.5490	

Set up a proportion to determine k.

$$\frac{0.004}{0.010} = \frac{k}{0.0012} \quad \text{and} \quad \frac{4}{10} = \frac{k}{0.0012}$$

$$10k = 0.0048 \quad \text{and} \quad k = 0.00048 \approx 0.0005$$

(rounded off to four places to comply with Appendix Table 3). Therefore,

$$\log 3.544 = 0.5490 + 0.0005 = 0.5495$$

Suppose $\log N = 1 + 0.4343$. In determining N, we see that 0.4343 is not an entry in Appendix Table 3; however, it is between 0.4330 and 0.4346. Using the interpolation process, we obtain

	x			$\log x$	
0.010	2.720			0.4346	0.0016
	$2.710 + k$	k	0.0013	0.4343	
	2.710			0.4330	

Setting up the proportion to determine k, we obtain

$$\frac{k}{0.010} = \frac{0.0013}{0.0016}$$

Simplifying gives

$$\frac{k}{0.01} = \frac{13}{16}$$

and

$$16k = 0.13$$

$$k = 0.0081 \approx 0.008$$

Therefore, $x = 2.710 + 0.008 = 2.718$, and

$$\log N = 1 + \log 2.718 = \log(2.718 \times 10^1) = \log 27.18$$
$$N = 27.18$$

COMMENT k was rounded off to 0.008 so that x would contain four significant digits because reverse interpolation in a four-place table yields, at most, four significant digits.

We now illustrate the use of logarithms involving products, quotients, power, and roots. In each of the following examples, we compute the value of N using logarithms and Theorems 1 to 3.

Example 13 Find N if $N = 0.836\sqrt{22.1}$.

Solution

$$\begin{aligned} \log N &= \log 0.836 + \tfrac{1}{2} \log 22.1 \\ \log 0.836 &= \log(8.36 \times 10^{-1}) = -1 + 0.9222 \\ \log 22.1 &= \log(2.21 \times 10^1) = 1 + 0.3444 \\ \tfrac{1}{2} \log 22.1 &= \tfrac{1}{2}(0 + 1.3444) = 0 + 0.6722 \end{aligned}$$

Note that 1 is *subtracted* from the characteristic and *added* to the mantissa, so that the resulting characteristic will be an integer.

$$\begin{aligned} \log N &= (-1 + 0.9222) + (0 + 0.6722) = 0 + 0.5944 \\ &= 0 + \log 3.93 = \log(3.93 \times 10^0) = \log 3.93 \\ N &= 3.93 \end{aligned}$$

Example 14 Find N if $N = \dfrac{\sqrt[3]{0.0327}}{86.9}$.

Solution

$$\begin{aligned} \log N &= \tfrac{1}{3} \log 0.0327 - \log 86.9 \\ \log 0.0327 &= \log(3.27 \times 10^{-2}) = -2 + 0.5145 \\ \log 86.9 &= \log(8.69 \times 10^1) = 1 + 0.9390 \\ \tfrac{1}{3} \log 0.0327 &= \tfrac{1}{3}(-3 + 1.5145) = -1 + 0.5048 \end{aligned}$$

Here 1 is subtracted from the characteristic and added to the mantissa so that division by 3 yields an integral characteristic. Now, since

$$\begin{aligned} \log N &= \tfrac{1}{3} \log .0327 - \log 86.9 \\ \log N &= (-1 + 0.5048) - (1 + 0.9390) \end{aligned}$$

Since the mantissa 0.9390 exceeds the mantissa 0.5048, the process of subtraction is simplified by writing $-1 + 0.5048$ as $-2 + 1.5048$ so that the resulting mantissa is always *positive*. Now

$$\begin{aligned}\log N &= (-2 + 1.5048) - (1 + 0.9390) = -3 + 0.5658\\ &= -3 + \log 3.68 = \log(3.68 \times 10^{-3}) = \log 0.00368\\ N &= 0.00368\end{aligned}$$

Example 15 Find N if $N = 0.225 \sqrt[3]{\dfrac{864}{(32.6)^2(0.0562)}}$

Solution

$$\begin{aligned}\log N &= \log 0.225 + \tfrac{1}{3}[\log 864 - (2 \log 32.6 + \log 0.0562)]\\ \log 0.225 &= -1 + 0.3522\\ \log 864 &= 2 + 0.9365\\ \log 32.6 &= 1 + 0.5132\\ \log 0.0562 &= -2 + 0.7497\end{aligned}$$

From Appendix Table 3

$$\begin{aligned}&2 \log 32.6 = 2(1 + 0.5132) = 2 + 1.0264 = 3 + 0.0264\\ &2 \log 32.6 + \log 0.0562 = (3 + 0.0264) + (-2 + 0.7497) = 1 + 0.7761\\ &\log 864 - (1 + 0.7761) = (2 + 0.9365) - (1 + 0.7761) = 1 + 0.1604\end{aligned}$$

The expression $1 + 0.1604$ is the logarithm of the combined operations under the radical. We must now determine the cube root of that number.

$$\tfrac{1}{3}(1 + 0.1604) = \tfrac{1}{3}(0 + 1.1604) = 0 + 0.3868$$

Finally, $\log N = \log 0.225 + (0 + 0.3868)$
$= (-1 + 0.3522) + (0 + 0.3868) = -1 + 0.7390$

Since the mantissa, 0.7390, is not an exact entry in Appendix Table 3, we interpolate.

	x			$\log x$	
0.01	5.49			0.7396	0.0008
	$5.48 + k$	k	0.0002	0.7390	
	5.48			0.7388	

$$\frac{k}{0.01} = \frac{0.0002}{0.0008} \Longrightarrow \frac{k}{0.01} = \frac{1}{4} \Longrightarrow k = \frac{0.01}{4} = 0.0025 \approx 0.003$$

Therefore, $x = 5.48 + 0.003 = 5.483$

$$\begin{aligned}\log N &= -1 + \log 5.483 = \log(5.483 \times 10^{-1})\\ &= \log 0.5483\\ N &= 0.5483\end{aligned}$$

In defining the logarithmic function in Section 10.4, we saw that the base b can be any positive real number other than 1. The following theorem can be used to find the logarithm of a number to the base b from the logarithm of that number to the same base a.

Theorem 5 $\log_b x = \dfrac{\log_a x}{\log_a b}$.

Proof Let $\log_b x = y$. Then, by the definition of a logarithm, $\log_b x = y$ if and only if $b^y = x$. Now, we take $\log_a$ of both sides of $b^y = x$:

$$\log_a b^y = \log_a x \qquad \text{or} \qquad y \log_a b = \log_a x$$

Dividing both sides by $\log_a b$, we obtain

$$y = \frac{\log_a x}{\log_a b}$$

$$\log_b x = \frac{\log_a x}{\log_a b} \qquad \text{since} \qquad y = \log_b x$$

Example 16 Find $\log_2 6$.
By Theorem 5

$$\log_2 6 = \frac{\log_a 6}{\log_a 2}$$

Suppose $a = 10$; then from Appendix Table 3

$$\log_2 6 = \frac{\log 6}{\log 2} = \frac{0.7782}{0.3010} = 2.585$$

Suppose $a = e$, then from Appendix Table 4

$$\log_2 6 = \frac{\ln 6}{\ln 2} = \frac{1.7918}{0.6931} = 2.585$$

The result indicates that $2^{2.585} = 6$.

COMMENT Recall that $\log_{10} N$ is simply $\log N$ and $\log_e N$ is simply $\ln N$. The solution is *independent of the base used.*

Example 17 Solve $3^{x+1} = 2^{3x-2}$ for x.
Taking the log of both sides and using Theorem 3, we have

$$(x + 1)\log 3 = (3x - 2)\log 2$$

$$x \log 3 + \log 3 = 3x \log 2 - 2 \log 2$$

$$\log 3 + 2 \log 2 = 3x \log 2 - x \log 3$$

$$\log 3 + 2 \log 2 = x(3 \log 2 - \log 3)$$

$$x = \frac{\log 3 + 2 \log 2}{3 \log 2 - \log 3}$$

Using the logarithm tables, we get

$$x = \frac{(0 + 0.4771) + 2(0 + 0.3010)}{3(0 + 0.3010) - (0 + 0.4771)} = \frac{1 + 0.0791}{0 + 0.4259} = \frac{0 + 1.0791}{0 + 0.4259} \approx 2.53$$

Logarithms can be used to facilitate solutions to many practical problems. The examples that follow are applications from physics. The first example involves the analysis of the motion of a pendulum and will be solved using logarithms to the base 10 (Appendix Table 3). The second example involves the cooling process when an object is placed in a medium whose temperature is lower than the object itself. This problem will be solved using logarithms to the base e (Appendix Table 4).

Other interesting applications such as compound interest, population growth, biological reproduction, electric circuits, effects of gravity, and geometric formulas are given in the exercises.

Example 18 The period T of the pendulum of a clock is given by

$$T = 2\pi \sqrt{\frac{L}{g}}$$

where T = time, seconds
g = acceleration due to gravity = 32.2 feet per second per second
L = length of pendulum, feet
$\pi = 3.14$

If the length of the pendulum is 0.661 feet, how long does the pendulum take for one oscillation?

Solution

$$\begin{aligned} \log T &= \log 2\pi + \tfrac{1}{2}(\log L - \log G) = \log 6.28 + \tfrac{1}{2}(\log 0.661 - \log 32.2) \\ &= (0 + 0.7980) + \tfrac{1}{2}[(-1 + 0.8202) - (1 + 0.5079)] \\ &= 0.7980 + \tfrac{1}{2}[(-1 + 0.8202) - (1 + 0.5079)] \\ &= 0.7980 + \tfrac{1}{2}(-2 + 0.3123) \end{aligned}$$

now $\tfrac{1}{2}(-2 + 0.3123) = -1 + 0.1562,$ hence

$$\begin{aligned} \log T &= (0 + 0.7980) + (-1 + 0.1562) = -1 + 0.9542 = -1 + \log 9.0 \\ &= \log(9 \times 10^{-1}) = \log 0.9 \end{aligned}$$

$$T = 0.9 \text{ second}$$

Example 19 If you want to increase the length of the pendulum so that the period is exactly 1 second, how long must the pendulum be?

Solution

$$1 = 2\pi\sqrt{\frac{L}{32.2}}$$

$$1 = \frac{(2\pi)^2 L}{32.2} \qquad \text{Squaring both sides}$$

$$L = \frac{32.2}{(6.28)^2} \qquad \text{Solving for } L$$

$$\log L = \log 32.2 - 2\log 6.28 = (1 + 0.5079) - 2(0 + 0.7980)$$
$$= (1 + 0.5079) - (1 + 0.5960)$$

or $$\log L = (0 + 1.5079) - (1 + 0.5960) = -1 + 0.9119$$

$$\log L = -1 + \log 8.164 \qquad \text{Using interpolation}$$
$$= \log(8.164 \times 10^{-1}) = \log 0.8164$$

$$L = 0.8164 \text{ feet}$$

Example 20 An object changes temperature slowly if the difference between its temperature and the surrounding medium is small and rapidly if this difference is large. The greater the difference, the faster the temperature change. (In other words, the temperature change is proportional to the difference in temperature.) This thermodynamic phenomenon is explained by Newton's law of cooling.

A refrigerator has a constant temperature of 20°F. A can of beer brought home from the local store on a hot summer day is at 80°F. The ideal drinking temperature of beer is 40°F. If the can of beer is cooled from 80 to 60°F in 8 minutes, how long (in minutes) will the thirsty shopper have to wait to drink this beer at its best?

The cooling process obeys Newton's law of cooling

$$T = T_0 + Ce^{-kt}$$

where T_0 = temperature of surrounding medium, °F
t = time, minutes
C = constant to be solved from initial conditions
k = constant of proportionality
T = temperature at any given time, °F

STEP 1 Determine C. Since $T_0 = 20$ and $T = 80$ when $t = 0$, therefore,

$$80 = 20 + Ce^{-k(0)} \qquad 60 = Ce^0 = C$$

STEP 2 The formula is now $T = 20 + 60e^{-kt}$. To find k, we use the fact that when $t = 8$, $T = 60$. Hence, $60 = 20 + 60e^{-8k}$.

$$\frac{40}{60} = e^{-8k} \qquad \text{or} \qquad 2e^{8k} = 3$$

Now, taking the natural logarithm of both sides (using Appendix Table 4), we obtain

$$\ln 2 + \ln e^{8k} = \ln 3$$

Note that $\ln e^{8k} = 8k \ln e = 8k$ (recall that $\ln e = 1$). Then

$$8k = \ln 3 - \ln 2 \qquad \text{and} \qquad k = \frac{\ln 3 - \ln 2}{8} \approx 0.05$$

The formula is now $T = 20 + 60e^{-0.05t}$.

STEP 3 We can now solve for t, the total time the shopper must wait. Substitute $T = 40$ and solve for t.

$$40 = 20 + 60e^{-0.05t}$$

$$20 = 60e^{-0.05t}$$

$$\frac{1}{3} = \frac{1}{e^{0.05t}} \qquad \text{and} \qquad e^{0.05t} = 3$$

$$0.05t \ln e = \ln 3 \qquad \text{and} \qquad 0.05t = \ln 3$$

$$t = \frac{\ln 3}{0.05} \approx 22 \text{ minutes}$$

EXERCISES

Graph the functions in Exercises 1 to 5.

1 $y = \log \sqrt{x}$ **2** $y = \log x^2$ **3** $y = \log 10x$
4 $y = \log(x - 1)$ **5** $y = \ln x^3$
6 Prove Theorem 2.
7 Prove Theorem 3.
8 Given $\log 2 = a$ and $\log 3 = b$. In terms of a and b, what is $\log 6$?
9 Show that $b^{\log_b x} = x$. *Hint:* Take the logarithm of both sides.
10 Simplify **(a)** $\pi^{\log_\pi \pi}$ and **(b)** $4^{\log_2 8}$
11 Show that

$$\log_e 10 = \frac{\log 10}{\log e} = \frac{1}{\log e}$$

Use tables in the Appendix to verify.
12 Show that

$$\log_e 10 = \frac{\ln 10}{\ln e} = \ln 10$$

Use tables in the Appendix to verify.
13 Suppose a gambler's lucky numbers are 7 and 11, and he demands that all his logarithms be expressed to one of these bases. Compute **(a)** $\log_7 11$ and **(b)** $\log_{11} 7$.
14 Solve for x (using logarithms): **(a)** $2^x = 10$ and **(b)** $10^x = 2$.
15 Solve for x (using logarithms): $2^{2x-1} = 3^{x+1}$.

16 A student simplifies an expression as follows:

$$\frac{\log x}{\log y} = \frac{x}{y}$$

Did he make a mistake? Explain.

17 A student simplifies the following expression by applying Theorem 2.

$$\frac{\log x}{\log y} = \log x - \log y$$

Did he make a mistake? Explain.

18 What is wrong with the following computation?

$$\begin{aligned} (-a)^2 &= a^2 \\ \log(-a)^2 &= \log a^2 \\ 2\log(-a) &= 2\log a \\ -a &= a \end{aligned}$$

19 The number n of the Methuselah bacteria (which causes the aging process) tends to increase according to the formula

$$n = n_0 2^{t/2}$$

where n_0 = number of bacteria present at age 25
t = number of years beyond 25

If there are approximately 2^4 bacteria present at age 25, find how many will be present at:

(a) Age 30 **(b)** Age 40 **(c)** Age 80

For Exercises 20 to 24, the compound-interest formula is

$$a = p\left(1 + \frac{r}{n}\right)^{nt}$$

where p = principal invested at r percent *annual* interest expressed as a decimal
n = number of times interest is compounded yearly
a = amount that accumulates after t years

20 A tax-free municipal bond pays 6 percent compounded semiannually. If an investor purchases a \$5000 bond, find the approximate value of the bond after it matures in 20 years.

21 Two banks offer different investment plans:

(a) 5 percent compounded yearly

(b) 4 percent compounded quarterly

An investor wishes to deposit \$4000 as a college fund for his newborn son. Into which bank, a or b, should he put his money to maximize his accumulated amount after 18 years? (Show all calculations.)

22 How much should be invested at 6 percent compounded yearly to amount to \$10,000 in 20 years?

23 How long does it take \$1000 invested at 6 percent compounded semiannually to double?

24 How long will it take \$1000 invested at 8 percent compounded quarterly to double?

In Exercises 25 to 50 solve for x using logarithms

25 $x = (24.3)(8.23)$

26 $x = (0.618)(0.055)$

27 $x = \dfrac{17.6}{2.12}$

28 $x = \dfrac{2.52}{16.8}$

29 $x = \dfrac{0.824}{0.0402}$

30 $x = \sqrt[3]{924}$

31 $x = (238)^{1.2}$

32 $x = 4\pi(1.32)^3$

33 $x = \sqrt[4]{0.346}$

34 $x = \sqrt[5]{0.985}$

35 $x = \sqrt[4]{0.00423}$

36 $x = (3.96)^3$

37 $x = (0.0805)^2$

38 $x = \sqrt[4]{\dfrac{46.8}{286}}$

39 $x = \dfrac{5.88(26.1)}{0.834}$

40 $x = \dfrac{0.608}{(.216)(12.4)}$

41 $x = \sqrt{\dfrac{(24.8)(5.2)}{42.6}}$

42 $x = \sqrt{\dfrac{80.8}{(6.16)(266)}}$

43 $x = \sqrt[3]{\dfrac{9030\sqrt{0.056}}{850}}$

44 $x = \sqrt{\dfrac{76.6}{(4.25)^3\sqrt{0.0814}}}$

45 $x = 0.225\sqrt[3]{\dfrac{(64.8)(3.4)^2}{824}}$

46 $2^{4x+1} = 3^{3x}$

47 $x = 0.677\sqrt[3]{\dfrac{64.3}{(2.1)^2(425)}}$

48 $x = \sqrt{\dfrac{(1.2)(465)}{(16.3)(0.9511)}}$

49 $x = \dfrac{(354.8)(4.835)^3}{(16.14)^2}$

50 $x = \dfrac{(0.6432)^2\sqrt[3]{700.4}}{(10.45)^3\sqrt{0.000824}}$

51 Assume that the acceleration due to gravity is 5.2 feet per second per second on the moon and 85.3 feet per second per second on Jupiter.

(a) From a model problem we know that a pendulum 0.8164 feet long on the earth will oscillate once in 1 second. Find the period of the same pendulum on the moon and on Jupiter.

(b) How long would the pendulum have to be on the moon and on Jupiter for the period to be 1 second?

52 In a triangle with sides a, b, and c, the area K is given by Heron's formula

$$K = \sqrt{s(s-a)(s-b)(s-c)}$$

where $s = \frac{1}{2} \times$ perimeter. Using logarithms, find the area of a triangle whose sides are 2.4, 4.2, and 5.8.

53 The radius of the inscribed circle on a triangle is given by

$$R = \sqrt{\frac{(s-a)(s-b)(s-c)}{s}}$$

Using the triangle in Exercise 52, find the radius and the area of the inscribed circle (use logarithms).

54 The volume of a cone is given by

$$V = \frac{\pi R^2 H}{3}$$

Find the height of a cone whose volume is 195 cubic inches and radius 2.58 inches.

55 Solve for x: $4^{2x-1} = 5^{x+2}$.

56 A piece of paper 1/1000 inch thick is torn in half, and the one piece is placed on top of the other. These are then torn in half, and the four pieces put together in a pile. If this process of tearing in half and piling is done 50 times, will the final pile of paper be more or less than 1 mile high?

57 The formula $a = pe^{rt}$ gives the amount a of money compounded *continuously* at rate r for t years, on a deposit of p dollars. Compare the amount after 1 year for \$1000 compounded at 6 percent continuously with \$1000 invested at 6 percent compounded semiannually. Use the compound-interest formula from Exercise 20.

58 The formula is $a = a_0 2^{-kt}$. Given: The quantity a of a radioactive substance remaining from an initial quantity a_0, after t years. If $k = 0.06$ for a new radioactive element "hillium," how much remains of 60 g of the element after 5 years?

59 When the voltage is turned off in an electric circuit, the current I after t seconds is given by

$$I = I_0 e^{-kt}$$

where I_0 = current at time voltage is terminated

amperes k = a physical constant depending upon circuit

If a current of 20 amperes drops to $\frac{1}{2}$ ampere in 0.05 second, find k.

60 If t is the time in hours needed for x_0 to become x_1 and d is the doubling period, then $x_1 = x_0 2^{t/d}$. The amount of a certain type of bacteria will become 16 times its original number in 3 hours. Find the doubling period.

61 Carbon 14 has a half-life of approximately 6000 years. A biblical scroll is believed to be 2000 years old. What percentage of the carbon 14 will the parchment have lost if this date is correct?

$$x = x_0 2^{-t/h}$$

where h is the half-life and t is the time in years. Let $x_0 = 1$ denote 100 percent.

62 In an attempt to freeze an antibiotic serum for experimentation, a physician places the liquid in a storage area of constant temperature -20°F. The liquid was at the laboratory temperature of 60°F and dropped to 30°F in 2.35 minutes.

(a) When will the temperature of the serum reach 20°F? 0°F?

(b) Make up a chart and compute the times it takes to go from 60 to 40°F, 40 to 20°F, and 20 to 0°F. Are the time intervals the same? Comment. Refer to the cooling formula used in Example 20.

63 If the population of a city doubles in 35 years, in how many years will it triple (the rate of increase is proportional to the population)?

$$x = x_0 e^{kt}$$

where k = constant of proportionality

x = population at time T years

x_0 = population at time $T = 0$

Hint: At $T = 35$, $x = 2x_0$.

64 **(a)** An object is heated to 180°F to test its resilience at high temperatures. It is then allowed to cool under a constant laboratory environment of 60°F. If the object cools to 160°F in 3 minutes, how long (in minutes) does it take the object to reach the temperatures of 140, 120, 100, and 80°F?

(b) Construct a graph which relates T in degrees Fahrenheit in 20-degree intervals from 180 to 60°F and t, the time in minutes.

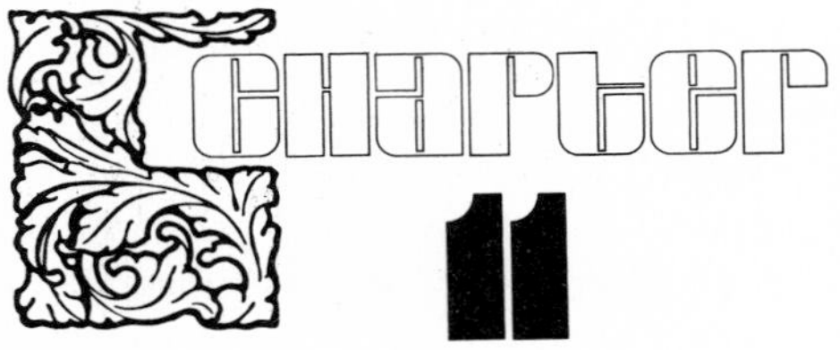

The Circle and the Parabola

11.1 INTRODUCTION

In Chapter 1 we represented a point in a plane by an ordered pair of real numbers (x, y). When there is a specific relationship between the variables x and y, that is, they are related by an equation, these points will not be scattered randomly throughout the plane but will lie in a definite pattern. The pattern (or curve) will depend on the equation, and only those points (x, y) which satisfy the equation will be on the curve.

DEFINITION 1 A set of points which satisfies specific algebraic or geometric conditions is called a *locus.* The locus of an equation is the curve which contains all the points that satisfy a given equation and which contains no other points.

The fundamental problem of analytic geometry is twofold: (1) to find the graph (locus) of a given equation and (2) for a given graph (locus) to find its equation. This second problem is the reverse of the first and usually more difficult. The graph is usually given as a figure with specific properties or as the path of a point which moves under specific conditions.

In this chapter we analyze the relationship between a curve and its equation and find the properties of the curve from its equation.

11.2 THE CIRCLE

DEFINITION A circle is the locus of all points in a plane a given distance from a given point in the plane.

The given distance r is the *radius* of the circle. The given point (h, k) is the *center* of the circle. We can derive the equation of this locus directly from the definition.

Theorem 1 A circle with center (h, k) and radius r has the equation

$$(x - h)^2 + (y - k)^2 = r^2$$

Proof

Let $P(x, y)$ be any point on the circle. Then using the distance formula, we can express the fact that r is the constant distance from $C(h, k)$ and $P(x, y)$ (Diagram 1).

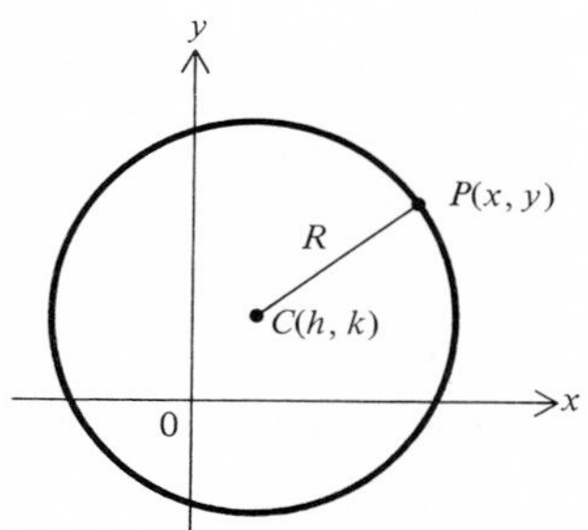

Diagram 1

$$\sqrt{(x - h)^2 + (y - k)^2} = r$$

$$(x - h)^2 + (y - k)^2 = r^2 \qquad \text{Squaring both sides} \qquad (1)$$

Equation (1) is called the *standard form* of the circle. If the center of the circle is the origin, $h = 0$ and $k = 0$ and the equation becomes $x^2 + y^2 = r^2$.

Example 1 Write the equation of the circle with center $(-3, 4)$ and radius 5 (Diagram 2).

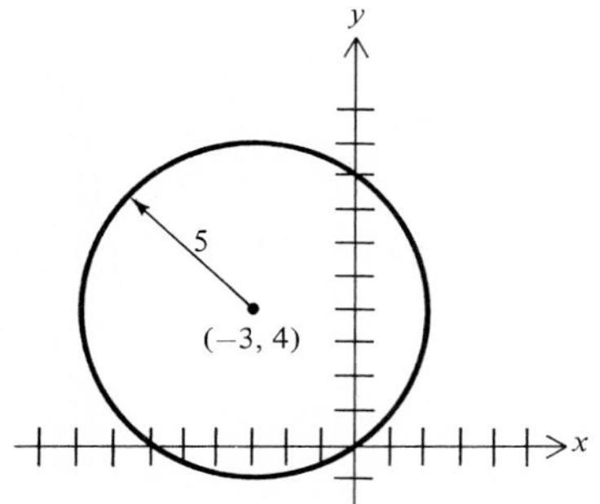

Diagram 2

$[x - (-3)]^2 + [y - (4)]^2 = 5^2$	Referring to (1)
$(x + 3)^2 + (y - 4)^2 = 25$	Simplifying
$x^2 + 6x + 9 + y^2 - 8y + 16 = 25$	Performing the indicated algebra
$x^2 + y^2 + 6x - 8y = 0$	Combining the constant terms

This is the same circle but expressed in a different form.

Theorem 2 All circles can be written in the *general form*

$$x^2 + y^2 + Ax + By + C = 0 \qquad (2)$$

Let us change from standard form (1) into general form (2).

$(x - h)^2 + (y - k)^2 = r^2$	Equation (1)
$x^2 - 2hx + h^2 + y^2 - 2ky + k^2 = r^2$	Expanding term on left side
$x^2 + y^2 - 2hx - 2ky + (h^2 + k^2 - r^2) = 0$	Combining constant terms

If we now compare Equations (1) and (2),

$$A = -2h \qquad B = -2k \qquad C = h^2 + k^2 - r^2$$

Both forms are equivalent in content yet different in appearance.

Example 2 Express $x^2 + y^2 - 4x + 8y - 16 = 0$ in standard form. Write the equation grouped as

$$(x^2 - 4x \qquad) + (y^2 + 8y \qquad) = 16$$

Make each of the terms on the left a perfect square. Recall the method of completing the square in Chapter 3.

$$(x^2 - 4x + 4) + (y^2 + 8y + 16) = 16 + 20$$

We added 20 to the left, and to preserve the equality we must add 20 to the right. Then

$$(x - 2)^2 + (y + 4)^2 = 36$$

is the equation of a circle with center $(2, -4)$ and radius $\sqrt{36}$ or 6.

Example 3 Suppose that in Example 2 the constant term was 20.

Solution

$$x^2 + y^2 - 4x + 8y + 20 = 0$$

$$(x^2 - 4x + 4) + (y^2 + 8y + 16) = -20 + 20 \qquad \text{Completing the square}$$

$$(x - 2)^2 + (y + 4)^2 = 0 \qquad \text{Factoring}$$

This is a circle with center $(2, -4)$ and radius 0. This will simply represent the point $(2, -4)$.

Example 4 Suppose that in Example 3 the constant term was 24.

Solution

$$x^2 + y^2 - 4x + 8y + 24 = 0$$

$$(x^2 - 4x + 4) + (y^2 + 8y + 16) = -24 + 20 \qquad \text{Completing the square}$$

$$(x - 2)^2 + (y + 4)^2 = -4 \qquad \text{Factoring}$$

There are no real numbers (x, y) that satisfy this equation, since $(x - 2)^2 \geq 0$ and $(y + 4)^2 \geq 0$ and therefore $(x - 2)^2 + (y + 4)^2 \geq 0$. In this case there is no graph. Some refer to it as an imaginary circle.

COMMENT In the general form (2) of a circle, the coefficients of x^2 and y^2 are both 1. Suppose in a given equation they are *not* both 1. For example,

$$3x^2 + 3y^2 - 18x + 6y + 2 = 0$$

$$x^2 + y^2 - 6x + 2y + \tfrac{2}{3} = 0 \qquad \text{Dividing by 3}$$

$$(x - 3)^2 + (y + 1)^2 = \tfrac{28}{3} \qquad \text{Expressing results in proper form}$$

which is a circle with center $(3, -1)$ and radius $\sqrt{\tfrac{28}{3}} \approx 3.055$.

Example 5 $y = x + 4$ is tangent to a circle with center $(3, 2)$. Find the equation of the circle and the point of tangency (Diagram 3).

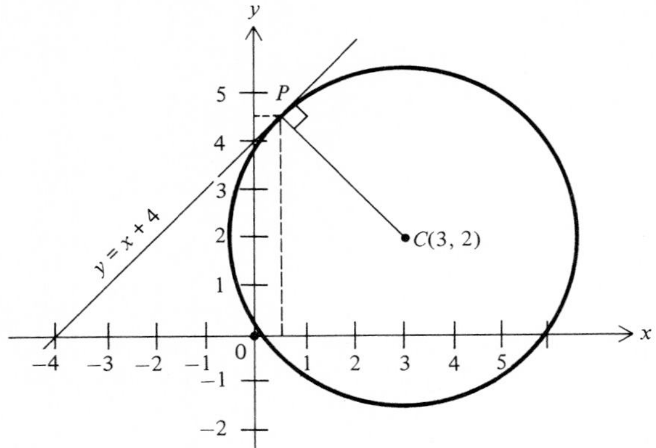

Diagram 3

STEP 1 From geometry, a line drawn from the center of a circle to a point of tangency is perpendicular to the tangent at that point. Hence line segment CP is perpendicular to $y = x + 4$. The slope of line segment CP is the negative reciprocal of the slope of $y = x + 4$.

$$m_{CP} = -1$$

STEP 2 Using the point-slope form for the equation of a straight line (see Chapter 1), we can find the equation of line segment CP passing through (3, 2) with slope $= -1$.

$$(y - 2) = -1(x - 3)$$
$$y - 2 = -x + 3$$
$$y = -x + 5$$

STEP 3 To find the coordinates of P, we solve $y = x + 4$ and $y = -x + 5$ simultaneously:

$$x + 4 = -x + 5$$
$$2x = 1$$
$$x = \tfrac{1}{2}$$

When $x = \frac{1}{2}$, $y = \frac{9}{2}$; $P(\frac{1}{2}, \frac{9}{2})$ is the point of intersection.

STEP 4 To find the radius of the circle, use the distance formula

$$r = \sqrt{(3 - \tfrac{1}{2})^2 + (2 - \tfrac{9}{2})^2} = \sqrt{(\tfrac{5}{2})^2 + (-\tfrac{5}{2})^2}$$
$$= \sqrt{\frac{50}{4}} = \frac{5\sqrt{2}}{2}$$

STEP 5 The equation of the required circle is

$$(x - 3)^2 + (y - 2)^2 = \tfrac{25}{2} \qquad \text{in standard form}$$

COMMENT An alternate (and much quicker) method of solution proceeds as follows. Express the equation of the tangent line in the form

$$Ax + By + C = 0 \qquad -x + y - 4 = 0$$

Use the formula for the distance from a point to a line to find the required radius, eliminating steps 1 to 3 from the previous method.

$$r = \frac{|(-1)(3) + (1)(2) - 4|}{\sqrt{1^2 + 1^2}} = \frac{|-3 + 2 - 4|}{\sqrt{2}} = \frac{5\sqrt{2}}{2}$$

We find the equation of the circle now directly from r and the fact that the center is given.

EXERCISES

Reduce each circle in Exercises 1 to 8 to the standard form $(x - h)^2 + (y - k)^2 = r^2$ and determine its center and radius.

1 $x^2 - 4x + y^2 - 6y - 7 = 0$

2 $x^2 + 12x + y^2 - 8y + 3 = 0$

3 $36x^2 - 108x + 36y^2 + 96y + 1 = 0$

4 $x^2 + 5x + y^2 - y - 16 = 0$

5 $x^2 + y^2 - 2x + 4y + 5 = 0$

6 $x^2 + y^2 - 2x + 4y + 6 = 0$

7 $2x^2 + 2y^2 - 2x + 6y - 3 = 0$

8 $5x^2 + 5y^2 - 8x - 4y - 121 = 0$

9 Find the points where $x^2 + y^2 - x + 4y - 12 = 0$ intersects the x and y axes. Sketch the circle.

10 Find the equation of each of the two lines with slope 2 tangent to the circle $(x - 3)^2 + (y + 4)^2 = 20$. Sketch the system.

11 Verify that if $y = mx + b$ is tangent to $x^2 + y^2 = r^2$, then $b = r\sqrt{1 + m^2}$.

12 Using the results of Exercise 11, show that $y = x + 6$ is tangent to $x^2 + y^2 = 18$ and find the point of intersection.

13 Show that $x^2 + y^2 = 45$ and $x = 15 - 2y$ are tangent to each other, and find the point of tangency.

14 Find the greatest and smallest distance between the point $(5, -2)$ and the circle $x^2 + y^2 + 2x - 8y - 1 = 0$.

15 A circle is inscribed in a square as shown in the diagram.

(a) Find the equation of the circle.

(b) Find the area of the regions outside the circle but inside the square.

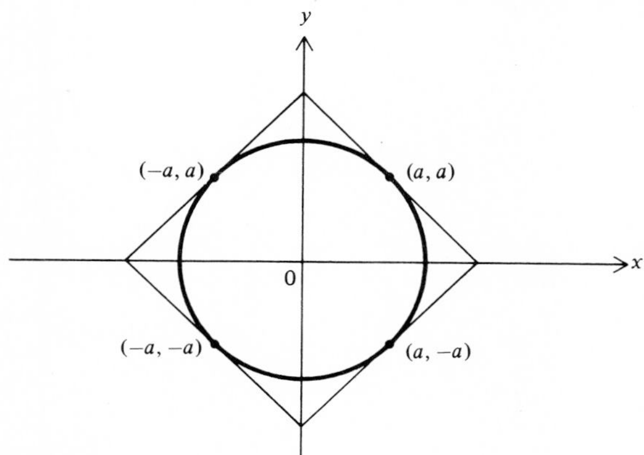

16 Given: $x^2 + y^2 + 8x - 2 = 0$ and $x^2 + y^2 - 4x - 14 = 0$.

(a) Find the point(s) of intersection of the two circles.

(b) Find the equation(s) of the tangent to *each* circle at the intersection point(s).

(c) Show that the tangents are perpendicular to each other. (This property is called *orthogonality* and implies that the circles intersect at right angles.)

17 Sketch the following system and determine the points of intersection:

$$\begin{aligned} x^2 + y^2 - 2x &= 3 \\ 3x + y &= 1 \end{aligned}$$

18 Given:

$$\begin{aligned} x^2 + y^2 + 2ax + 2by + c &= 0 \\ x^2 + y^2 + 2px + 2qy + r &= 0 \end{aligned}$$

Under what conditions are the two circles **(a)** concentric and **(b)** equal in area.

19 Find the equation of the line which passes through the center of $x^2 + y^2 - 4x + 12y - 12 = 0$ and the midpoint of the segment joining $(-2, -3)$ and $(-6, 7)$.

20 Show that $y = -4x/3 + 2$ is tangent to $(x - 1)^2 + (y - 4)^2 = 4$.

21 Repeat Exercise 16 for the two circles $x^2 + y^2 + 8x - 4 = 0$ and $x^2 + y^2 - 6y + 4 = 0$.

It can be shown using the methods of the calculus that an equation of the tangent to the circle $x^2 + y^2 + 2ax + 2by + c = 0$ at the point (x_1, y_1) on the circle is $xx_1 + yy_1 + a(x + x_1) + b(y + y_1) + c = 0$.

22 Verify this theorem for the tangent to $x^2 + y^2 - 4x + 2y = 0$ at the point $(3, 1)$.

23 Check Exercise 20 using this fact.

24 A chord of $x^2 + y^2 - 6x + 4y - 12 = 0$ is parallel to $y = -2x - 8$ and $2\sqrt{5}$ units long. Find the equation of the chord and coordinates of its end points (two solutions). Sketch the system.

25 Find the equations of the four lines which are tangent to both circles

$$\begin{aligned} x^2 + y^2 - 4x - 2y + 4 &= 0 \\ x^2 + y^2 + 4x + 2y - 4 &= 0 \end{aligned}$$

These are referred to as *common tangents.*

11.3 SOME CONDITIONS THAT DETERMINE A CIRCLE

The standard form of a circle is $(x - h)^2 + (y - k)^2 = r^2$. If we could determine the values of the *parameters* h, k, and r, we would determine the circle. Note that the *general form* of the circle, $x^2 + y^2 + Ax + By + C = 0$, also contains three parameters, A, B, and C.

Example 1 Find the equation of the circle passing through the points $(1, -2)$, $(3, 0)$, and $(-1, 4)$ (Diagram 4).

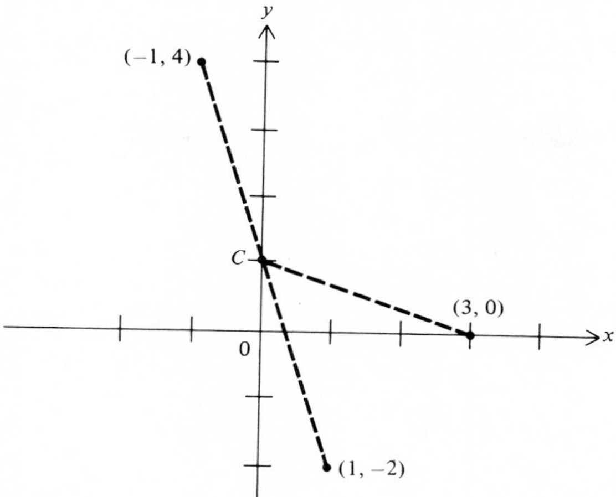

Diagram 4

Using the *general form* of the circle, we must determine the values of the parameters *A, B,* and *C*. Since the three points must satisfy the equation of the circle we have

$$\begin{aligned} 1 + 4 + A - 2B + C &= 0 \quad \text{For } (1, -2) \\ 9 + 0 + 3A + C &= 0 \quad \text{For } (3, 0) \\ 1 + 16 - A + 4B + C &= 0 \quad \text{For } (-1, 4) \end{aligned}$$

This gives the system

$$\begin{aligned} A - 2B + C &= -5 \\ 3A + C &= -9 \\ -A + 4B + C &= -17 \end{aligned}$$

Solving simultaneously for *A, B,* and *C,* we find $A = 0, B = -2,$ and $C = -9$; then the equation of the circle can be expressed

$$x^2 + y^2 - 2y - 9 = 0 \qquad \text{or} \qquad x^2 + (y - 1)^2 = 10$$

which is a circle with center (0, 1) and radius $\sqrt{10}$.

Geometrical Method The intersection of the perpendicular bisectors of the sides of a triangle is the *circumcenter,* which is the center of the circumscribed circle about the triangle. Since the three given points also determine a triangle, we will solve the example using this method (Diagram 5).

STEP 1 Find the midpoints of line segments *AB* and *BC*. $P = (1, 2)$ and $Q = (2, -1)$.

STEP 2 Find the equations of the lines perpendicular to *AB* and *BC* and passing through their midpoints. These are the perpendicular bisectors of

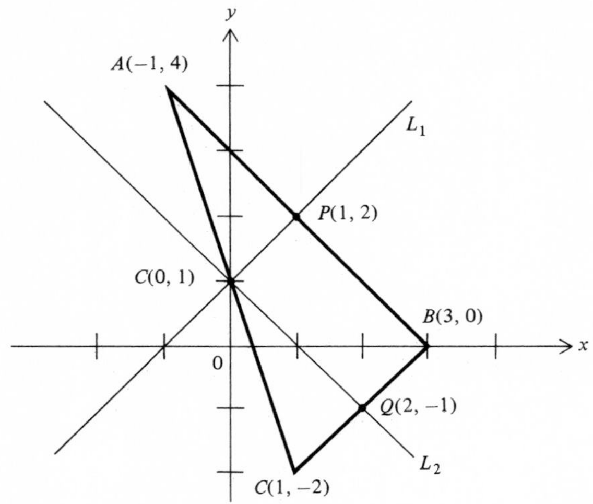

Diagram 5

the sides. Call these lines L_1 and L_2, respectively. Since the slope of AB is -1, the slope of L_1 is 1. Therefore the equation of L_1 is

$$y - 2 = 1(x - 1) \Longrightarrow y = x + 1$$

Since the slope of BC is 1, the slope of L_2 is -1. Therefore the equation of L_2 is

$$y - (-1) = -1(x - 2) \Longrightarrow y = -x + 1$$

STEP 3 Their point of intersection is the center of the circumscribed circle. Solving simultaneously

$$\begin{aligned} -x + 1 &= x + 1 \\ -2x &= 0 \\ x &= 0 \end{aligned}$$

When $x = 0$, $y = 1$. (0, 1) is the point of intersection.

STEP 4 The radius equals the distance from (0, 1) to *any* vertex, say $B(3, 0)$.

$$r = \sqrt{(0 - 3)^2 + (1 - 0)^2} = \sqrt{10}$$

and $\qquad (x - 0)^2 + (y - 1)^2 = 10 \qquad$ Standard form of a circle

Example 2 Find the equation of the circle which passes through (6, 2) and is tangent to $x + 2y - 1 = 0$ at $(3, -1)$ (Diagram 6).

STEP 1 Let $C(h, k)$ denote the center of the circle and r the radius; then $AC = CB = r$.

$$\sqrt{(h - 3)^2 + (k + 1)^2} = \sqrt{(h - 6)^2 + (k - 2)^2}$$

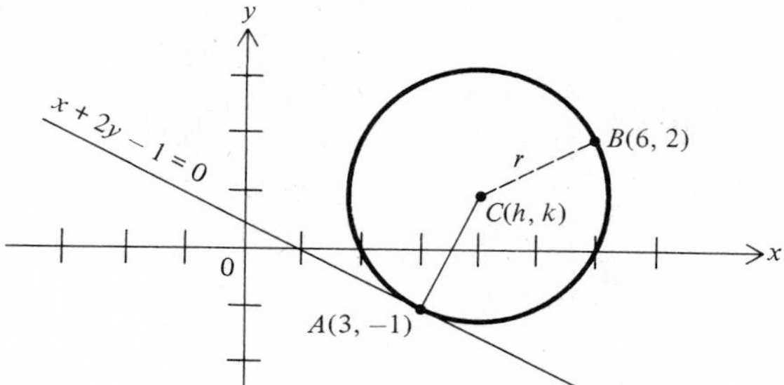

Diagram 6

Squaring both sides and expanding gives

$$h^2 - 6h + 9 + k^2 + 2k + 1 = h^2 - 12h + 36 + k^2 - 4k + 4$$

$$6h + 6k - 30 = 0 \qquad \text{Combining like terms}$$

$$h + k = 5 \qquad (1)$$

STEP 2 Also, AC is perpendicular to $x + 2y - 1 = 0$ due to the property of tangents. The slope of AC is 2 since $y = -\frac{1}{2}x + \frac{1}{2}$. Expressing the slope of AC in terms of h and k, we have

$$\frac{k + 1}{h - 3} = 2$$

Multiplying both sides by $h - 3$ gives

$$k + 1 = 2h - 6$$

$$2h - k = 7 \qquad (2)$$

STEP 3 Solving Equations (1) and (2) simultaneously, we find $h = 4$ and $k = 1$. Then the radius of the circle is $\sqrt{(4 - 3)^2 + (1 + 1)^2} = \sqrt{5}$, and the equation of the required circle is

$$(x - 4)^2 + (y - 1)^2 = 5$$

Example 3 (*Optional*) Find the equation of the circle (or circles) whose center is on the line $y = 2x - 2$, which has a radius of 5, and which passes through $(-1, 1)$ (Diagram 7).

STEP 1 We let $C(h, k)$ represent the center of the required circle. Since $r = 5$ with $(-1, 1)$ a point on the circle, we have

$$(h + 1)^2 + (k - 1)^2 = r^2 = 25$$

$$h^2 + 2h + 1 + k^2 - 2k + 1 = 25 \qquad \text{Squaring}$$

$$h^2 + 2h + k^2 - 2k - 23 = 0 \qquad \text{Combining like terms} \qquad (3)$$

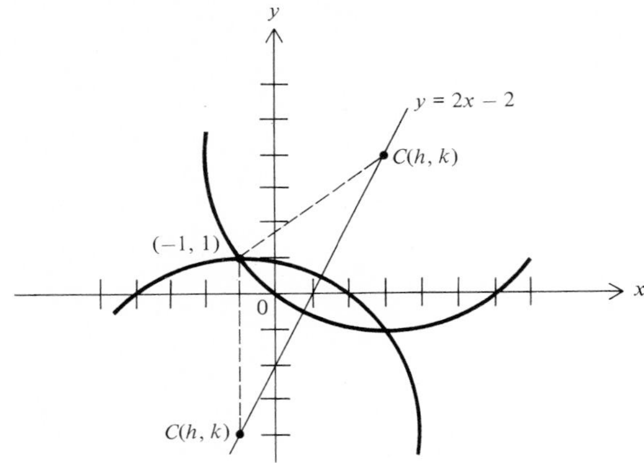

Diagram 7

STEP 2 Since $C\,(h,\ k)$ is also a point on the line $y = 2x - 2$,

$$k = 2h - 2 \qquad \text{Substituting } y = k \text{ and } x = h \tag{4}$$

STEP 3 Substituting Equation (4) into (3) for k gives

$$\begin{aligned} h^2 + 2h + (2h - 2)^2 - 2(2h - 2) - 23 &= 0 \\ h^2 + 2h + 4h^2 - 8h + 4 - 4h + 4 - 23 &= 0 \\ 5h^2 - 10h - 15 &= 0 \\ h^2 - 2h - 3 &= 0 \\ (h - 3)(h + 1) &= 0 \Longrightarrow h = 3, -1 \end{aligned}$$

STEP 4 When $h = 3$, $k = 4$; and when $h = -1$, $k = -4$.

CIRCLE 1	CIRCLE 2
$(x - 3)^2 + (y - 4)^2 = 25$	$(x + 1)^2 + (y + 4)^2 = 25$

EXERCISES

1 Find the equation of the circle passing through (6, 2) and $(-3, -1)$ whose center is on the line $y = 2x + 2$.

2 Find the equation of the circle:

(a) Which has its center at $(3, -2)$ and is tangent to the x axis.

(b) Which has its center at $(-4, -3)$ and is tangent to the y axis.

(c) Which is tangent to both axes and has its center at (a, a), $a > 0$.

3 Determine the equation of the circle passing through $(-1, 5)$, $(2, 6)$, and $(-4, 6)$. Sketch. Find the center and radius.

4 Find the equation of the circle passing through $(-1, 2)$, $(4, -3)$, and $(0, 3)$. Sketch. Find the center and radius.

5 Find the equation of the circle passing through (6, 2), (8, −2), and (−1, 1). Sketch. Find the center and radius.

6 Find the equation of the circle passing through (9, −7), (−3, −1), and (6, 2). Sketch. Find the center and radius.

7 Find the equation of the circle passing through (1, 6), (2, 5), and (−6 −1). Sketch. Find the center and radius.

8 Find the equation of the circle that is concentric with $x^2 + y^2 - 4x + 10y = 0$ and passes through the point (−2, 3).

9 Find the equation of the circle tangent to $x - 4y + 3 = 0$ at (5, 2) and tangent to $4x + y - 5 = 0$ at (2, −3).

10 Find an equation of the circle which is tangent to both axes and passes through the point (1, 2).

11 **(a)** Find the point(s) of intersection of the circle $x^2 + y^2 - 2x - 4y - 3 = 0$ and the line $2x - y - 2 = 0$.

(b) Find the equation of the tangent line to the given circle at each of the intersection points in part (*a*).

12 Find an equation of the line tangent to the circle $(x + 2)^2 + (y - 2)^2 = 45$ at the point (1, 8).

13 A circle has center (−2, −3). Find the equation of the circle if it is tangent to $x + y - 3 = 0$. Sketch the system.

14 Repeat Exercise 13 for a circle with center (−2, 1) tangent to $y = -2x + 7$.

15 Find an equation of the line tangent to $x^2 + y^2 + 4x - 6y - 19 = 0$ at the point (2, −1).

16 Find the equation of the circle tangent to $y = x - 1$ at (2, 1) if its center is on the y axis.

17 **(a)** Find the equation of the circle circumscribing the triangle formed by the intersection of the lines

$$y = -2x + 3 \qquad 3y = -x + 4 \qquad 2y = x - 4$$

(b) Find the center and radius of the circle.

(c) Show that the center of the circle is the intersection of the perpendicular bisectors of the sides of the triangle.

18 Find the equations of lines tangent to the circle $(x - 1)^2 + (y + 3)^2 = 10$ from the point (−4, 2).

19 Find the equations of the lines tangent to the circle $x^2 + y^2 - 2x + 4y = 0$ from the point (−2, −1).

20 Two tangent lines are drawn from the point (2, 2) to the circle $(x + 2)^2 + y^2 = 4$. Find the coordinates of the points where the lines intersect the circle.

21 Find the equation of the tangents from the origin to

$$x^2 + y^2 - 6x - 2y + 8 = 0$$

22 Find the equations of the tangents to the circle $x^2 + y^2 - 5x - 7y + 6 = 0$ at the points where the circle intersects the axes.

23 Find the equations of all the circles having their center on the line $2y = 3x - 6$ and tangent to both axes.

24 Find the equation of the circle(s) tangent to the lines $y = -x - 4$ and $y = 7x + 4$ and passing through the point (−1, 3). Sketch an accurate diagram.

11.4 THE PARABOLA

DEFINITION 1 A *parabola* is the locus of all points in a plane equidistant from a fixed point and a fixed line. The fixed point is called the *focus*. The fixed line is called the *directrix*.

Let us develop the standard equation by setting up a specialized coordinate system as follows. Let the focus be at $F(0, p)$. Let the directrix be parallel to the x axis. Let the parabola pass through $V(0, 0)$ and open upward. Since V is on the parabola, the distance from V to F *must* be the same as the distance from V to the directrix, by the definition. Therefore the equation of the directrix is $y = -p$ (Diagram 8).

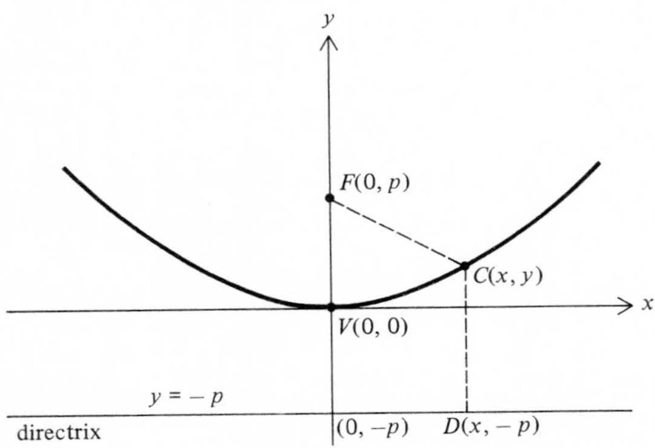

Diagram 8

According to the definition, $CF = CD$. Note that the coordinates of D are $(x, -p)$ and CD is perpendicular to the directrix. Hence

$$\sqrt{(x - 0)^2 + (y - p)^2} = \sqrt{(x - x)^2 + [y - (-p)]^2}$$

$$x^2 + y^2 - 2py + p^2 = y^2 + 2py + p^2 \qquad \text{Squaring both sides}$$

$$x^2 = 4py \qquad \text{Simplifying}$$

COMMENT Remember that p is the distance from the vertex to the focus and also from the vertex to the directrix.

Theorem 1 The equation of a parabola with focus $(0, p)$ and directrix $y = -p$ is given by $x^2 = 4py$.

Theorem 2 The equation of a parabola with focus $(p, 0)$ and directrix $x = -p$ is given by $y^2 = 4px$.

Theorem 2 can be developed in a similar manner to Theorem 1.

COMMENT If $p > 0$ in the first part of the theorem, the curve opens *upward;* if $p < 0$, the curve opens *downward.* If $p > 0$ in the second part of the theorem, the curve opens to the *right;* if $p < 0$, the curve opens to the *left.*

It should be clear from the discussion that the point V (called the *vertex* or *turning point*) is midway between the focus and directrix. The parabola is *symmetric* about the line through the focus and perpendicular to the directrix, and the equation of the axis of symmetry is $x = 0$ (or $y = 0$, according as the focus is $(0, p)$ or $(p, 0)$.

Example 1 Analyze $x^2 = 16y$.

$$4p = 16$$
$$p = 4$$

According to the theorem, the focus is at (0, 4) and the directrix is $y = -4$. Since $p > 0$, the curve opens upward (Diagram 9).

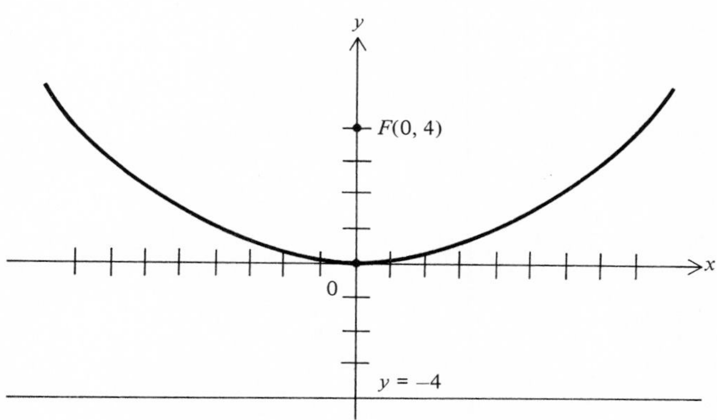

Diagram 9

Example 2 Analyze $y^2 = -8x$

$$4p = -8$$
$$p = -2$$

According to the theorem, the focus is $(-2, 0)$, the equation of the directrix is $x = 2$. Since $p < 0$, the curve opens to the left (Diagram 10).

Example 3 When the weight of a suspension bridge is evenly distributed, the cable hangs in the form of a parabolic arc. Two towers are 220 feet above the road surface and are separated by 800 feet. The turning point of the cable is at the center of the bridge and 20 feet above the roadway at this point. Abnormal oscillations during periods of high winds make it necessary to install an auxiliary vertical support cable 200 feet on either side of the center of the bridge. What will the height of this vertical cable be if it is in-

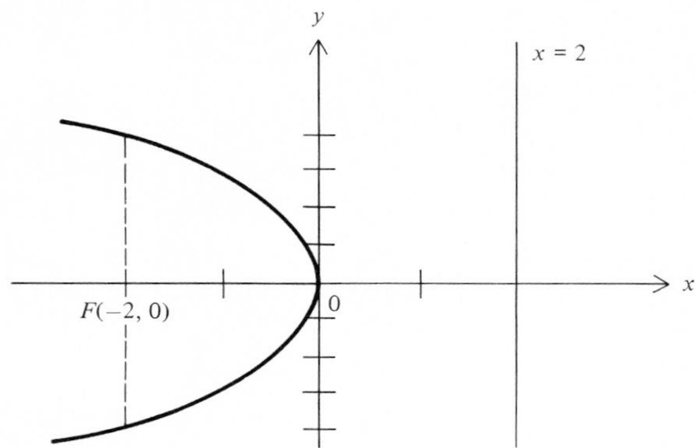

Diagram 10

stalled at these points? This problem can easily be expressed in coordinates (Diagram 11).

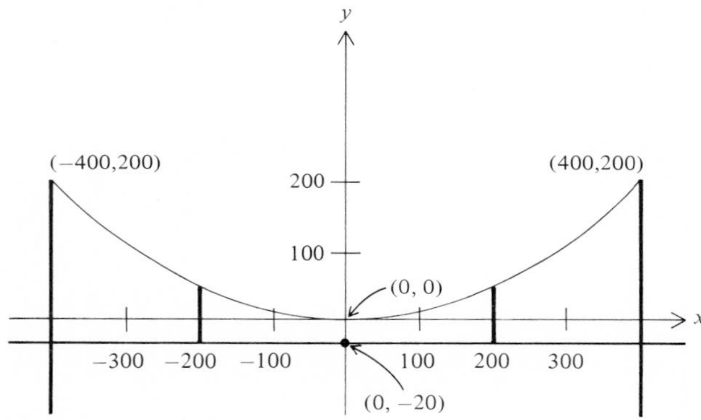

Diagram 11

Let the roadway be represented 20 feet *below* the x axis. Hence the top of the towers will be represented as (400, 200) and (−400, 200). The standard form of the parabola is $x^2 = 4py$. Since (400, 200) is on the parabola,

$$(400)^2 = 4p(200)$$

$$800 = 4p$$

$$x^2 = 800y \qquad \text{Substituting } 4p = 800$$

When $x = 200$ (the position of the vertical support cable),

$$200^2 = 800y \qquad \text{and} \qquad y = 50$$

Therefore, since the cables will be anchored to the roadway, their height will be $50 + 20 = 70$ feet.

EXERCISES

In Exercises 1 to 6 sketch and determine the focus and directrix.

1 $y^2 = 8x$ **2** $y = \dfrac{-x^2}{4}$ **3** $y^2 + 12 = 0$

4 $4x^2 = 9y$ **5** $x^2 = -2y$ **6** $2y^2 = x$

7 Find an equation of the parabola with vertex at the origin, focus on the x axis, and passing through $(-2, 4)$.

8 Find an equation of the parabola with vertex at the origin, focus on the y axis, and passing through $(2, 8)$.

9 A parabola has its vertex at the origin and axis parallel to the y axis. If the parabola passes through $(-2, 6)$, find the equation of the parabola. Find the focus and directrix.

10 Find the equation of the parabola:

(a) Whose focus is $(8, 0)$ and directrix is $x = -8$.

(b) Whose focus is $(0, -2)$ and directrix is $y = 2$.

11 Find the intersection of $x^2 + y^2 = 8$ and $y^2 = 2x$. Sketch the system.

12 A *focal chord* of a parabola is a line segment which passes through the focus and contains two points on the parabola.

(a) For the parabola $x^2 = 2y$, find the equation of the focal chord perpendicular to $3x + 4y + 2 = 0$.

(b) Find the points of intersection of the chord and the parabola.

13 A restaurant would like to install a parabolic entranceway with a wooden door placed symmetrically within the arch so that the top of the door passes through the focus of the parabola. The arch is 9 feet high and 12 feet wide.

(a) Represent the problem graphically and label your points accordingly. (One method would be to let the vertex of the arch be the origin.)

(b) Find the equation of the arch (see the diagram).

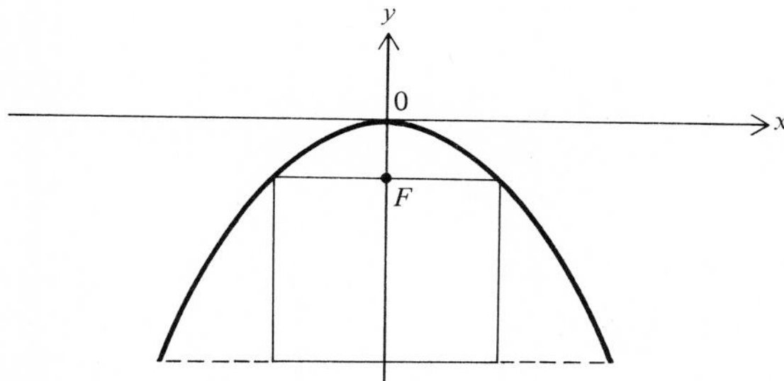

(c) Find the dimension of the door.

(d) If the restaurant wanted a door twice as wide as in part (*c*), disregarding the focus condition, how high would it be?

14 Given the parabola $y = x^2$, where (a, a^2) is a point on the parabola.

(a) Find the equation of the line passing through (a, a^2) with a slope of $2a$.

(b) Find the equation of the line passing through (a, a^2) and perpendicular to the line in part (*a*).

(c) What *seems* to happen to the y intercept of the line from part (*b*) as $a \to 0$?

15 A theorem states that $y = mx + p/m$ is tangent to the parabola $y^2 = 4px$ $(m \neq 0)$, where m = slope of the line. Verify this theorem for $m = 2, p = -4$.

Using the methods of the calculus, it can be shown that given a parabola $y^2 = 4px$ and (x_1, y_1), a point on the parabola, then $yy_1 = 2p(x + x_1)$ is the equation of the tangent line at (x_1, y_1).

16 (a) Find an equation of the line tangent to $y^2 = -8x$ at the point $(-2, 4)$.

(b) Graphically verify the results of part (*a*).

17 The path of a projectile shot horizontally from a point y feet above the ground with a speed of v feet per second is given by the formula $x^2 = -2v^2y/g$, where x is the horizontal distance from the point of projection and $g = -32$ feet per second per second is the acceleration due to gravity. A major league pitcher throws his fastball 93 miles per hour (or approximately 136 feet per second). If he releases the ball 6 feet above the ground and throws it horizontally, how far will the ball travel before it hits the ground?

11.5
THE PARABOLA WITH VERTEX AT $V(h, k)$

Suppose the vertex of the parabola was not $V(0, 0)$ but *translated* to $V(h, k)$ and the directrix remained parallel to the x axis. Examine Diagram 12 to see the effect of this translation on the standard form of the parabola.

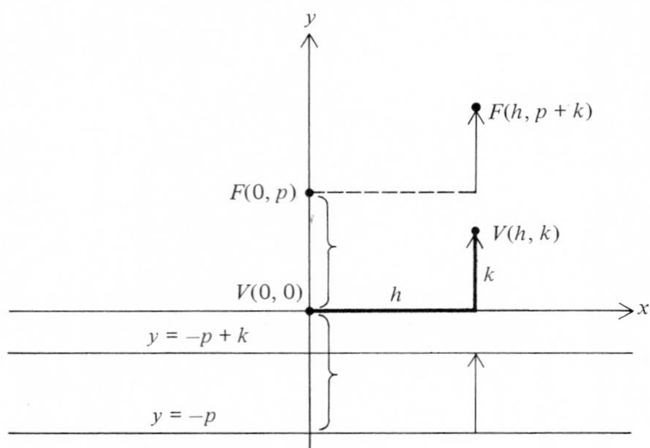

Diagram 12

Theorem 1 The equation of a parabola with focus $F(h, k + p)$ and directrix $y = -p + k$ is

$$(x - h)^2 = 4p(y - k) \tag{1}$$

(a) The vertex is $V(h, k)$.
(b) For $p > 0$ ($p < 0$) the parabola opens upward (downward).
(c) The axis of symmetry is $x = h$.

Theorem 2 The equation of a parabola with focus $F(h + p, k)$ and directrix $x = -p + h$ is

$$(y - k)^2 = 4p(x - h)$$

(a) The vertex is $V(h, k)$.
(b) For $p > 0$ ($p < 0$) the parabola opens to the right (left).
(c) The axis of symmetry is $y = k$.

Example 1 Analyze $x^2 - 4x - 3y - 5 = 0$.

In order to use the preceding theorems we must put the equation into standard form.

STEP 1 Separate the variables:

$$x^2 - 4x = 3y + 5$$

STEP 2 Complete the square on the left side of the equation:

$$x^2 - 4x + 4 = 3y + 5 + 4$$

STEP 3 Simplify: $(x - 2)^2 = 3y + 9$

This is *almost* in required form. From our theorem, the coefficients of x and y within the parentheses are both 1; hence

$$(x - 2)^2 = 3(y + 3) \qquad \text{standard form of the parabola}$$

FINAL ANALYSIS (See Diagram 13).

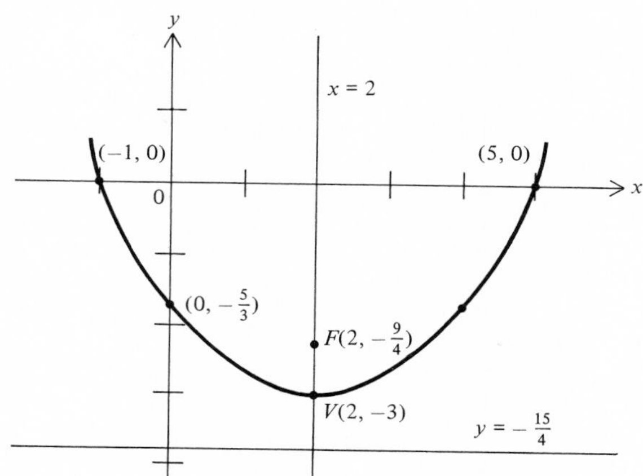

Diagram 13

$h = 2$	$k = -3$	therefore	$V(2, -3)$	
$4p = 3$	$p = \frac{3}{4}$	and	$p > 0$	curve opens upward

focus: $(2, -3 + \frac{3}{4}) = (2, -\frac{9}{4})$ directrix: $y = -3 - \frac{3}{4} \Rightarrow y = -\frac{15}{4}$

Additional points to aid in the sketch. When $x = 0$,

$$3y = -5 \qquad y = -\tfrac{5}{3} \qquad \text{y intercept is } (0, -\tfrac{5}{3})$$

When $y = 0$,

$$x^2 - 4x - 5 = 0 \qquad \text{Substituting } y = 0$$

$$(x - 5)(x + 1) = 0 \qquad \text{Factoring}$$

$$x = 5, -1 \qquad \text{Solving for } x$$

$(5, 0)$ and $(-1, 0)$ are the coordinates of the x intercepts of the parabola.

Example 2 Expressing the parabola in the required form may be more of an algebraic exercise than in the previous example. Let us analyze

$$3y^2 - 9y + 5x - 12 = 0$$

STEP 1 Separate the variables:

$$3y^2 - 9y = -5x + 12$$

STEP 2 In the standard form $(y - k)^2$ has a coefficient of 1 for the y term. Therefore divide both sides of the equation by 3

$$y^2 - 3y = -\tfrac{5}{3}x + 4$$

STEP 3 Complete the square on the left side of the equation:

$$y^2 - 3y + \tfrac{9}{4} = -\tfrac{5}{3}x + 4 + \tfrac{9}{4}$$

Simplifying gives

$$(y - \tfrac{3}{2})^2 = -\tfrac{5}{3}x + \tfrac{25}{4}$$

STEP 4 Now factor $-\frac{5}{3}$ from the right-hand member so that we have the form $-\frac{5}{3}(x - h)$. We have to find h so that $-\frac{5}{3}(-h) = \frac{25}{4}$. Solving gives $h = \frac{15}{4}$ and $(y - \frac{3}{2})^2 = -\frac{5}{3}(x - \frac{15}{4})$.

FINAL ANALYSIS (See Diagram 14.)

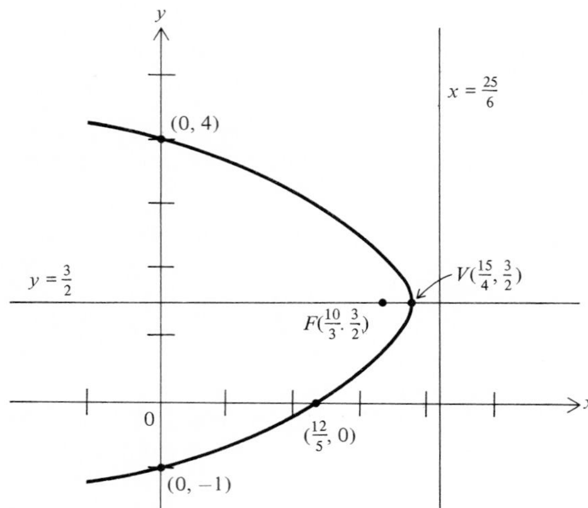

Diagram 14

$h = \frac{15}{3}$ $\quad k = \frac{3}{2}$ $\quad$ therefore $\quad V(\frac{15}{4}, \frac{3}{2})$
$4p = -\frac{5}{3}$ $\quad p = -\frac{5}{12}$ $\quad$ and $\quad p < 0$ curve opens left
axis of symmetry: $y = \frac{3}{2}$
focus: $(\frac{15}{4} - \frac{5}{12}, \frac{3}{2}) = (\frac{10}{3}, \frac{3}{2})$
directrix: $x = \frac{5}{12} + \frac{15}{4} = \frac{25}{6}$

Additional points to aid the sketch. When $y = 0$

$5x - 12 = 0 \qquad x = \frac{12}{5}$	x intercept
$3y^2 - 9y - 12 = 0$	Substituting $x = 0$
$y^2 - 3y - 4 = 0$	Dividing both sides by 3
$(y - 4)(y + 1) = 0$	Factoring
$y = 4, -1$	Solving for y

$(0, -1)(0, 4)$, are the coordinates of the y intercepts.

COMMENT In the standard form of the parabola

$$(x - h)^2 = 4p(y - k) \tag{1}$$

h, p, and k are the parameters that determine the parabola. But $(y - k)^2 = 4p(x - h)$ determines a different parabola with the same three parameters. Therefore, if we wish to specify a particular parabola with parameters h, p, and k, we also have to specify the axis of the parabola.

COMMENT A parabola with axis parallel to the y axis can be written in the form $y = Ax^2 + Bx + C$. A parabola with axis parallel to the x axis can be written in the form $x = Ay^2 + By + C$.

Example 3 Determine the equation of the parabola with axis parallel to the y axis and passing through the three points $(-1, 4)$, $(3, 0)$, and $(1, -2)$. Note that these are the same three points that we used in Section 11.3. Three noncollinear points do not require the locus to be a circle. Both the circle and the parabola depend on three parameters.

Using the general form $y = Ax^2 + Bx + C$, we obtain

$4 = A - B + C$	when	$x = -1$ and $y = 4$
$0 = 9A + 3B + C$	when	$x = 3$ and $y = 0$
$-2 = A + B + C$	when	$x = 1$ and $y = -2$

Solving this system of three equations and three unknowns, we obtain the solution, $A = 1, B = -3$, and $C = 0$. Therefore the equation of the parabola which satisfies the given conditions is

$$x^2 - 3x - y = 0$$

or

$$(x - \tfrac{3}{2})^2 = y + \tfrac{9}{4}$$

(see Diagram 15). An excellent exercise, which we recommend, would be to determine the sister parabola passing through the same three points but with axis parallel to the x axis. Check your solution with ours:

$$y^2 - y + 3x - 9 = 0$$

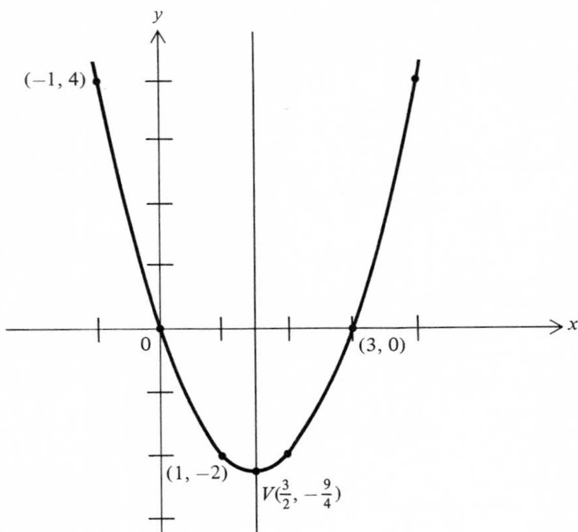

Diagram 15

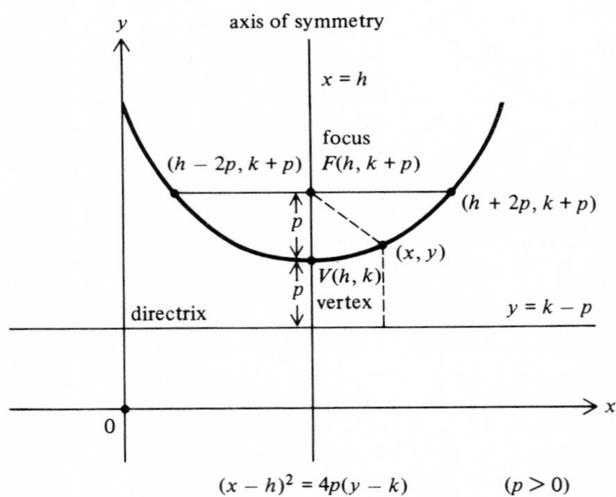

Diagram 16

Graphical Summary of the Parabola (See Diagrams 16 and 17.)

FINAL REMARKS Given the general equation of the parabola, $(x - h)^2 = 4p(y - k)$, if $p = 0$, we obtain $(x - h)^2 = 0$ or $x^2 - 2hx + h^2 = 0$. This represents a line, a pair of parallel lines, or no graph. These alternatives are referred to as *degenerate cases* of the parabola.

Example 4 Analyze **(a)** $x^2 - x - 6 = 0$, **(b)** $y^2 - 2y + 1 = 0$, and **(c)** $x^2 - 2x + 2 - 0$. Note that none of the equations contain *both* variables x and y. This is because $p = 0$. All qualify as degenerate parabolas.

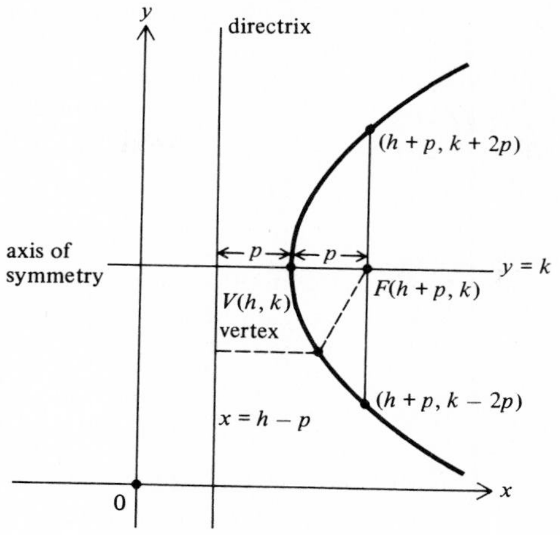

Diagram 17 $(y - k)^2 = 4p(x - h) \quad (p > 0)$

(a) $(x - 3)(x + 2) = 0$

therefore $x = 3, x = -2 \implies$ (two parallel lines).

(b) $(y - 1)(y - 1) = 0$

therefore $y = 1, \implies$ (one straight line).

(c) $x^2 - 2x + 2 = 0$

cannot be rectified since $b^2 - 4ac < 0 \implies$ (no graph exists).

EXERCISES

1 Find the vertex of $2x^2 + 5y = 3x - 4$.
2 Find the vertex of $4ay = -x^2 - ax$.

In Exercises 3 to 20, analyze, graph, and find the vertex, focus, and directrix.

3 $x^2 - 4x - 4y - 2 = 0$
4 $(x - 1)^2 = -2(y + 4)$
5 $y^2 + 4y + 8x = 0$
6 $x^2 - 4x - 3y + 5 = 0$
7 $y^2 - 4y - 4x - 2 = 0$
8 $y = -2x^2 + 3x + 2$
9 $2x^2 + 3y - 4x + 4 = 0$
10 $3y^2 + 12y - 20x + 42 = 0$
11 $3x^2 - 8y - 12x = 4$
12 $y^2 + 8y + 6x + 1 = 0$
13 $2x^2 - 8x = 6y + 1$
14 $x^2 - 6x + 8y + 25 = 0$
15 $y^2 - 6y - 12x + 3 = 0$
16 $2x^2 - 24x + 3y + 78 = 0$
17 $4y^2 + 12y - 20x + 49 = 0$
18 $y = 2x^2 - 6x + 1$
19 $y^2 + 2y + 6x - 11 = 0$
20 $x^2 + 4x = 6y - 22$

21 Show, in general, that the vertex of $x = py^2 + qy + r$ is $(r - q^2/4p, -q/2p)$.
22 Using the results of Exercise 21, find the vertex of $x = 3y^2 - 9y + 1$.
23 In general, find the coordinates of the vertex of $y = px^2 + qx + r$ and verify the results of Exercises 8 and 18.
24 Find an equation of a parabola with:
(a) vertex $(-2, 1)$ and focus $(2, 1)$ **(b)** vertex $(2, -1)$ and focus $(2, 1)$

(c) vertex $(1, -2)$ and directrix $y = -4$
(d) vertex $(-1, 2)$ and directrix $x = 2$
(e) vertex $(1, -2)$ and focus $(-4, -2)$
(f) vertex $(-2, 1)$ and focus $(-2, -3)$
(g) vertex $(-2, 1)$ and directrix $y = 4$
(h) vertex $(2, -1)$ and directrix $x = -3$

25 Find the equation of the parabola opening to the left and passing through the point $(0, -1)$ with focus $(0, 3)$.

26 Find the equation of the parabola opening upward, passing through the point $(-2, -4)$, and having focus $(2, -4)$.

27 Find the equation of the parabola with axis parallel to the y axis passing through $(-1, -1)$, $(1, 1)$, and $(3, 4)$.

28 Find the equation of the parabola with axis parallel to the x axis passing through $(0, 0)$, $(1, 1)$, and $(-2, 4)$.

29 Answer Exercise 28 if the axis of the parabola is parallel to the y axis.

30 How high is a parabolic arch 24 feet wide and 18 feet high at a point 8 feet from the center of the arch?

31 Find the equation of the line passing through the vertex of $x^2 - 4x - 8y = 0$ and perpendicular to the line containing the focus and the point $(-4, 4)$ on the parabola.

32 On the same axes, sketch $y^2 = 64 - 16x$ and $y^2 = 16 + 8x$.
(a) Find the focus and vertex of each.
(b) Find the coordinates of their point(s) of intersection.

33 Show that $y = x^2 + 4x + 8$ and $y = -4x - x^2$ are tangent to each other. Sketch the system and determine the point of tangency.

34 Find the equation of the parabola whose axis is vertical and passes through the points of intersection of the lines

$$y = -\tfrac{4}{3}x + 4 \qquad y = \tfrac{2}{3}x \qquad y = -2x + 4$$

Find the vertex, focus, and directrix. Sketch the system.

35 Find the equation of the parabola passing the points $(-1, 2)$, $(1, 0)$, and $(5, 4)$ with **(a)** a horizontal axis and **(b)** a vertical axis. Sketch and analyze.

36 **(a)** Find the equation of the parabolic arch that is 12 feet wide and 8 feet high. Place point A at the origin $(0, 0)$ (see the diagram).

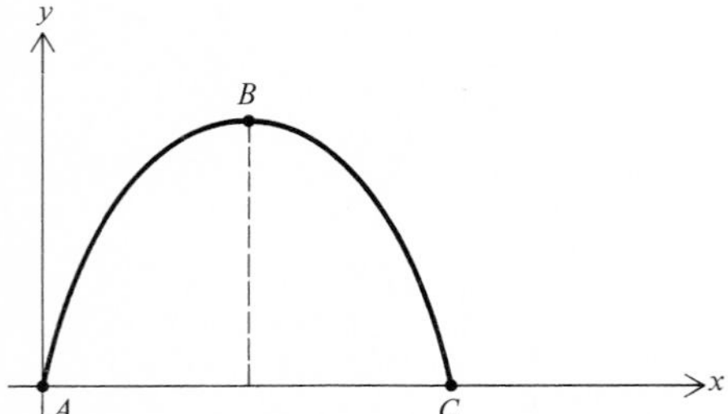

(b) In general, find the equation of any parabolic arch twice as wide as it is high. (Let H = height.) Specify the initial position of the parabola.

37 Using the definition of the parabola, derive an equation for a parabola whose focus is the origin and whose vertex is $(a, 0)$.

38 A parabolic arch crosses a highway 40 feet wide, with the edges of the arch at the edges of the highway. The highest point of the arch is above the center of the road, 20 feet high. An 8-foot-high truck wishes to pass beneath the arch 4 feet from the edge of the roadway.

(a) Is the clearance possible?

(b) What is the closest the truck can come to the edge of the road and still pass safely under the arch?

39 Suppose a projectile is shot from a gun (at ground level) at an angle of elevation of α radians. If v_0 is the muzzle velocity of the projectile, x is the horizontal distance of the projectile from the starting point (called the *range*), and y is the height of the projectile for any allowable x, then

$$y = x(\tan\alpha) - \frac{16x^2}{v_0^2\cos^2\alpha}$$

If a projectile is shot from a gun at an angle of 60° with a muzzle velocity of 640 feet per second find:

(a) An equation for the path of the projectile

(b) The range of the projectile

(c) The maximum height attained by the projectile

40 Answer the three questions in Exercise 39 under the given conditions:

(a) $\alpha = 30°$, $v_0 = 480$ **(b)** $\alpha = 45°$, $v_0 = 2500$

41 Referring to Exercise 39, the equation of the path of the projectile can be expressed in terms of time t using *parametric equations* (see Chapter 7):

$$x = v_0t\cos\alpha \qquad \text{and} \qquad y = v_0t\sin\alpha - 16t^2$$

(a) Find the time of flight in Exercise 39

(b) Find the time of flight in parts (*a*) and (*b*) of Exercise 40.

42 A projectile is shot from the ground at an angle of 30° with a muzzle velocity of 1600 feet per second. Find:

(a) Time of flight

(b) Range of the projectile

(c) Maximum height of the projectile

The Ellipse and the Hyperbola

12.1 THE ELLIPSE

DEFINITION 1 An *ellipse* is the locus of all points in a plane such that the sum of the distances from two fixed points remains constant. The two fixed points are the *foci* of the ellipse.

To develop the standard form of the ellipse, let us proceed in a manner similar to that for the parabola and set up a specialized coordinate system (Diagram 1).

STEP 1 Let the foci be on the x axis, labeled $(-c, 0)$ and $(c, 0)$ with the origin midway between the foci, for simplicity. The origin in this case is referred to as the center of the ellipse.

STEP 2 Let $P(x, y)$ represent any point on the graph of the ellipse.

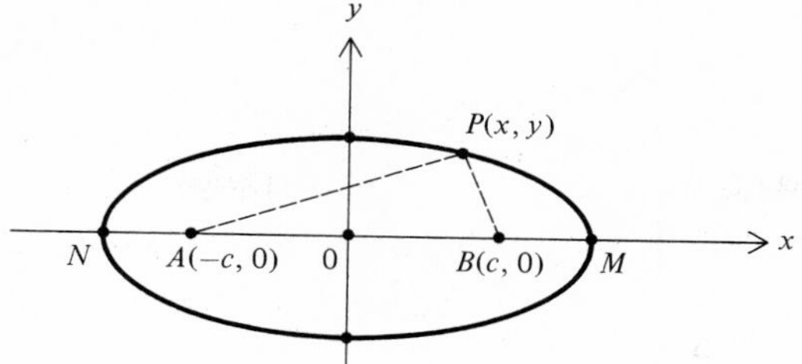

Diagram 1

STEP 3 According to the definition, $PA + PB =$ a constant. Let this constant be $2a$.

STEP 4 Note that in Diagram 2, when $P(x, y)$ is on the y axis,

$$PA = PB = a \qquad OB = c$$

The length of OP will be called b.

STEP 5 From Diagram 2, it can be seen that $a^2 = b^2 + c^2$.

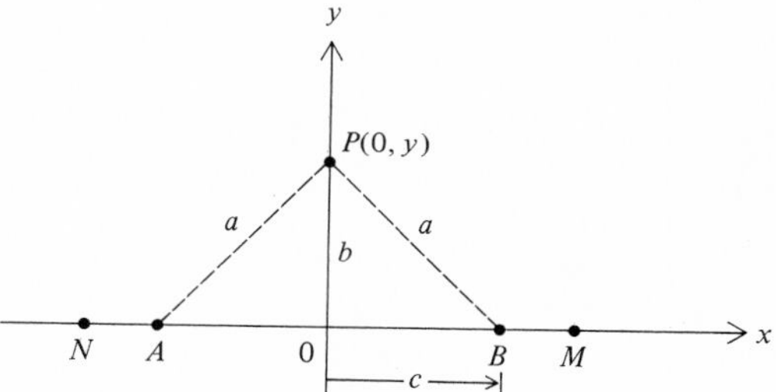

Diagram 2

STEP 6 The coordinates of the end points of the ellipse M and N are $(a, 0)$ and $(-a, 0)$, respectively (see Diagram 3). To verify the above statement, when $P(x, y)$ is on the x axis at M, $y = 0$.

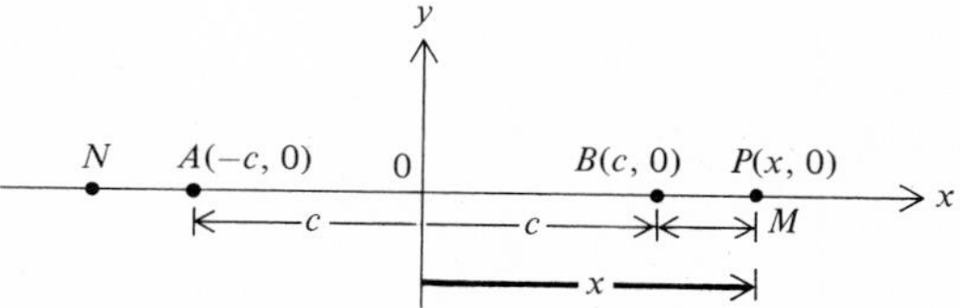

Diagram 3

$PA + PB = 2a$	Definition of the ellipse	(1)
$PA = 2c + PB$	Diagram 3	(2)
$2c + PB + PB = 2a$	Substituting (2) into (1)	(3)
$PB = a - c$	Solving for PB	(4)

Since $OP = c + PB$,

$$OP = c + a - c = a \tag{5}$$

Therefore, $x = a$, and the coordinates of P are $(a, 0)$.

STEP 7 Now that we have the ellipse expressed in coordinates, we follow through the algebraic computations to develop the equation of the ellipse.

$\sqrt{(x+c)^2+y^2} + \sqrt{(x-c)^2+y^2} = 2a$	From definition; $PA + PB = 2a$
$\sqrt{(x-c)^2+y^2} = 2a - \sqrt{(x+c)^2+y^2}$	Transposing one radical
$(x-c)^2 + y^2 = 4a^2 - 4a\sqrt{(x+c)^2+y^2} + (x+c)^2 + y^2$	Squaring both sides
$4a^2 + 4cx = 4a\sqrt{(x+c)^2+y^2}$	Expanding and regrouping
$a + \dfrac{cx}{a} = \sqrt{(x+c)^2+y^2}$	Dividing by $4a$
$a^2 + 2cx + \dfrac{c^2x^2}{a^2} = x^2 + 2cx + c^2 + y^2$	Squaring both sides *again*
$a^2 - c^2 = x^2 - \dfrac{c^2x^2}{a^2} + y^2 = x^2\left(\dfrac{a^2-c^2}{a^2}\right) + y^2$	Regrouping *again*
$1 = \dfrac{x^2}{a^2} + \dfrac{y^2}{a^2-c^2}$	Dividing by $a^2 - c^2$
$1 = \dfrac{x^2}{a^2} + \dfrac{y^2}{b^2}$	From step 5, $a^2 - c^2 = b^2$

COMMENT (see Diagram 4). MN is called the major (long) axis, and a is one-half the length of the major axis. QR is called the minor (short) axis, and b is one-half the length of the minor axis. We are assuming here that $a > b$. If $a < b$, MN would be the minor axis and QR the major axis. If $a = b$, the locus would be a circle. $(a, 0)$, $(-a, 0)$, $(0, b)$, and $(0, -b)$ are called the vertices of the ellipse. $(c, 0)$ and $(-c, 0)$ are the foci of the ellipse. Note that the foci always lie on the major axis.

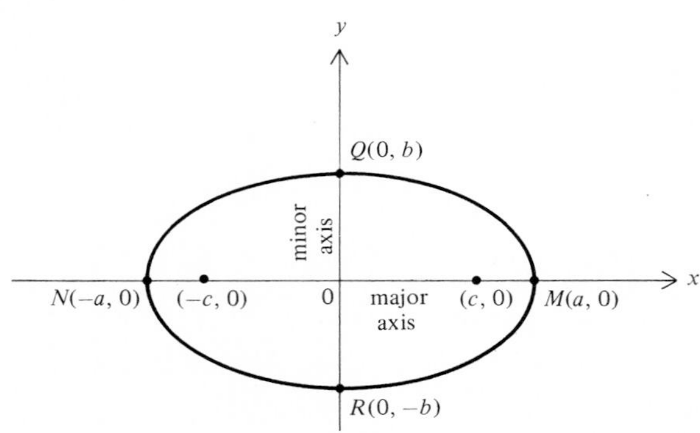

Diagram 4

Theorem 1 The equation of an ellipse with foci $(c, 0)$ and $(-c, 0)$ and the major axis the x axis is

$$\frac{x^2}{a^2} + \frac{y^2}{b^2} = 1 \qquad \text{where } c^2 = a^2 - b^2 \text{ and } a > b$$

Theorem 2 The equation of an ellipse with foci $(0, c)$ and $(0, -c)$ and the major axis the y axis is

$$\frac{y^2}{a^2} + \frac{x^2}{b^2} = 1 \qquad \text{where } c^2 = a^2 - b^2 \text{ and } a > b$$

COMMENT The *larger* denominator is labeled a^2 so that $a^2 - b^2 > 0$ and hence $c^2 > 0$.

Example 1 Analyze $36x^2 + 100y^2 = 3600$.

Putting the equation into standard form, we divide both sides of the equation by 3600:

$$\frac{x^2}{100} + \frac{y^2}{36} = 1 \qquad \begin{aligned} a^2 &= 100 \\ a &= 10 \end{aligned} \qquad \begin{aligned} b^2 &= 36 \\ b &= 6 \end{aligned} \qquad \begin{aligned} c^2 &= a^2 - b^2 \\ &= 100 - 36 = 64 \\ c &= 8 \end{aligned}$$

From Diagram 5 the center is $(0, 0)$ the foci are $(8, 0)$ and $(-8, 0)$, and the vertices are $(10, 0)$, $(-10, 0)$, $(0, 6)$, and $(0, -6)$.

Once the key points are determined, sketching the ellipse is rather simple. The most important consideration is putting the original expression into *perfect form.*

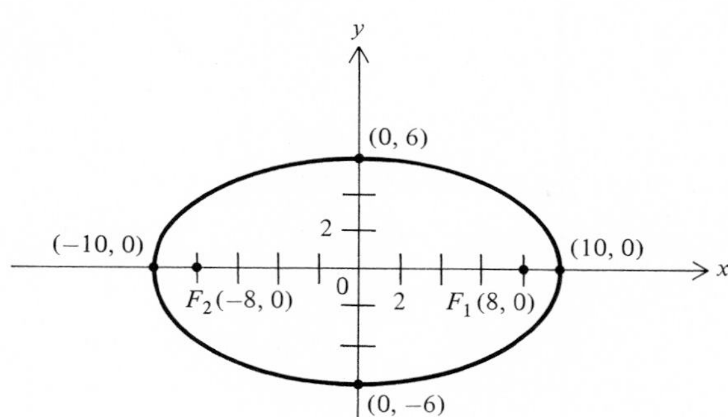

Diagram 5

Example 2 Analyze $\dfrac{x^2}{4} + \dfrac{y^2}{16} = 1$.

Since $16 > 4$, the focus will lie on the y axis, which is the major axis.

$$\frac{y^2}{16} + \frac{x^2}{4} = 1 \qquad \begin{aligned} a^2 &= 16 \\ a &= 4 \end{aligned} \qquad \begin{aligned} b^2 &= 4 \\ b &= 2 \end{aligned} \qquad \begin{aligned} c^2 &= 12 \\ c &= 2\sqrt{3} \end{aligned}$$

The center is (0, 0). The foci are $(0, 2\sqrt{3})$ and $(0, -2\sqrt{3})$. The vertices are (0, 4), (0, −4), (2, 0), and (−2, 0) (Diagram 6).

To determine whether the major axis is the x or y axis, we simply choose the variable which is above the larger denominator.

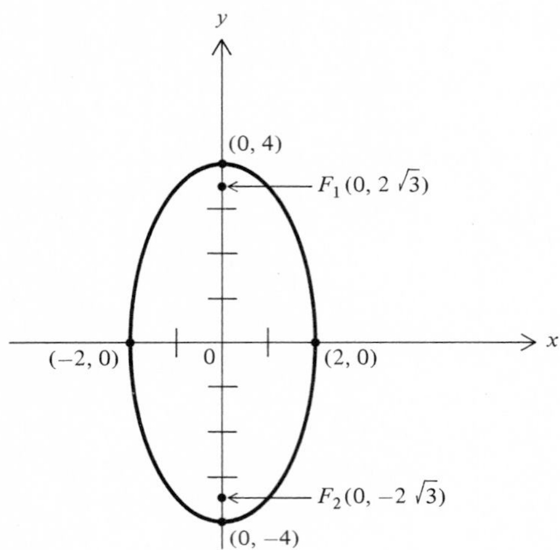

Diagram 6

Example 3 Find the equation of the ellipse with center at the origin and major axis on the x axis and passing through points (6, 1) and (2, 3).

The standard form is

$$\frac{x^2}{a^2} + \frac{y^2}{b^2} = 1$$

Since (6, 1) and (2, 3) must satisfy the equation of the required ellipse,

$$\frac{36}{a^2} + \frac{1}{b^2} = 1 \qquad \text{When } x = 6 \text{ and } y = 1 \qquad (1)$$

$$\frac{4}{a^2} + \frac{9}{b^2} = 1 \qquad \text{When } x = 2 \text{ and } y = 3 \qquad (2)$$

$$\frac{1}{b^2} = 1 - \frac{36}{a^2} = \frac{a^2 - 36}{a^2} \qquad \text{From (1)}$$

$$\frac{9}{b^2} = \frac{9a^2 - 324}{a^2} \qquad \text{Multiplying both sides by 9}$$

$$\frac{4}{a^2} + \frac{9a^2 - 324}{a^2} = 1 \qquad \text{Substituting into (2)}$$

$$4 + 9a^2 - 324 = a^2$$

$$8a^2 = 320 \qquad \text{Solving for } a^2$$

$$a^2 = 40$$

$$\frac{4}{40} + \frac{9}{b^2} = 1 \qquad \text{Substituting } a^2 = 40 \text{ into (2)}$$

$$b^2 = 10 \qquad \text{Solving for } b^2$$

$$\frac{x^2}{40} + \frac{y^2}{10} = 1 \qquad \text{The required equation}$$

(See Diagram 7.)

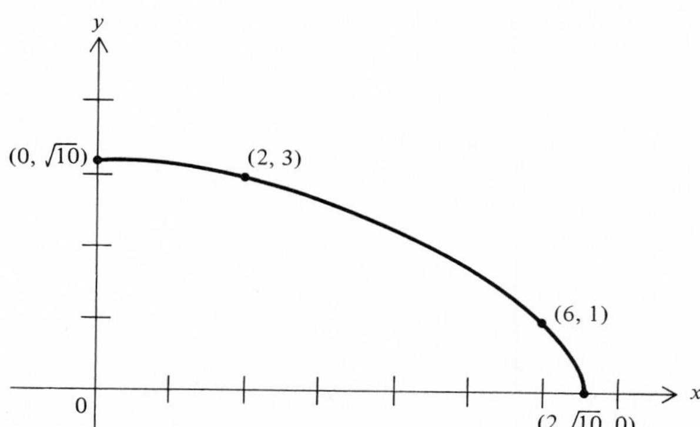

Diagram 7

Example 4 A series of semielliptic arches is being constructed to form the roof of a truck tunnel. The tunnel is to be 96 feet wide at the base and 16 feet high at its highest point. The tunnel will have a six-lane roadway, 72 feet wide, with three equal lanes on each side of a center divider. What will be the minimum clearance for truck traffic in each of the three lanes (Diagram 8)?

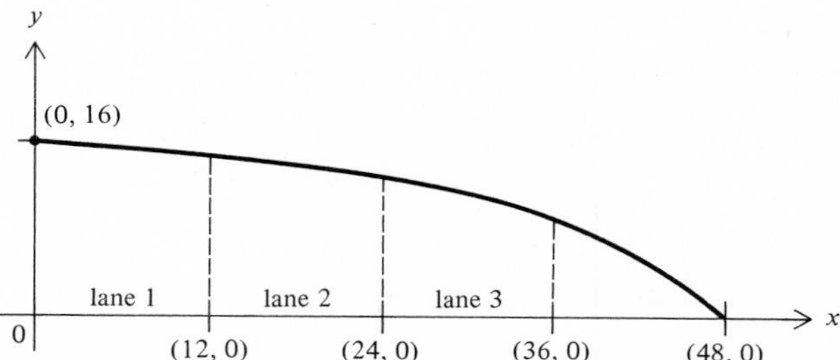

Diagram 8

We can use coordinates in this problem thanks to its elliptic properties. From the diagram, $a = 48$, $b = 16$, and the equation of the ellipse is

$$\frac{x^2}{2304} + \frac{y^2}{256} = 1$$

Since $y > 0$, it is a semiellipse. To find the heights at $x = 12$, 24, and 36 we solve for y:

$$y = \sqrt{256 - \frac{x^2}{9}}$$

When $x = 12$, $y = \sqrt{240} \approx 15$ feet 6 inches. When $x = 24$, $y = \sqrt{192} \approx 13$ feet 10 inches. When $x = 36$, $y = \sqrt{112} \approx 10$ feet 7 inches. These are the minimum clearances in each lane.

EXERCISES

Sketch Exercises 1 to 6, and find the coordinates of the foci and vertices.

1 $\dfrac{x^2}{25} + \dfrac{y^2}{9} = 1$

2 $9x^2 + 4y^2 - 144 = 0$

3 $x^2 + 4y^2 = 16$

4 $\dfrac{x^2}{8} + \dfrac{y^2}{12} = 1$

5 $\dfrac{9x^2}{4} + \dfrac{4y^2}{9} = 1$

6 $225x^2 + 289y^2 = 65{,}025$

7 Find the equation of the ellipse with foci $(0, \pm 2)$ and passing through $(-3, 2)$.

8 A semielliptic wooden arch 40 feet wide and 10 feet high is supported by two vertical beams each $5\sqrt{3}$ feet high. How far from the center of the arch are the beams located?

9 A focal radius of an ellipse is a line drawn from the focus to a point on the ellipse. Find the equations of the focal radii containing $(-5, 3)$ on the ellipse $x^2/40 + y^2/24 = 1$. Sketch the system. If the shorter radius were extended, at what point would it intersect the ellipse?

10 Using the methods of the calculus, it can be shown that if $p(x_1, y_1)$ is a point on $x^2/a^2 + y^2/b^2 = 1$, then $b^2xx_1 + a^2yy_1 = a^2b^2$ is an equation of the line tangent to the ellipse at (x_1, y_1). Given $x^2/100 + y^2/25 = 1$ and point $(-8, 3)$ on the ellipse, sketch the ellipse and determine an equation of the tangent line at the given point.

11 Find the equation of the ellipse with center at the origin and major axis on the x axis and passing through $(8, 2)$ and $(4, -4)$. Sketch the ellipse.

12 An elliptic oil tank with a false bottom is 8 feet wide and 4 feet high. To determine when the tank is empty, a gauge is inserted into a hole directly above the focus and lowered perpendicularly until it touches the bottom directly below the focus. The numbers on the gauge represent the fuel level (in feet) from the bottom. What number on the gauge will correspond to the half-full position?

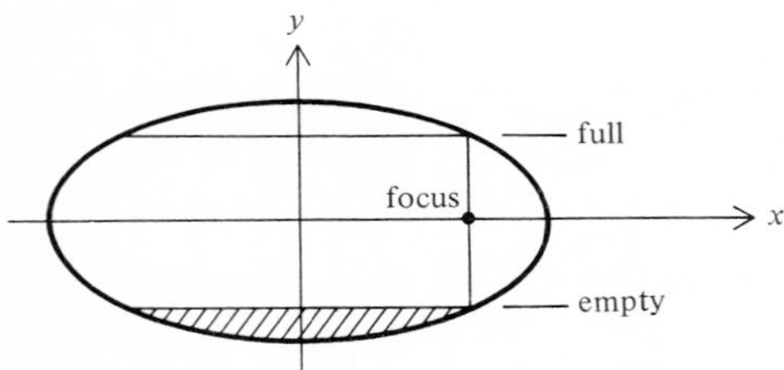

13 A chord drawn through the focus of an ellipse perpendicular to the major axis is called the latus rectum. Show that the length of this chord is $2b^2/a$.

14 An ellipse has its center at (0, 0) and its foci on the x axis. A point m is on the ellipse so that the lines from m to each of the foci are perpendicular to each other. The lengths of these lines are 2 and 4.

(a) Find the equation of the ellipse.

(b) Can you determine the actual point on the ellipse in part (*a*) which corresponds to point m?

An ellipse $x^2/a^2 + y^2/b^2 = 1$ may vary in shape. If $c = 0$ (remember that $c^2 = a^2 - b^2, a > b$), then $a^2 = b^2$ and the ellipse is a circle. In fact, as c increases from 0 to a, the ellipse becomes *flatter* and approaches a straight-line segment, with its foci as end points. The ratio $e = c/a$, called the *eccentricity* of the ellipse, varies from 0 to 1.

0 ← circular | flat → 1

15 Find the eccentricity of the ellipses in Exercises 1 to 4.

16 Find the equation of the ellipse with foci (6, 0) and (−6, 0) and eccentricity $\frac{2}{3}$.

17 The orbit of the earth about the sun as a focus is elliptical. If the major axis of the ellipse is approximately 180 million miles and the eccentricity is approximately 1/60, what are the greatest and smallest distances between the earth and the sun?

18 Find a more precise answer to Exercise 17 by using the more accurate data of 185.5 million miles and $e = 17/1000$.

19 Given the circle $25x^2 + 25y^2 = 144$ and the ellipse $9x^2 + 16y^2 = 144$. A tangent line to the circle parallel to the y axis is drawn and intersects the ellipse in points M and N. Prove:

(a) MN is equal to the diameter of the circle.

(b) $\angle MON = 90°$

20 An ellipse has its center at the origin and its foci on the x axis. P and Q represent the foci. Point $R(4, 3)$ is on the ellipse, and $\angle PRQ = 90°$. Find the equation of the ellipse.

21 **(a)** Find the area of the rectangle circumscribing an ellipse whose equation is

$$\frac{x^2}{25} + \frac{y^2}{81} = 1$$

(b) What is the area of the parallelogram inscribed in the ellipse from part (*a*) if the vertices of the parallelogram are the end points of the major and minor axes of the ellipse?

(c) Estimate the areas of the ellipse by averaging the areas found in part (*a*) and part (*b*).

(d) If the exact area of an ellipse is πab, how does the answer to part (*c*) compare?

12.2 THE ELLIPSE WITH CENTER $C(h, k)$

Theorem 1 The equation of an ellipse with center $C(h, k)$ and major axis parallel to the x axis is

$$\frac{(x-h)^2}{a^2} + \frac{(y-k)^2}{b^2} = 1 \qquad \text{where } a > b \tag{1}$$

$$c^2 = a^2 - b^2$$

The coordinates of the foci are $(h + c, k)$ and $(h - c, k)$; the coordinates of the end points of the major axis are $(h + a, k)$ and $(h - a, k)$. And the coordinates of the end points of the minor axis are $(h, k + b)$ and $(h, k - b)$ (see Diagram 9).

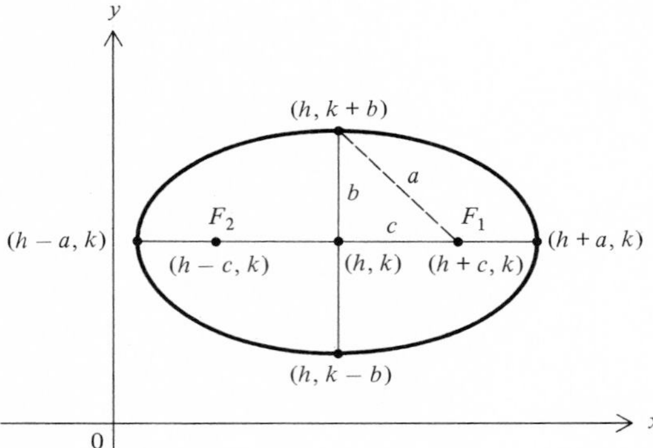

Diagram 9

Theorem 2 The equation of an ellipse with center $C(h, k)$ and major axis parallel to the y axis is

$$\frac{(y - k)^2}{a^2} + \frac{(x - h)^2}{b^2} = 1 \qquad \text{where } a > b \tag{2}$$

$$c^2 = a^2 - b^2$$

The coordinates of the foci are $(h, k + c)$ and $(h, k - c)$. The coordinates of the end points of the major axis are $(h, k + a)$ and $(h, k - a)$. And the coordinates of the end points of the minor axis are $(h + b, k)$ and $(h - b, k)$ (see Diagram 10).

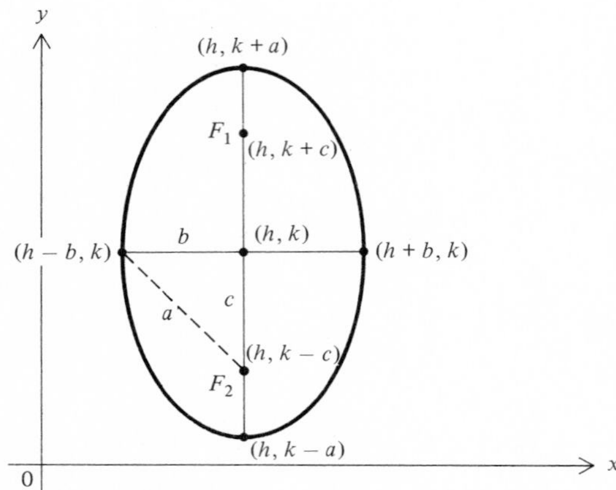

Diagram 10

Example 1 Analyze $\frac{(x-2)^2}{16} + \frac{(y+3)^2}{9} = 1.$

Since $16 > 9$, the major axis is parallel to the x axis

$$a^2 = 16 \qquad b^2 = 9 \qquad c^2 = a^2 - b^2 = 16 - 9 = 7$$
$$a = 4 \qquad b = 3 \qquad c = \sqrt{7}$$

The center is $C(2, -3)$, and the foci are $(2 + \sqrt{7}, -3)$ and $(2 - \sqrt{7}, -3)$ (Diagram 11).

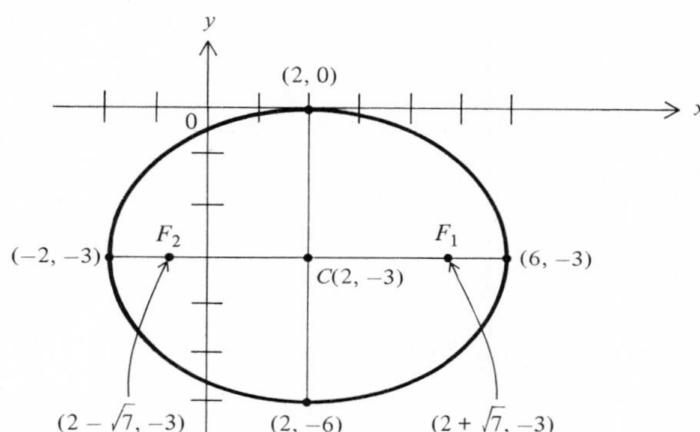

Diagram 11

Example 2 Analyze $9x^2 + 4y^2 - 72x - 48y + 144 = 0$.
This equation must be put into standard form before analysis:

$$9(x^2 - 8x \qquad) + 4(y^2 - 12y \qquad) = -144$$

Complete the square within *each* of the parentheses.

$$9(x^2 - 8x + 16) + 4(y^2 - 12y + 36) = -144 + 9(16) + 4(36)$$

In effect we have added $9(16) + 4(36) = 288$ to both sides, reducing the equation to

$$9(x-4)^2 + 4(y-6)^2 = 144$$

To generate the proper form according to Theorem 2, we divide both sides by 144:

$$\frac{(x-4)^2}{16} + \frac{(y-6)^2}{36} = 1$$

Since $36 > 16$, the major axis is parallel to the y axis

$$a^2 = 36 \qquad b^2 = 16 \qquad c^2 = a^2 - b^2 = 36 - 16 = 20$$
$$a = 6 \qquad b = 4 \qquad c = \sqrt{20} = 2\sqrt{5}$$

Examine Diagram 12 carefully and see why the coordinates of the foci are $(4, 6 + 2\sqrt{5})$ and $(4, 6 - 2\sqrt{5})$.

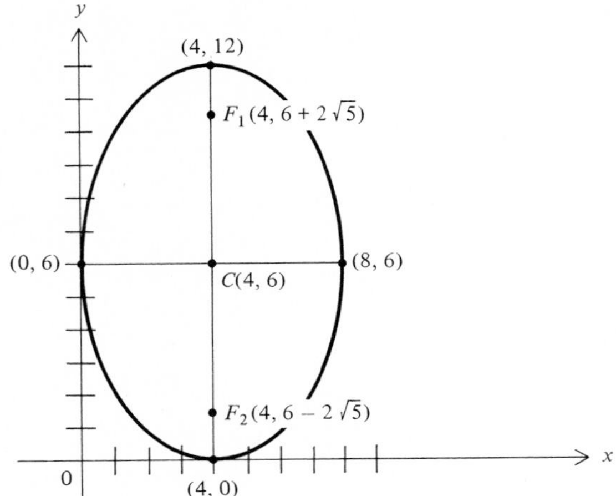

Diagram 12

Example 3 The end points of the major and minor axes of an ellipse are $(-6, 3)$, $(4, 3)$, $(-1, 6)$, $(-1, 0)$. After sketching the ellipse find the coordinates of the foci (Diagram 13).

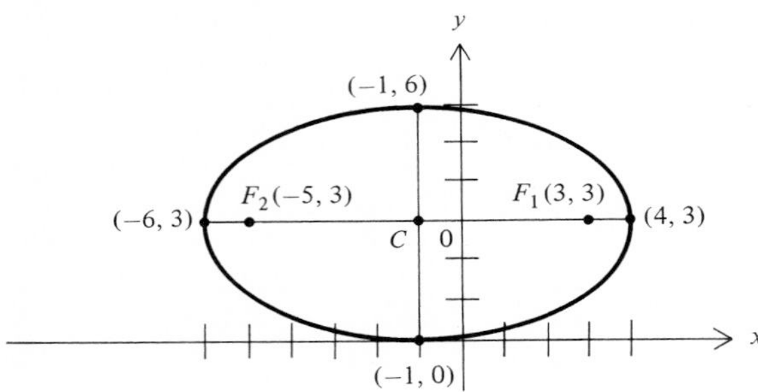

Diagram 13

From the diagram $C(h, k)$ is the midpoint of the minor axis and $C(-1, 3)$ satisfies the condition. By careful examination of the diagram, $a = 5$, $b = 3$, and the equation of the ellipse is:

$$\frac{(x + 1)^2}{25} + \frac{(y - 3)^2}{9} = 1$$

Then $$c^2 = a^2 - b^2 = 16 \qquad c = 4$$

The coordinates of the foci are $(-5, 3)$ and $(3, 3)$.

COMMENT Always analyze these problems with an accurate graph. This prevents a careless error with coordinates: just verify them on your diagram.

Theorem 3 An ellipse can be written in the *general form*

$$x^2 + Ay^2 + Bx + Cy + D = 0 \qquad \text{where} \qquad A > 0$$

In the standard form we have been using

$$\frac{(x-h)^2}{a^2}+\frac{(y-k)^2}{b^2}=1$$

h, k, a, and b are the parameters that determine the ellipse. For the circle and the parabola we needed only three conditions to determine their locus. In order to find an ellipse we need only analyze certain given conditions. If from these conditions we can find the values of A, B, C, D in the general form or h, k, a, b in the standard form, we will have determined an ellipse.

Example 4 Find the equation of the ellipse passing through $(-4, 1)$, $(2, 2)$, $(-6, 0)$, and $(-8, -3)$.

Since each of the four points must satisfy the equation of the required ellipse, using the *general form,*

$$16 + A - 4B + C + D = 0 \qquad \text{for } (-4, 1) \qquad (3)$$

$$4 + 4A + 2B + 2C + D = 0 \qquad \text{for } (2, 2) \qquad (4)$$

$$36 - 6B + D = 0 \qquad \text{for } (-6, 0) \qquad (5)$$

$$64 + 9A - 8B - 3C + D = 0 \qquad \text{for } (-8, -3) \qquad (6)$$

In order to solve the four equations and four unknowns, a simple algebraic solution will be to eliminate one of the variables (D in this case) and form three equations and three unknowns. In eliminating the variable D, we must be careful to use *each* of the four equations in our system. It is possible to eliminate D from the first three equations and never use the last equation. It will be an interesting exercise for you to try this and see what happens.

$$3A + 6B + C = 12 \qquad \text{Combining (3) and (4)}$$

$$A + 2B + C = 20 \qquad \text{Combining (3) and (5)}$$

$$9A - 2B - 3C = -20 \qquad \text{Combining (6) and (5)}$$

Solving the system, we have

$$A = 4 \qquad B = -4 \qquad C = 24 \qquad D = -60$$

which yields,

$$x^2 + 4y^2 - 4x + 24y - 60 = 0 \qquad \text{or} \qquad \frac{(x-2)^2}{100}+\frac{(y+3)^2}{25}=1$$

(See Diagram 14.)

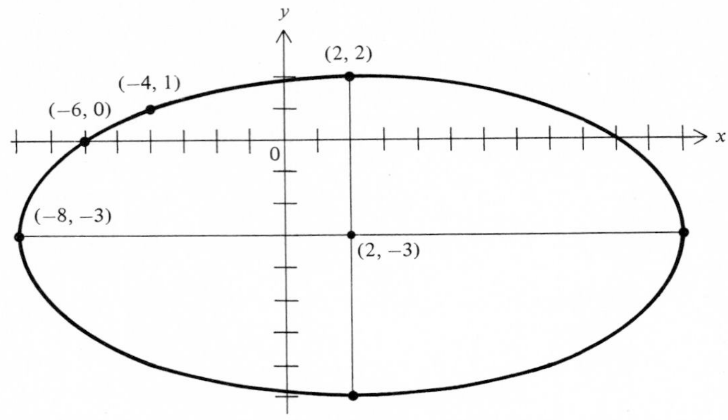

Diagram 14

EXERCISES

Analyze and sketch each of the equations in Exercises 1 to 11.

1 $x^2 + 4y^2 - 4x = 0$

2 $\dfrac{(x+1)^2}{25} + \dfrac{(y-2)^2}{16} = 1$

3 $x^2 - 8x + 2y^2 + 8y + 20 = 0$

4 $4x^2 + y^2 - 16x + 4y + 4 = 0$

5 $16x^2 + 64x + 9y^2 - 54y + 1 = 0$

6 $9x^2 + 4y^2 - 36x + 24y + 36 = 0$

7 $4x^2 - 48x + 9y^2 + 72y + 144 = 0$

8 $4(x-1)^2 + (y+2)^2 = 100$

9 $8x^2 + 4y^2 - 64x - 8y + 68 = 0$

10 $25x^2 + 9y^2 - 36y - 189 = 0$

11 $\dfrac{(x-4)^2}{49} + \dfrac{(y+3)^2}{16} = 1$

12 Find equation of ellipse with end points $(-8, 3)$, $(4, 3)$, $(-2, 5)$, and $(-2, 1)$.

13 Find the foci of the ellipse whose end points are $(4, 7)$, $(4, -1)$, $(2, 3)$, and $(6, 3)$.

14 On the same axes, draw the graphs of

$$\frac{(x+3)^2}{48} + \frac{(y-2)^2}{4} = 1 \qquad \text{and} \qquad (x+3)^2 = -12(y-4)$$

(a) Determine algebraically their points of intersection.
(b) Find the equation of the circle passing through these points.

15 The system of lines $2y = x + 6$ $2y = x - 6$ $2y = -x - 10$ $2y = -x + 2$ forms a parallelogram, whose vertices are the end points of the axes of an ellipse.
(a) Find the equation of the ellipse.
(b) Find the area of the parallelogram.
(c) Find the area of the rectangle circumscribing the ellipse.
(d) The area of an ellipse is πab; does this satisfy part (*b*) and part (*c*)?

16 Find the equation of the ellipse whose major axis is parallel to the y axis and is the diameter of $x^2 + y^2 + 4x - 6y + 1 = 0$ and whose minor axis is equal in length to the radius of the circle.

17 Find the equation of the ellipse passing through $(-8, 2)$, $(-6, 4)$, $(2, -2)$, and $(4, 0)$ and express in the form $x^2 + Ay^2 + Bx + Cy + D = 0$.

18 Find the equation of an ellipse which has center $(4, -1)$ and focus at $(1, -1)$ and which passes through the point $(8, 0)$.

19 An ellipse passes through the points (8, 18), (14, 2), (11, −10), and (−1, −18). If the equation of the ellipse is written in the form

$$\frac{(x-h)^2}{a^2} + \frac{(y-k)^2}{b^2} = 1$$

find the value of a, b, h, and k.

12.3
THE HYPERBOLA

DEFINITION 1 A *hyperbola* is the locus of all points in a plane such that the difference of the distances from two fixed points remains constant. The two fixed points are called the *foci.*

Theorem 1 The equation of a hyperbola with center $C(0, 0)$, foci $(c, 0)$ and $(-c, 0)$, and vertices $(a, 0)$ and $(-a, 0)$ is given by

$$\frac{x^2}{a^2} - \frac{y^2}{b^2} = 1 \qquad \text{where} \qquad c^2 = a^2 + b^2 \tag{1}$$

and the equations of the asymptotes are $y = bx/a$ and $y = -bx/a$ (refer to Diagram 15).

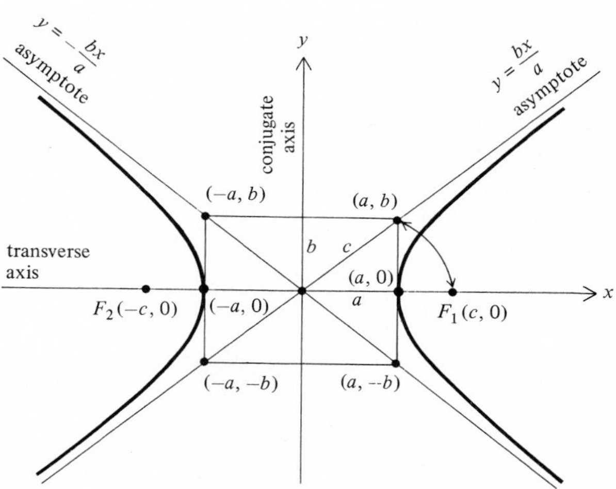

Diagram 15

Verification of this standard form is like that of Section 11.1 for the ellipse. We leave this problem as an exercise.

Refer to Diagram 15 for a graphic interpretation of the theorem.

COMMENTS

1 The major axis on which the foci are located is called the *transverse axis.* The minor axis is called the *conjugate axis.*

2 If we set $x = 0$, then $-y^2/b^2 = 1$ and $y^2 = -b^2$. Hence y is imaginary, and there is no value of y that corresponds to $x = 0$ in the real-number system. Therefore there is no y intercept.

3 The indicated rectangle is called the *reference rectangle,* and is the most important aid to sketching the hyperbola accurately. Note the significance of a and b in the equation as represented by the rectangle:

$2a$ = length of major axis
$2b$ = length of minor axis

4 For every value of x in the interval $(-a, a)$, y is imaginary.

$$y^2 = \frac{b^2(x^2 - a^2)}{a^2}$$

Since $y^2 \geq 0$ and $b^2/a^2 > 0$, $(x^2 - a^2) \geq 0$ and $x^2 \geq a^2$. Therefore we can conclude that $x \geq a$ or $x \leq -a$.

5 The hyperbola possesses a pair of asymptotes. They pass through the center of the reference rectangle and are the extended diagonals of the rectangle. The equations of the asymptotes are $y = bx/a$ and $y = -bx/a$.

6 There is no intersection between the hyperbola and its asymptote. When $y = bx/a$,

$$\frac{x^2}{a^2} - \frac{(bx/a)^2}{b^2} = 1 \Longrightarrow \frac{x^2}{a^2} - \frac{x^2}{a^2} \stackrel{?}{=} 1 \Longrightarrow \text{no solution}$$

7 As $|x| \to \infty$, the hyperbola approaches its asymptotes. Symbolically, this means that as $|x| \to \infty$, $D \to 0$ (where D is the distance between the curve and its asymptote) (Diagram 16).

$$D = \frac{bx}{a} - \frac{b}{a}\sqrt{x^2 - a^2} = \frac{b}{a}(x - \sqrt{x^2 - a^2})$$

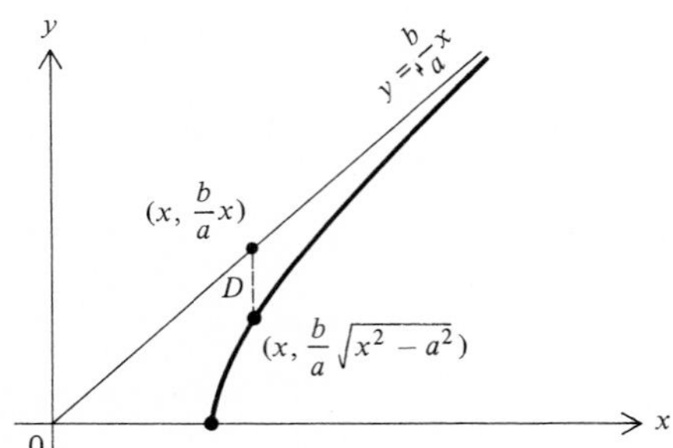

Diagram 16

Multiplying the numerator and denominator of the right-hand member by $x + \sqrt{x^2 - a^2}$, we get

$$D = \frac{ab}{x + \sqrt{x^2 - a^2}}$$

Now, as $|x| \to \infty$, the denominator $x + \sqrt{x^2 - a^2} \to \infty$; we can conclude that $D \to 0$.

Theorem 2 The equation of the hyperbola with center $C(0, 0)$, foci $(0, c)$ and $(0, -c)$, and vertices $(0, a)$ and $(0, -a)$ is

$$\frac{y^2}{a^2} - \frac{x^2}{b^2} = 1 \qquad \text{where} \qquad c^2 = a^2 + b^2$$

and the equations of the asymptotes are $y = ax/b$ and $y = -ax/b$ (refer to Diagram 17).

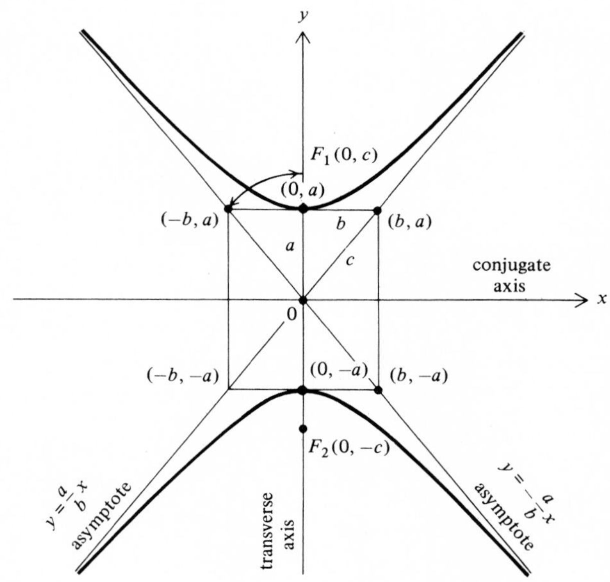

Diagram 17

Example 1 Analyze $\frac{x^2}{64} - \frac{y^2}{36} = 1$.

Solution

$$a^2 = 64 \qquad b^2 = 36$$
$$a = 8 \qquad b = 6$$

The center is $(0, 0)$.

The end points of the major axis are (8, 0) and (−8, 0), and the end points of the minor axis are (0, 6) and (0, −6).

Construct a reference rectangle using the above information as your guide and draw the asymptotes through the vertices of the rectangle (Diagram 18).

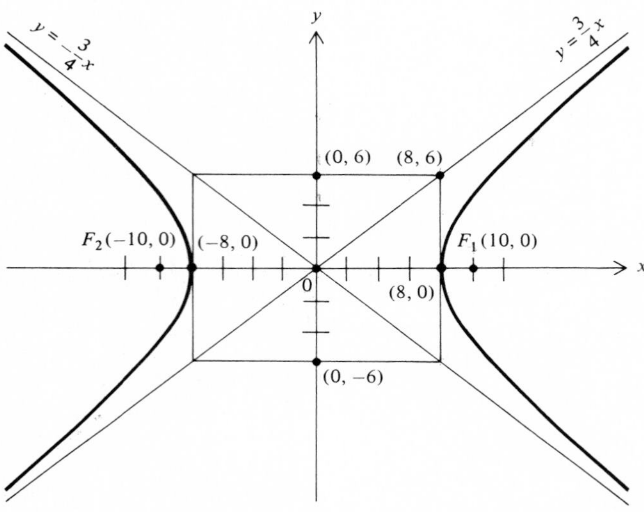

Diagram 18

From

$$c^2 = 64 + 36 = 100$$
$$c = 10$$

The foci are (10, 0) and (−10, 0).

The equations of the asymptotes are

$$y = \frac{3x}{4} \quad \text{and} \quad y = -\frac{3x}{4}$$

COMMENT Note the importance of the reference rectangle in drawing an accurate graph of the hyperbola. Use it carefully. It will be an invaluable aid in Section 12.4 also.

Example 2 Analyze $\frac{y^2}{25} - \frac{x^2}{16} = 1$.

The major axis is now the y axis, and since $a = 5$, $b = 4$, and $c^2 = 25 + 16 = 41$, we have that $c = \sqrt{41}$. The foci are $(0, \sqrt{41})$ and $(0, -\sqrt{41})$; the vertices are (0, 5) and (0, −5). The equations of the asymptotes are

$$y = \tfrac{5}{4}x \quad \text{and} \quad y = -\tfrac{5}{4}x$$

(See Diagram 19.)

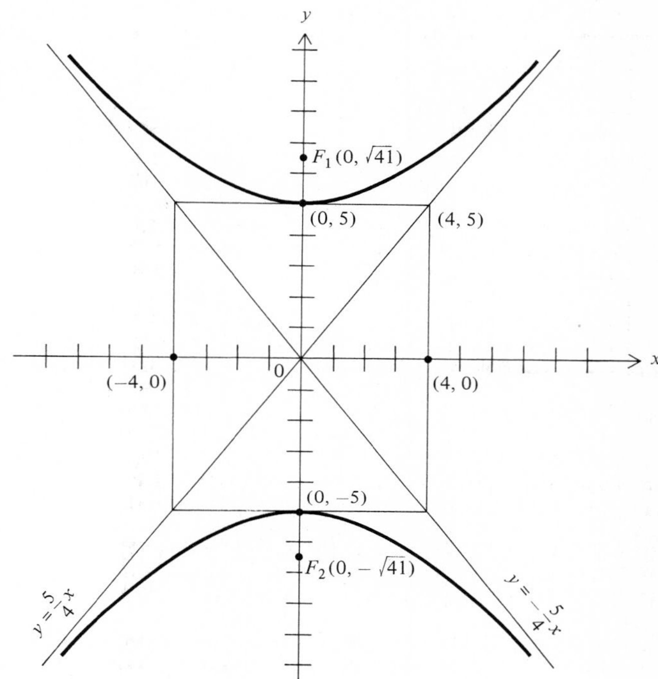

Diagram 19

EXERCISES

Sketch and analyze each of the hyperbolas in Exercises 1 to 9.

1 $\frac{x^2}{16} - \frac{y^2}{9} = 1$ **2** $\frac{x^2}{9} - \frac{y^2}{16} = 1$ **3** $3y^2 - x^2 = 27$

4 $x^2 - 4y^2 = 4$ **5** $4x^2 - y^2 = 4$ **6** $\frac{x^2}{36} - \frac{y^2}{64} = 1$

7 $7y^2 - 9x^2 = 252$ **8** $x^2 - y^2 = 16$ **9** $0.4x^2 - 0.6y^2 = 14.4$

10 Find the equations of the asymptotes of $25x^2 - 9y^2 = 225$.

11 Find the equations of the asymptotes of $4x^2 - 5y^2 + 20 = 0$.

12 Find the equation of the hyperbola with axes along the coordinate axes and passing through $(-4, -3)$ and $(-6, 5)$.

13 Find the equation of the hyperbola with asymptotes $y = \pm 3x/2$ and vertices $(\pm 4, 0)$.

14 Find the equation of the hyperbola with asymptotes $y = \pm 3x/2$ and vertices $(0, \pm 6)$.

15 Find the equation of the hyperbola with asymptotes $y = \pm 4x/3$ passing through $(-5, 4)$.

16 Find the equation of the hyperbola with center at (0, 0), axes on the coordinate axes, passing through (3, 1) and (9, 5).

17 Find the equation of the hyperbola with center at (0, 0) axes on the coordinate axes, passing through (4, 6) and (1, −3).

18 Using the methods of the calculus, it can be shown that the equation of the line tangent to the hyperbola

$$\frac{x^2}{a^2} - \frac{y^2}{b^2} = 1$$

at the point $P(x_1, y_1)$ is given by

$$\frac{x_1x}{a^2} - \frac{y_1y}{b^2} = 1$$

Find the equation of the line tangent to $x^2 - y^2 = 16$ at the point (−5, 3).

19 Sketch the hyperbolas $x^2/16 - y^2/4 = 1$ and $y^2/4 - x^2/16 = 1$ on the same axes. Find the equations of their asymptotes.

20 Find the equation of the hyperbola with foci (±3, 0) and vertices (±2, 0).

21 Find the equation of the hyperbola with foci (0, ±1) and vertices (0, $\pm\frac{1}{2}$).

22 Using the results of Exercise 18, find the equation of the line tangent to $x^2/36 - y^2/9 = 1$ at the point (10, −4).

23 Find the equation of the hyperbola passing through the point $(2\sqrt{2}, 4)$ and having asymptotes $y = \pm 2x$.

24 An ellipse and a hyperbola have their vertices at the foci of the other. The equation of the ellipse is $x^2/25 + y^2/16 = 1$. Find the equation of the hyperbola and sketch the system.

25 Find the points of intersection of $x^2/36 + y^2/20 = 1$ and $x^2/18 - y^2/10 = 1$. Sketch the system.

12.4
THE HYPERBOLA WITH CENTER $C(h, k)$

Theorem 1 The equation of a hyperbola with center at $C(h, k)$ and foci $(h + c, k)$ and $(h - c, k)$ is

$$\frac{(x-h)^2}{a^2} - \frac{(y-k)^2}{b^2} = 1 \qquad \text{where} \qquad c^2 = a^2 + b^2 \tag{1}$$

The vertices are $(h + a, k)$ and $(h - a, k)$. The equation of transverse axis is $y = k$. The equation of the conjugate axis is $x = h$. The equations of the asymptotes are

$$y = \pm\frac{b}{a}(x - h) + k$$

(See Diagram 20.)

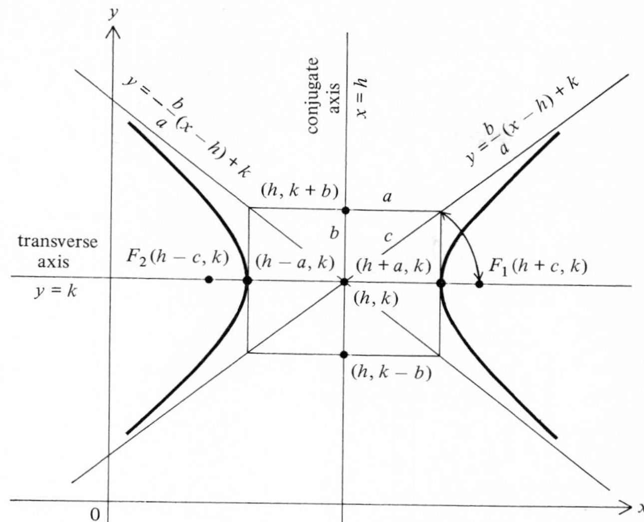

Diagram 20

Theorem 2 The equation of a hyperbola with center at $C(h, k)$ and foci $(h, k + c)$ and $(h, k - c)$ is

$$\frac{(y - k)^2}{a^2} - \frac{(x - h)^2}{b^2} = 1 \quad \text{where} \quad c^2 = a^2 + b^2 \tag{2}$$

The vertices are $(h, k + a)$ and $(h, k - a)$. The equation of the transverse axis is $x = h$. The equation of the conjugate axis is $y = k$. The equations of the asymptotes are

$$y = \pm \frac{a}{b}(x - h) + k$$

(See Diagram 21.)

Example 1 Analyze $\dfrac{(x + 2)^2}{16} - \dfrac{(y - 3)^2}{9} = 1$.

Solution

$$\begin{aligned} a^2 &= 16 \\ a &= 4 \\ b^2 &= 9 \\ b &= 3 \\ c^2 &= 16 + 9 = 25 \\ c &= 5 \end{aligned}$$

The center is $(-2, 3)$; the vertices are $(-2 + 4, 3)$ and $(-2 - 4, 3)$ or $(2, 3)$ and $(-6, 3)$. The foci are $(-2 + 5, 3)$ and $(-2 - 5, 3)$ or $(3, 3)$ and $(-7, 3)$.

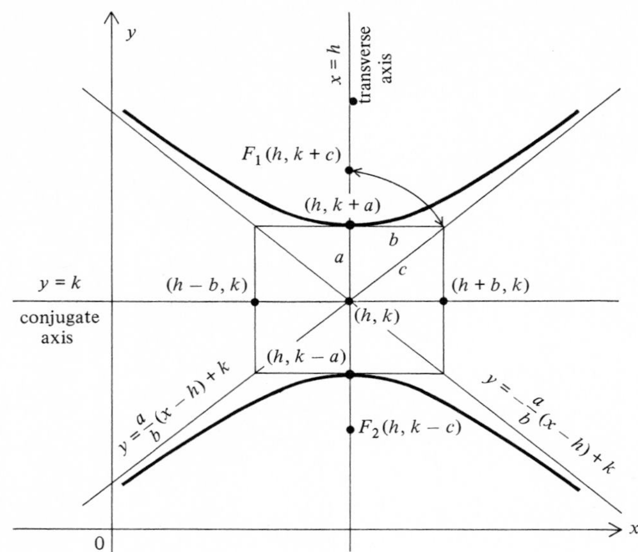

Diagram 21

The asymptotes are

$y = \pm\frac{3}{4}(x + 2) + 3 \Longrightarrow y = \frac{3}{4}x + \frac{9}{2},\ y = -\frac{3}{4}x + \frac{3}{2}$ (See Diagram 22.)

Example 2 Analyze $4y^2 - 16y - 9x^2 - 18x - 29 = 0$.
First express the equation in standard form by completing the square.

$$4(y^2 - 4y \quad) - 9(x^2 + 2x \quad) = 29$$
$$4(y^2 - 4y + 4) - 9(x^2 + 2x + 1) = 29 + 16 - 9$$
$$4(y - 2)^2 - 9(x + 1)^2 = 36$$

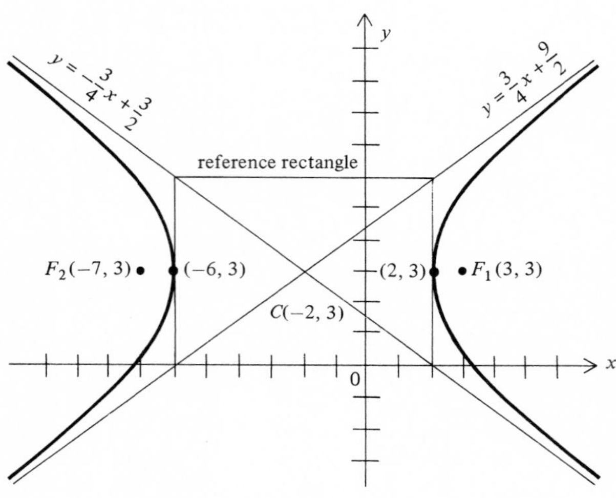

Diagram 22

Divide both sides by 36, forcing the equation into the form specified in Theorem 2.

$$\frac{(y-2)^2}{9} - \frac{(x+1)^2}{4} = 1$$

This is an example of form 2. The center is $C(-1, 2)$. The vertices are $(-1, 2+3)$ and $(-1, 2-3)$ or $(-1, 5)$ and $(-1, -1)$. The end points of the minor axis, since $b = 2$, are $(1, 2)$ and $(-3, 2)$. Since $c^2 = 9 + 4 = 13$ and $c = \sqrt{13}$, the foci are $(-1, 2+\sqrt{13})$ and $(-1, 2-\sqrt{13})$. The equations of the asymptotes are

$$y = \tfrac{3}{2}x + \tfrac{7}{2} \qquad \text{and} \qquad y = -\tfrac{3}{2}x + \tfrac{1}{2}. \qquad \text{(See Diagram 23.)}$$

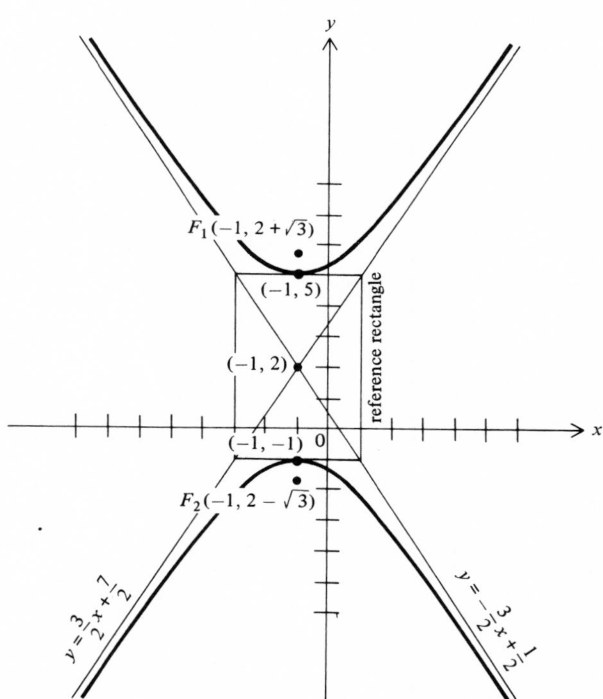

Diagram 23

COMMENT Under certain circumstances, the hyperbola in standard form may appear as

$$\frac{(x-h)^2}{a^2} - \frac{(y-k)^2}{b^2} = 0$$

In this case $\quad \dfrac{(x-h)^2}{a^2} = \dfrac{(y-k)^2}{b^2}$

$$b^2(x-h)^2 = a^2(y-k)^2 \qquad \text{Multiplying both sides by } a^2b^2$$

$$\pm b(x - h) = \pm a(y - k) \qquad \text{Taking the square root of both sides}$$

$$\pm \frac{b}{a}(x - h) + k = y \qquad \text{Solving for } y$$

which represents two lines intersecting at (h, k).

Example 3 Show that $4x^2 - 9y^2 + 16x + 72y - 128 = 0$ represents two intersecting lines.

Solution

$$\begin{aligned} 4(x^2 + 4x \qquad) - 9(y^2 - 8y \qquad) &= 128 \\ 4(x^2 + 4x + 4) - 9(y^2 - 8y + 16) &= 128 + 16 - 144 \\ 4(x + 2)^2 - 9(y - 4)^2 &= 0 \\ 4(x + 2)^2 &= 9(y - 4)^2 \\ \pm 2(x + 2) &= 3(y - 4) \\ \pm \tfrac{2}{3}(x + 2) + 4 &= y \end{aligned}$$

The equations of the lines are

$$y = \tfrac{2}{3}x + \tfrac{16}{3} \qquad \text{and} \qquad y = -\tfrac{2}{3}x + \tfrac{8}{3}$$

intersecting at $(-2, 4)$ (Diagram 24).

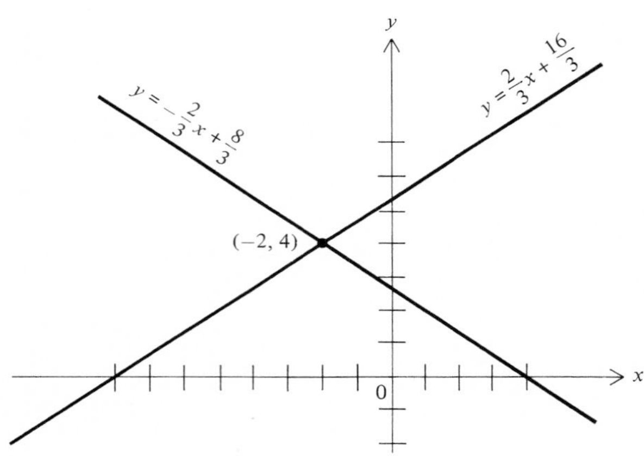

Diagram 24

Example 4 Find the equation of the hyperbola passing through the point $(7, \frac{9}{2})$ with asymptotes $y = \pm\frac{3}{4}(x + 3)$.

Draw the intersecting asymptotes (Diagram 25). They intersect at the center $C(h, k)$. Since $4y = 3x + 9$ and $4y = -3x - 9$ are the required asymptotes, solving simultaneously gives $x = -3$ and $y = 0$. $(-3, 0)$ is the center.

Since $(7, \frac{9}{2})$ is located in quadrant I and positioned as shown in the dia-

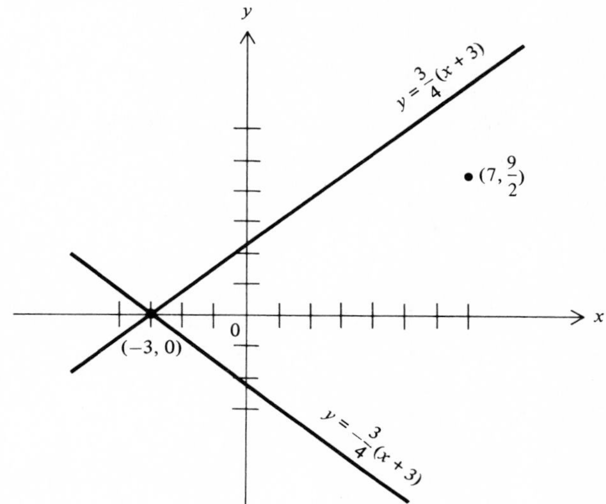

Diagram 25

gram, the major axis of the hyperbola is parallel to the x axis, and its standard form is

$$\frac{(x+3)^2}{a^2} - \frac{y^2}{b^2} = 1$$

The general equation of an asymptote is

$$y = \frac{b}{a}(x - h) + k$$

In this problem $h = -3$, $k = 0$, and $y = (b/a)(x - 3) + 0$. Since we have matched the general equations exactly, we can conclude that $b/a = \frac{3}{4}$.

Caution: This does not necessarily imply that $b = 3$ and $a = 4$. It merely implies that the ratio of b to a is 3 to 4; for example, $b = 12$ and $a = 16$.

To find the exact values of b and a, we note that $b = \frac{3}{4}a$ and $(7, \frac{9}{2})$ must satisfy the equation of the hyperbola. Therefore

$$\frac{(7+3)^2}{a^2} - \frac{(\frac{9}{2})^2}{9a^2/16} = 1$$

$$\frac{100}{a^2} - \frac{324}{9a^2} = 1 \qquad \text{Simplifying terms}$$

$$9a^2 = 576$$

$$a^2 = 64$$

$$a = 8 \qquad \text{Solving for } a \text{ and } b$$

$$b = \tfrac{3}{4}(8) = 6$$

$$\frac{(x+3)^2}{64} - \frac{y^2}{36} = 1 \qquad \text{Solution}$$

EXERCISES

Sketch and analyze each of the hyperbolas in Exercises 1 to 13.

1 $\frac{(x-2)^2}{9} - \frac{(y+3)^2}{16} = 1$

2 $\frac{(x+4)^2}{9} - \frac{(y-2)^2}{4} = 1$

3 $\frac{(y-4)^2}{64} - \frac{(x+5)^2}{36} = 1$

4 $3x^2 - y^2 + 12x = 0$

5 $x^2 - 3y^2 + 6x + 6y + 18 = 0$

6 $x^2 - 4y^2 - 4x = 0$

7 $x^2 - 2x - 4y^2 - 16y + 1 = 0$

8 $y^2 - 4x^2 + 6y + 16x - 11 = 0$

9 $4x^2 - y^2 - 16x + 6y - 29 = 0$

10 $16x^2 - 9y^2 + 96x + 36y - 36 = 0$

11 $x^2 - 4y^2 - 16y - 6x - 23 = 0$

12 $4x^2 - 9y^2 - 36y - 8x - 68 = 0$

13 $36x^2 - 25y^2 + 216x + 100y - 676 = 0$

14 Find an equation of the hyperbola whose foci are $(-3, 2)$ and $(5, 2)$, and vertices $(-1, 2)$ and $(3, 2)$.

15 Find an equation of the hyperbola with transverse axis 8 units long and foci at $(-4, 2)$ $(6, 2)$.

16 Find the equation of the hyperbola passing through $(0, -4)$ with asymptotes $y - 2 = \pm\sqrt{2}(x + 4)$.

17 Find an equation of the hyperbola with conjugate axis 8 units long and foci at $(-4, -1)$ $(6, -1)$.

18 Show that $x^2 - 2y^2 - 6x + 8y + 1 = 0$ is a degenerate hyperbola.

12.5 ROTATION OF AXES

The circle, parabola, ellipse, and hyperbola are all special cases of the *general second-degree equation*

$$Ax^2 + Bxy + Cy^2 + Dx + Ey + F = 0 \tag{1}$$

The terms Ax^2, Bxy, and Cy^2 are called the *second-degree terms.* However, the Bxy term (called the *cross-product* term) has not yet appeared in our analyses. We have examined loci only under the specific conditions that $B = 0$. The summary below shows the existence of each locus when $B = 0$.

$$Ax^2 + Cy^2 + Dx + Ey + F = 0$$

Locus	Conditions
Straight line	$A = C = 0$ (D and E not both zero)
Parabola	A or C (not both) $= 0$
Circle	$A = C \neq 0$
Ellipse	$AC > 0$
Hyperbola	$AC < 0$

In certain cases, each of these loci may *degenerate* into a special case.

Locus	May degenerate into:
Parabola	One line
	Two parallel lines
	No graph
Circle	One point
	No graph
Ellipse	Circle
	One point
	No graph
Hyperbola	Two intersecting lines

QUESTION What is the significance of the Bxy term?

Let us examine $xy = 8$. Since we have not yet developed a method to handle the cross-product term, we simply generate a set of points and graph the results.

x	y	x	y
1	8	-1	-8
2	4	-2	-4
3	$\frac{8}{3}$	-3	$-\frac{8}{3}$
4	2	-4	-2
6	$\frac{4}{3}$	-6	$-\frac{4}{3}$
8	1	-8	-1
16	$\frac{1}{2}$	-16	$-\frac{1}{2}$
64	$\frac{1}{8}$	-64	$-\frac{1}{8}$
800	$\frac{1}{100}$	-800	$-\frac{1}{100}$

(See Diagram 26.)

The graph appears to be a *tilted* hyperbola, with the x and y axes as asymptotes. However, $xy = 8$ does not *fit* one of the standard forms presented in this chapter. The *dark line* (transverse axis of the hyperbola) is *not* parallel to either the x or the y axis. The axis of the hyperbola has been rotated through an angle α. If we examine this graph *as if* this dark line were the x axis, the graph is identical to the graph of $x^2 - y^2 = 16$ (see Diagram 27).

The hyperbola $x^2 - y^2 = 16$ can be expressed as

$$\frac{x^2}{16} - \frac{y^2}{16} = 1$$

and then analyzed using the methods of Section 12.3.

The reason why we have not encountered the Bxy term in previous problems is that at least one of the coordinate axes was always parallel to an axis of symmetry of the curve.

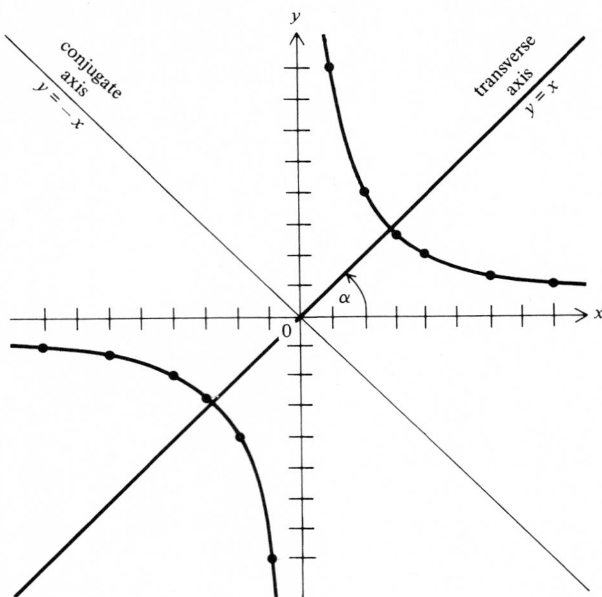

Diagram 26

Graphically, the presence of the Bxy term involves a rotation of the x and y axes. If we were to rotate the axes counterclockwise, through an angle α, every point in the *original* xy system can also be expressed by coordinates in terms of a *new* system, which we call the $x'y'$ system (see Diagram 28).

Although the point P is in the same position in Diagram 28a and b, the coordinates of P depend on which system we refer to.

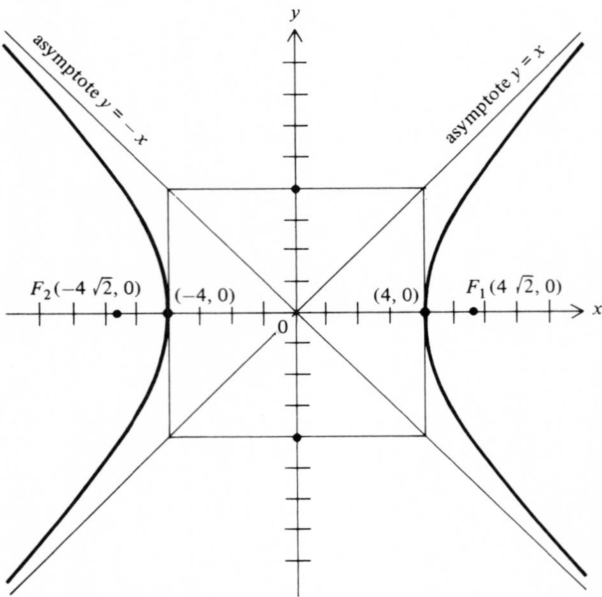

Diagram 27

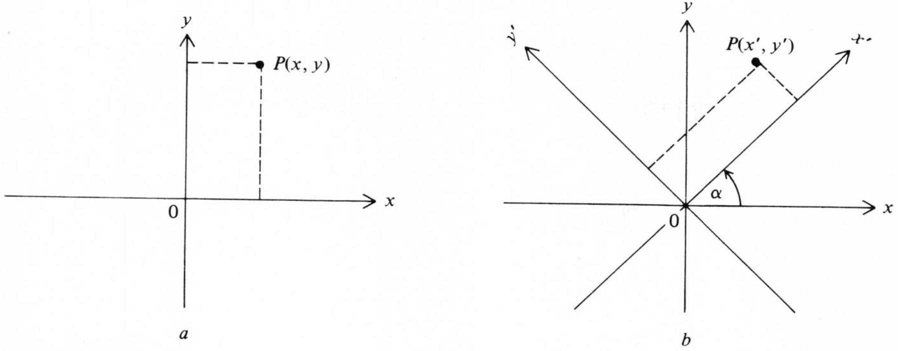

Diagram 28

If we could rotate the original x and y axes the proper angle α so that an axis of symmetry of the curve would be parallel to the rotated axes ($x'y'$ system), then in terms of x' and y', the cross-product term would be missing. If this term is not present, we can analyze the curve using the methods presented in this chapter and Chapter 11. For a graphical explanation, see Diagram 29.

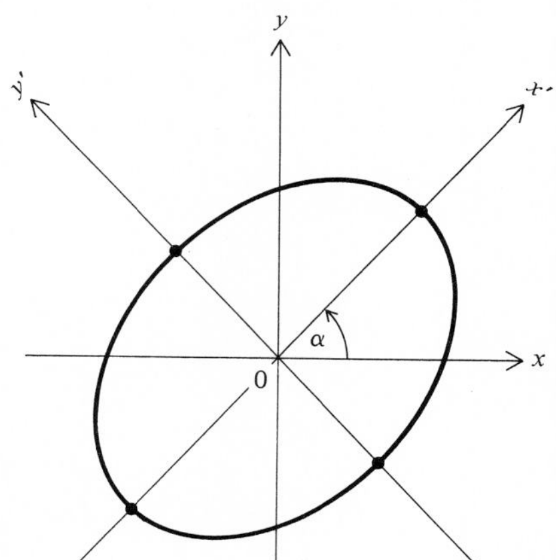

Diagram 29

The ellipse in Diagram 29 is tilted with respect to the x and y axes. The axes of symmetry are *not* parallel to either the x or y axis. Since the equation of this ellipse will possess an xy term, it cannot be analyzed in the usual manner.

If we could determine α, the axes of symmetry would be parallel to the x' and y' axes and the equation of the ellipse would not have a cross-product term in terms of x' and y'.

Let P be a point in quadrant I (see Diagram 30). Let α be an angle of rotation (measured in radians). Draw line segment OP and let β represent

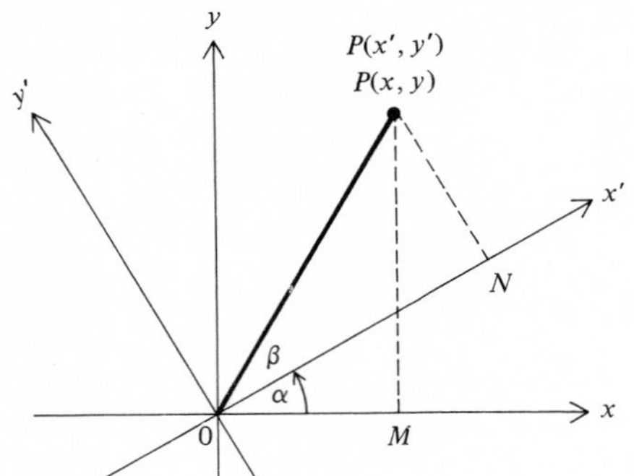

Diagram 30

angle PON. Line PM is drawn perpendicular to the x axis, and PN is drawn perpendicular to the x' axis.

Referring to triangle POM,

$$\cos(\alpha + \beta) = \frac{OM}{OP} \qquad \sin(\alpha + \beta) = \frac{PM}{OP}$$

In terms of x and y, $OM = x$ and $PM = y$. Then

$$x = OP\cos(\alpha + \beta) \qquad y = OP\sin(\alpha + \beta) \tag{2}$$

Referring to triangle PON,

$$\cos\beta = \frac{ON}{OP} \qquad \sin\beta = \frac{PN}{OP}$$

In terms of x' and y', $ON = x'$ and $PN = y'$. Then

$$x' = OP\cos\beta \qquad \text{and} \qquad y' = OP\sin\beta \tag{3}$$

From the formula development in Chapter 9,

$$\begin{aligned} \cos(\alpha + \beta) &= \cos\alpha\cos\beta - \sin\alpha\sin\beta \\ \sin(\alpha + \beta) &= \sin\alpha\cos\beta + \sin\beta\cos\alpha \end{aligned}$$

Using these equations in (2), we find

$$x = OP(\cos\alpha\cos\beta - \sin\alpha\sin\beta) = OP\cos\alpha\cos\beta - OP\sin\alpha\sin\beta$$

Since $\cos\beta = x'/OP$ and $\sin\beta = y'/OP$ [from Equations (3)], we have

$$x = x'\cos\alpha - y'\sin\alpha \tag{4}$$

Similarly, from $y = OP\sin(\alpha + \beta)$, it follows that

$$y = x'\sin\alpha + y'\cos\alpha \tag{5}$$

Equations (4) and (5) are the *formulas* for the rotation of the axes through an angle α.

Example 1 Determine the equation of the parabola

$$x^2 - 2xy + y^2 - 8\sqrt{2}x + 16 = 0$$

when $\alpha = \pi/4$ is the angle of rotation.

Solution Referring to the equations of rotating Equations (4) and (5),

$$x = x' \cos\frac{\pi}{4} - y' \sin\frac{\pi}{4} = \frac{x' - y'}{\sqrt{2}} \qquad y = x' \sin\frac{\pi}{4} + y' \cos\frac{\pi}{4} = \frac{x' + y'}{\sqrt{2}}$$

Substituting these values into the given equation yields

$$\left(\frac{x' - y'}{\sqrt{2}}\right)^2 - 2\left(\frac{x' - y'}{\sqrt{2}}\right)\left(\frac{x' + y'}{\sqrt{2}}\right) + \left(\frac{x' + y'}{\sqrt{2}}\right)^2 - 8\sqrt{2}\left(\frac{x' - y'}{\sqrt{2}}\right) + 16 = 0$$

$$\tfrac{1}{2}(x'^2 - 2x'y' + y'^2) - (x'^2 - y'^2) + \tfrac{1}{2}(x'^2 + 2x'y' + y'^2) - 8x' + 8y' + 16 = 0$$

Combining terms gives

$$2y'^2 + 8y' - 8x' + 16 = 0 \qquad \text{or} \qquad y'^2 + 4y' - 4x' + 8 = 0$$

This equation is a parabola, in the form $y'^2 + Ay' + Bx' + C = 0$ and axis of symmetry parallel to the new x' axis.

COMMENT To simplify your visual interpretation of the rotated graph, rotate the paper an angle α in a *clockwise direction.* The new $x'y'$ axes will appear normal, and your graphical analysis will be considerably simplified.

$$y'^2 + 4y' = 4x' - 8$$
$$y'^2 + 4y' + 4 = 4x' - 8 + 4$$
$$(y' + 2)^2 = 4(x' - 1)$$

The vertex is $(1, -2)$ (in the $x'y'$ system). The equation of the axis of symmetry is $y' = -2$

$4P = 4 \Longrightarrow P = 1$ focus $(2, -2)$ equation of the directrix $x' = 0$

(See Diagram 31.)

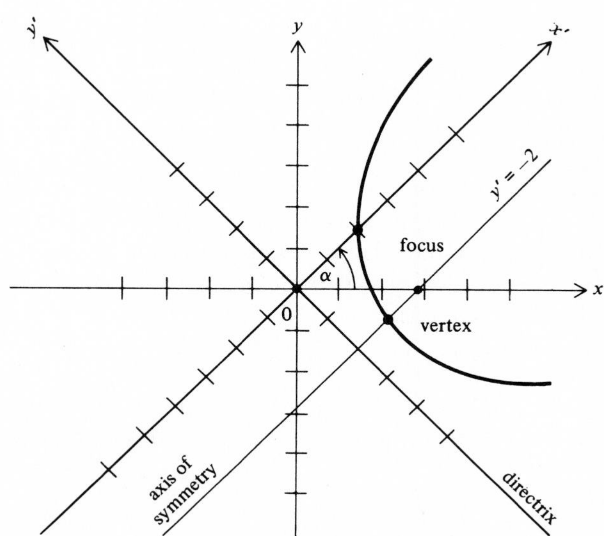

Diagram 31

COMMENT When $\alpha = \pi/4$, the axes are rotated so that the axis of symmetry, $y' = -2$, is parallel to the x' axis. The equation of the parabola in terms of x' and y' *does not contain* a cross-product term, and therefore it was analyzed using the methods of Chapter 11.

The coordinates of the focus, $(2, -2)$, in the $x'y'$ system can be expressed in the original xy system. Using the Equations (4) and (5), we have

$$x = 2\frac{1}{\sqrt{2}} - (-2)\frac{1}{\sqrt{2}} = 2\sqrt{2} \qquad y = 2\frac{1}{\sqrt{2}} + (-2)\frac{1}{\sqrt{2}} = 0$$

Therefore, $(2, -2)$ in the *new* $x'y'$ system corresponds to $(2\sqrt{2}, 0)$ in the *old* xy system.

The coordinates $(1, -2)$ in the $x'y'$ system are equivalent to $(\frac{3}{2}\sqrt{2}, -\sqrt{2}/2)$ in the xy system.

Equations (4) and (5) express x and y in terms of x' and y'. Below, Equations (6) and (7) express x' and y' in terms of x and y:

$$x' = x\cos\alpha + y\sin\alpha \tag{6}$$

$$y' = y\cos\alpha - x\sin\alpha \tag{7}$$

The equation of the directrix of the parabola is $x' = 0$. Using Equation (6), we have

$$0 = x\frac{1}{\sqrt{2}} + y\frac{1}{\sqrt{2}}$$

$$0 = x + y$$

$$y = -x$$

This is the equation of the directrix in the xy system.

The equation of the axis of symmetry of the parabola is $y' = -2$. Using Equation (7), we get

$$-2 = y\frac{1}{\sqrt{2}} - x\frac{1}{\sqrt{2}}$$

$$-2\sqrt{2} = y - x$$

$$y = x - 2\sqrt{2}$$

This is the equation of the axis of symmetry in the xy system.

In the previous example, $\alpha = \pi/4$ just *happened* to be the proper angle of rotation which eliminated the cross-product term $-2xy$. The angle α, through which it is necessary to rotate the axes to eliminate the xy term, can be found in general by analyzing

$$Ax^2 + Bxy + Cy^2 + Dx + Ey + F = 0 \tag{1}$$

If we substitute the values of x and y from the rotation Equations (4) and (5) in (1), we will form another general second-degree equation in the $x'y'$ system:

$$A^*x'^2 + B^*x'y' + C^*y'^2 + D^*x' + E^*y' + F^* = 0 \tag{8}$$

The *new* starred coefficients are related to the *old* coefficients as follows:

$$
\begin{aligned}
A^* &= A \cos^2 \alpha + B \sin \alpha \cos \alpha + C \sin^2 \alpha \\
B^* &= (C - A)\sin 2\alpha + B \cos 2\alpha \\
C^* &= A \sin^2 \alpha - B \sin \alpha \cos \alpha + C \cos^2 \alpha \\
D^* &= D \cos \alpha + E \sin \alpha \\
E^* &= E \cos \alpha - D \sin \alpha \\
F^* &= F
\end{aligned}
\tag{9}
$$

When $B^* = 0$, the cross-product term vanishes.

$$(C - A)\sin 2\alpha + B \cos 2\alpha = 0$$

$$B \cos 2\alpha = -(C - A)\sin 2\alpha$$

$$\frac{B}{A - C} = \frac{\sin 2\alpha}{\cos 2\alpha} \qquad \frac{B}{A - C} = \tan 2\alpha \qquad \text{if } A \neq C$$

or

$$\alpha = \frac{\pi}{4} \qquad \text{if } A = C$$

Note that α is an *acute angle* (counterclockwise rotation).

Example 2 Analyze $x^2 + xy + y^2 = 6$.

Solution $\quad A = 1 \qquad B = 1 \qquad C = 1$

Since $A = C$, $\alpha = \pi/4$ is the angle of rotation. Using Equations (4) and (5), we have

$$x = \frac{x'}{\sqrt{2}} - \frac{y'}{\sqrt{2}} \qquad y = \frac{x'}{\sqrt{2}} + \frac{y}{\sqrt{2}}$$

Substituting into the given equation

$$\left(\frac{x' - y'}{\sqrt{2}}\right)^2 + \frac{(x' - y')}{\sqrt{2}}\frac{(x' + y')}{\sqrt{2}} + \left(\frac{x' + y'}{\sqrt{2}}\right)^2 = 6$$

we obtain

$$\frac{3x'^2}{2} + \frac{y'^2}{2} = 6 \qquad \text{or} \qquad \frac{x'^2}{4} + \frac{y'^2}{12} = 1$$

This curve can be identified as an ellipse with its focus on the y' axis.

$$
\begin{aligned}
a^2 &= 12 & b^2 &= 4 & c^2 &= 12 - 4 = 8 \\
a &= 2\sqrt{3} & b &= 2 & c &= 2\sqrt{2}
\end{aligned}
$$

The center is (0, 0), and the foci are $(0, 2\sqrt{2})$ and $(0, -2\sqrt{2})$ (see Diagram 32).

The foci $(0, 2\sqrt{2})$ and $(0, 2\sqrt{2})$ in the $x'y'$ system correspond to the foci $(-2, 2)$ and $(2, -2)$ in the xy system.

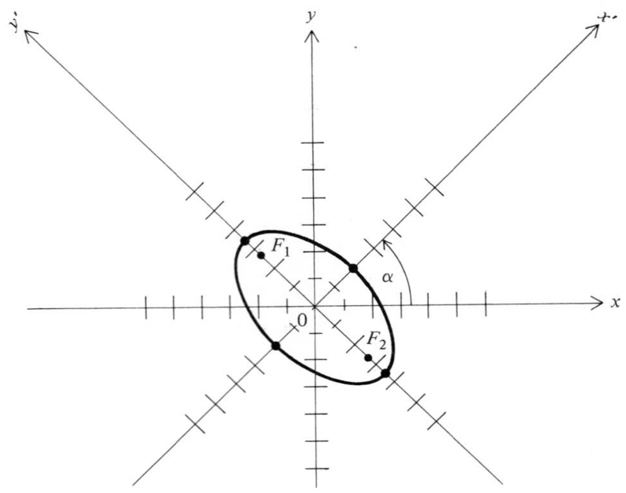

Diagram 32

Example 3 Analyze $2x^2 - \sqrt{3}xy + y^2 = 10$.

Solution

$$A = 2 \qquad B = -\sqrt{3} \qquad C = 1$$

$$\tan 2\alpha = -\frac{\sqrt{3}}{2 - 1} = -\sqrt{3}$$

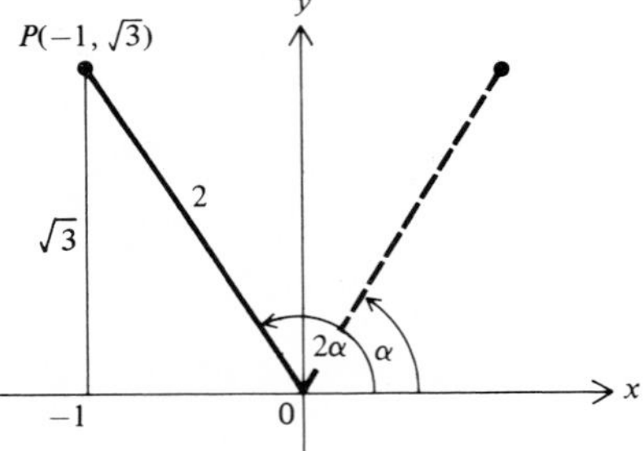

Diagram 33

Since α is an acute angle (see Diagram 33), and $\tan 2\alpha < 0$, then 2α is the second quadrant and since $\tan(2\pi/3) = -\sqrt{3}$, then

$$2\alpha = \frac{2\pi}{3} \Longrightarrow \alpha = \frac{\pi}{3}$$

$$\cos \alpha = \tfrac{1}{2} \qquad \sin \alpha = \frac{\sqrt{3}}{2}$$

Substituting these results into the rotation Equations (4) and (5) gives

$$x = \frac{x'}{2} - \frac{\sqrt{3}y'}{2} \qquad y = \frac{\sqrt{3}x'}{2} + \frac{y'}{2}$$

Replacing x and y in the original equation and simplifying gives

$$\frac{x'^2}{2} + \frac{5y'^2}{2} = 10 \qquad \text{or} \qquad \frac{x'^2}{20} + \frac{y'^2}{4} = 1$$

This equation in x' and y' is an ellipse, with center at the origin. The major axis is along the x' axis.

$$a = 2\sqrt{5} \qquad b = 2$$

$$c^2 = 20 - 4 = 16 \qquad \text{and} \qquad c = 4$$

The coordinates of the foci are (4, 0) and (−4, 0) (see Diagram 34).

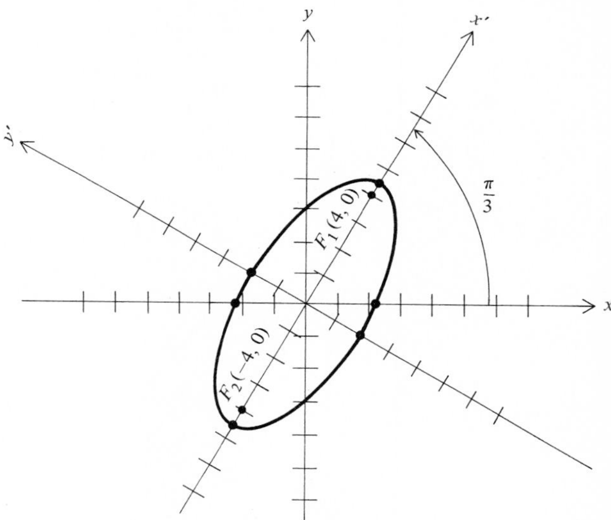

Diagram 34

SUMMARY

1 A second-degree equation in x and y represents a circle, parabola, ellipse, or hyperbola (except for special cases when the locus degenerates into a point, a line, a pair of lines or fails to exist).

2 To graph any second-degree equation, we must first rotate the axes (*if necessary*) to eliminate the cross-product term. Then we must express the equation in one of the standard forms presented in this chapter and Chapter 11.

Theorem 1 The general equation

$$Ax^2 + Bxy + Cy^2 + Dx + Ey + F = 0$$

represents:

(a) A parabola if $B^2 - 4AC = 0$.

(b) An ellipse if $B^2 - 4AC < 0$.

(c) A hyperbola if $B^2 - 4AC > 0$.

(d) One of the *degenerate cases* of the locus.

Example 4 Analyze

$$x^2 - 4xy + y^2 + \sqrt{2}x - 5\sqrt{2}y = 13 \tag{10}$$

STEP 1 Classify the conic.

$$A = 1 \qquad B = -4 \qquad C = 1$$

$$B^2 - 4AC = 16 - 4 = 12 \Longrightarrow \text{hyperbola}$$

STEP 2 Determine angle of rotation. Since $A = C$,

$$\alpha = \frac{\pi}{4}$$

STEP 3 The equations of rotation are

$$x = \frac{x' - y'}{\sqrt{2}} \qquad y = \frac{x' + y'}{\sqrt{2}}$$

STEP 4 Substituting into the given Equation (10), we have

$$3y'^2 - 6y' - x'^2 - 4x' = 13 \tag{11}$$

STEP 5 Expressing Equation (11) in standard form, we get

$$3(y'^2 - 2y' + 1) - (x'^2 + 4x' + 4) = 13 + 3 - 4$$

$$3(y' - 1)^2 - (x' + 2)^2 = 12$$

$$\frac{(y' - 1)^2}{4} - \frac{(x' + 2)^2}{12} = 1 \tag{12}$$

STEP 6 Refer to Section 12.4 and analyze Equation (12):

$$a^2 = 4 \qquad b^2 = 12 \qquad c^2 = a^2 + b^2 = 16$$
$$a = 2 \qquad b = 2\sqrt{3} \qquad c = 4$$

The transverse (major) axis is $x' = -2$. The conjugate (minor) axis is $y' = 1$. The center is $(-2, 1)$; the vertices are $(-2, 3)$ and $(-2, -1)$; and the foci are $(-2, 5)$ and $(-2, -3)$. We find the asymptotes as follows:

$$y' = \pm \frac{1}{\sqrt{3}}(x' - 2) + 1$$

$$y' = \frac{x'}{\sqrt{3}} - \frac{2}{\sqrt{3}} + 1 \qquad y' = \frac{-x'}{\sqrt{3}} + \frac{2}{\sqrt{3}} + 1$$

(See Diagram 35.)

In each of the previous examples, the angle of rotation was one of the common angles, for example, $\pi/4$, $\pi/6$, $\pi/3$, ... However, in order to

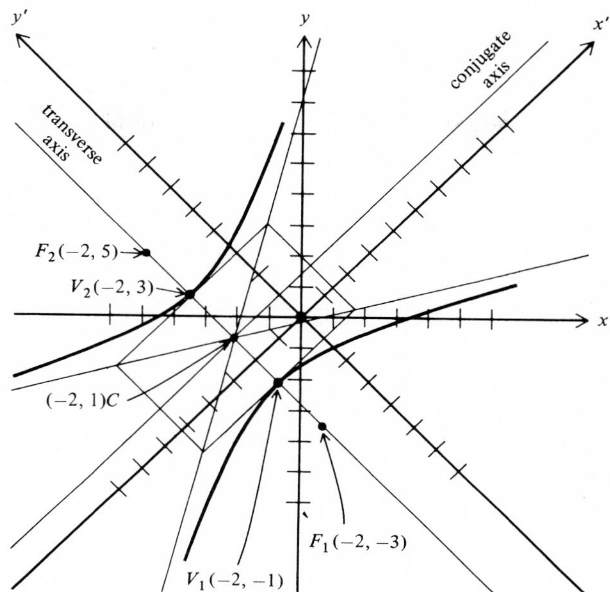

Diagram 35

apply the rotation equations, $x = x' \cos \alpha - y' \sin \alpha$ and $y = x' \sin \alpha + y' \cos \alpha$, it will *not be necessary* to determine the *exact* value of α as long as we can determine $\cos \alpha$ and $\sin \alpha$.

Using the methods of Chapter 9, we can state and prove the following theorem.

Theorem 2 If $0 < \alpha < \pi/2$, then

$$\sin \alpha = \sqrt{\frac{1 - \cos 2\alpha}{2}} \quad \text{and} \quad \cos \alpha = \sqrt{\frac{1 + \cos 2\alpha}{2}}$$

Proof

$$\begin{aligned} \cos 2\alpha &= 1 - 2 \sin^2 \alpha \\ &= 2 \cos^2 \alpha - 1 \end{aligned} \qquad \text{Double-angle formula (Chapter 9)} \qquad (13)$$

$$2 \sin^2 \alpha = 1 - \cos 2\alpha$$

$$\sin^2 \alpha = \frac{1 - \cos 2\alpha}{2} \qquad \text{Solving for } \sin^2 \alpha \qquad (14)$$

$$\sin \alpha = \sqrt{\frac{1 - \cos 2\alpha}{2}} \qquad \text{Solving for } \sin \alpha$$

$$2\cos^2\alpha = 1 + \cos 2\alpha$$

$$\cos^2\alpha = \frac{1 + \cos 2\alpha}{2} \qquad \text{Solving for } \cos^2\alpha$$

$$\cos\alpha = \sqrt{\frac{1 + \cos 2\alpha}{2}} \qquad \text{Solving for } \cos\alpha$$

Since $0 < \alpha < \pi/2$, $\sin\alpha > 0$ and $\cos\alpha > 0$.

We illustrate the usefulness of this theorem in rotation problems in an example.

Example 5 Analyze

$$x^2 + 6xy + 9y^2 - 16\sqrt{10}x + 12\sqrt{10}y - 110 = 0$$

Solution $A = 1 \quad B = 6 \quad C = 9$

$$B^2 - 4AC = 36 - 4(9)(1) = 0 \Longrightarrow \text{parabola}$$

$$\tan 2\alpha = \frac{6}{1 - 9} = -\frac{6}{8} = -\frac{3}{4}$$

Since $\tan 2\alpha < 0$ and $0 < \alpha < \pi/2$, 2α is in quadrant II. Diagram 36 gives a graphical representation of angle 2α. From Diagram 36, it can be determined that $\cos 2\alpha = -\frac{4}{5}$, which is required in using the formulas developed in the previous theorem.

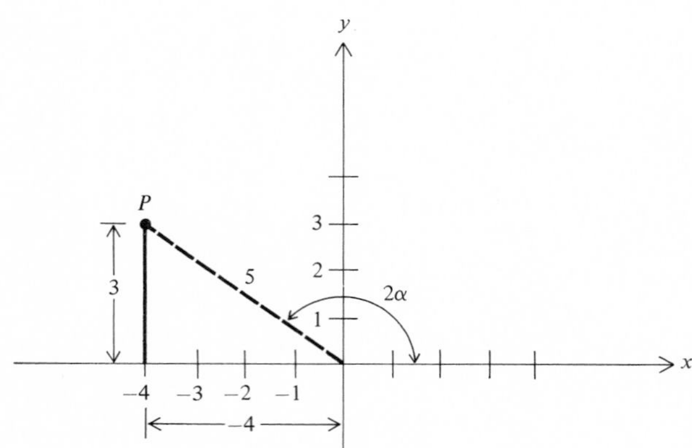

Diagram 36

Returning to the original problem, since $\cos 2\alpha = -\frac{4}{5}$, then

$$\sin\alpha = \sqrt{\frac{1 - (-\frac{4}{5})}{2}} = \sqrt{\frac{9}{10}} = \frac{3}{\sqrt{10}}$$

$$\cos\alpha = \sqrt{\frac{1 + (-\frac{4}{5})}{2}} = \sqrt{\frac{1}{10}} = \frac{1}{\sqrt{10}}$$

Substituting these values into the rotation equations, we obtain

$$x = \frac{x'}{\sqrt{10}} - \frac{3y'}{\sqrt{10}} \qquad \text{and} \qquad y = \frac{3x'}{\sqrt{10}} + \frac{y'}{\sqrt{10}}$$

Substituting the values of x and y into the original equation and simplifying yields

$$10x'^2 - 16x' + 48y' + 36x' + 12y' - 110 = 0$$

Grouping like terms gives

$$10x'^2 + 20x' + 60y' = 110 \qquad \text{or} \qquad x'^2 + 2x' + 6y' - 11 = 0$$

This is a parabola in the form $x'^2 + Ax' + By' + C = 0$. This can now be analyzed using the methods of Chapter 11.

$$\begin{aligned} x'^2 + 2x' &= -6y' + 11 \\ x'^2 + 2x' + 1 &= -6y' + 12 \\ (x' + 1)^2 &= -6(y' - 2) \end{aligned}$$

In terms of the $x'y'$ system the coordinates of the vertex are $(-1, 2)$. Since $4P = -6$, $P = -\frac{3}{2}$. The coordinates of the focus are $(-1, \frac{1}{2})$. The equation of the axis of symmetry is $x' = -1$. The equation of the directrix is $y' = \frac{7}{2}$.

COMMENT The technique presented in Example 5 demonstrates why we need not determine the actual angle of rotation α but merely $\sin\alpha$ and $\cos\alpha$ to be used in the rotation equations. In order to construct the new x' and y' axes, we can refer to the *original xy* system and construct a *reference right triangle.* In the previous example, $\sin\alpha = 3/\sqrt{10}$ and $\cos\alpha = 1/\sqrt{10}$, and angle α is in quadrant I. The reference triangle is constructed as shown in Diagram 37. By extending the hypotenuse in Diagram 37, we determine the x' axis. To construct the y' axis perpendicular to the x' axis we should note that the slope of the x' axis is $\frac{3}{1}$. Therefore, we construct another reference triangle in quadrant II such that the slope of the y' axis is the negative reciprocal, or $-\frac{1}{3}$

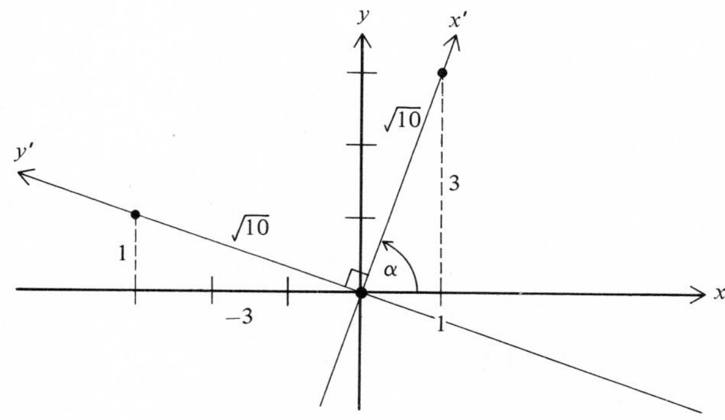

Diagram 37

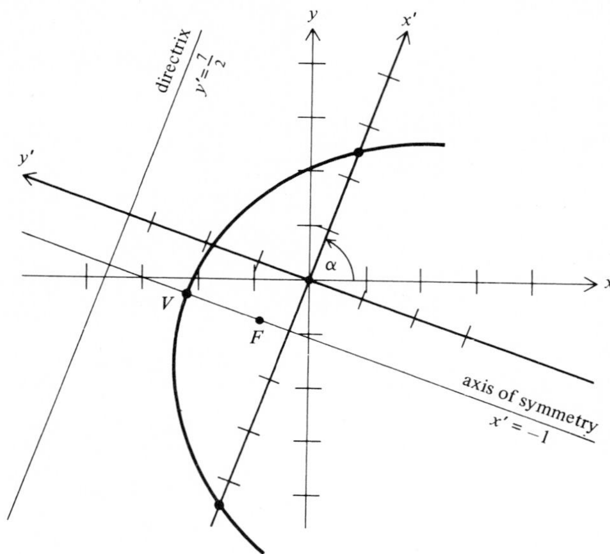

Diagram 38

(see Diagram 37). With this procedure for constructing the new x' and y' axes, Diagram 38 represents the graph of

$$x^2 + 6xy + 9y^2 - 16\sqrt{10}x + 12\sqrt{10}y - 110 = 0$$

EXERCISES

1 Using the methods of this section, show that $2xy = a^2$ in the xy system becomes $x'^2 - y'^2 = a^2$ in the $x'y'$ system.

2 Find the equation of $y = x^2$ after the axes have been rotated $\pi/2$ radians.

3 Determine the equation of $x^2 + y^2 - 4x = 0$ after the axes have been rotated π, $\pi/2$, and $\pi/4$ radians.

4 Determine the equation of $\sqrt{3}y^2 - 2xy - \sqrt{3}x^2 - 4 = 0$ after the axes have been rotated $\pi/3$ radians.

5 Through what angle must the axes be rotated so that $x^2 + xy = .1$ contains *no xy* term?

Simplify each of the equations in Exercises 6 to 20 by rotating the axes and eliminating the xy term. Sketch a graph of each equation showing both the old and new axes.

6 $x^2 + 2xy + y^2 - 8y = 0$

7 $3x^2 - 2\sqrt{3}xy + 5y^2 = 24$

8 $xy - \sqrt{2}y = 2$

9 $6x^2 + 2\sqrt{3}xy + 8y^2 = 45$

10 $7x^2 - 8xy + y^2 = 9$

11 $3x^2 + xy + 3y^2 = 0$

12 $x^2 - 6xy - 6\sqrt{2}x + 10\sqrt{2}y = 2$

13 $2x^2 + 4\sqrt{3}xy + 6y^2 + \sqrt{3}x - y = 0$

14 $73x^2 - 72xy + 52y^2 + 100x - 200y + 100 = 0$

15 $4x^2 + 12xy + 9y^2 + 8\sqrt{13}x + 12\sqrt{13}y - 65 = 0$

16 $16x^2 - 24xy + 9y^2 - 80x - 190y + 425 = 0$

17 $4x^2 - 4xy + y^2 - 8\sqrt{5}x - 16\sqrt{5}y = 0$
18 $x^2 + 2xy - y^2 + 8x + 4y - 8 = 0$
19 $2x^2 + 3xy - 2y^2 = 25$
20 $12x^2 - 3xy + 8y^2 = 3000$
21 Examine each of the following equations under the proper rotation:
(a) $x^2 + 2xy + y^2 = 0$ **(b)** $x^2 + 2xy + y^2 = 8$
(c) $x^2 + 2xy + y^2 = -4$
22 Derive Equations (6) and (7) in this section.
23 Derive the values of A^*, B^*, C^*, D^*, E^*, F^* in Equation (9) of this section.

Appendix Tables

Table 1 VALUES OF e^x AND e^{-x}

x	e^x	e^{-x}	x	e^x	e^{-x}
0.00	1.0000	1.00000	2.10	8.1662	0.12246
0.01	1.0101	0.99005	2.20	9.0250	0.11080
0.02	1.0202	0.98020	2.30	9.9742	0.10026
0.03	1.0305	0.97045	2.40	11.023	0.09072
0.04	1.0408	0.96079	2.50	12.182	0.08208
0.05	1.0513	0.95123	2.60	13.464	0.07427
0.06	1.0618	0.94176	2.70	14.880	0.06721
0.07	1.0725	0.93239	2.80	16.445	0.06081
0.08	1.0833	0.92312	2.90	18.174	0.05502
0.09	1.0942	0.91393	3.00	20.086	0.04979
0.10	1.1052	0.90484	3.10	22.198	0.04505
0.20	1.2214	0.81873	3.20	24.533	0.04076
0.30	1.3499	0.74082	3.30	27.113	0.03688
0.40	1.4918	0.67032	3.40	29.964	0.03337
0.50	1.6487	0.60653	3.50	33.115	0.03020
0.60	1.8221	0.54881	3.60	36.598	0.02732
0.70	2.0138	0.49659	3.70	40.447	0.02472
0.80	2.2255	0.44933	3.80	44.701	0.02237
0.90	2.4596	0.40657	3.90	49.402	0.02024
1.00	2.7183	0.36788	4.00	54.598	0.01832
1.10	3.0042	0.33287	4.10	60.340	0.01657
1.20	3.3201	0.30119	4.20	66.686	0.01500
1.30	3.6693	0.27253	4.30	73.700	0.01357
1.40	4.0552	0.24660	4.40	81.451	0.01228
1.50	4.4817	0.22313	4.50	90.017	0.01111
1.60	4.9530	0.20190	4.60	99.484	0.01005
1.70	5.4739	0.18268	4.70	109.95	0.00910
1.80	6.0496	0.16530	4.80	121.51	0.00823
1.90	6.6859	0.14957	4.90	134.29	0.00745
2.00	7.3891	0.13534	5.00	148.41	0.00674

Table 2 TRIGONOMETRIC FUNCTIONS OF REAL NUMBERS

x	Sin x	Tan x	Cot x	Cos x	x	Sin x	Tan x	Cot x	Cos x
.00	.00000	.00000	∞	1.00000	**.50**	.47943	.54630	1.8305	.87758
.01	.01000	.01000	99.997	0.99995	.51	.48818	.55936	1.7878	.87274
.02	.02000	.02000	49.993	.99980	.52	.49688	.57256	1.7465	.86782
.03	.03000	.03001	33.323	.99955	.53	.50553	.58592	1.7067	.86281
.04	.03999	.04002	24.987	.99920	.54	.51414	.59943	1.6683	.85771
.05	.04998	.05004	19.983	.99875	.55	.52269	.61311	1.6310	.85252
.06	.05996	.06007	16.647	.99820	.56	.53119	.62695	1.5950	.84726
.07	.06994	.07011	14.262	.99755	.57	.53963	.64097	1.5601	.84190
.08	.07991	.08017	12.473	.99680	.58	.54802	.65517	1.5263	.83646
.09	.08988	.09024	11.081	.99595	.59	.55636	.66956	1.4935	.83094
.10	.09983	.10033	9.9666	.99500	**.60**	.56464	.68414	1.4617	.82534
.11	.10978	.11045	9.0542	.99396	.61	.57287	.69892	1.4308	.81965
.12	.11971	.12058	8.2933	.99281	.62	.58104	.71391	1.4007	.81388
.13	.12963	.13074	7.6489	.99156	.63	.58914	.72911	1.3715	.80803
.14	.13954	.14092	7.0961	.99022	.64	.59720	.74454	1.3431	.80210
.15	.14944	.15144	6.6166	.98877	.65	.60519	.76020	1.3154	.79608
.16	.15932	.16138	6.1966	.98723	.66	.61312	.77610	1.2885	.78999
.17	.16918	.17166	5.8256	.98558	.67	.62099	.79225	1.2622	.78382
.18	.17903	.18197	5.4954	.98384	.68	.62879	.80866	1.2366	.77757
.19	.18886	.19232	5.1997	.98200	.69	.63654	.82534	1.2116	.77125
.20	.19867	.20271	4.9332	.98007	**.70**	.64422	.84229	1.1872	.76484
.21	.20846	.21314	4.6917	.97803	.71	.65183	.85953	1.1634	.75836
.22	.21823	.22362	4.4719	.97590	.72	.65938	.87707	1.1402	.75181
.23	.22798	.23414	4.2709	.97367	.73	.66687	.89492	1.1174	.74517
.24	.23770	.24472	4.0864	.97134	.74	.67429	.91309	1.0952	.73847
.25	.24740	.25534	3.9163	.96891	.75	.68164	.93160	1.0734	.73169
.26	.25708	.26602	3.7591	.96639	.76	.68892	.95045	1.0521	.72484
.27	.26673	.27676	3.6133	.96377	.77	.69614	.96967	1.0313	.71791
.28	.27636	.28755	3.4776	.96106	.78	.70328	.98926	1.0109	.71091
.29	.28595	.29841	3.3511	.95824	.79	.71035	1.0092	.99084	.70385
.30	.29552	.30934	3.2327	.95534	**.80**	.71736	1.0296	.97121	.69671
.31	.30506	.32033	3.1218	.95233	.81	.72429	1.0505	.95197	.68950
.32	.31457	.33139	3.0176	.94924	.82	.73115	1.0717	.93309	.68222
.33	.32404	.34252	2.9195	.94604	.83	.73793	1.0934	.91455	.67488
.34	.33349	.35374	2.8270	.94275	.84	.74464	1.1156	.89635	.66746
.35	.34290	.36503	2.7395	.93937	.85	.75128	1.1383	.87848	.65998
.36	.35227	.37640	2.6567	.93590	.86	.75784	1.1616	.86091	.65244
.37	.36162	.38786	2.5782	.93233	.87	.76433	1.1853	.84365	.64483
.38	.37092	.39941	2.5037	.92866	.88	.77074	1.2097	.82668	.63715
.39	.38019	.41105	2.4328	.92491	.89	.77707	1.2346	.80998	.62941
.40	.38942	.42279	2.3652	.92106	**.90**	.78333	1.2602	.79355	.62161
.41	.39861	.43463	2.3008	.91712	.91	.78950	1.2864	.77738	.61375
.42	.40776	.44657	2.2393	.91309	.92	.79560	1.3133	.76146	.60582
.43	.41687	.45862	2.1804	.90897	.93	.80162	1.3409	.74578	.59783
.44	.42594	.47078	2.1241	.90475	.94	.80756	1.3692	.73034	.58979
.45	.43497	.48306	2.0702	.90045	.95	.81342	1.3984	.71511	.58168
.46	.44395	.49545	2.1084	.89605	.96	.81919	1.4284	.70010	.57352
.47	.45289	.50797	1.9686	.89157	.97	.82489	1.4592	.68531	.56530
.48	.46178	.52061	1.9208	.88699	.98	.83050	1.4910	.67071	.55702
.49	.47063	.53339	1.8748	.88233	.99	.83603	1.5237	.65631	.54869
.50	.47943	.54630	1.8305	.87758	**1.00**	.84147	1.5574	.64209	.54030
x	Sin x	Tan x	Cot x	Cos x	x	Sin x	Tan x	Cot x	Cos x

Table 2 TRIGONOMETRIC FUNCTIONS OF REAL NUMBERS (*continued*)

x	Sin x	Tan x	Cot x	Cos x	x	Sin x	Tan x	Cot x	Cos x
1.00	.84147	1.5574	.64209	.54030	**1.50**	.99749	14.101	.07091	.07074
1.01	.84683	1.5922	.62806	.53186	1.51	.99815	16.428	.06087	.06076
1.02	.85211	1.6281	.61420	.52337	1.52	.99871	19.670	.05084	.05077
1.03	.85730	1.6652	.60051	.51482	1.53	.99917	24.498	.04082	.04079
1.04	.86240	1.7036	.58699	.50622	1.54	.99953	32.461	.03081	.03079
1.05	.86742	1.7433	.57362	.49757	1.55	.99978	48.078	.02080	.02079
1.06	.87236	1.7844	.56040	.48887	1.56	.99994	92.621	.01080	.01080
1.07	.87720	1.8270	.54734	.48012	1.57	1.00000	1255.8	.00080	.00080
1.08	.88196	1.8712	.53441	.47133	1.58	.99996	−108.65	−.00920	−.00920
1.09	.88663	1.9171	.52162	.46249	1.59	.99982	−52.067	−.01921	−.01920
1.10	.89121	1.9648	.50897	.45360	**1.60**	.99957	−34.233	−.02921	−.02920
1.11	.89570	2.0143	.49644	.44466	1.61	.99923	−25.495	−.03922	−.03919
1.12	.90010	2.0660	.48404	.43568	1.62	.99879	−20.307	−.04924	−.04918
1.13	.90441	2.1198	.47175	.42666	1.63	.99825	−16.871	−.05927	−.05917
1.14	.90863	2.1759	.45959	.41759	1.64	.99761	−14.427	−.06931	−.06915
1.15	.91276	2.2345	.44753	.40849	1.65	.99687	−12.599	−.07937	−.07912
1.16	.91680	2.2958	.43558	.39934	1.66	.99602	−11.181	−.08944	−.08909
1.17	.92075	2.3600	.42373	.39015	1.67	.99508	−10.047	−.09953	−.09904
1.18	.92461	2.4273	.41199	.38092	1.68	.99404	− 9.1208	−.10964	−.10899
1.19	.92837	2.4979	.40034	.37166	1.69	.99290	− 8.3492	−.11977	−.11892
1.20	.93204	2.5722	.38878	.36236	**1.70**	.99166	− 7.6966	−.12993	−.12884
1.21	.93562	2.6503	.37731	.35302	1.71	.99033	− 7.1373	−.14011	−.13875
1.22	.93910	2.7328	.36593	.34365	1.72	.98889	− 6.6524	−.15032	−.14865
1.23	.94249	2.8198	.35463	.33424	1.73	.98735	− 6.2281	−.16056	−.15853
1.24	.94578	2.9119	.34341	.32480	1.74	.98572	− 5.8535	−.17084	−.16840
1.25	.94898	3.0096	.33227	.31532	1.75	.98399	− 5.5204	−.18115	−.17825
1.26	.95209	3.1133	.32121	.30582	1.76	.98215	− 5.2221	−.19149	−.18808
1.27	.95510	3.2236	.31021	.29628	1.77	.98022	− 4.9534	−.20188	−.19789
1.28	.95802	3.3413	.29928	.28672	1.78	.97820	− 4.7101	−.21231	−.20768
1.29	.96084	3.4672	.28842	.27712	1.79	.97607	− 4.4887	−.22278	−.21745
1.30	.96356	3.6021	.27762	.26750	**1.80**	.97385	− 4.2863	−.23330	−.22720
1.31	.96618	3.7471	.26687	.25785	1.81	.97153	− 4.1005	−.24387	−.23693
1.32	.96872	3.9033	.25619	.24818	1.82	.96911	− 3.9294	−.25449	−.24663
1.33	.97115	4.0723	.24556	.23848	1.83	.96659	− 3.7712	−.26517	−.25631
1.34	.97348	4.2556	.23498	.22875	1.84	.96398	− 3.6245	−.27590	−.26596
1.35	.97572	4.4552	.22446	.21901	1.85	.96128	− 3.4881	−.28669	−.27559
1.36	.97786	4.6734	.21398	.20924	1.86	.95847	− 3.3608	−.29755	−.28519
1.37	.97991	4.9131	.20354	.19945	1.87	.95557	− 3.2419	−.30846	−.29476
1.38	.98185	5.1774	.19315	.18964	1.88	.95258	− 3.1304	−.31945	−.30430
1.39	.98370	5.4707	.18279	.17981	1.89	.94949	− 3.0257	−.33051	−.31381
1.40	.98545	5.7979	.17248	.16997	**1.90**	.94630	− 2.9271	−.34164	−.32329
1.41	.98710	6.1654	.16220	.16010	1.91	.94302	− 2.8341	−.35284	−.33274
1.42	.98865	6.5811	.15195	.15023	1.92	.93965	− 2.7463	−.36413	−.34215
1.43	.99010	7.0555	.14173	.14033	1.93	.93618	− 2.6632	−.37549	−.35153
1.44	.99146	7.6018	.13155	.13042	1.94	.93262	− 2.5843	−.38695	−.36087
1.45	.99271	8.2381	.12139	.12050	1.95	.92896	− 2.5095	−.39849	−.37018
1.46	.99387	8.9886	.11125	.11057	1.96	.92521	− 2.4383	−.41012	−.37945
1.47	.99492	9.8874	.10114	.10063	1.97	.92137	− 2.3705	−.42185	−.38868
1.48	.99588	10.983	.09105	.09067	1.98	.91744	− 2.3058	−.43368	−.39788
1.49	.99674	12.350	.08097	.08071	1.99	.91341	− 2.2441	−.44562	−.40703
1.50	.99749	14.101	.07081	.07074	**2.00**	.90930	− 2.1850	−.45766	−.41615
x	Sin x	Tan x	Cot x	Cos x	x	Sin x	Tan x	Cot x	Cos x

Table 3 COMMON LOGARITHMS (BASE 10)

N.	0	1	2	3	4	5	6	7	8	9
10	0000	0043	0086	0128	0170	0212	0253	0294	0334	0374
11	0414	0453	0492	0531	0569	0607	0645	0682	0719	0755
12	0792	0828	0864	0899	0934	0969	1004	1038	1072	1106
13	1139	1173	1206	1239	1271	1303	1335	1367	1399	1430
14	1461	1492	1523	1553	1584	1614	1644	1673	1703	1732
15	1761	1790	1818	1847	1875	1903	1931	1959	1987	2014
16	2041	2068	2095	2122	2148	2175	2201	2227	2253	2279
17	2304	2330	2355	2380	2405	2430	2455	2480	2504	2529
18	2553	2577	2601	2625	2648	2672	2695	2718	2742	2765
19	2788	2810	2833	2856	2878	2900	2923	2945	2967	2989
20	3010	3032	3054	3075	3096	3118	3139	3160	3181	3201
21	3222	3243	3263	3284	3304	3324	3345	3365	3385	3404
22	3424	3444	3464	3483	3502	3522	3541	3560	3579	3598
23	3617	3636	3655	3674	3692	3711	3729	3747	3766	3784
24	3802	3820	3838	3856	3874	3892	3909	3927	3945	3962
25	3979	3997	4014	4031	4048	4065	4082	4099	4116	4133
26	4150	4166	4183	4200	4216	4232	4249	4265	4281	4298
27	4314	4330	4346	4362	4378	4393	4409	4425	4440	4456
28	4472	4487	4502	4518	4533	4548	4564	4579	4594	4609
29	4624	4639	4654	4669	4683	4698	4713	4728	4742	4757
30	4771	4786	4800	4814	4829	4843	4857	4871	4886	4900
31	4914	4928	4942	4955	4969	4983	4997	5011	5024	5038
32	5051	5065	5079	5092	5105	5119	5132	5145	5159	5172
33	5185	5198	5211	5224	5237	5250	5263	5276	5289	5302
34	5315	5328	5340	5353	5366	5378	5391	5403	5416	5428
35	5441	5453	5465	5478	5490	5502	5514	5527	5539	5551
36	5563	5575	5587	5599	5611	5623	5635	5647	5658	5670
37	5682	5694	5705	5717	5729	5740	5752	5763	5775	5786
38	5798	5809	5821	5832	5843	5855	5866	5877	5888	5899
39	5911	5922	5933	5944	5955	5966	5977	5988	5999	6010
40	6021	6031	6042	6053	6064	6075	6085	6096	6107	6117
41	6128	6138	6149	6160	6170	6180	6191	6201	6212	6222
42	6232	6243	6253	6263	6274	6284	6294	6304	6314	6325
43	6335	6345	6355	6365	6375	6385	6395	6405	6415	6425
44	6435	6444	6454	6464	6474	6484	6493	6503	6513	6522
45	6532	6542	6551	6561	6571	6580	6590	6599	6609	6618
46	6628	6637	6646	6656	6665	6675	6684	6693	6702	6712
47	6721	6730	6739	6749	6758	6767	6776	6785	6794	6803
48	6812	6821	6830	6839	6848	6857	6866	6875	6884	6893
49	6902	6911	6920	6928	6937	6946	6955	6964	6972	6981
50	6990	6998	7007	7016	7024	7033	7042	7050	7059	7067
51	7076	7084	7093	7101	7110	7118	7126	7135	7143	7152
52	7160	7168	7177	7185	7193	7202	7210	7218	7226	7235
53	7243	7251	7259	7267	7275	7284	7292	7300	7308	7316
54	7324	7332	7340	7348	7356	7364	7372	7380	7388	7396
N.	0	1	2	3	4	5	6	7	8	9

Table 3 COMMON LOGARITHMS (BASE 10) (*continued*)

N.	0	1	2	3	4	5	6	7	8	9
55	7404	7412	7419	7427	7435	7443	7451	7459	7466	7474
56	7482	7490	7497	7505	7513	7520	7528	7536	7543	7551
57	7559	7566	7574	7582	7589	7597	7604	7612	7619	7627
58	7634	7642	7649	7657	7664	7672	7679	7686	7694	7701
59	7709	7716	7723	7731	7738	7745	7752	7760	7767	7774
60	7782	7789	7796	7803	7810	7818	7825	7832	7839	7846
61	7853	7860	7868	7875	7882	7889	7896	7903	7910	7917
62	7924	7931	7938	7945	7952	7959	7966	7973	7980	7987
63	7993	8000	8007	8014	8021	8028	8035	8041	8048	8055
64	8062	8069	8075	8082	8089	8096	8102	8109	8116	8122
65	8129	8136	8142	8149	8156	8162	8169	8176	8182	8189
66	8195	8202	8209	8215	8222	8228	8235	8241	8248	8254
67	8261	8267	8274	8280	8287	8293	8299	8306	8312	8319
68	8325	8331	8338	8344	8351	8357	8363	8370	8376	8382
69	8388	8395	8401	8407	8414	8420	8426	8432	8439	8445
70	8451	8457	8463	8470	8476	8482	8488	8494	8500	8506
71	8513	8519	8525	8531	8537	8543	8549	8555	8561	8567
72	8573	8579	8585	8591	8597	8603	8609	8615	8621	8627
73	8633	8639	8645	8651	8657	8663	8669	8675	8681	8686
74	8692	8698	8704	8710	8716	8722	8727	8733	8739	8745
75	8751	8756	8762	8768	8774	8779	8785	8791	8797	8802
76	8808	8814	8820	8825	8831	8837	8842	8848	8854	8859
77	8865	8871	8876	8882	8887	8893	8899	8904	8910	8915
78	8921	8927	8932	8938	8943	8949	8954	8960	8965	8971
79	8976	8982	8987	8993	8998	9004	9009	9015	9020	9025
80	9031	9036	9042	9047	9053	9058	9063	9069	9074	9079
81	9085	9090	9096	9101	9106	9112	9117	9122	9128	9133
82	9138	9143	9149	9154	9159	9165	9170	9175	9180	9186
83	9191	9196	9201	9206	9212	9217	9222	9227	9232	9238
84	9243	9248	9253	9258	9263	9269	9274	9279	9284	9289
85	9294	9299	9304	9309	9315	9320	9325	9330	9335	9340
86	9345	9350	9355	9360	9365	9370	9375	9380	9385	9390
87	9395	9400	9405	9410	9415	9420	9425	9430	9435	9440
88	9445	9450	9455	9460	9465	9469	9474	9479	9484	9489
89	9494	9499	9504	9509	9513	9518	9523	9528	9533	9538
90	9542	9547	9552	9557	9562	9566	9571	9576	9581	9586
91	9590	9595	9600	9605	9609	9614	9619	9624	9628	9633
92	9638	9643	9647	9652	9657	9661	9666	9671	9675	9680
93	9685	9689	9694	9699	9703	9708	9713	9717	9722	9727
94	9731	9736	9741	9745	9750	9754	9759	9763	9768	9773
95	9777	9782	9786	9791	9795	9800	9805	9809	9814	9818
96	9823	9827	9832	9836	9841	9845	9850	9854	9859	9863
97	9868	9872	9877	9881	9886	9890	9894	9899	9903	9908
98	9912	9917	9921	9926	9930	9934	9939	9943	9948	9952
99	9956	9961	9965	9969	9974	9978	9983	9987	9991	9996
N.	0	1	2	3	4	5	6	7	8	9

Table 4 NATURAL LOGARITHMS (BASE *e*)

	.00	.01	.02	.03	.04	.05	.06	.07	.08	.09
1.0	0.0000	0.0100	0.0198	0.0296	0.0392	0.0488	0.0583	0.0677	0.0770	0.0862
1.1	0.0953	0.1044	0.1133	0.1222	0.1310	0.1398	0.1484	0.1570	0.1655	0.1740
1.2	0.1823	0.1906	0.1989	0.2070	0.2151	0.2231	0.2311	0.2390	0.2469	0.2546
1.3	0.2624	0.2700	0.2776	0.2852	0.2927	0.3001	0.3075	0.3148	0.3221	0.3293
1.4	0.3365	0.3436	0.3507	0.3577	0.3646	0.3716	0.3784	0.3853	0.3920	0.3988
1.5	0.4055	0.4121	0.4187	0.4253	0.4318	0.4383	0.4447	0.4511	0.4574	0.4637
1.6	0.4700	0.4762	0.4824	0.4886	0.4947	0.5008	0.5068	0.5128	0.5188	0.5247
1.7	0.5306	0.5365	0.5423	0.5481	0.5539	0.5596	0.5653	0.5710	0.5766	0.5822
1.8	0.5878	0.5933	0.5988	0.6043	0.6098	0.6152	0.6206	0.6259	0.6313	0.6366
1.9	0.6419	0.6471	0.6523	0.6575	0.6627	0.6678	0.6729	0.6780	0.6831	0.6881
2.0	0.6932	0.6981	0.7031	0.7080	0.7129	0.7178	0.7227	0.7275	0.7324	0.7372
2.1	0.7419	0.7467	0.7514	0.7561	0.7608	0.7655	0.7701	0.7747	0.7793	0.7839
2.2	0.7885	0.7930	0.7975	0.8020	0.8065	0.8109	0.8154	0.8198	0.8242	0.8286
2.3	0.8329	0.8373	0.8416	0.8459	0.8502	0.8544	0.8587	0.8629	0.8671	0.8713
2.4	0.8755	0.8796	0.8838	0.8879	0.8920	0.8961	0.9002	0.9042	0.9083	0.9123
2.5	0.9163	0.9203	0.9243	0.9282	0.9322	0.9361	0.9400	0.9439	0.9478	0.9517
2.6	0.9555	0.9594	0.9632	0.9670	0.9708	0.9746	0.9783	0.9821	0.9858	0.9895
2.7	0.9933	0.9969	1.0006	1.0043	1.0080	1.0116	1.0152	1.0188	1.0225	1.0260
2.8	1.0296	1.0332	1.0367	1.0403	1.0438	1.0473	1.0508	1.0543	1.0578	1.0613
2.9	1.0647	1.0682	1.0716	1.0750	1.0784	1.0818	1.0852	1.0886	1.0919	1.0953
3.0	1.0986	1.1019	1.1053	1.1086	1.1119	1.1151	1.1184	1.1217	1.1249	1.1282
3.1	1.1314	1.1346	1.1378	1.1410	1.1442	1.1474	1.1506	1.1537	1.1569	1.1600
3.2	1.1632	1.1663	1.1694	1.1725	1.1756	1.1787	1.1817	1.1848	1.1878	1.1909
3.3	1.1939	1.1969	1.2000	1.2030	1.2060	1.2090	1.2119	1.2149	1.2179	1.2208
3.4	1.2238	1.2267	1.2296	1.2326	1.2355	1.2384	1.2413	1.2442	1.2470	1.2499
3.5	1.2528	1.2556	1.2585	1.2613	1.2641	1.2669	1.2698	1.2726	1.2754	1.2782
3.6	1.2809	1.2837	1.2865	1.2892	1.2920	1.2947	1.2975	1.3002	1.3029	1.3056
3.7	1.3083	1.3110	1.3137	1.3164	1.3191	1.3218	1.3244	1.3271	1.3297	1.3324
3.8	1.3350	1.3376	1.3403	1.3429	1.3455	1.3481	1.3507	1.3533	1.3558	1.3584
3.9	1.3610	1.3635	1.3661	1.3686	1.3712	1.3737	1.3762	1.3788	1.3813	1.3838
4.0	1.3863	1.3888	1.3913	1.3938	1.3962	1.3987	1.4012	1.4036	1.4061	1.4085
4.1	1.4110	1.4134	1.4159	1.4183	1.4207	1.4231	1.4255	1.4279	1.4303	1.4327
4.2	1.4351	1.4375	1.4398	1.4422	1.4446	1.4469	1.4493	1.4516	1.4540	1.4563
4.3	1.4586	1.4609	1.4633	1.4656	1.4679	1.4702	1.4725	1.4748	1.4771	1.4793
4.4	1.4816	1.4839	1.4861	1.4884	1.4907	1.4929	1.4951	1.4974	1.4996	1.5019
4.5	1.5041	1.5063	1.5085	1.5107	1.5129	1.5151	1.5173	1.5195	1.5217	1.5239
4.6	1.5261	1.5282	1.5304	1.5326	1.5347	1.5369	1.5390	1.5412	1.5433	1.5454
4.7	1.5476	1.5497	1.5518	1.5539	1.5560	1.5581	1.5602	1.5623	1.5644	1.5665
4.8	1.5686	1.5707	1.5728	1.5748	1.5769	1.5790	1.5810	1.5831	1.5851	1.5872
4.9	1.5892	1.5913	1.5933	1.5953	1.5974	1.5994	1.6014	1.6034	1.6054	1.6074
5.0	1.6094	1.6114	1.6134	1.6154	1.6174	1.6194	1.6214	1.6233	1.6253	1.6273
5.1	1.6292	1.6312	1.6332	1.6351	1.6371	1.6390	1.6409	1.6429	1.6448	1.6467
5.2	1.6487	1.6506	1.6525	1.6544	1.6563	1.6582	1.6601	1.6620	1.6639	1.6658
5.3	1.6677	1.6696	1.6715	1.6734	1.6752	1.6771	1.6790	1.6808	1.6827	1.6845
5.4	1.6864	1.6882	1.6901	1.6919	1.6938	1.6956	1.6974	1.6993	1.7011	1.7029

Table 4 NATURAL LOGARITHMS (BASE e) (*continued*)

	.00	.01	.02	.03	.04	.05	.06	.07	.08	.09
5.5	1.7047	1.7066	1.7084	1.7102	1.7120	1.7138	1.7156	1.7174	1.7192	1.7210
5.6	1.7228	1.7246	1.7263	1.7281	1.7299	1.7317	1.7334	1.7352	1.7370	1.7387
5.7	1.7405	1.7422	1.7440	1.7457	1.7475	1.7492	1.7509	1.7527	1.7544	1.7561
5.8	1.7579	1.7596	1.7613	1.7630	1.7647	1.7664	1.7681	1.7699	1.7716	1.7733
5.9	1.7750	1.7766	1.7783	1.7800	1.7817	1.7834	1.7851	1.7868	1.7884	1.7901
6.0	1.7918	1.7934	1.7951	1.7967	1.7984	1.8001	1.8017	1.8034	1.8050	1.8066
6.1	1.8083	1.8099	1.8116	1.8132	1.8148	1.8165	1.8181	1.8197	1.8213	1.8229
6.2	1.8245	1.8262	1.8278	1.8294	1.8310	1.8326	1.8342	1.8358	1.8374	1.8390
6.3	1.8405	1.8421	1.8437	1.8453	1.8469	1.8485	1.8500	1.8516	1.8532	1.8547
6.4	1.8563	1.8579	1.8594	1.8610	1.8625	1.8641	1.8656	1.8672	1.8687	1.8703
6.5	1.8718	1.8733	1.8749	1.8764	1.8779	1.8795	1.8810	1.8825	1.8840	1.8856
6.6	1.8871	1.8886	1.8901	1.8916	1.8931	1.8946	1.8961	1.8976	1.8991	1.9006
6.7	1.9021	1.9036	1.9051	1.9066	1.9081	1.9095	1.9110	1.9125	1.9140	1.9155
6.8	1.9169	1.9184	1.9199	1.9213	1.9228	1.9242	1.9257	1.9272	1.9286	1.9301
6.9	1.9315	1.9330	1.9344	1.9359	1.9373	1.9387	1.9402	1.9416	1.9430	1.9445
7.0	1.9459	1.9473	1.9488	1.9502	1.9516	1.9530	1.9544	1.9559	1.9573	1.9587
7.1	1.9601	1.9615	1.9629	1.9643	1.9657	1.9671	1.9685	1.9699	1.9713	1.9727
7.2	1.9741	1.9755	1.9769	1.9782	1.9796	1.9810	1.9824	1.9838	1.9851	1.9865
7.3	1.9879	1.9892	1.9906	1.9920	1.9933	1.9947	1.9961	1.9974	1.9988	2.0001
7.4	2.0015	2.0028	2.0042	2.0055	2.0069	2.0082	2.0096	2.0109	2.0122	2.0136
7.5	2.0149	2.0162	2.0176	2.0189	2.0202	2.0215	2.0229	2.0242	2.0255	2.0268
7.6	2.0281	2.0295	2.0308	2.0321	2.0334	2.0347	2.0360	2.0373	2.0386	2.0399
7.7	2.0412	2.0425	2.0438	2.0451	2.0464	2.0477	2.0490	2.0503	2.0516	2.0528
7.8	2.0541	2.0554	2.0567	2.0580	2.0592	2.0605	2.0618	2.0631	2.0643	2.0656
7.9	2.0669	2.0681	2.0694	2.0707	2.0719	2.0732	2.0744	2.0757	2.0769	2.0782
8.0	2.0794	2.0807	2.0819	2.0832	2.0844	2.0857	2.0869	2.0882	2.0894	2.0906
8.1	2.0919	2.0931	2.0943	2.0956	2.0968	2.0980	2.0992	2.1005	2.1017	2.1029
8.2	2.1041	2.1054	2.1066	2.1078	2.1090	2.1102	2.1114	2.1126	2.1138	2.1150
8.3	2.1163	2.1175	2.1187	2.1199	2.1211	2.1223	2.1235	2.1247	2.1259	2.1270
8.4	2.1282	2.1294	2.1306	2.1318	2.1330	2.1342	2.1353	2.1365	2.1377	2.1389
8.5	2.1401	2.1412	2.1424	2.1436	2.1448	2.1459	2.1471	2.1483	2.1494	2.1506
8.6	2.1518	2.1529	2.1541	2.1552	2.1564	2.1576	2.1587	2.1599	2.1610	2.1622
8.7	2.1633	2.1645	2.1656	2.1668	2.1679	2.1691	2.1702	2.1713	2.1725	2.1736
8.8	2.1748	2.1759	2.1770	2.1782	2.1793	2.1804	2.1815	2.1827	2.1838	2.1849
8.9	2.1861	2.1872	2.1883	2.1894	2.1905	2.1917	2.1928	2.1939	2.1950	2.1961
9.0	2.1972	2.1983	2.1994	2.2006	2.2017	2.2028	2.2039	2.2050	2.2061	2.2072
9.1	2.2083	2.2094	2.2105	2.2116	2.2127	2.2138	2.2148	2.2159	2.2170	2.2181
9.2	2.2192	2.2203	2.2214	2.2225	2.2235	2.2246	2.2257	2.2268	2.2279	2.2289
9.3	2.2300	2.2311	2.2322	2.2332	2.2343	2.2354	2.2364	2.2375	2.2386	2.2396
9.4	2.2407	2.2418	2.2428	2.2439	2.2450	2.2460	2.2471	2.2481	2.2492	2.2502
9.5	2.2513	2.2523	2.2534	2.2544	2.2555	2.2565	2.2576	2.2586	2.2597	2.2607
9.6	2.2618	2.2628	2.2638	2.2649	2.2659	2.2670	2.2680	2.2690	2.2701	2.2711
9.7	2.2721	2.2732	2.2742	2.2752	2.2762	2.2773	2.2783	2.2793	2.2803	2.2814
9.8	2.2824	2.2834	2.2844	2.2854	2.2865	2.2875	2.2885	2.2895	2.2905	2.2915
9.9	2.2925	2.2935	2.2946	2.2956	2.2966	2.2976	2.2986	2.2996	2.3006	2.3016

Table 4 NATURAL LOGARITHMS (BASE e) (*continued*)

N	Nat Log	N	Nat Log	N	Nat Log	N	Nat Log	N	Nat Log
0	$-\infty$	40	3.68 888	80	4.38 203	120	4.78 749	160	5.07 517
1	0.00 000	41	3.71 357	81	4.39 445	121	4.79 579	161	5.08 140
2	0.69 315	42	3.73 767	82	4.40 672	122	4.80 402	162	5.08 760
3	1.09 861	43	3.76 120	83	4.41 884	123	4.81 218	163	5.09 375
4	1.38 629	44	3.78 419	84	4.43 082	124	4.82 028	164	5.09 987
5	1.60 944	45	3.80 666	85	4.44 265	125	4.82 831	165	5.10 595
6	1.79 176	46	3.82 864	86	4.45 435	126	4.83 628	166	5.11 199
7	1.94 591	47	3.85 015	87	4.46 591	127	4.84 419	167	5.11 799
8	2.07 944	48	3.87 120	88	4.47 734	128	4.85 203	168	5.12 396
9	2.19 722	49	3.89 182	89	4.48 864	129	4.85 981	169	5.12 990
10	2.30 259	50	3.91 202	90	4.49 981	130	4.86 753	170	5.13 580
11	2.39 790	51	3.93 183	91	4.51 086	131	4.87 520	171	5.14 166
12	2.48 491	52	3.95 124	92	4.52 179	132	4.88 280	172	5.14 749
13	2.56 495	53	3.97 029	93	4.53 260	133	4.89 035	173	5.15 329
14	2.63 906	54	3.98 898	94	4.54 329	134	4.89 784	174	5.15 906
15	2.70 805	55	4.00 733	95	4.55 388	135	4.90 527	175	5.16 479
16	2.77 259	56	4.02 535	96	4.56 435	136	4.91 265	176	5.17 048
17	2.83 321	57	4.04 305	97	4.57 471	137	4.91 998	177	5.17 615
18	2.89 037	58	4.06 044	98	4.58 497	138	4.92 725	178	5.18 178
19	2.94 444	59	4.07 754	99	4.59 512	139	4.93 447	179	5.18 739
20	2.99 573	60	4.09 434	100	4.60 517	140	4.94 164	180	5.19 296
21	3.04 452	61	4.11 087	101	4.61 512	141	4.94 876	181	5.19 850
22	3.09 104	62	4.12 713	102	4.62 497	142	4.95 583	182	5.20 401
23	3.13 549	63	4.14 313	103	4.63 473	143	4.96 284	183	5.20 949
24	3.17 805	64	4.15 888	104	4.64 439	144	4.96 981	184	5.21 494
25	3.21 888	65	4.17 439	105	4.65 396	145	4.97 673	185	5.22 036
26	3.25 810	66	4.18 965	106	4.66 344	146	4.98 361	186	5.22 575
27	3.29 584	67	4.20 469	107	4.67 283	147	4.99 043	187	5.23 111
28	3.33 220	68	4.21 951	108	4.68 213	148	4.99 721	188	5.23 644
29	3.36 730	69	4.23 411	109	4.69 135	149	5.00 395	189	5.24 175
30	3.40 120	70	4.24 850	110	4.70 048	150	5.01 064	190	5.24 702
31	3.43 399	71	4.26 268	111	4.70 953	151	5.01 728	191	5.25 227
32	3.46 574	72	4.27 667	112	4.71 850	152	5.02 388	192	5.25 750
33	3.49 651	73	4.29 046	113	4.72 739	153	5.03 044	193	5.26 269
34	3.52 636	74	4.30 407	114	4.73 620	154	5.03 695	194	5.26 786
35	3.55 535	75	4.31 749	115	4.74 493	155	5.04 343	195	5.27 300
36	3.58 352	76	4.33 073	116	4.75 359	156	5.04 986	196	5.27 811
37	3.61 092	77	4.34 381	117	4.76 217	157	5.05 625	197	5.28 320
38	3.63 759	78	4.35 671	118	4.77 068	158	5.06 260	198	5.28 827
39	3.66 356	79	4.36 945	119	4.77 912	159	5.06 890	199	5.29 330
40	3.68 888	80	4.38 203	120	4.78 749	160	5.07 517	200	5.29 832

Table 5 POWERS OF 2 AND 3

x	2^x	2^{-x}	3^x	3^{-x}
0.05	1.035	0.966	1.056	0.946
0.10	1.072	0.933	1.116	0.896
0.20	1.149	0.871	1.246	0.803
0.30	1.231	0.812	1.390	0.719
0.40	1.320	0.758	1.552	0.644
0.50	1.414	0.707	1.732	0.577
0.60	1.516	0.660	1.933	0.517
0.70	1.625	0.616	2.158	0.463
0.80	1.741	0.574	2.408	0.415
0.90	1.866	0.536	2.688	0.372
1.00	2.000	0.500	3.000	0.333
1.10	2.144	0.467	3.348	0.299
1.20	2.297	0.435	3.737	0.266
1.30	2.462	0.406	4.171	0.240
1.40	2.639	0.379	4.656	0.215
1.50	2.828	0.354	5.196	0.192
1.60	3.031	0.330	5.800	0.172
1.70	3.249	0.308	6.473	0.154
1.80	3.482	0.287	7.225	0.138
1.90	3.732	0.268	8.064	0.124
2.00	4.000	0.250	9.000	0.111
2.10	4.287	0.233	10.045	0.096
2.20	4.595	0.218	11.212	0.089
2.30	4.925	0.203	12.514	0.080
2.40	5.278	0.189	13.967	0.072
2.50	5.657	0.177	15.589	0.064
2.60	6.063	0.165	17.399	0.057
2.70	6.498	0.154	19.419	0.051
2.80	6.964	0.144	21.674	0.046
2.90	7.464	0.134	24.191	0.041
3.00	8.000	0.125	27.000	0.037
3.50	11.314	0.088	46.765	0.021
4.00	16.000	0.063	81.000	0.012
4.50	22.627	0.044	140.296	0.007
5.00	32.000	0.031	243.000	0.004

Table 6 TRIGONOMETRIC FUNCTIONS $\theta°$

→	Sin	Cos	Tan	Cot	Sec	Csc	
0°00′	.0000	1.000	.0000	——	1.000	——	**90°00′**
10′	029	000	029	343.8	000	343.8	89°50′
20′	058	000	058	171.9	000	171.9	40′
30′	.0087	1.000	.0087	114.6	1.000	114.6	30′
40′	116	.9999	116	85.94	000	85.95	20′
0°50′	145	999	145	68.75	000	68.76	10′
1°00′	.0175	.9998	.0175	57.29	1.000	57.30	**89°00′**
10′	204	998	204	49.10	000	49.11	88°50′
20′	233	997	233	42.96	000	42.98	40′
30′	.0262	.9997	.0262	38.19	1.000	38.20	30′
40′	291	996	291	34.37	000	34.38	20′
1°50′	320	995	320	31.24	001	31.26	10′
2°00′	.0349	.9994	.0349	28.64	1.001	28.65	**88°00′**
10′	378	993	378	26.43	001	26.45	87°50′
20′	407	992	407	24.54	001	24.56	40′
30′	.0436	.9990	.0437	22.90	1.001	22.93	30′
40′	465	989	466	21.47	001	21.49	20′
2°50′	494	988	495	20.21	001	20.23	10′
3°00′	.0523	.9986	.0524	19.08	1.001	19.11	**87°00′**
10′	552	985	553	18.07	002	18.10	86°50′
20′	581	983	582	17.17	002	17.20	40′
30′	.0610	.9981	.0612	16.35	1.002	16.38	30′
40′	640	980	641	15.60	002	15.64	20′
3°50′	669	978	670	14.92	002	14.96	10′
4°00′	.0698	.9976	.0699	14.30	1.002	14.34	**86°00′**
10′	727	974	729	13.73	003	13.76	85°50′
20′	756	971	758	13.20	003	13.23	40′
30′	.0785	.9969	.0787	12.71	1.003	12.75	30′
40′	814	967	816	12.25	003	12.29	20′
4°50′	843	964	846	11.83	004	11.87	10′
5°00′	.0872	.9962	.0875	11.43	1.004	11.47	**85°00′**
10′	901	959	904	11.06	004	11.10	84°50′
20′	929	957	934	10.71	004	10.76	40′
30′	.0958	.9954	.0963	10.39	1.005	10.43	30′
40′	.0987	951	.0992	10.08	005	10.13	20′
5°50′	.1016	948	.1022	9.788	005	9.839	10′
6°00′	.1045	.9945	.1051	9.514	1.006	9.567	**84°00′**
	Cos	Sin	Cot	Tan	Csc	Sec	←

Table 6 TRIGONOMETRIC FUNCTIONS θ° (*continued*)

⟶	Sin	Cos	Tan	Cot	Sec	Csc	
6°00′	.1045	.9945	.1051	9.514	1.006	9.567	**84°00′**
10′	074	942	080	255	006	309	83°50′
20′	103	939	110	9.010	006	9.065	40′
30′	.1132	.9936	.1139	8.777	1.006	8.834	30′
40′	161	932	169	556	007	614	20′
6°50′	190	929	198	345	007	405	10′
7°00′	.1219	.9925	.1228	8.144	1.008	8.206	**83°00′**
10′	248	922	257	7.953	008	8.016	82°50′
20′	276	918	287	770	008	7.834	40′
30′	.1305	.9914	.1317	7.596	1.009	7.661	30′
40′	334	911	346	429	009	496	20′
7°50′	363	907	376	269	009	337	10′
8°00′	.1392	.9903	.1405	7.115	1.010	7.185	**82°00′**
10′	421	899	435	6.968	010	7.040	81°50′
20′	449	894	465	827	011	6.900	40′
30′	.1478	.9890	.1495	6.691	1.011	6.765	30′
40′	507	886	524	561	012	636	20′
8°50′	536	881	554	435	012	512	10′
9°00′	.1564	.9877	.1584	6.314	1.012	6.392	**81°00′**
10′	593	872	614	197	013	277	80°50′
20′	622	868	644	6.084	013	166	40′
30′	.1650	.9863	.1673	5.976	1.014	6.059	30′
40′	679	858	703	871	014	5.955	20′
9°50′	708	853	733	769	015	855	10′
10°00′	.1736	.9848	.1763	5.671	1.015	5.759	**80°00′**
10′	765	843	793	576	016	665	79°50′
20′	794	838	823	485	016	575	40′
30′	.1822	.9833	.1853	5.396	1.017	5.487	30′
40′	851	827	883	309	018	403	20′
10°50′	880	822	914	226	018	320	10′
11°00′	.1908	.9816	.1944	5.145	1.019	5.241	**79°00′**
10′	937	811	.1974	5.066	019	164	78°50′
20′	965	805	.2004	4.989	020	089	40′
30′	.1994	.9799	.2035	4.915	1.020	5.016	30′
40′	.2022	793	065	843	021	4.945	20′
11°50′	051	787	095	773	022	876	10′
12°00′	.2079	.9781	.2126	4.705	1.022	4.810	**78°00′**
	Cos	Sin	Cot	Tan	Csc	Sec	⟵

Table 6 TRIGONOMETRIC FUNCTIONS θ° (continued)

⟶	Sin	Cos	Tan	Cot	Sec	Csc	
12°00′	.2079	.9781	.2126	4.705	1.022	4.810	**78°00′**
10′	108	775	156	638	023	745	77°50′
20′	136	769	186	574	024	682	40′
30′	.2164	.9763	.2217	4.511	1.024	4.620	30′
40′	193	757	247	449	025	560	20′
12°50′	221	750	278	390	026	502	10′
13°00′	.2250	.9744	.2309	4.331	1.026	4.445	**77°00′**
10′	278	737	339	275	027	390	76°50′
20′	306	730	370	219	028	336	40′
30′	.2334	.9724	.2401	4.165	1.028	4.284	30′
40′	363	717	432	113	029	232	20′
13°50′	391	710	462	061	030	182	10′
14°00′	.2419	.9703	.2493	4.011	1.031	4.134	**76°00′**
10′	447	696	524	3.962	031	086	75°50′
20′	476	689	555	914	032	4.039	40′
30′	.2504	.9681	.2586	3.867	1.033	3.994	30′
40′	532	674	617	821	034	950	20′
14°50′	560	667	648	776	034	906	10′
15°00′	.2588	.9659	.2679	3.732	1.035	3.864	**75°00′**
10′	616	652	711	689	036	822	74°50′
20′	644	644	742	647	037	782	40′
30′	.2672	.9636	.2773	3.606	1.038	3.742	30′
40′	700	628	805	566	039	703	20′
15°50′	728	621	836	526	039	665	10′
16°00′	.2756	.9613	.2867	3.487	1.040	3.628	**74°00′**
10′	784	605	899	450	041	592	73°50′
20′	812	596	931	412	042	556	40′
30′	.2840	.9588	.2962	3.376	1.043	3.521	30′
40′	868	580	.2994	340	044	487	20′
16°50′	896	572	.3026	305	045	453	10′
17°00′	.2924	.9563	.3057	3.271	1.046	3.420	**73°00′**
10′	952	555	089	237	047	388	72°50′
20′	.2979	546	121	204	048	356	40′
30′	.3007	.9537	.3153	3.172	1.049	3.326	30′
40′	035	528	185	140	049	295	20′
17°50′	062	520	217	108	050	265	10′
18°00′	.3090	.9511	.3249	3.078	1.051	3.236	**72°00′**
	Cos	Sin	Cot	Tan	Csc	Sec	⟵

Table 6 TRIGONOMETRIC FUNCTIONS θ° (*continued*)

⟶	Sin	Cos	Tan	Cot	Sec	Csc	
18°00′	.3090	.9511	.3249	3.078	1.051	3.236	**72°00′**
10′	118	502	281	047	052	207	71°50′
20′	145	492	314	3.018	053	179	40′
30′	.3173	.9483	.3346	2.989	1.054	3.152	30′
40′	201	474	378	960	056	124	20′
18°50′	228	465	411	932	057	098	10′
19°00′	.3256	.9455	.3443	2.904	1.058	3.072	**71°00′**
10′	283	446	476	877	059	046	70°50′
20′	311	436	508	850	060	3.021	40′
30′	.3338	.9426	.3541	2.824	1.061	2.996	30′
40′	365	417	574	798	062	971	20′
19°50′	393	407	607	773	063	947	10′
20°00′	.3420	.9397	.3640	2.747	1.064	2.924	**70°00′**
10′	448	387	673	723	065	901	69°50′
20′	475	377	706	699	066	878	40′
30′	.3502	.9367	.3739	2.675	1.068	2.855	30′
40′	529	356	772	651	069	833	20′
20°50′	557	346	805	628	070	812	10′
21°00′	.3584	.9336	.3839	2.605	1.071	2.790	**69°00′**
10′	611	325	872	583	072	769	68°50′
20′	638	315	906	560	074	749	40′
30′	.3665	.9304	.3939	2.539	1.075	2.729	30′
40′	692	293	.3973	517	076	709	20′
21°50′	719	283	.4006	496	077	689	10′
22°00′	.3746	.9272	.4040	2.475	1.079	2.669	**68°00′**
10′	773	261	074	455	080	650	67°50′
20′	800	250	108	434	081	632	40′
30′	.3827	.9239	.4142	2.414	1.082	2.613	30′
40′	854	228	176	394	084	595	20′
22°50′	881	216	210	375	085	577	10′
23°00′	.3907	.9205	.4245	2.356	1.086	2.559	**67°00′**
10′	934	194	279	337	088	542	66°50′
20′	961	182	314	318	089	525	40′
30′	.3987	.9171	.4348	2.300	1.090	2.508	30′
40′	.4014	159	383	282	092	491	20′
23°50′	041	147	417	264	093	475	10′
24°00′	.4067	.9135	.4452	2.246	1.095	2.459	**66°00′**
	Cos	Sin	Cot	Tan	Csc	Sec	⟵

Table 6 TRIGONOMETRIC FUNCTIONS θ° (*continued*)

⟶	Sin	Cos	Tan	Cot	Sec	Csc	
24°00′	.4067	.9135	.4452	2.246	1.095	2.459	**66°00′**
10′	094	124	487	229	096	443	65°50′
20′	120	112	522	211	097	427	40′
30′	.4147	.9100	.4557	2.194	1.099	2.411	30′
40′	173	088	592	177	100	396	20′
24°50′	200	075	628	161	102	381	10′
25°00′	.4226	.9063	.4663	2.145	1.103	2.366	**65°00′**
10′	253	051	699	128	105	352	64°50′
20′	279	038	734	112	106	337	40′
30′	.4305	.9026	.4770	2.097	1.108	2.323	30′
40′	331	013	806	081	109	309	20′
25°50′	358	.9001	841	066	111	295	10′
26°00′	.4384	.8988	.4877	2.050	1.113	2.281	**64°00′**
10′	410	975	913	035	114	268	63°50′
20′	436	962	950	020	116	254	40′
30′	.4462	.8949	.4986	2.006	1.117	2.241	30′
40′	488	936	.5022	1.991	119	228	20′
26°50′	514	923	059	977	121	215	10′
27°00′	.4540	.8910	.5095	1.963	1.122	2.203	**63°00′**
10′	566	897	132	949	124	190	62°50′
20′	592	884	169	935	126	178	40′
30′	.4617	.8870	.5206	1.921	1.127	2.166	30′
40′	643	857	243	907	129	154	20′
27°50′	669	843	280	894	131	142	10′
28°00′	.4695	.8829	.5317	1.881	1.133	2.130	**62°00′**
10′	720	816	354	868	134	118	61°50′
20′	746	802	392	855	136	107	40′
30′	.4772	.8788	.5430	1.842	1.138	2.096	30′
40′	797	774	467	829	140	085	20′
28°50′	823	760	505	816	142	074	10′
29°00′	.4848	.8746	.5543	1.804	1.143	2.063	**61°00′**
10′	874	732	581	792	145	052	60°50′
20′	899	718	619	780	147	041	40′
30′	.4924	.8704	.5658	1.767	1.149	2.031	30′
40′	950	689	696	756	151	020	20′
29°50′	.4975	675	735	744	153	010	10′
30°00′	.5000	.8660	.5774	1.732	1.155	2.000	**60°00′**
	Cos	Sin	Cot	Tan	Csc	Sec	⟵

Table 6 TRIGONOMETRIC FUNCTIONS θ° (*continued*)

→	Sin	Cos	Tan	Cot	Sec	Csc	
30°00′	.5000	.8660	.5774	1.732	1.155	2.000	**60°00′**
10′	025	646	812	720	157	1.990	59°50′
20′	050	631	851	709	159	980	40′
30′	.5075	.8616	.5890	1.698	1.161	1.970	30′
40′	100	601	930	686	163	961	20′
30°50′	125	587	.5969	675	165	951	10′
31°00′	.5150	.8572	.6009	1.664	1.167	1.942	**59°00′**
10′	175	557	048	653	169	932	58°50′
20′	200	542	088	643	171	923	40′
30′	.5225	.8526	.6128	1.632	1.173	1.914	30′
40′	250	511	168	621	175	905	20′
31°50′	275	496	208	611	177	896	10′
32°00′	.5299	.8480	.6249	1.600	1.179	1.887	**58°00′**
10′	324	465	289	590	181	878	57°50′
20′	348	450	330	580	184	870	40′
30′	.5373	.8434	.6371	1.570	1.186	1.861	30′
40′	398	418	412	560	188	853	20′
32°50′	422	403	453	550	190	844	10′
33°00′	.5446	.8387	.6494	1.540	1.192	1.836	**57°00′**
10′	471	371	536	530	195	828	56°50′
20′	495	355	577	520	197	820	40′
30′	.5519	.8339	.6619	1.511	1.199	1.812	30′
40′	544	323	661	501	202	804	20′
33°50′	568	307	703	492	204	796	10′
34°00′	.5592	.8290	.6745	1.483	1.206	1.788	**56°00′**
10′	616	274	787	473	209	781	55°50′
20′	640	258	830	464	211	773	40′
30′	.5664	.8241	.6873	1.455	1.213	1.766	30′
40′	688	225	916	446	216	758	20′
34°50′	712	208	.6959	437	218	751	10′
35°00′	.5736	.8192	.7002	1.428	1.221	1.743	**55°00′**
10′	760	175	046	419	223	736	54°50′
20′	783	158	089	411	226	729	40′
30′	.5807	.8141	.7133	1.402	1.228	1.722	30′
40′	831	124	177	393	231	715	20′
35°50′	854	107	221	385	233	708	10′
36°00′	.5878	.8090	.7265	1.376	1.236	1.701	**54°00′**
	Cos	Sin	Cot	Tan	Csc	Sec	←

Table 6 TRIGONOMETRIC FUNCTIONS θ° (*continued*)

⟶	Sin	Cos	Tan	Cot	Sec	Csc	
36°00′	.5878	.8090	.7265	1.376	1.236	1.701	**54°00′**
10′	901	073	310	368	239	695	53°50′
20′	925	056	355	360	241	688	40′
30′	.5948	.8039	.7400	1.351	1.244	1.681	30′
40′	972	021	445	343	247	675	20′
36°50′	.5995	.8004	490	335	249	668	10′
37°00′	.6018	.7986	.7536	1.327	1.252	1.662	**53°00′**
10′	041	969	581	319	255	655	52°50′
20′	065	951	627	311	258	649	40′
30′	.6088	.7934	.7673	1.303	1.260	1.643	30′
40′	111	916	720	295	263	636	20′
37°50′	134	898	766	288	266	630	10′
38°00′	.6157	.7880	.7813	1.280	1.269	1.624	**52°00′**
10′	180	862	860	272	272	618	51°50′
20′	202	844	907	265	275	612	40′
30′	.6225	.7826	.7954	1.257	1.278	1.606	30′
40′	248	808	.8002	250	281	601	20′
38°50′	271	790	050	242	284	595	10′
39°00′	.6293	.7771	.8098	1.235	1.287	1.589	**51°00′**
10′	316	753	146	228	290	583	50°50′
20′	338	735	195	220	293	578	40′
30′	.6361	.7716	.8243	1.213	1.296	1.572	30′
40′	383	698	292	206	299	567	20′
39°50′	406	679	342	199	302	561	10′
40°00′	.6428	.7660	.8391	1.192	1.305	1.556	**50°00′**
10′	450	642	441	185	309	550	49°50′
20′	472	623	491	178	312	545	40′
30′	.6494	.7604	.8541	1.171	1.315	1.540	30′
40′	517	585	591	164	318	535	20′
40°50′	539	566	642	157	322	529	10′
41°00′	.6561	.7547	.8693	1.150	1.325	1.524	**49°00′**
10′	583	528	744	144	328	519	48°50′
20′	604	509	796	137	332	514	40′
30′	.6626	.7490	.8847	1.130	1.335	1.509	30′
40′	648	470	899	124	339	504	20′
41°50′	670	451	.8952	117	342	499	10′
42°00′	.6691	.7431	.9004	1.111	1.346	1.494	**48°00′**
	Cos	Sin	Cot	Tan	Csc	Sec	⟵

Table 6 TRIGONOMETRIC FUNCTIONS $\theta°$ (*continued*)

⟶	Sin	Cos	Tan	Cot	Sec	Csc	
42°00′	.6691	.7431	.9004	1.111	1.346	1.494	**48°00′**
10′	713	412	057	104	349	490	47°50′
20′	734	392	110	098	353	485	40′
30′	.6756	.7373	.9163	1.091	1.356	1.480	30′
40′	777	353	217	085	360	476	20′
42°50′	799	333	271	079	364	471	10′
43°00′	.6820	.7314	.9325	1.072	1.367	1.466	**47°00′**
10′	841	294	380	066	371	462	46°50′
20′	862	274	435	060	375	457	40′
30′	.6884	.7254	.9490	1.054	1.379	1.453	30′
40′	905	234	545	048	382	448	20′
43°50′	926	214	601	042	386	444	10′
44°00′	.6947	.7193	.9657	1.036	1.390	1.440	**46°00′**
10′	967	173	713	030	394	435	45°50′
20′	.6988	153	770	024	398	431	40′
30′	.7009	.7133	.9827	1.018	1.402	1.427	30′
40′	030	112	884	012	406	423	20′
44°50′	050	092	.9942	006	410	418	10′
45°00′	.7071	.7071	1.000	1.000	1.414	1.414	**45°00′**
	Cos	Sin	Cot	Tan	Csc	Sec	⟵

Answers to Odd-Numbered Exercises

CHAPTER 1

Section 1.2 (Page 8)

3 **(a)** $\Delta x = 0$, $\Delta y = 11$; **(b)** $\Delta x = 0$, $\Delta y = -11$
5 $|\Delta x| = 0$, $|\Delta y| = 11$

Section 1.3 (Page 12)

1 10
3 $3\sqrt{5}$
5 $\sqrt{10}$
7 $2x\sqrt{x^2 + 1}$
9 $\sqrt{9a^2 + 4b^2}$
11 $(\frac{1}{2}, 0)$
13 $(-\frac{1}{2}, -2)$
15 $(-0.3, -0.05)$
17 $[(3b - 2a)/2, (a - 2b)]$
19 $\sqrt{34}/2$, $\sqrt{34}/2$, $4\sqrt{2}$
21 $R(4, -7)$
23 1:2
25 10 or -6
27 **(a)** $(\frac{1}{2}, 0)$; **(b)** $(-\frac{1}{4}, -1)$ or $(\frac{5}{4}, 1)$

Section 1.4 (Page 16)

1 2
3 1
5 $\frac{3}{4}$
7 $\frac{3}{2}$
9 undefined
11 yes, $m = -2$
13 yes, $m = -\frac{5}{2}$
15 not on the same line

Section 1.5 (Page 21)

1 $y = 3x + 9$ **3** $y = -3x/2 - 1$ **5** $y = -4x + 2$
7 $y = 3x/2 + 3$ **9** $y = 9x/5 + 32$ **11** $x = 2$
13 **(a)** $m = 2$, y intercept $= 4$ **(b)** $m = \frac{3}{2}$, y intercept $= -\frac{5}{2}$ **(c)** $m = -\frac{3}{2}$, y intercept $= 2$ **(d)** $m = \frac{1}{2}$, y intercept $= 2$ **(e)** $m = -\frac{1}{2}$, y intercept $= 2$ **(f)** $m = -\frac{3}{2}$, y intercept $= 3$
17 **(a)** $x/2 + y/3 = 1$ **(b)** $2x/5 + y/5 = 1$ **(c)** $x/3 - y/4 = 1$ **(d)** $x/(\frac{4}{3}) - y/2 = 1$ **(e)** cannot be expressed in intercept form
19 $P = Q = R = 1$ **21** $y = 5x/2$, yes
23 **(a)** $m = -1$ **(b)** $m = -\frac{1}{2}$ **(c)** $m = -2$ **(d)** $m = 1$

Section 1.6 (Page 25)

1 $y = \frac{5}{2}x + \frac{7}{4}$ **3** $y = 8x/9 + \frac{59}{9}$
5 $y = 2x + 10$ **7** **(a)** $y = -7x + 1$ **(b)** $y = -5x/4 + 3$
9 $M(2, 8)$, $N(-4, 0)$ **11** $y = -2x/c^2 + 1/c$
13 $y = -3x/2 + 18$ **15** **(a)** -4 **(b)** 1
17 $T = -\frac{28}{3}$ **19** $C = \frac{4}{5}$
21 **(a)** no triangle **(b)** $(3, 3)$ or $(-1, -3)$

Section 1.7 (Page 34)

1 $(\frac{5}{2}, -2)$ **3** $(\frac{1}{2}, -\frac{3}{2})$ **5** $(\frac{3}{2}, -2)$
7 $(2, -3)$ **9** $(-4, -\frac{2}{3})$ **11** $(0.5, 0.2)$
13 -4 **15** $y = 3x - 7$ **17** $\pm\sqrt{2}$
19 $(\frac{10}{3}, \frac{11}{3})$ **21** $(2, \frac{19}{6})$ **23** $(\frac{11}{5}, \frac{6}{5})$
25 $A = -2$ $B = 3$
27 federal tax = \$200,000, state tax = \$50,000
29 **(a)** $y = x$, $y = 2x - 10$, $y = -2x + 30$ **(b)** $(10, 10)$
31 -2 **33** 30 **35** 42
37 0 **39** -53 **41** $(0, 5)$
43 $(\frac{9}{2}, \frac{27}{5})$ **45** $\left(\frac{ac - bd}{a^2 - b^2}, \frac{ad - bc}{a^2 - b^2}\right)$ **47** $(1, 0)$
49 $(\frac{3}{2}, \frac{5}{3})$ **51** $(-\frac{7}{3}, \frac{7}{19})$ **53** 20 and -8
55 $(\frac{1}{2}, 1, -\frac{1}{2})$ **57** $(3, -1, 2)$ **59** $(2, -1, -3)$
61 $(\frac{2}{3}, \frac{3}{4}, -1)$ **63** $a = 8$, $b = -12$, $c = 14$ **65** 30°, 40°, 110°

Section 1.8 (Page 39)

1 **(a)** $16x + 12y - 1 = 0$ **(b)** $(1, -\frac{5}{4})$ **(c)** $3x - 4y - 8 = 0$ **(d)** $(-\frac{1}{2}, \frac{3}{4})$
3 $\frac{12}{5}$ **5** $L = 5\sqrt{2}$, $A = 50$ **7** $\frac{27}{2}$ **9** $C = 1$ or 27
11 **(a)** $y = 2x + 11$ **(b)** $(-4, 3)$ **(c)** $\sqrt{5}$
13 $(0, \frac{27}{4})$, $(0, -\frac{3}{4})$, $(9, 0)$, $(-1, 0)$
15 **(a)** $2\sqrt{2}$, $2\sqrt{10}$, $2\sqrt{10}$; area = 10 **(b)** $9/\sqrt{5}$, $36/\sqrt{50}$, $36/\sqrt{26}$; area = 18

CHAPTER 2

Section 2.2 (Page 43)

3 all real numbers that avoid division by 0 or taking an even root of a negative quantity
5 **(a)** **(1)** 0, 0 **(2)** -1, 1 **(3)** 0, 4 **(b)** $a^2 - b^2 \neq (a - b)^2$
7 **(a)** **(1)** -1, -1 **(2)** 0, 0 **(3)** 7, 1 **(4)** -2, -8 **(b)** $x^3 - 1^3 \neq (x - 1)^3$
9 -16 **11** 2 **13** 0
15 0 **17** π **19** π
21 $\frac{51}{50}$ **23** undefined **25** -9
27 81 **29** $1 + 2/(x + c)$
31 $f(-3)$ is interpreted as f times -3.
33 No: there is no possible domain.
35 **(a)** $c - 1$ **(b)** 1 **(c)** 1
37 **(a)** $(c - 1)^2$ **(b)** $x - 2 + c$ **(c)** $x - 2$
39 **(a)** $(x + h)^2$ **(b)** $2x + h$ **(c)** $2x$
41 **(a)** $(x + h)^2 - (x + h)$ **(b)** $2x + h - 1$ **(c)** $2x - 1$
43 **(a)** $a(x + h)^2 + b(x + h) + c$ **(b)** $2ax + ah + b$ **(c)** $2ax + b$
45 **(a)** $1/(x + h)^2$ **(b)** $(-2x - h)/x^2(x + h)^2$ **(c)** $-2/x^3$
47 **(a)** **(1)** to **(3)** 2 **(c)** **(1)** to **(3)** m
49 -1

Section 2.4 (Page 52)

1 V, W
3 Diagram A1

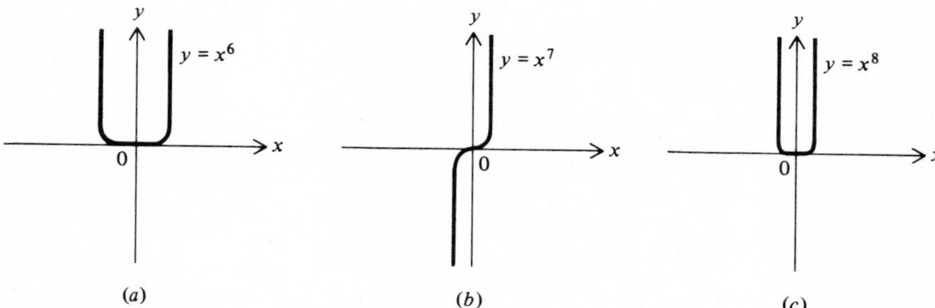

Diagram A1

5 domain: all real numbers; range: all nonpositive real numbers
7 domain: all real numbers; range: 2
9 domain: all real numbers less than or equal to 1; range: all nonnegative real numbers
11 domain: all real numbers; range: all real numbers
13 **(a)** e^3 **(b)** $(S/6)^{3/2}$ **(c)** $(L/12)^3$ **(d)** $d^3/(3)^{3/2}$
15 $f(r) = r^2(\pi + 4)$
17 **(a)** $r^2(\pi - 2)$ **(b)** $s^2(\pi/2 - 1)$ **(c)** $d^2(\pi/4 - \frac{1}{2})$
19 **(a)** $f(d) = 5\pi d^3/24$ **(b)** $f(h) = \pi h^3/75$

21 **(a)** $f(x) = c$ **(b)** $f(x) = \begin{cases} c & x \leq 0 \\ -c & x > 0 \end{cases}$

23 Diagram A2 **25** Diagram A3 **27** Diagram A4 **29** Diagram A5

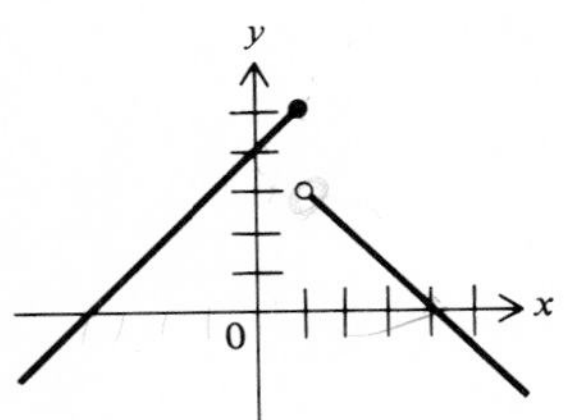

Diagram A2

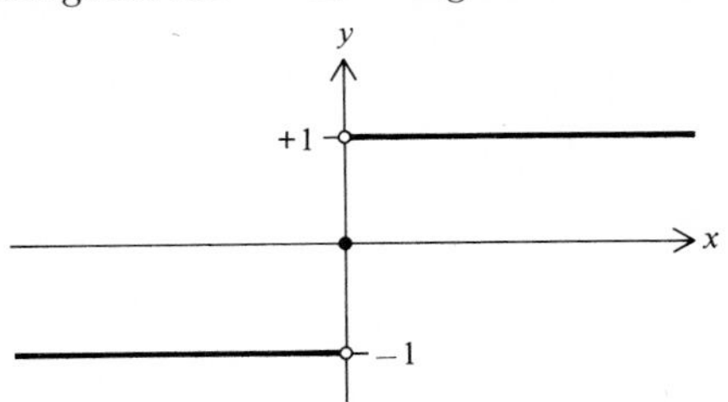

Diagram A3

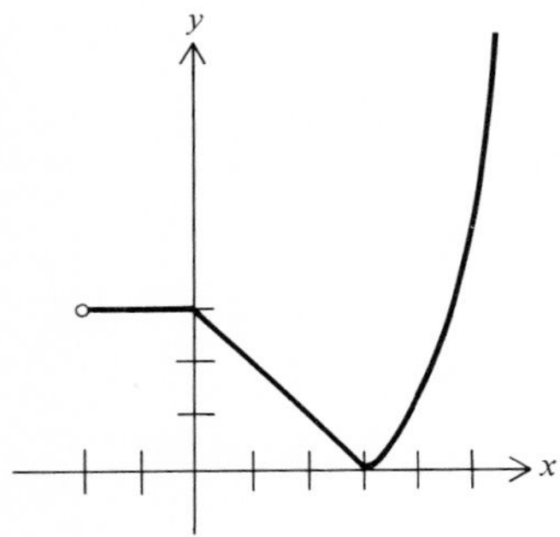

Diagram A4

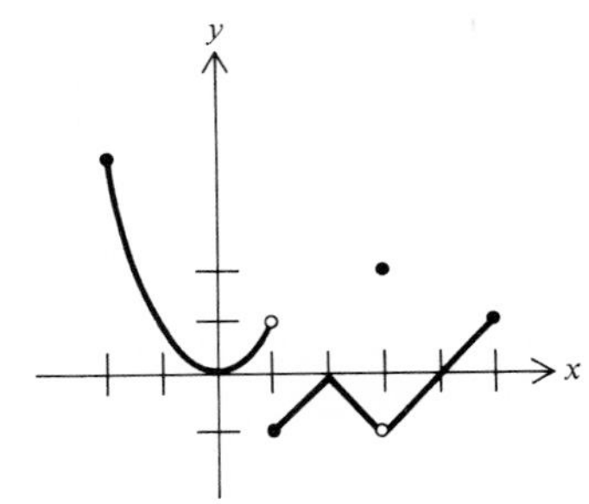

Diagram A5

31 **(a)** $f(x) = \begin{cases} 10 & 0 \leq x < 30 \\ 20 & 30 \leq x < 60 \\ 30 & 60 \leq x < 100 \\ 10x - 950 & x \geq 100 \end{cases}$ **(b)** Diagram A6

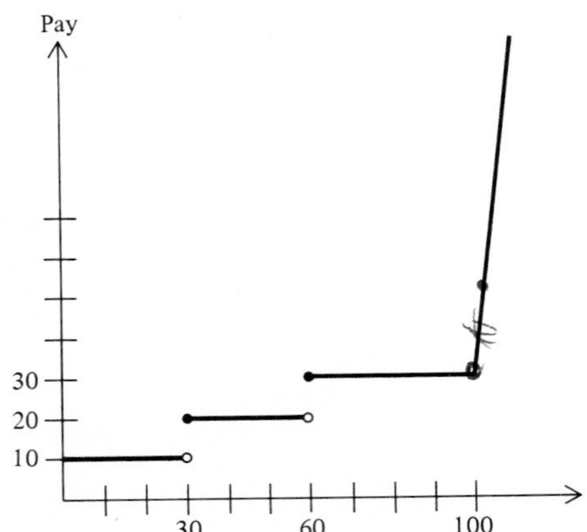

Diagram A6

33 **(a)** $x = 2, x \neq 2$, no value **(b)** $f(x) = \begin{cases} x - 2 & x > 2 \\ 0 & x = 2 \\ 2 - x & x < 2 \end{cases}$

35 **(a)** $s(t) = 16t^2 + 60t, v(t) = 32t + 60$ **(b)** Diagram A7
(c) $s(2) = 184, s(3) = 324, s(5) = 700$ **(d)** same as part **(c)**

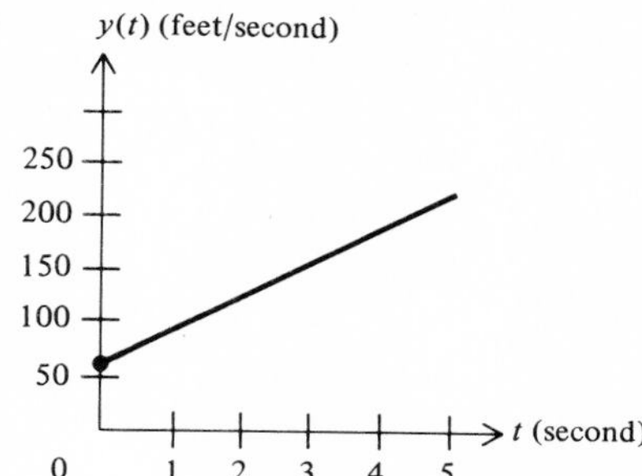

Diagram A7

37 **(a)** F **(b)** T **(c)** T **(d)** F **(e)** T **(f)** T

CHAPTER 3

Section 3.1 (Page 64)

1 Diagram A8 **3** Diagram A9 **5** Diagram A10 **7** Diagram A11
9 Diagram A12 **11** Diagram A13

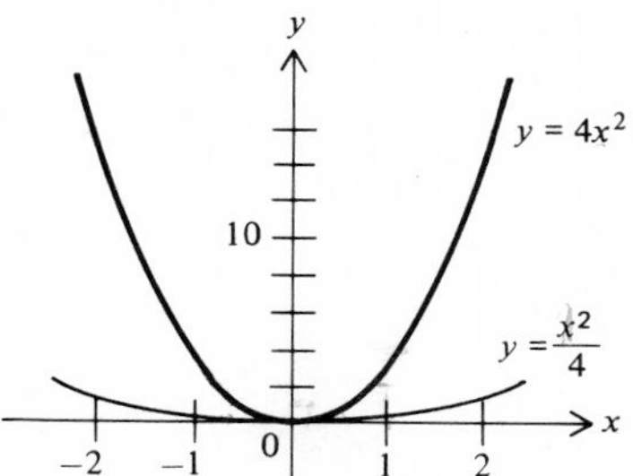

Diagram A8

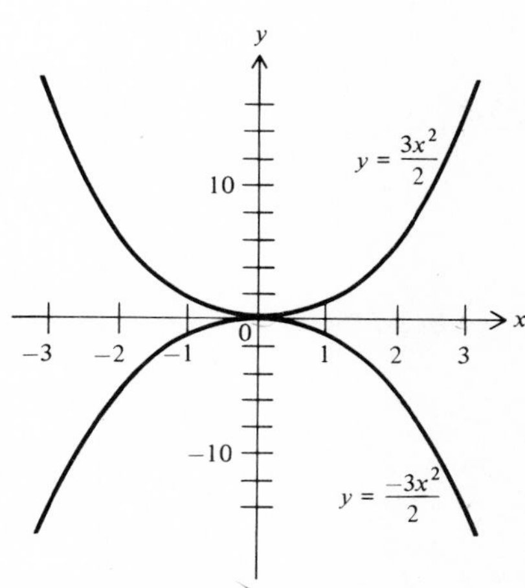

Diagram A9

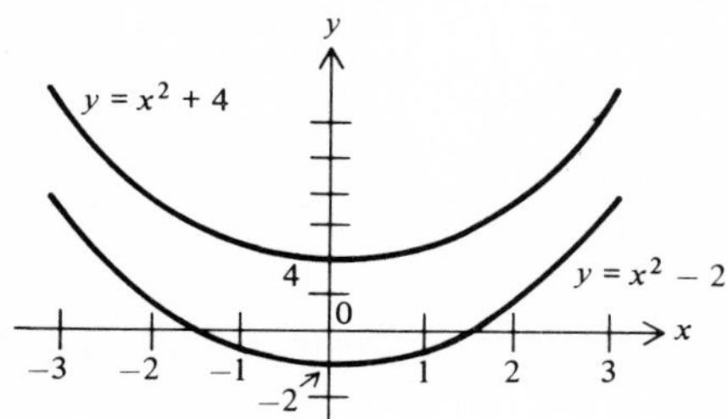

Diagram A10

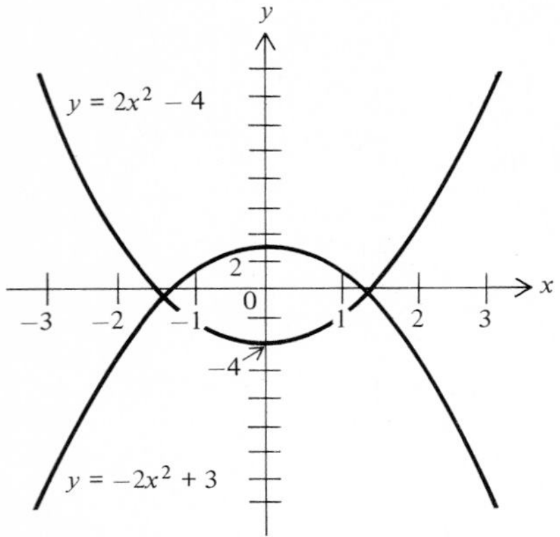

Diagram A11

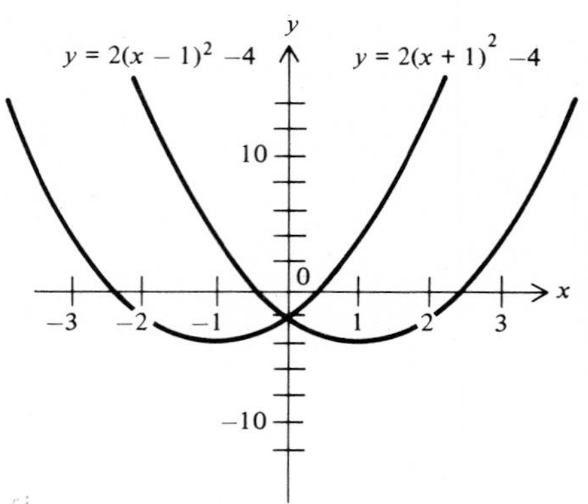

Diagram A12

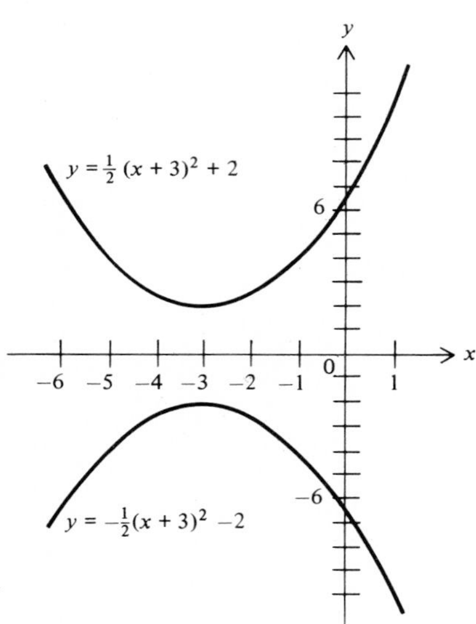

Diagram A13

13 $y = (x - 1)^2 + 1$ **(a)** $x = 1$ **(b)** $(1, 1)$
15 $y = -2(x - 1)^2 + 1$ **(a)** $x = 1$ **(b)** $(1, 1)$
17 $y = -3(x - \frac{2}{3})^2 - \frac{2}{3}$ **(a)** $x = \frac{2}{3}$ **(b)** $(\frac{2}{3}, -\frac{2}{3})$
19 $y = -(x + 4)^2 + 16$ **(a)** $x = -4$ **(b)** $(-4, 16)$
21 $y = \frac{3}{2}(x - 1)^2 + \frac{8}{3}$ **(a)** $x = 1$ **(b)** $(1, \frac{8}{3})$
23 $y = -\frac{2}{3}(x - \frac{3}{2})^2 - \frac{33}{6}$ **(a)** $x = \frac{3}{2}$ **(b)** $(\frac{3}{2}, -\frac{33}{6})$
25 $y = -13(x - 0)^2 - 13$ **(a)** $x = 0$ **(b)** $(0, -13)$
27 $k = \frac{3}{2}$

Section 3.2 (Page 70)

1 roots: $-2, -4$; axis of symmetry: $x = -3$; turning point: $(-3, -1)$
3 roots: $+3, +\frac{3}{2}$; axis of symmetry: $x = \frac{9}{4}$; turning point: $(\frac{9}{4}, \frac{9}{8})$
5 roots: $4, -2$; axis of symmetry: $x = 1$; turning point: $(1, 9)$
7 roots: $-\frac{3}{4}, \frac{1}{2}$; axis of symmetry: $x = -\frac{1}{8}$; turning point: $(-\frac{1}{8}, -\frac{25}{8})$
9 roots: $-\frac{4}{3}$ (double root); axis of symmetry: $x = -\frac{4}{3}$; turning point: $(-\frac{4}{3}, 0)$
11 $2, -3$ **13** $-2 \pm \sqrt{2}$ **15** $\frac{1}{3}, -\frac{5}{2}$
17 $-\frac{5}{4}$ **19** $1 \pm \sqrt{2}$ **21** $2, -\frac{6}{5}$
23 $-1 \pm \sqrt{3}$ **25** $(-3 \pm \sqrt{2})/2$ **27** $\frac{5}{3}, \frac{3}{2}$
29 $-\frac{4}{3}$ **31** $\frac{3}{2}, -\frac{2}{5}$ **33** $(9 \pm \sqrt{321})/24$

Section 3.3 (Page 74)

1 $(-1 \pm 2i)/4$ **3** $(4 \pm 3i)/2$ **5** $(2 \pm i\sqrt{6})/2$
7 **(a)** $x = -\frac{3}{2}$ **(b)** $(-\frac{3}{2}, -\frac{9}{2})$ **(c)** $0, -3$
9 **(a)** $x = 2$ **(b)** $(2, 3)$ **(c)** $(4 \pm \sqrt{6})/2$
11 **(a)** $x = 2$ **(b)** $(2, 0)$ **(c)** 2
13 **(a)** $x = 2$ **(b)** $(2, \frac{3}{2})$ **(c)** $2 \pm i\sqrt{3}$
15 **(a)** $x = -2$ **(b)** $(-2, -3)$ **(c)** $1, -5$
17 **(a)** $x = \frac{3}{2}$ **(b)** $(\frac{3}{2}, -\frac{11}{2})$ **(c)** $(3 \pm i\sqrt{33})/2$
19 $b^2 - 4ac = 169$; two real, unequal, rational roots
21 $b^2 - 4ac = 0$; two real, equal, rational roots

Section 3.4 (Page 80)

1 $-2, 4$
3 $y = 4x + 9$
5 $\pm\frac{2}{3}$
7 $\frac{3}{2}$; $(21 - 3\sqrt{37})/4 \approx 0.69$
9 $R_1 = R_2$
11 **(a)** $-8 + 4\sqrt{6} \approx 1.8$ **(b)** $8 - 4\sqrt{2} \approx 2.34$
13 $\frac{11}{2}, \frac{15}{2}$
15 15
17 $\frac{5}{2}$
19 7×8 or 3×12
21 **(c)** $T = 4$ **(d)** $V = 0$
23 $\frac{9}{2}$
25 $4, -4$
27 200×300
29 120×300
31 350×350
33 300×600
35 **(a)** 1083.75 **(b)** 12
37 **(a)** 1125 **(b)** 12
39 12
41 **(b)** 30 **(c)** 20 **(d)** 25

CHAPTER 4

Section 4.4 (Page 94)

1 **(a)** $2 < 7$ **(b)** $\sqrt{3} < \pi\sqrt{\pi}$ **(c)** $a - b < a + b$
15 $x \leq 2$ **17** $x \leq -3$ **19** $x < -12$ **21** $x \leq 6$
23 $x < 27$ **25** $2 < x < 6$ **27** $x < 5$ **29** $-2 \leq x \leq -1$
31 $-3 < x < \frac{1}{4}$ **33** $x \leq 2$ **35** $-4 \leq x < 2$ **37** $-1 \leq x < 5$

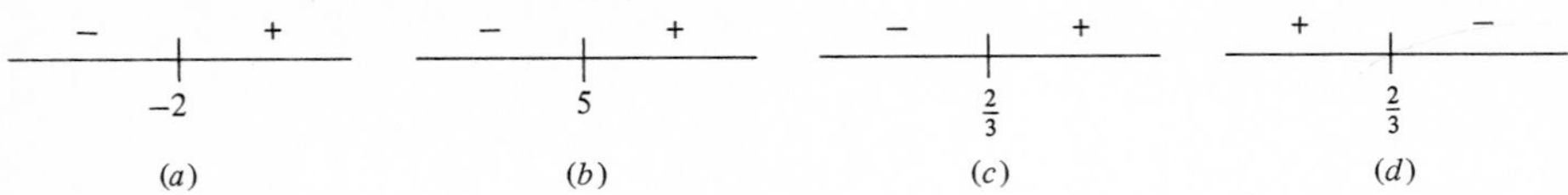

Diagram A14

Section 4.5 (Page 100)

1 Diagram A14
3 **(a)** $x \geq -2$ **(b)** $x < \frac{2}{3}$ **(c)** $x < \frac{2}{3}$
5 $t > 2$ or $t < 0$
7 $1 \leq x \leq 4$
9 $-4 \leq x \leq \frac{3}{2}$
11 $-\frac{1}{3} < x < 6$
13 $-5 < x < -\frac{3}{5}$
15 $x < -4$ or $x > 3$
17 $x \geq 2$ or $-2 \leq x \leq 0$
19 $x \leq 1$ or $x \geq 3$
21 $x \leq -2$ or $x \geq -\frac{1}{2}$
23 $x > 3$
25 $-5 \leq x < 4$
27 $0 < x < 1$ or $2 < x < 4$
29 $x < -2$ or $x \geq 1$
31 $x < -1$
33 $x < 0$
35 $-2 < x < -1$ or $x > 1$
37 $x \leq 0$ or $1 < x < 4$
39 $-2 < x < -1$ or $2 < x < 4$
41 $x < 3$ or $x > 4$

Section 4.6 (Page 106)

1 Diagram A15
3 $\sqrt{a^2/b^2}$
5 5
7 $x = \frac{5}{2}$ or $-\frac{3}{2}$
9 $-\frac{4}{3} < x < 2$
11 $2 < x < 4$
13 $1.999 < x < 2.001$
15 $x \leq -3$ or $x \geq 0$
17 $2 < x < 3$
19 $\frac{1}{3} < x < 9$
21 all real x
23 $x < 4$
25 $4 < x < 5$ or $-2 < x < -1$
27 $-\frac{4}{5} < x < \frac{16}{3}$
29 $-4 \leq x \leq 1$ (excluding $x = -\frac{3}{2}$)
31 $1 \leq x \leq 3$ (excluding $x = 2$)
33 $-1, 2, 3, 6$
35 $x = \frac{3}{2}$
37 $x \leq 2$
39 $x > \frac{1}{2}$ or $x < \frac{1}{4}$
41 $x = -\frac{5}{2}, -\frac{1}{4}$
43 $x < -1$
49 Diagram A16; **(a)** $d = 2$

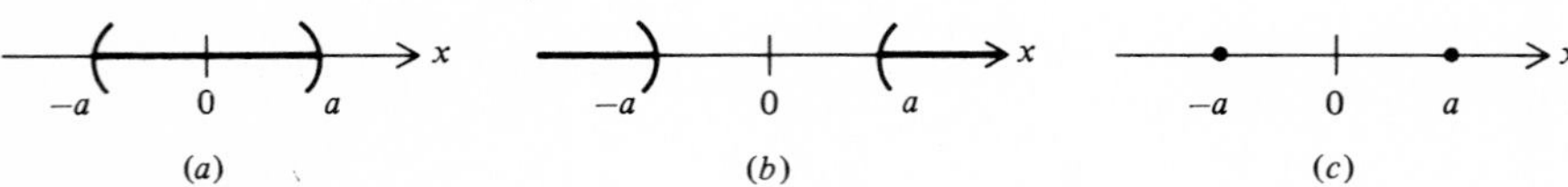

Diagram A15

CHAPTER 5

Section 5.2 (Page 111)

1 **(a)** -4 **(b)** -26 **(c)** 80 **(d)** -1228
3 **(a)** 6 **(b)** 2 **(c)** 8 **(d)** -30
5 **(a)** 0.65 **(b)** -25
7 **(a)** $-ab - a^2b$ **(b)** $b^2 - 2ab - ab^2$ **(c)** $b^2 - ab$ **(d)** $1 - a - 2ab$

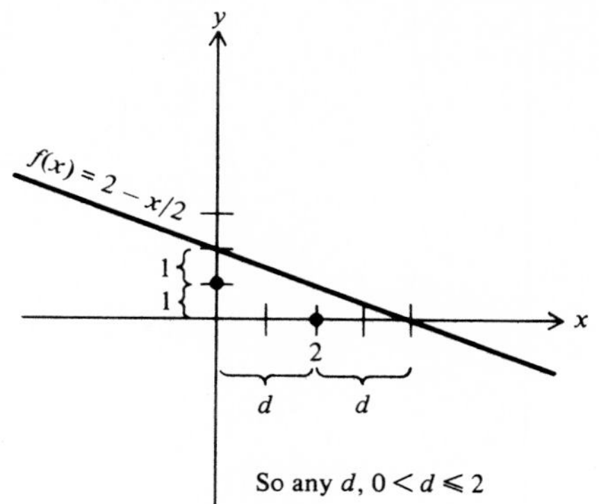

Diagram A16

9 **(a)** $\frac{3}{2}$ **(b)** -1
11 **(a)** 110 **(b)** $\frac{5}{16}$ **(c)** $-\frac{8}{81}$ **(d)** 342
13 $a = 2, b = -3, c = -4$

Section 5.3 (Page 114)

1 $Q(x) = 3x - 1, R(x) = 5x - 1$ **3** $Q(x) = x^6 + x^4 + x^2 + 1, R(x) = 0$
5 $Q(x) = 3x^2 + 14x + 88, R(x) = 513x - 84$
7 $Q(x) = 3x/2 + \frac{1}{4}, R(x) = -11x/4 + \frac{3}{4}$
9 $Q(x) = \frac{2}{3}x^3 + x^2 + \frac{1}{3}, R(x) = 1$ **11** $Q(x) = 4x^2 + 2x + 1, R(x) = 0$
13 $Q(x) = 3x^2 - 4x + 13, R(x) = 0$ **15** $Q(x) = 2x^2 + 2x - 2, R(x) = -10$

Section 5.4 (Page 119)

1 $Q(x) = 3x^2 + x - 14, R(x) = -16$ **3** $Q(x) = x^3 - 6x^2 + 12, R(x) = -18$
5 $Q(x) = x^4 + x^3 - 6, R(x) = -3$
7 $Q(x) = x^5 - 2x^4 + 4x^3 - 8x^2 + 16x - 32, R(x) = 128$
9 $Q(x) = 2x^2 - 4x + 2, R(x) = -1$
15 $d = 2$ **17** Multiplicity 2 **19** $x + 1$ is not a factor
21 $P(0) = 1, P(1) = 2, P(0.2) = 1.0912, P(0.4) = 1.2752, P(2) = 13, P(0.6) = 1.4752$
23 14 **25** -2 **27** $k = -2$ **29** $k = -3$

Section 5.5 (Page 129)

1 $\frac{2}{3}, -1 \pm \sqrt{2}$ **3** $3, 2, -\frac{1}{2}, -3$ **5** $-1, \frac{1}{3}, (1 \pm \sqrt{13})/3$
7 $4, \pm \sqrt{2}/2$ **9** $6, -6, \pm \sqrt{3}$ **11** -2
13 $-\frac{3}{4}$ **15** $-\frac{2}{3}, -\frac{1}{2}, \frac{3}{2}, 3$ **17** $-\frac{8}{3}, \pm 1/\sqrt{2}$
19 $-\frac{1}{3}, -\frac{2}{3}, -1$ **21** $3, -\frac{3}{2}, 1, \frac{1}{2}$ **23** $1, \frac{2}{3}, -\frac{1}{5}, -3$
25 $\frac{9}{2}, -\frac{2}{5}, \pm \sqrt{2}$ **27** $\frac{5}{2}, -\frac{7}{8}, -1 \pm \sqrt{2}$ **31** $x \le -\frac{2}{3}$ or $\frac{1}{2} \le x \le 3$
33 $x < -2$ or $\frac{3}{4} < x < 2$ **35** $\frac{7}{2}$ or $\frac{1}{2}$ **37** $x = 2$

Section 5.6 (Page 134)

1 $-1, \frac{1}{3}, \pm i$ **3** $-\frac{1}{2}, (1 \pm 2i)/2$ **5** $-\frac{3}{4}, 2, -1 \pm i\sqrt{3}$
7 $1, 1, \pm i/\sqrt{2}$ **9** $\frac{7}{2}, \pm\sqrt{2}$ **11** $3, \frac{1}{2}, -1 \pm i$

13 $3, \frac{1}{2}, -5 \pm i\sqrt{15}$ **15** $-\frac{1}{24}, 1 \pm i\sqrt{7}$ **17** $\frac{5}{3}, -\frac{7}{2}, 1 \pm i$
19 $-\frac{5}{24}, (1 \pm i\sqrt{15})/2$ **21** $\frac{1}{3}, -\frac{5}{6}, \pm 2i$ **25** $-1 - i, 2 \pm i\sqrt{2}$
27 $1 + i, -1 + i$

Section 5.7 (Page 142)

1 2, 3 **3** -1.2 **5** **(a)** 0, 1 **(b)** 1, 2
7 **(a)** $-4, -3$ **(b)** $-2, -1$ **(c)** 1, 2 **9** Diagram A17
11 Diagram A18 **13** Diagram A19 **15** Diagram A20
17 Diagram A21 **19** Diagram A22 **21** Diagram A23

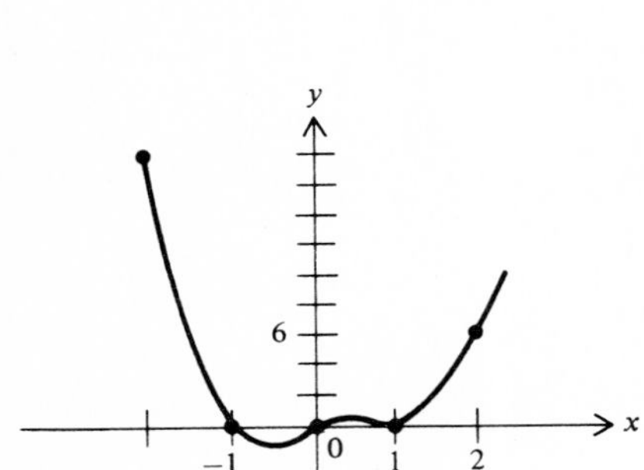

Diagram A17

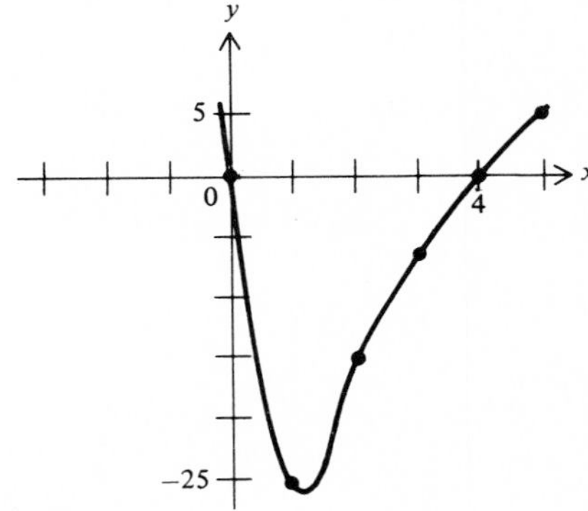

Diagram A18

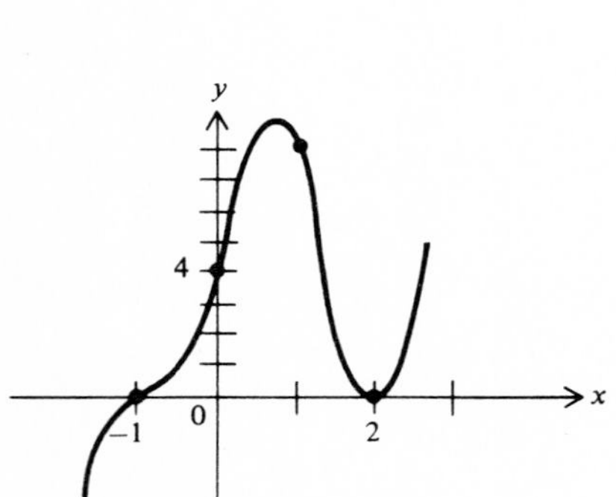

Diagram A19

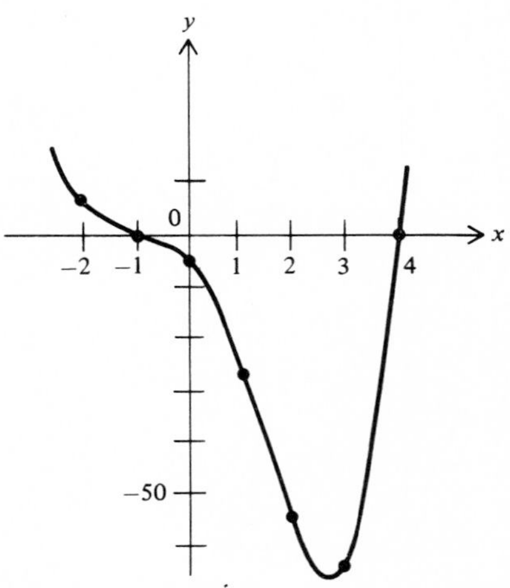

Diagram A20

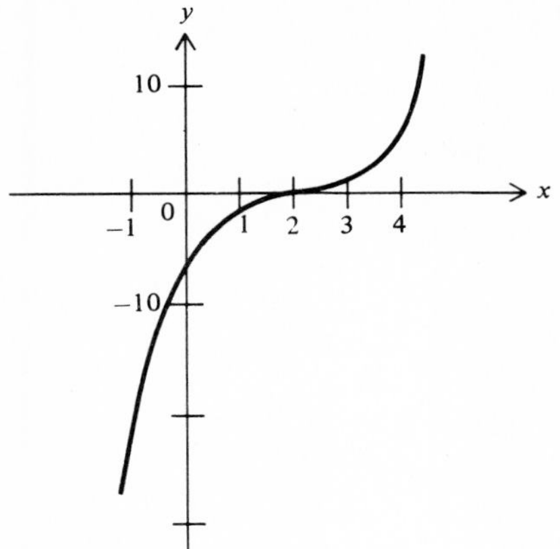

Diagram A21

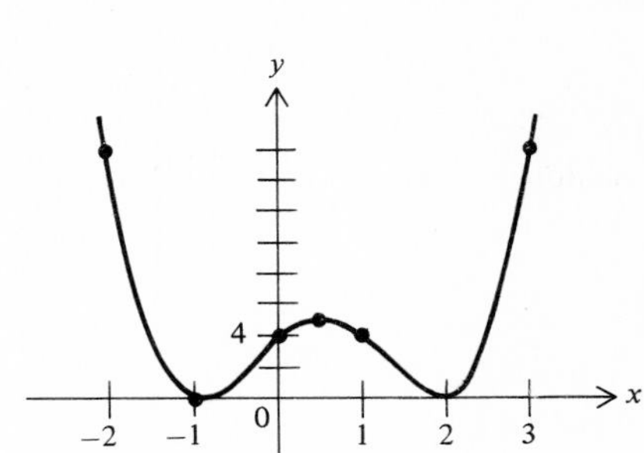

Diagram A22

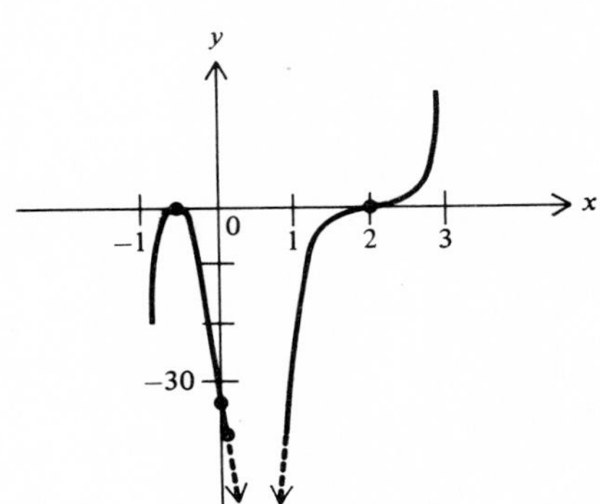

Diagram A23

Section 5.8 (Page 145)

1 1.5 **3** 2.44 **5** −2.2
7 2.62 **9** 0.217, 3.95, −1.17 **11** 3.85

Section 5.9 (Page 150)

1 **(a)** $B = -\frac{15}{2}$, $C = 18$ **(b)** $(3, \frac{27}{2})$
3 (1, 5)(2, 4)
5 (−1, 16)(3, −16)
7 no turning point
9 (1, −1)(−1, 3)
11 $(-1, 0)(\frac{1}{3}, -\frac{32}{27})$
13 (0, 17)(4, −15)
15 $(1, -1)(-\frac{2}{3}, \frac{98}{27})$
17 (0, 2)(2, 1)
19 108π
21 4×8
23 $14 \times 14 \times 28$
25 **(a)** $2 \times 4 \times 8$ **(b)** $\frac{5}{3} \times \frac{14}{3} \times \frac{35}{6}$

27 top: $5 \times 20 \times 8$; no top: $5\sqrt{2} \times 20\sqrt{2} \times 4\sqrt{2}$
29 **(a)** max $(-1, 4)$; min $(-3, -16)$ **(b)** Diagram A24

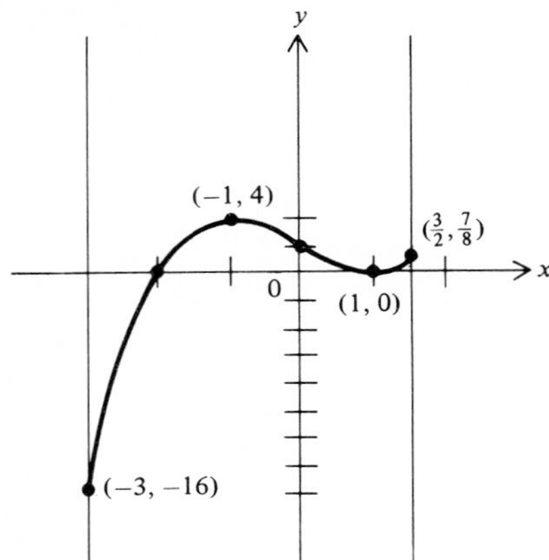

Diagram A24

31 max $(0, 0)$ or $(3, 0)$; min $(2, -4)$ or $(-1, 4)$
33 $r = 4, h = 8$ **35** 2048/27 **37** $8 \times 16 \times \frac{16}{3}$

CHAPTER 6

Section 6.2 (Page 164)

1 domain: all reals, $x \neq -4$; vertical asymptote: $x = -4$; horizontal asymptote: $y = 0$; y intercept at $(0, \frac{1}{2})$ (Diagram A25)
3 domain: all reals; horizontal asymptote: $y = 0$; y intercept at $(0, 4)$ (Diagram A26)

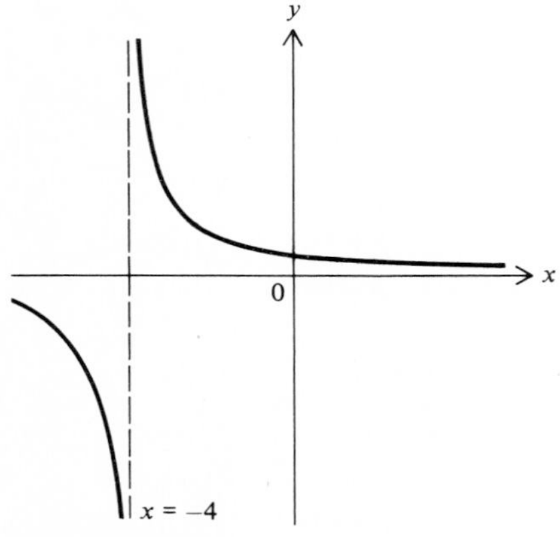

Diagram A25

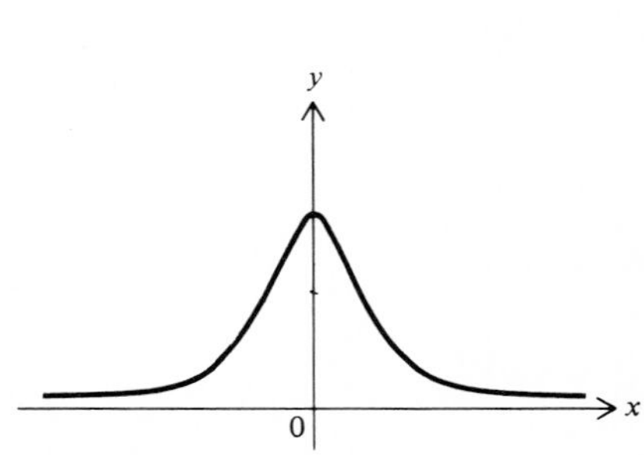

Diagram A26

5 domain: all reals, $x \neq \frac{2}{3}$; vertical asymptote: $x = \frac{2}{3}$; horizontal asymptote: $y = \frac{2}{3}$; zeros at $x = -\frac{1}{2}$; y intercept at $(0, -\frac{1}{2})$ (Diagram A27)

7 domain: all reals, $x \neq -2, 1$; vertical asymptote: $x = -2, x = 1$; horizontal asymptote: $y = 2$; intersection of the graph of $y = f(x)$ with $y = 2$ at $(2, 2)$; double zero at $x = 0$ (Diagram A28)

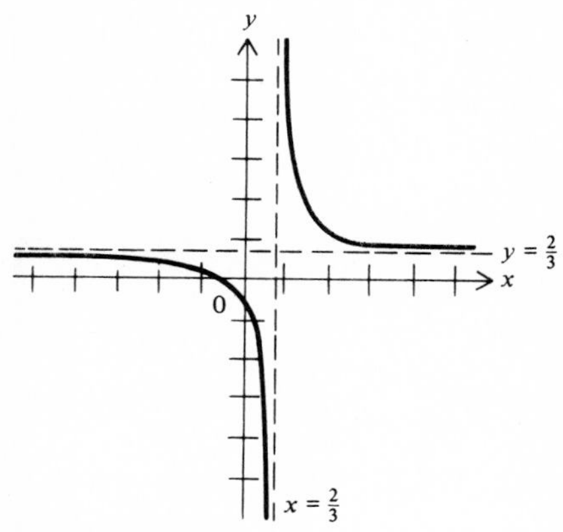

Diagram A27

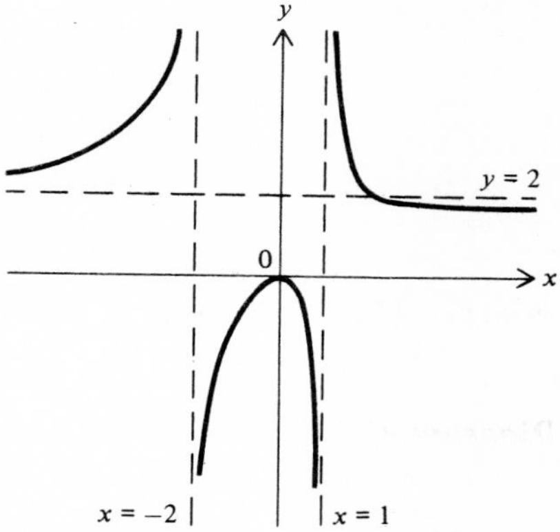

Diagram A28

9 domain: all reals, $x \neq \pm 2$; vertical asymptote: $x = 2, x = -2$; horizontal asymptote: $y = 0$; zero at $x = \frac{2}{3}$; y intercept at $(0, \frac{1}{2})$ (Diagram A29)

11 domain: all reals; horizontal asymptote: $y = 0$; zero at $x = -\frac{3}{2}$; y intercept at $(0, \frac{3}{2})$ (Diagram A30)

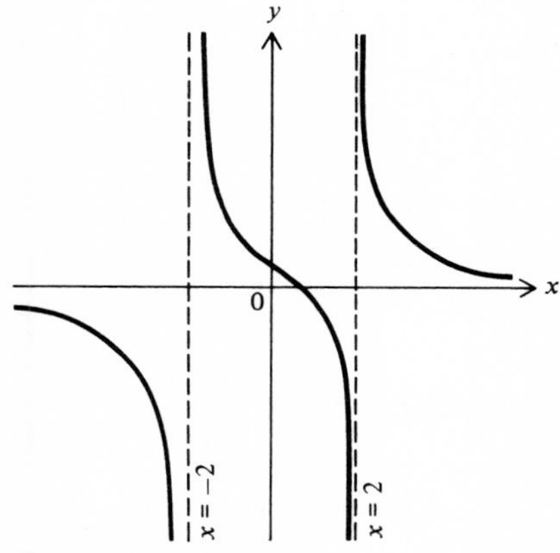

Diagram A29

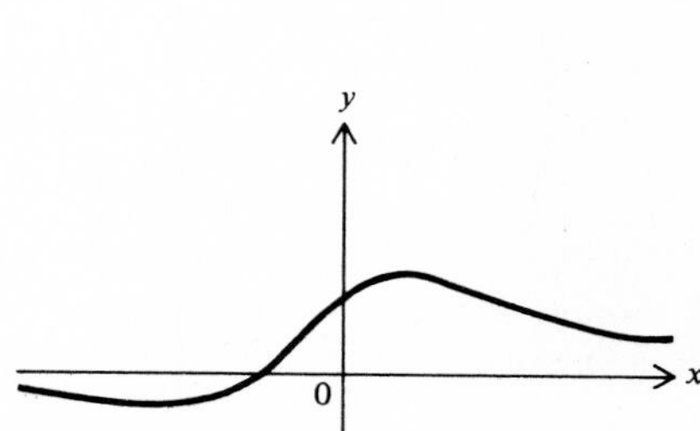

Diagram A30

13 domain: all reals, $x \neq \pm 1$; vertical asymptote: $x = 1, x = -1$; horizontal asymptote: $y = 4$; zero of multiplicity four at $x = 0$; y intercept at $(0, 0)$ (Diagram A31)

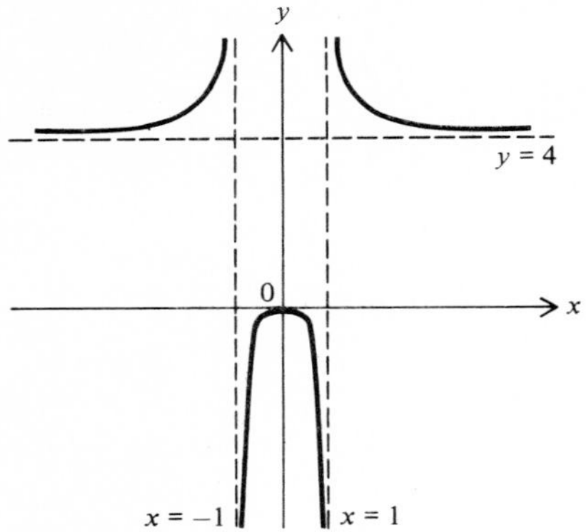

Diagram A31

15 domain: all reals, $x \neq -2 \pm \sqrt{3}$; vertical asymptote: $x = -2 + \sqrt{3}$, $x = -2\sqrt{3}$; horizontal asymptote: $y = 1$; intersection of the graph of $y = f(x)$ with $y = 1$ at $(-\frac{1}{4}, 1)$; double zero at $x = 0$ (Diagram A32)

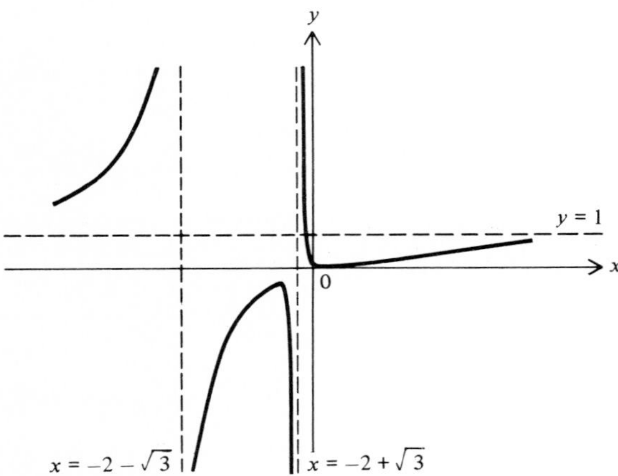

Diagram A32

17 domain: all reals, $x \neq -1, -2$; vertical asymptote: $x = -1, x = -2$; horizontal asymptote: $y = 2$; intersection of the graph of $y = f(x)$ with $y = 2$ at $(-\frac{2}{3}, 2)$; double zero at $x = 0$ (Diagram A33)

19 domain: all reals, $x \neq 2$; vertical asymptote: $x = 2$; horizontal asymptote: $y = 0$; zero at $x = 0$ (Diagram A34)

21 domain: all reals; horizontal asymptote: $y = 1$; no vertical asymptote; zeros at $x = -1$ and 1 (Diagram A35)

23 domain: all reals, $x \neq 0$; vertical asymptote: $x = 0$ horizontal asymptote: $y = 0$; zero at $x = -\frac{1}{2}$ (Diagram A36)

25 $y = (2x + 4)/(x + 1)$

31 domain: all reals, $x \neq -3, 1$; vertical asymptote: $x = -3, x = 1$; horizontal asymptote: $y = -2$; intersection of the graph of $y = f(x)$ with $y = -2$ at $(-\frac{1}{2}, -2)$; zeros at $x = \pm 2$; y intercept at $(0, -\frac{8}{3})$ (Diagram A37)

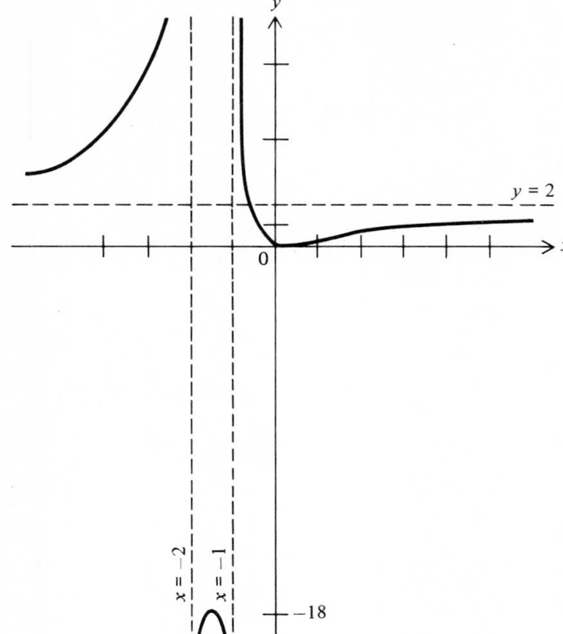

Diagram A33

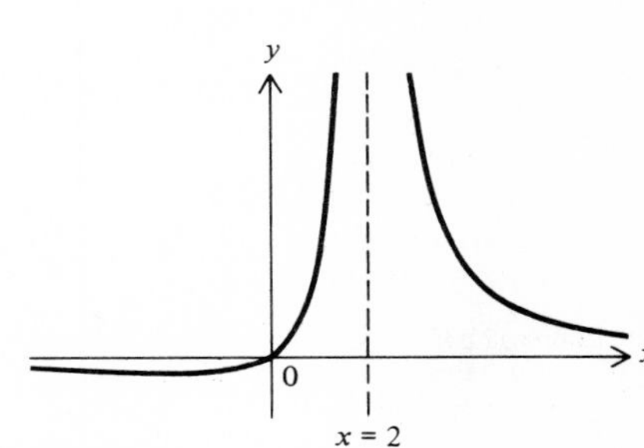

Diagram A34

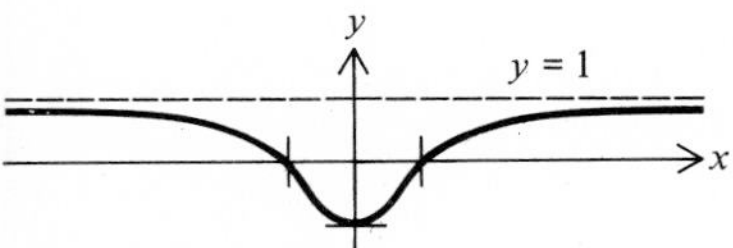

Diagram A35

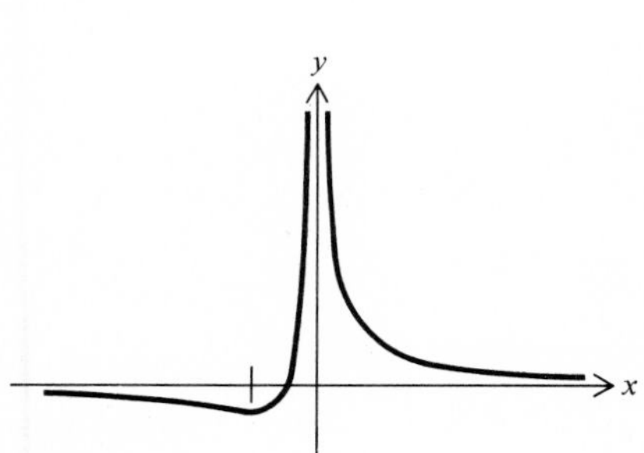

Diagram A36

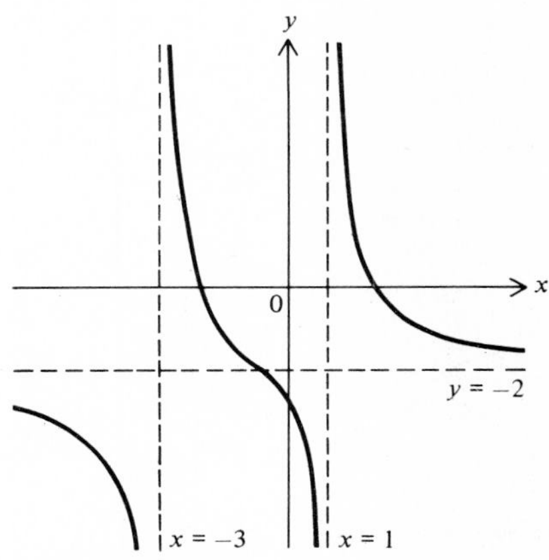

Diagram A37

33 **(a)** domain: all reals; horizontal asymptote: $y = 0$; zero at $x = 0$ (Diagram A38)
(b) domain: all reals, $x \neq \pm\sqrt{-b}$; vertical asymptote: $x = \sqrt{-b}$, $x = -\sqrt{-b}$; horizontal asymptote: $y = 0 = 0$; zeros at $x = 0$ (Diagram A39)

35 The function has its maximum value at $x = 2ab/(a + b) = \frac{8}{3}$ (Diagram A40)

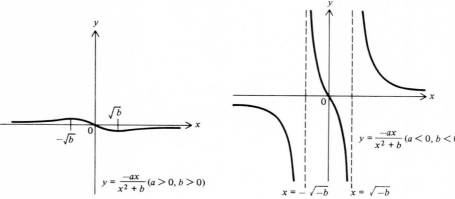

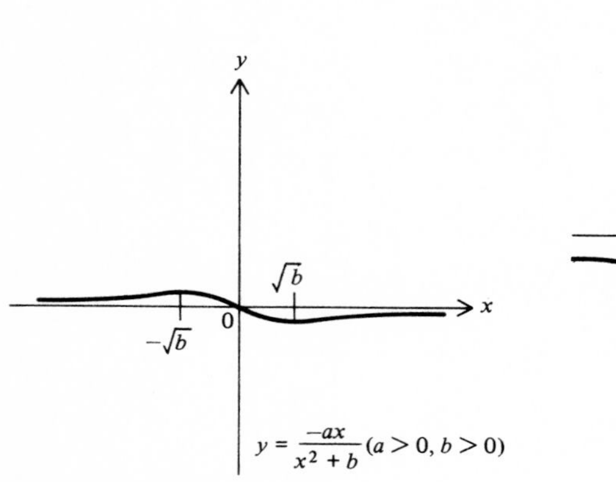

Diagram A38

Diagram A39

Section 6.3 (Page 173)

1 domain: all reals, $x \neq -1$; vertical asymptote: $x = -1$; oblique asymptote: $y = x - 3$; double zero at $x = 1$; y intercept at $(0, 1)$ (Diagram A41)

3 domain: all reals, $x \neq -1$; vertical asymptote: $x = -1$; oblique asymptote: $y = x + 1$; zeros at $x = -4, 2$; y intercept at $(0, -8)$ (Diagram A42)

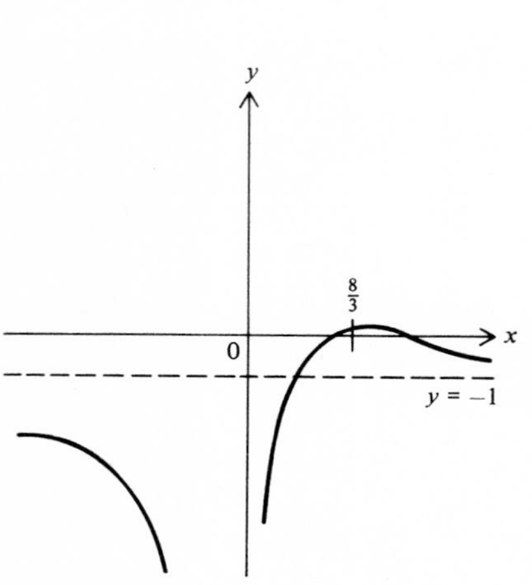

Diagram A40

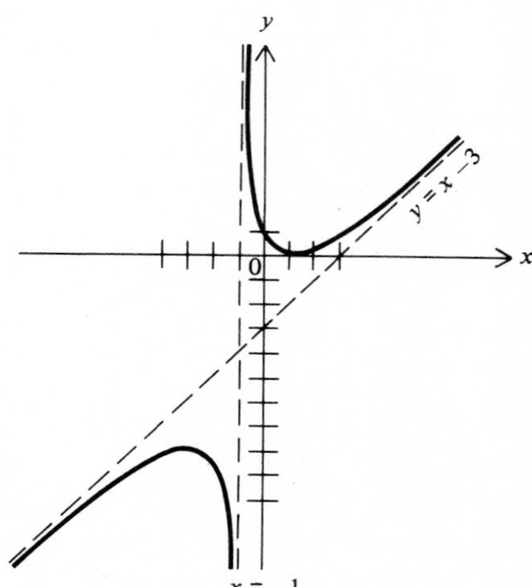

Diagram A41

5 domain: all reals, $x \neq 0$; vertical asymptote: $x = 0$; oblique asymptote $y = x^2$; zeros at $x = 1$ (Diagram A43)

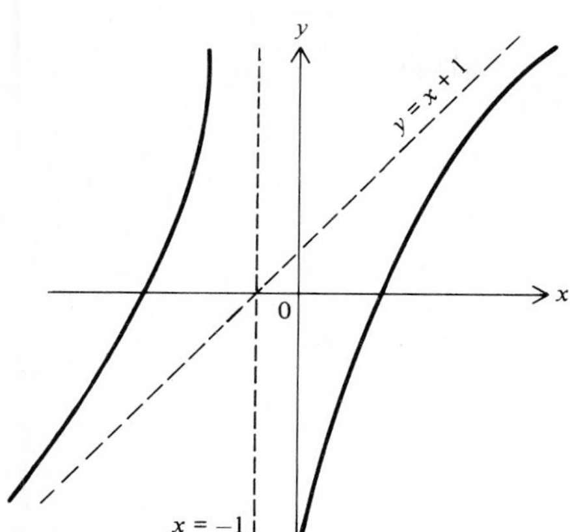

Diagram A42

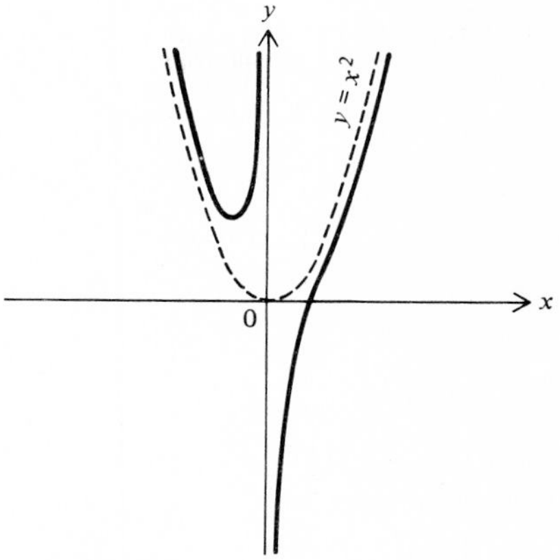

Diagram A43

7 domain: all reals, $x \neq 4$; vertical asymptote: $x = 4$; oblique asymptote: $y = x^2 + 1$; zeros at $x = -1, 2, 3$; y intercept at $(0, -\frac{3}{2})$ (Diagram A44)

9 domain: all reals, $x \neq 1$; vertical asymptote: $x = 1$; oblique asymptote: $y = -x$; zeros at $x = -1, 2$; y intercept at $(0, -2)$ (Diagram A45)

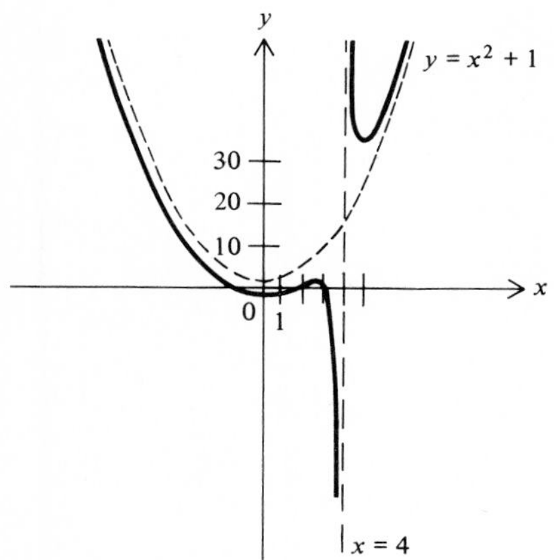

Diagram A44

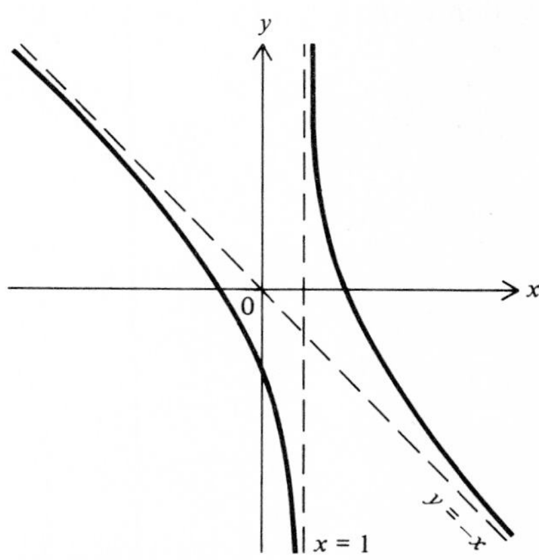

Diagram A45

11 domain: all reals, $x \neq -1$; vertical asymptote: $x = -1$; oblique asymptote: $y = x - 2$; zeros at $x = 0, 1$ (Diagram A46)

13 domain: all reals, $x \neq -2, 3$; vertical asymptote: $x = -2, x = 3$; oblique asymptote: $y = x + 4$; intersection of $y = f(x)$ with $y = x + 4$ at $(-\frac{7}{3}, \frac{5}{3})$; zeros at $x = -3, -1, 1$; y intercept at $(0, \frac{1}{2})$ (Diagram A47)

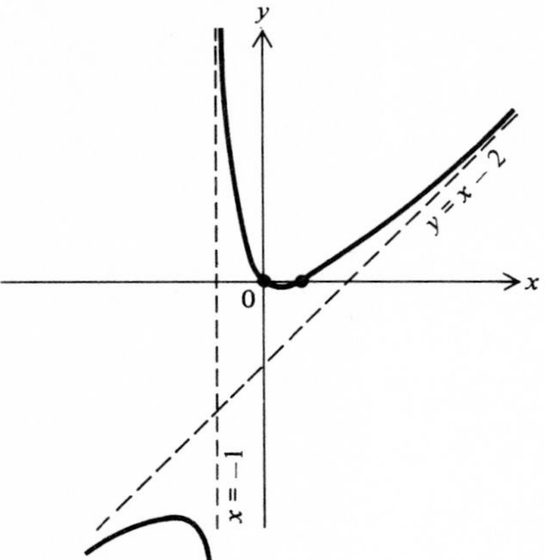

Diagram A46

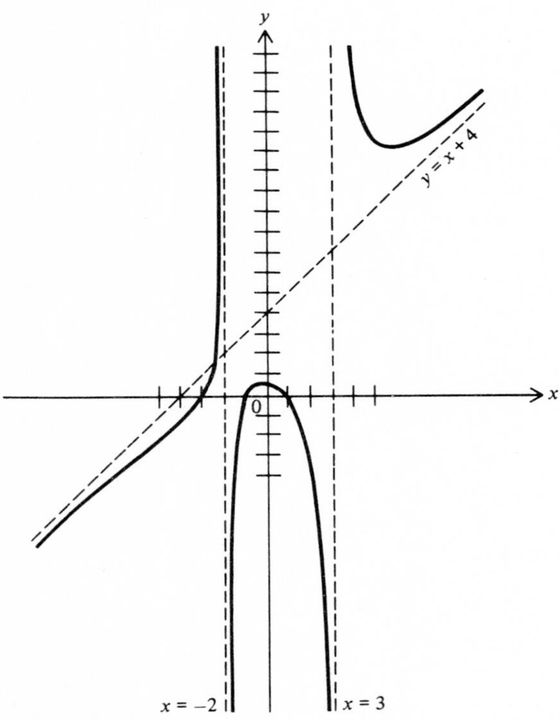

Diagram A47

15 For $a = -27$, $y = (x^3 - 27)/x$; domain: all reals, $x \neq 0$; vertical asymptote: $x = 0$; oblique asymptote: $y = x^2$; zeros at $x = 3$. For $a = 8$, $y = (x^3 + 8)/x$; domain: all reals, $x \neq 0$; vertical asymptote: $x = 0$; oblique asymptote: $y = x^2$; zeros at $x = -2$. For $a = -1$, $y = (x^3 - 1)/x$; domain: all reals, $x \neq 0$; vertical asymptote: $x = 0$; oblique asymptote: $y = x^2$; zeros at $x = 1$ (Diagram A48)

17 $y = g(x)$ is an oblique asymptote of $y = f(x)$.

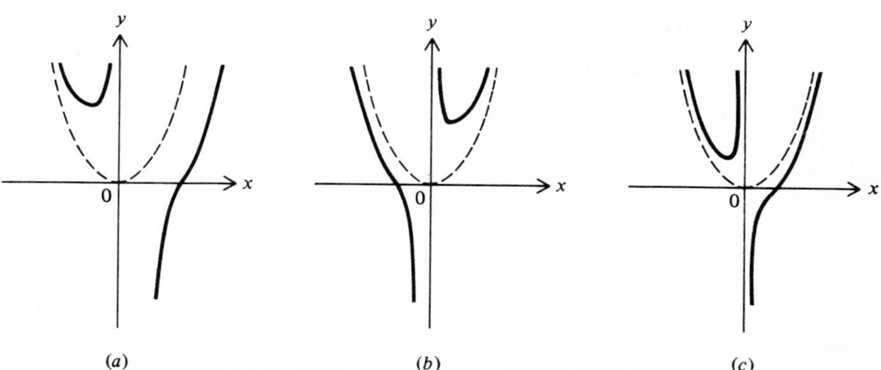

Diagram A48

Section 6.4 (Page 180)

1 $\frac{1}{x-1} - \frac{1}{2x} - \frac{1}{2(x+2)}$

3 $\frac{2}{2x-1} - \frac{x+2}{x^2+1}$

5 $x + 1 + \frac{4}{x-2} + \frac{2}{x+1}$

7 $\frac{3}{x-1} - \frac{1}{(x-1)^2} - \frac{2}{x}$

9 $\frac{2x-6}{x^2-2x+2} - \frac{1}{x+1}$

11 $\frac{2}{x-1} + \frac{1}{(x-1)^2} + \frac{2}{(x-1)^3} - \frac{1}{x}$

13 $\frac{1}{x-3} - \frac{3}{2(x+2)} + \frac{1}{2(x+4)}$

15 $\frac{4}{3x-2} + \frac{x-5}{x^2-2x-1}$

17 $\frac{3}{x-2} + \frac{4}{(x-2)^2} + \frac{5}{(x-2)^3}$

19 $\frac{2}{x} + \frac{3x-4}{x^2+2x-4}$

21 $\frac{11}{8(x+1)} - \frac{5}{4(x+3)} + \frac{15}{8(x-3)}$

CHAPTER 7

Section 7.2 (Page 187)

1 (a) $(f + g)(x) = x^2 + 1$ **(b)** $(f - g)(x) = 4x - x^2 - 1$
(c) $fg(x) = 2x^3 - 4x^2 + 2x$ **(d)** $(f/g)(x) = 2x/(x - 1)^2$
(e) domain for parts (*a*), (*b*), and (*c*): all reals; domain for part (*d*): all real, $x \neq 1$

3 (a) $(f + g)(x) = \sqrt{x-1} + 5$ **(b)** $(f - g)(x) = \sqrt{x-1} - 5$
(c) $fg(x) = 5\sqrt{x-1}$ **(d)** $(f/g)(x) = \sqrt{x-1}/5$
(e) domain for parts (*a*), (*b*), (*c*), and (*d*): $x \geq 1$

5 **(a)** $(f + g)(x) = (2x^2 - 2x + 1)/x(x - 1)$ **(b)** $(f - g)(x) = (2x - 1)/x(x - 1)$ **(c)** $fg(x) = 1$ **(d)** $(f/g)(x) = x^2/(x - 1)^2$ **(e)** domain for parts (*a*), (*b*), (*c*), and (*d*): all reals, $x \neq 0, 1$

7 **(a)** $(f + g)(x) = \sqrt{x} + |x|$ **(b)** $(f - g)(x) = \sqrt{x} - |x|$ **(c)** $fg(x) = \sqrt{x^3}$ **(d)** $(f/g)(x) = 1/\sqrt{x}$ **(e)** domain for parts (*a*), (*b*), and (*c*): $x \geq 0$; domain for $(f/g)(x)$: $x > 0$

9 **(a)** $(f + g)(x) = \sqrt{1 - x} + \sqrt{2 + x}$ **(b)** $(f - g)(x) = \sqrt{1 - x} - \sqrt{2 - x}$ **(c)** $fg(x) = \sqrt{2 - x - x^2}$ **(d)** $(f/g)(x) = \sqrt{2 - x - x^2}/(x + 2)$ **(e)** domain for parts (*a*), (*b*), and (*c*): $-2 \leq x \leq 1$; domain for $(f/g)(x)$: $-2 < x \leq 1$

11 **(a)** $\frac{5}{2}$; **(b)** undefined **13** **(a)** $(f + g)(x) = 0$ **(b)** all reals

15 **(a)** $f[g(x)] = 6(x - 4)$ **(b)** all reals

17 **(a)** $f[g(x)] = 2$ **(b)** $x \geq 0$ **19** **(a)** $f[g(x)] = 1/\sqrt{x}$ **(b)** $x > 0$

21 **(a)** $f[g(x)] = (2x + 1)/x$ **(b)** all reals, $x \neq 0, -1$

23 **(a)** $f[g(x)] = \sqrt{\dfrac{2\sqrt{x} + x + 1}{x - 1}}$ **(b)** $x > 1$

25 **(a)** mp **(b)** mp **(c)** no

Section 7.3 (Page 191)

1 $f^{-1}(x) = x/2$

3 $f^{-1}(x) = (x + 3)/4$

5 $f^{-1}(x) = (-\frac{1}{2})(1 + \sqrt{4x + 1})$

7 $f^{-1}(x) = x^2 \quad x \geq 0$

9 $f^{-1}(x) = (1 - 4x)/x$

11 $f^{-1}(x) = (2 - x)/(1 + x)$

23 See Diagram A48a

25 See Diagram A48b

27 See Diagram A48c

29 See Diagram A48d

31 See Diagram A48e

39 many answers possible, such as $x > 1$

41 many answers possible, such as $x < -3$

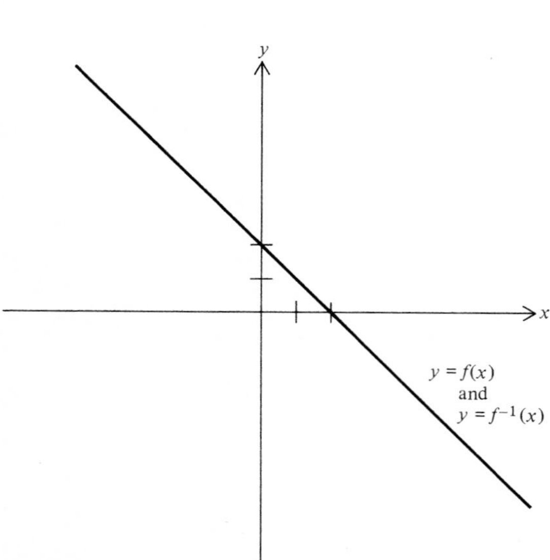

Diagram A48a

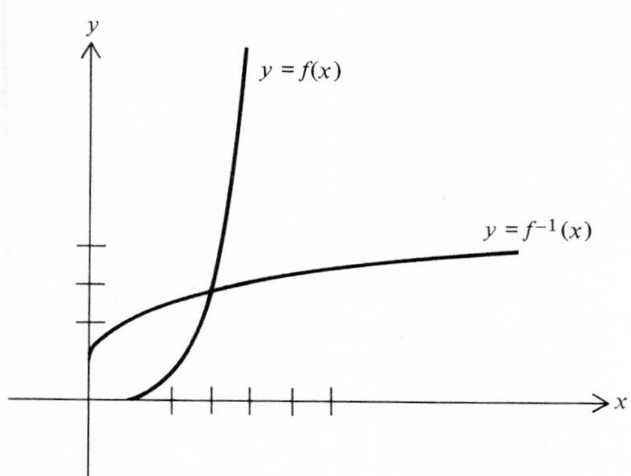

Diagram A48b

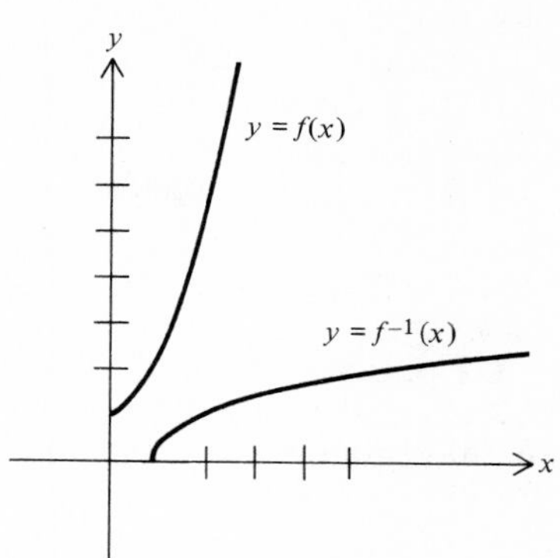

Diagram A48c

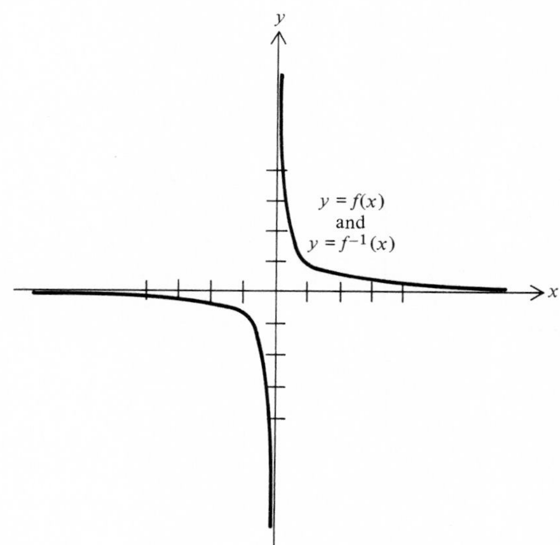

Diagram A48d

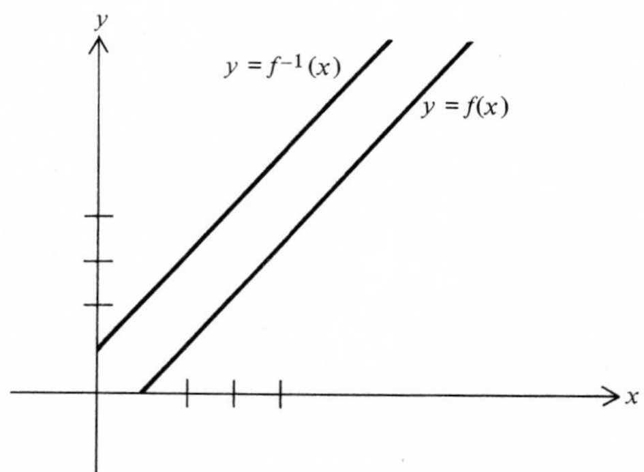

Diagram A48e

Section 7.4 (Page 194)

1 $y = x - 2$ **3** $y = x^4$ **5** $x = (1 - y)/y^2$
7 $y = x + 2$ **9** $y = (x + 1)/(3x + 2)$
11 **(a)** $y = x^2$ **(b)** $y = x^2, x \geq 0$ **(c)** No

Section 7.5 (Page 199)

1 $y = \sqrt{(x - 1)^2} = |x - 1|$ (Diagram A49)
3 domain: $x > 1$ or $x \leq -1$; zero at $x = -1$; vertical asymptote: $x = 1$; horizontal asymptote: $y = 1$ (Diagram A50)

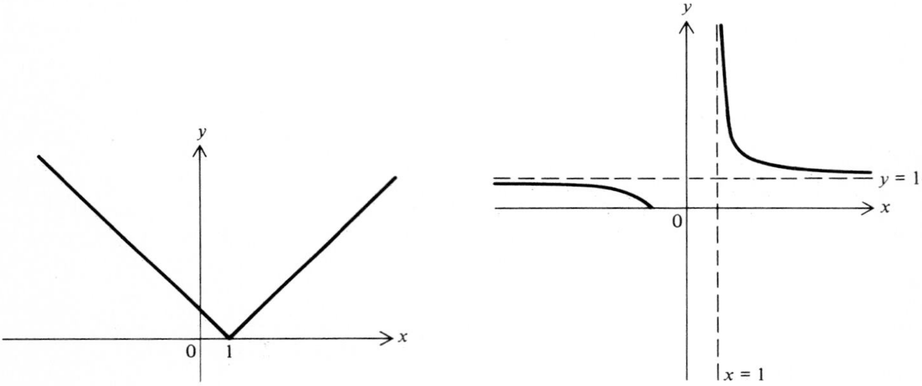

Diagram A49 *Diagram A50*

5 domain: $x \geq 0$ (Diagram A51)
7 domain: $x > -2$; vertical asymptote: $x = -2$ y intercept $3/\sqrt{2}$ (Diagram A52)

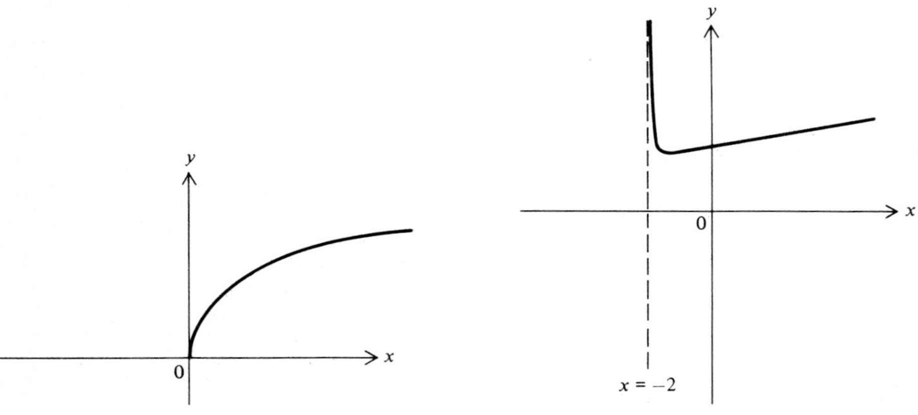

Diagram A51 *Diagram A52*

9 domain: $x > 1$ or $x < -1$; vertical asymptote: $x = \pm 1$; horizontal asymptote: $y = \pm 2$; intersection of the graph of $y = f(x)$ with $y = 2$ at $x = \frac{5}{4}$ (Diagram A53)
11 domain: $x \geq -1$; zeros at -1 and 0; range: $y \geq 0$ (Diagram A54)

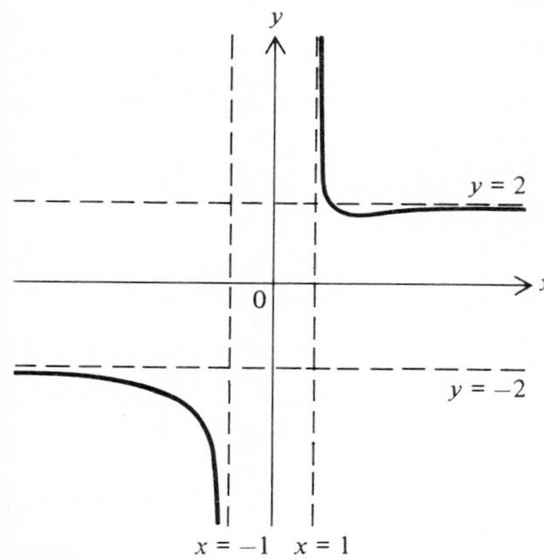

Diagram A53

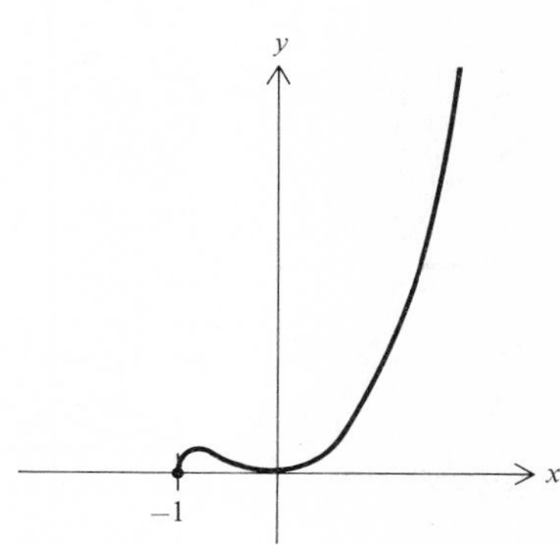

Diagram A54

13 domain: all reals, excluding $-1 \leq x \leq 1$; oblique asymptote: $y = 2x$; horizontal asymptote: $y = 0$ (Diagram A55)
15 domain: $0 \leq x < 4$; zero at 0; y intercept at 0; vertical asymptote: $x = 4$ (Diagram A56)

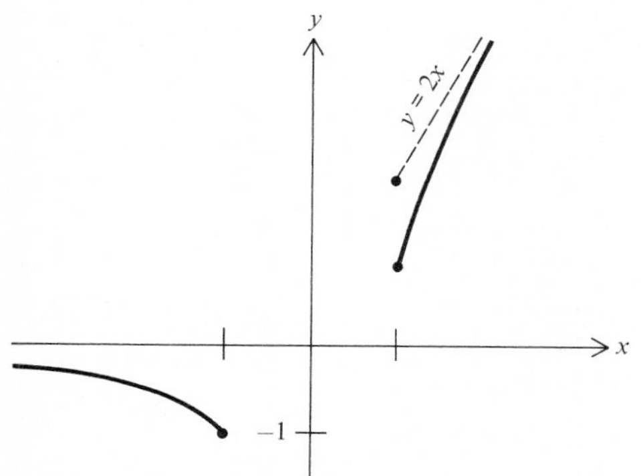

Diagram A55

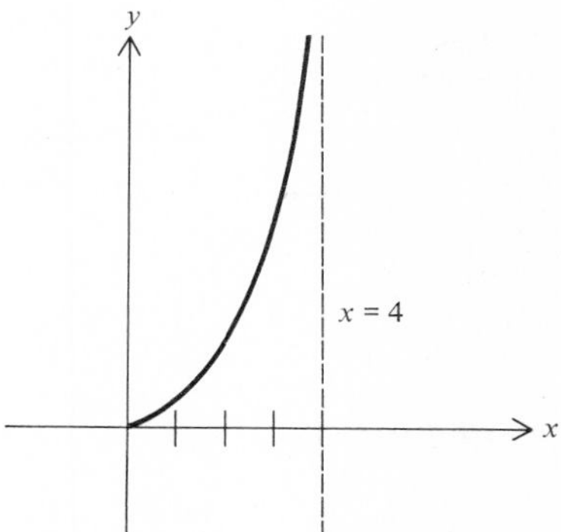

Diagram A56

17 domain: $x \geq 1$ or $x < -1$; zero at 1; horizontal asymptote: $y = 1$ and $y = -1$; vertical asymptote: $x = -1$ (Diagram A57)

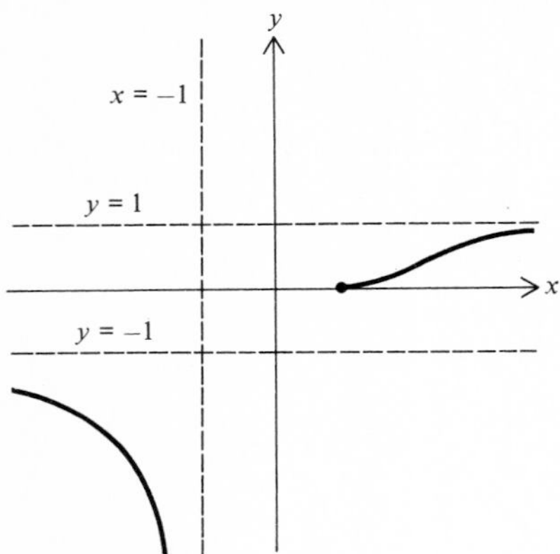

Diagram A57

CHAPTER 8

Section 8.2 (Page 211)

1 13 **3** 2 **5** 17.3 **7** 1.46
9 1.0724 **11** 0.1564 **13** 0.9228 **15** 0.2193
17 32°20′ **19** 66° **21** 61°50′ **23** 64°
25 26°10′ **27** 2.6 **29** 5.5 **31** 63.8
33 63.5 **35** $\frac{1}{2}$ **37** 68°40′

Section 8.3 (Page 219)

1 $a = 33, b = 22, B = 34°$ **3** $c = 17, A = 28°, B = 62°$
5 $b = 18, B = 66°, A = 24°$ **7** S26°30′E
9 N38°40′E **11** 47
13 **(a)** N53°30′E; **(b)** 53°30′ **15** **(a)** S56°20′E; **(b)** 303°40′
17 44.5 **19** 3.3
21 374 **23** 81
25 $w = 108, h = 97$ **27** 38
29 63.7 **31** 234
33 130

Section 8.4 (Page 229)

1 $-\frac{1}{2}$ **3** -1 **5** $2/\sqrt{3}$ **7** $-\sqrt{2}/2$
9 $-\sqrt{3}$ **11** $\sqrt{2}$ **13** $-\frac{1}{2}$ **15** $-1/\sqrt{3}$
17 $-2/\sqrt{3}$ **19** $-\sqrt{3}/2$ **21** $-\sin 50°$ **23** $\tan 32°$
25 $-\csc 36°$ **27** $\sin 6°$ **29** $-\tan 19°10'$ **31** $-\csc 88°$
33 $-\sin 18°$ **35** $-\tan 17°50'$ **37** $-\cos 32°$ **39** $\sec 44°30'$
41 $\cot 28°$ **43** $-\cos 1°$ **45** 26°, 154′ **47** 8°30′, 188°30′
49 1°, 179°

51 256°50′, 283°10′
53 104°, 284°
55 236°30′, 303°30′

In Exercises 57 to 67 the answers are specified in the following order: sin, cos, tan, csc, sec, cot.

57 $\frac{5}{13}, \frac{12}{13}, \frac{5}{12}, \frac{13}{5}, \frac{13}{12}, \frac{12}{5}$
59 $-\frac{24}{25}, -\frac{7}{25}, \frac{24}{7}, -\frac{25}{24}, -\frac{25}{7}, \frac{7}{24}$
61 $5/\sqrt{41}, 4/\sqrt{41}, \frac{5}{4}, \sqrt{41}/5, \sqrt{41}/4, \frac{4}{5}$
63 $-4/\sqrt{41}, -5/\sqrt{41}, \frac{4}{5}, -\sqrt{41}/4, -\sqrt{41}/5, \frac{5}{4}$
65 $1/\sqrt{17}, 4/\sqrt{17}, \frac{1}{4}, \sqrt{17}, \sqrt{17}/4, 4$
67 impossible data

Section 8.5 (Page 238)

1 $a = 10.7, b = 18.1$
3 $c = 6.6, A = 18°10'$
5 $a = 7.4, A = 85°10'$
7 no solution
9 $c = 34.4, B = 65°40', C = 60°20'$
11 14.9
13 101°30′
15 $A = 36°20', B = 62°40', C = 81°$
17 $c = 5.3, A = 26°10', B = 117°50'$
19 16.7
21 40 minutes
23 47
25 15°
27 10
29 **(a)** 11 miles **(b)** 43.2 second
31 **(a)** 111°20′ **(b)** 40°

Section 8.6 (Page 249)

1 **(a)** 20 **(b)** 30°
3 **(a)** 125 **(b)** 16°20′
5 **(a)** $25\sqrt{5}$ **(b)** 26°30′
7 60, 25
9 $15\sqrt{3}$, 15
11 **(a)** $20\sqrt{29}$ **(b)** 21°50′
13 87
15 15°40′, 56
17 N23°30′E
19 87 miles, 659 mph
21 479, 20°
23 7.1
25 $S = 146, T = 179$

CHAPTER 9

Section 9.2 (Page 259)

1 $(0, -1)$
3 $(1, 0)$
5 $\left(-\frac{\sqrt{3}}{2}, \frac{1}{2}\right)$
7 $(\sqrt{3}/2, -\frac{1}{2})$
9 $(-1/\sqrt{2}, -1/\sqrt{2})$
11 $(-\frac{1}{2}, \sqrt{3}/2)$
13 $(\frac{1}{2}, -\sqrt{3}/2)$
23 $(-x, -y)$
25 $(x, -y)$

Section 9.3 (Page 261)

1 See table on pages 438 and 439.
3 second quadrant
5 $-\frac{4}{5}$
7 0.84
9 0.73
11 0.10
13 0.47943
15 0.54030

Z	$W(Z)$	$\cos Z$	$\sin Z$
0	$(1, 0)$	1	0
$\frac{\pi}{3}$	$\left(\frac{1}{2}, \frac{\sqrt{3}}{2}\right)$	$\frac{1}{2}$	$\frac{\sqrt{3}}{2}$
$\frac{2\pi}{3}$	$\left(-\frac{1}{2}, \frac{\sqrt{3}}{2}\right)$	$-\frac{1}{2}$	$\frac{\sqrt{3}}{2}$
$\frac{4\pi}{3}$	$\left(-\frac{1}{2}, -\frac{\sqrt{3}}{2}\right)$	$-\frac{1}{2}$	$-\frac{\sqrt{3}}{2}$
$\frac{5\pi}{3}$	$\left(\frac{1}{2}, -\frac{\sqrt{3}}{2}\right)$	$\frac{1}{2}$	$-\frac{3}{2}$
$\frac{\pi}{6}$	$\left(\frac{\sqrt{3}}{2}, \frac{1}{2}\right)$	$\frac{\sqrt{3}}{2}$	$\frac{1}{2}$
$\frac{5\pi}{6}$	$\left(-\frac{\sqrt{3}}{2}, \frac{1}{2}\right)$	$-\frac{\sqrt{3}}{2}$	$\frac{1}{2}$
$\frac{7\pi}{6}$	$\left(-\frac{\sqrt{3}}{2}, -\frac{1}{2}\right)$	$-\frac{\sqrt{3}}{2}$	$-\frac{1}{2}$
$\frac{11\pi}{6}$	$\left(\frac{\sqrt{3}}{2}, -\frac{1}{2}\right)$	$\frac{\sqrt{2}}{2}$	$-\frac{1}{2}$
$\frac{\pi}{4}$	$\left(\frac{1}{\sqrt{2}}, \frac{1}{\sqrt{2}}\right)$	$\frac{1}{\sqrt{2}}$	$\frac{1}{\sqrt{2}}$
$\frac{3\pi}{4}$	$\left(-\frac{1}{\sqrt{2}}, \frac{1}{\sqrt{2}}\right)$	$-\frac{1}{\sqrt{2}}$	$\frac{1}{\sqrt{2}}$
$\frac{5\pi}{4}$	$\left(-\frac{1}{\sqrt{2}}, -\frac{1}{\sqrt{2}}\right)$	$-\frac{1}{\sqrt{2}}$	$-\frac{1}{\sqrt{2}}$
$\frac{7\pi}{4}$	$\left(\frac{1}{\sqrt{2}}, -\frac{1}{\sqrt{2}}\right)$	$\frac{1}{\sqrt{2}}$	$-\frac{1}{\sqrt{2}}$
$\frac{\pi}{2}$	$(0, 1)$	0	1
π	$(-1, 0)$	-1	0
$\frac{3\pi}{2}$	$(0, -1)$	0	-1
2π	$(1, 0)$	1	0
100π	$(1, 0)$	1	0
$\frac{22\pi}{3}$	$\left(-\frac{1}{2}, -\frac{\sqrt{3}}{2}\right)$	$-\frac{1}{2}$	$-\frac{\sqrt{3}}{2}$

Z	$W(Z)$	$\cos Z$	$\sin Z$
$\frac{43\pi}{4}$	$\left(-\frac{1}{\sqrt{2}}, \frac{1}{\sqrt{2}}\right)$	$-\frac{1}{\sqrt{2}}$	$\frac{1}{\sqrt{2}}$
$-\frac{9\pi}{4}$	$\left(\frac{1}{\sqrt{2}}, -\frac{1}{\sqrt{2}}\right)$	$\frac{1}{\sqrt{2}}$	$-\frac{1}{\sqrt{2}}$
$\frac{37\pi}{6}$	$\left(\frac{\sqrt{3}}{2}, \frac{1}{2}\right)$	$\frac{\sqrt{3}}{2}$	$\frac{1}{2}$
$-\frac{13\pi}{6}$	$\left(\frac{\sqrt{3}}{2}, -\frac{1}{2}\right)$	$\frac{\sqrt{3}}{2}$	$-\frac{1}{2}$

Section 9.4 (Page 268)

1 $(\sqrt{6} - \sqrt{2})/4$ **3** $(\sqrt{6} + \sqrt{2})/4$ **5** $-(\sqrt{6} + \sqrt{2})/4$
7 $(\sqrt{6} - \sqrt{2})/4$ **9** $(\sqrt{6} + \sqrt{2})/4$ **11** -1
13 $\frac{1}{2}$ **15** $\frac{1}{2}$ **35** false
37 $1/\sqrt{2}$

Section 9.6 (Page 272)

1 zeros: $x = -3\pi/2, -\pi/2$; maximum value: $x = -2\pi$; minimum value: $x = -\pi$ (Diagram A58)

3 zeros: $x = \pi/2$; maximum value: $x = 0$; minimum value: $x = \pi$; $f(4) \approx -0.65$ (Diagram A59)

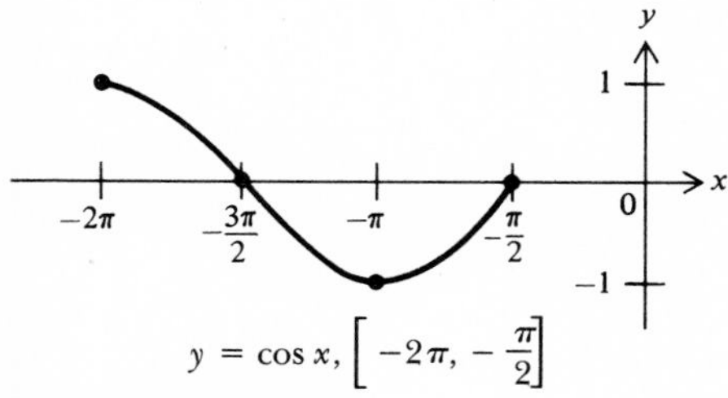

Diagram A58

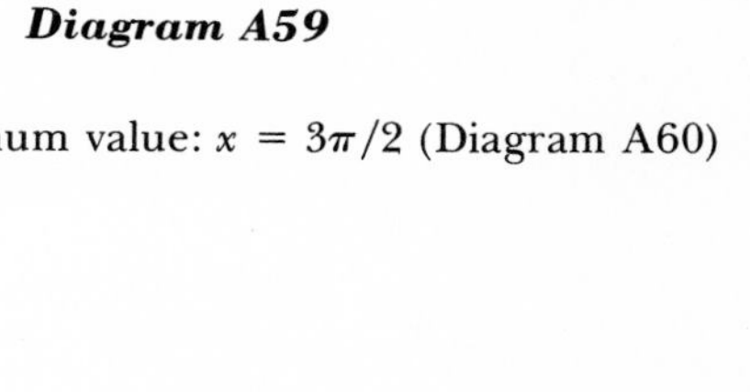

Diagram A59

5 zeros: none; maximum value: none; minimum value: $x = 3\pi/2$ (Diagram A60)

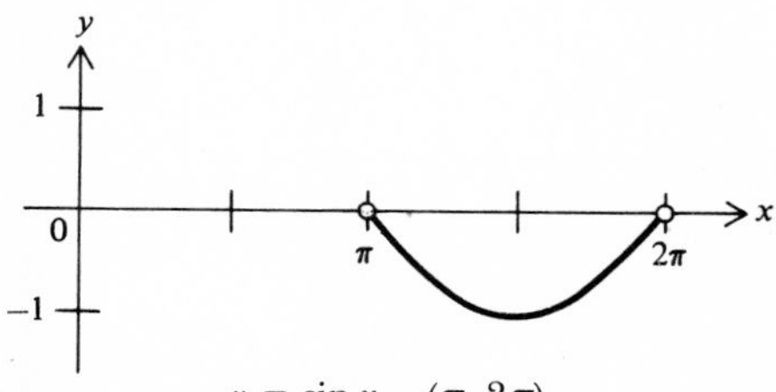

Diagram A60

7 zeros: $x = 0$; maximum value: $x = \pi/2$; minimum value: none (Diagram A61)
9 zeros: $x = -5\pi/2$; maximum value: $x = -2\pi$; minimum value: $x = -3\pi$ (Diagram A62)
11 $x = \pi/4$ and $5\pi/4$

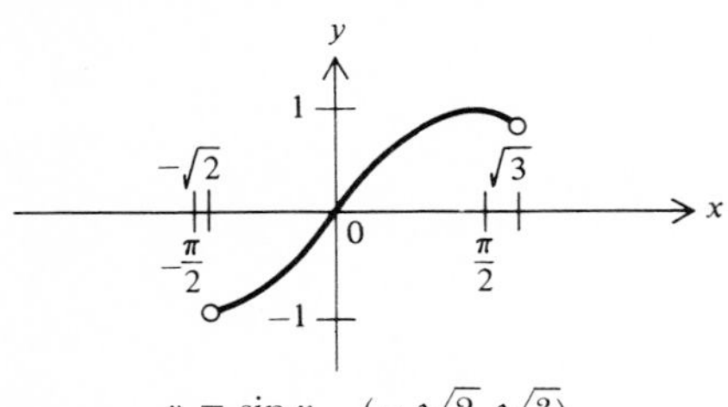

$y = \sin x, \quad (-\sqrt{2}, \sqrt{3})$

Diagram A61

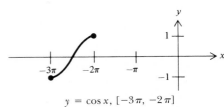

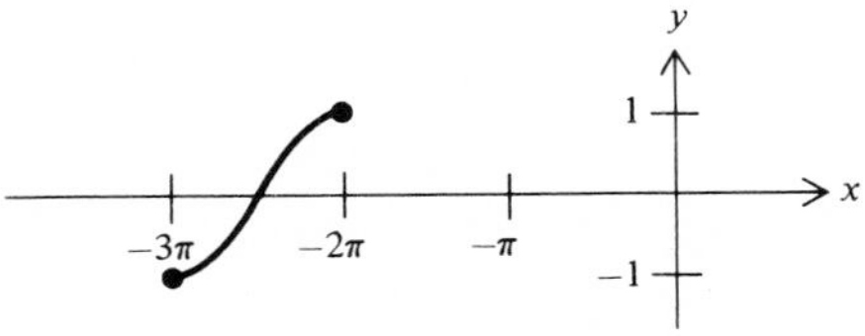

$y = \cos x, \; [-3\pi, -2\pi]$

Diagram A62

Section 9.7 (Page 284)

1 Diagram A63 **3** Diagram A64 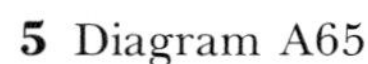 **5** Diagram A65

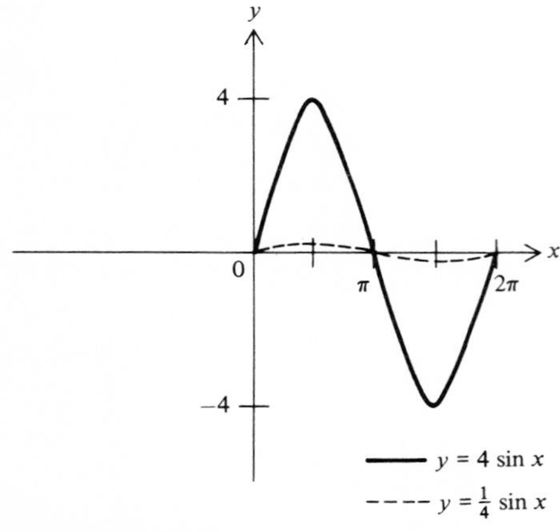

Diagram A63

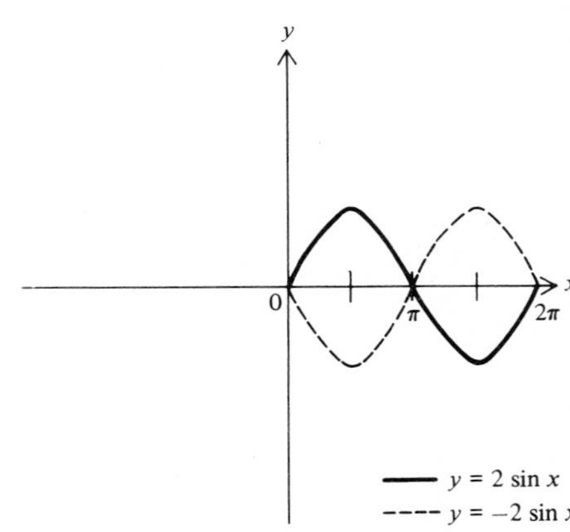

Diagram A64

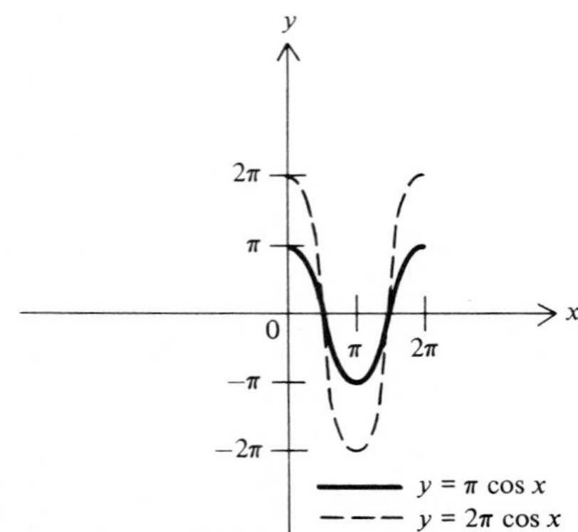

Diagram A65

7 $y = \frac{1}{2}\cos 2x$ and $y = \frac{1}{2}\sin 2x$
9 $y = 2\pi \cos 4x$ and $y = 2\pi \sin 4x$
11 Diagram A66
13 Diagram A67
15 Diagram A68
17 Diagram A69
19 Diagram A70
21 Diagram A71

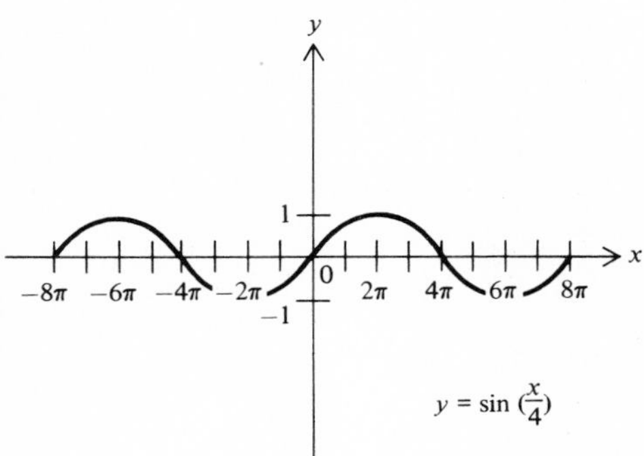

Diagram A66

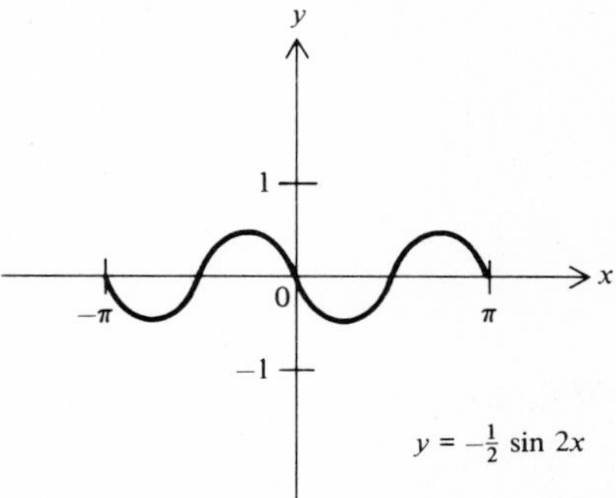

Diagram A67

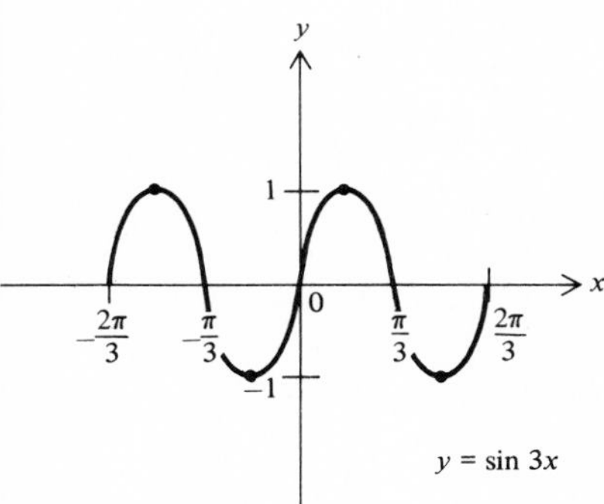

Diagram A68

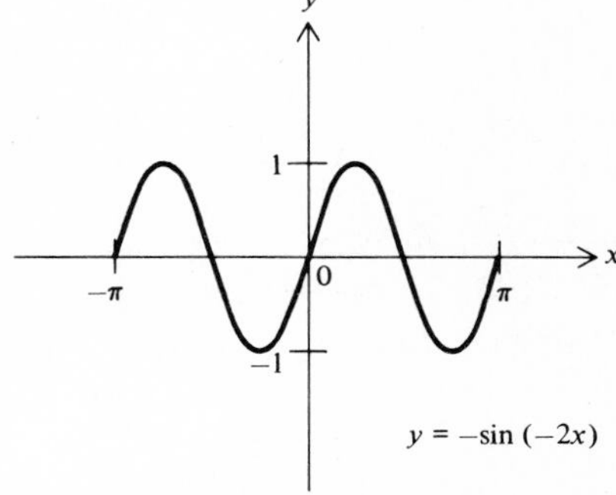

Diagram A69

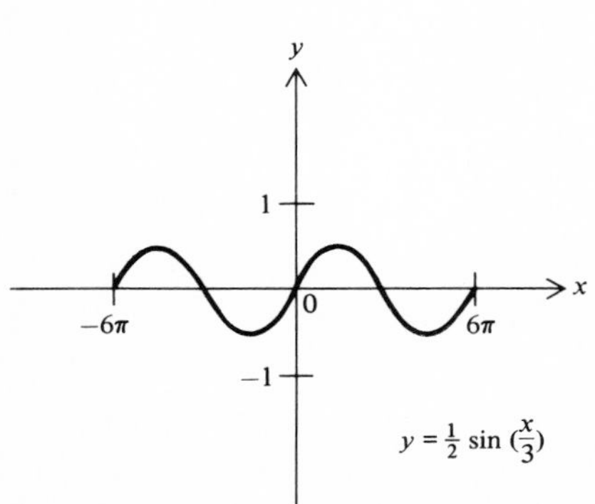

Diagram A70

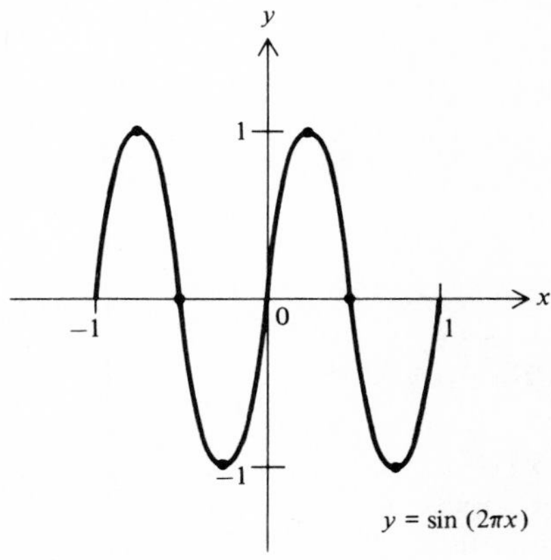

Diagram A71

23 Diagram A72 **25** Diagram A73 **27** Diagram A74
29 Diagram A75 **31** Diagram A76 **33** Diagram A77

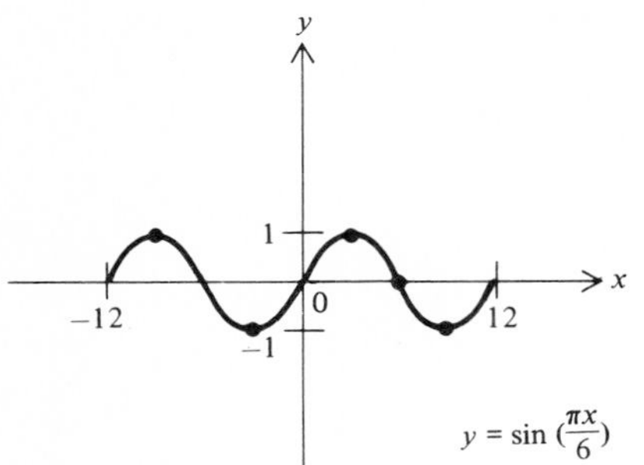

Diagram A72

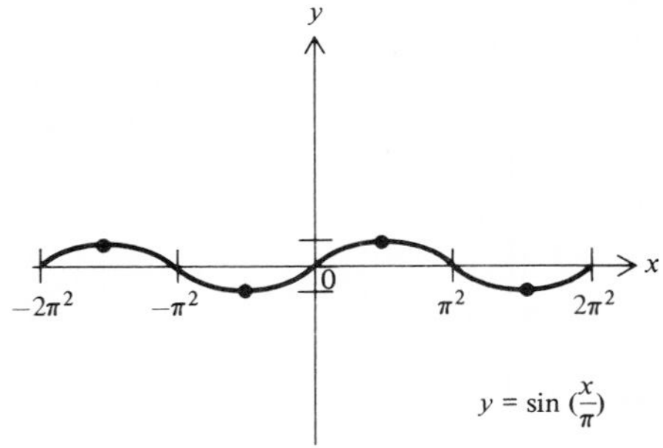

Diagram A73

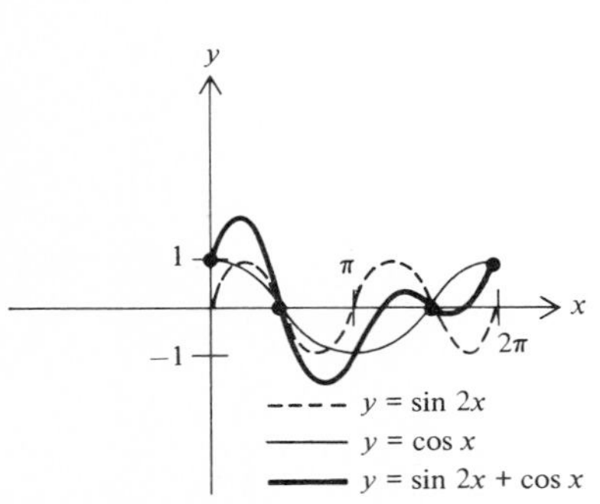

Diagram A74

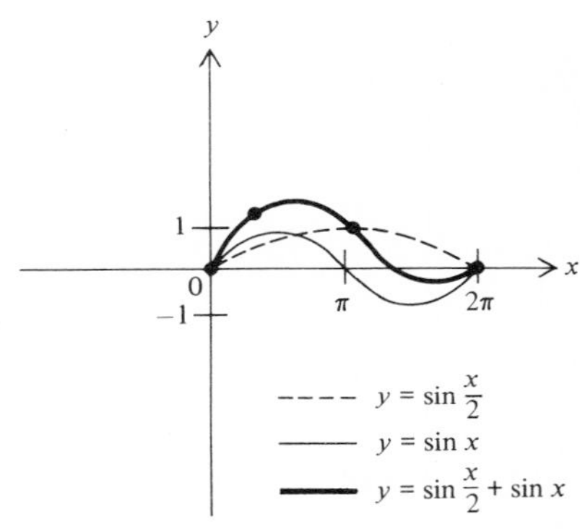

Diagram A75

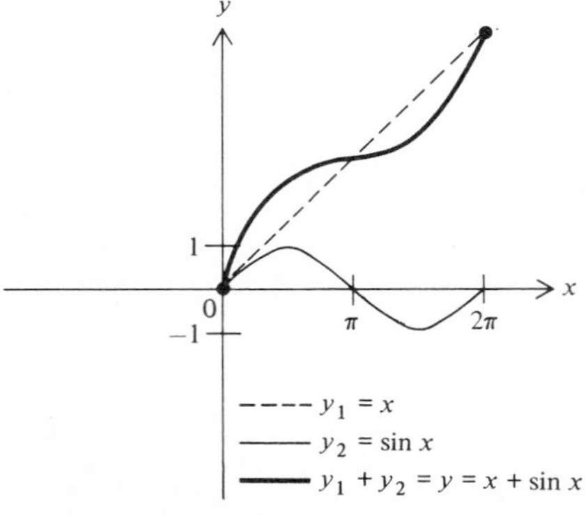

Diagram A76

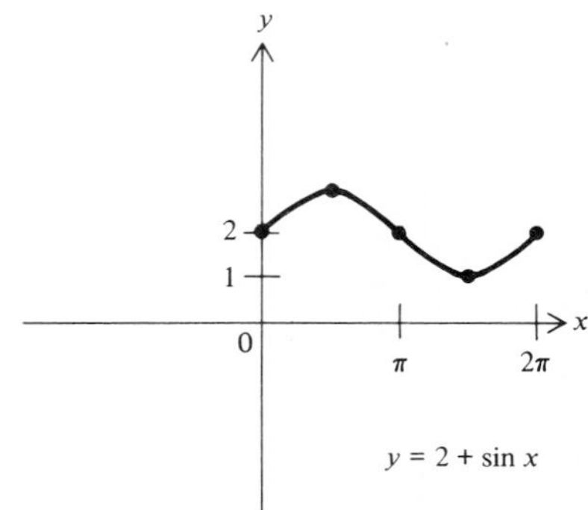

Diagram A77

35 If $k > 0$, then the graphs of these functions are translated k units above the x axis. If $k < 0$ then the graphs of these functions are translated k units below the x axis.

Section 9.8 (Page 290)

1 Diagram A78 **3** Diagram A79 **5** Diagram A80
7 Diagram A81 **9** Diagram A82 **11** Diagram A83

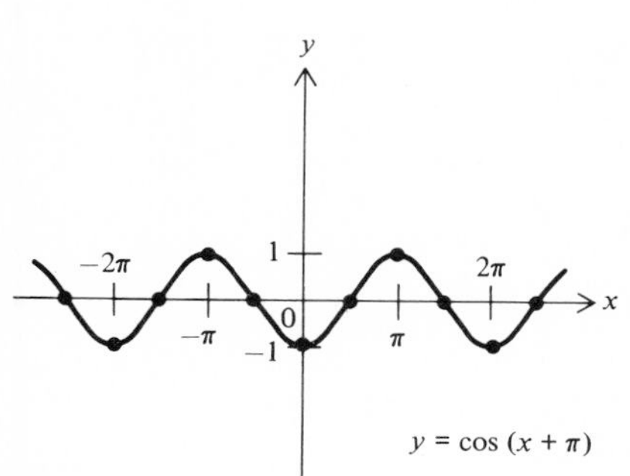

Diagram A78

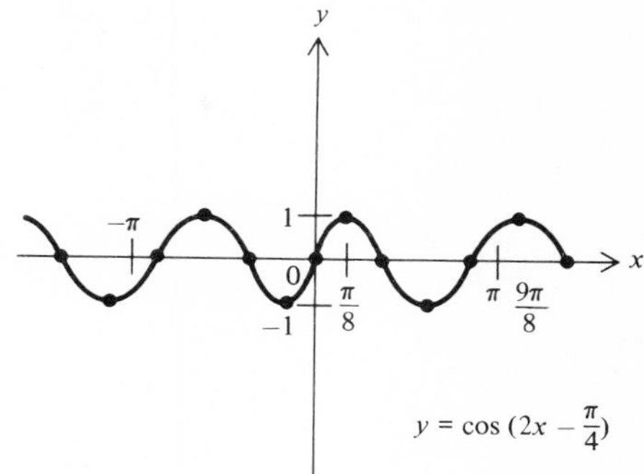

Diagram A79

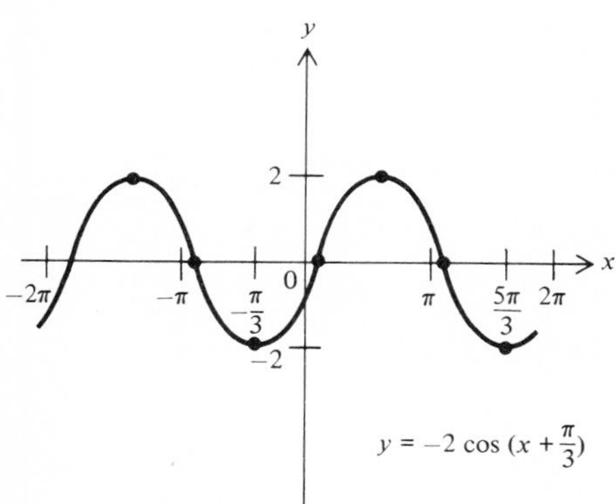

Diagram A80

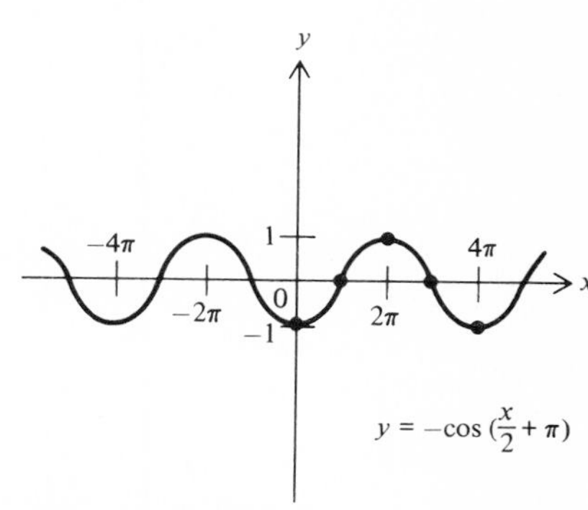

Diagram A81

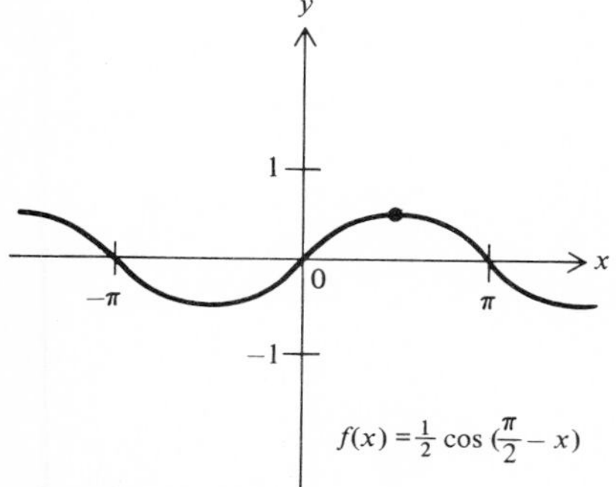

Diagram A82

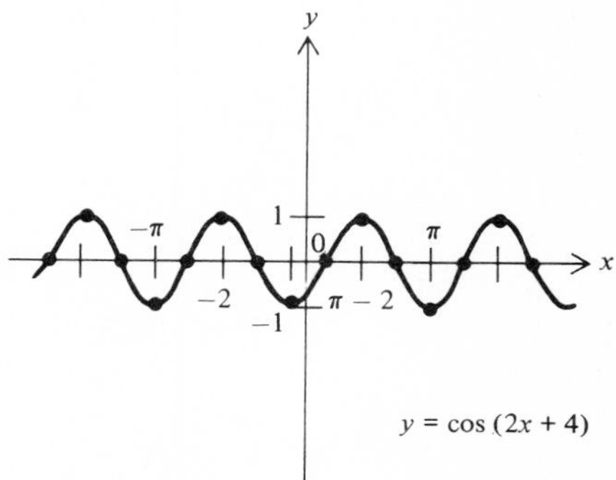

Diagram A83

13 Diagram A84 **15** Diagram A85 **17** Diagram A86

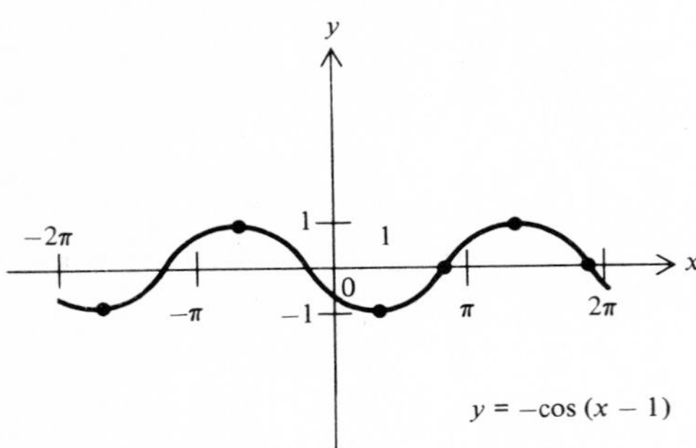

Diagram A84

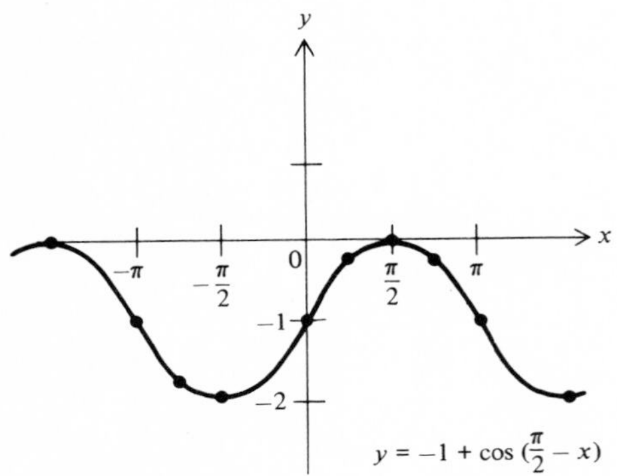

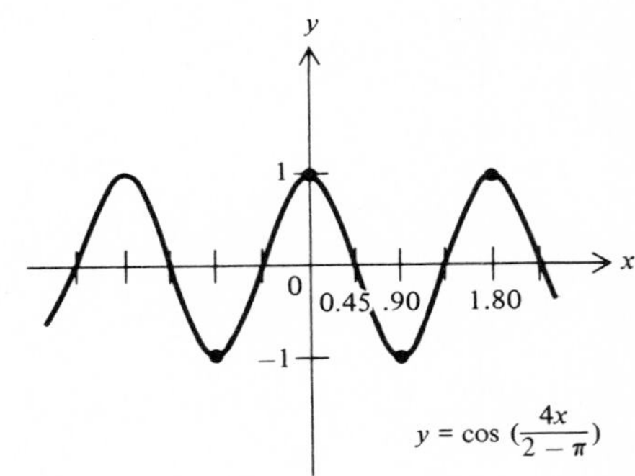

Diagram A85 ***Diagram A86***

Section 9.9 (Page 295)

1 no solution **3** $\pi/6$ **5** 0.21
7 $5\pi/6$ **9** 0.64 **11** $1\sqrt{2}$
13 $\pi/3$ **15** 0.96872 **17** 0
19 2 **23** $(\sqrt{6} + \sqrt{2})/4$

Section 9.10 (Page 299)

1 0 **3** $2/\sqrt{3}$
5 1 **7** -1
9 undefined **11** 1
13 $-2 - \sqrt{3}/2$ **15** $2 + \sqrt{3}/2$
19 Diagram A87 **(a)** $\pi/2$ and $3\pi/2$ **(b)** $0 \leq x \leq 2\pi, x \neq \pi/2, 3\pi/2$ **(c)** $y \geq 1$ or $y \leq -1$ **(d)** $x = \pi/2$ and $x = 3\pi/2$
21 Diagram A88 **(a)** 0, π and 2π **(b)** $0 < x < \pi$ or $\pi < x < 2\pi$ **(c)** all reals **(d)** $x = 0, x = \pi, x = 2\pi$.

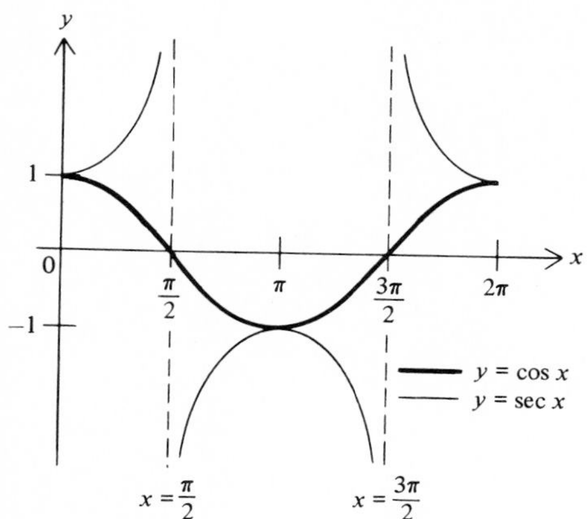

Diagram A87

Diagram A88

25 **(a)** derivation **(b)** $x \neq n\pi/2$, n is an integer
27 **(a)** $\sqrt{3} - 2$ **(b)** $2 + \sqrt{3}$ **(c)** $2 - \sqrt{3}$
29 $\pi/2$ and $\pi/3$, respectively
31 There are real numbers x, such that tan x is undefined. The function tends towards infinity as x nears such values (i.e., $\pi/2$). Hence, no maximum or minimum function values can be determined.

CHAPTER 10

Section 10.1 (Page 304)

1 3^6	**3** 4^6	**5** $8b^3y^3$	**7** 8^3	**9** 1
11 2π	**13** 1	**15** 2	**17** 24	**19** 64
21 27	**23** $2^{7/12}$	**25** $2x^{1/3}$	**27** 3	**29** 4
31 $\frac{1}{25}$	**33** 2	**35** $2^{3.3}$	**37** b^{2m+2}	**39** $b^{2n} + 1/b^{2n}$
41 2	**43** $1/b^k$	**45** 5	**47** -4	**49** $\frac{2}{3}$

Section 10.2 (Page 309)

1 Diagram A89 **3** Diagram A90 **(a)** 2; **(b)** (2, 4)
5 Diagram A91 **7** Diagram A92 **9** Diagram A93
11 **(a)** 0.6 **(b)** -0.4 **(c)** 0.9 **(d)** -0.1 **(e)** 0.8 **(f)** no value

Section 10.3 (Page 312)

1 1.649	**3** 1.350	**5** 0.905
9 Diagram A94	**11** Diagram A95	**15** 0
17 1.1752	**19** 0.5211	

23 Some selected ordered pairs $(x, f(x))$: (1, 2), (2, 2.25), (4, 2.44), (10, 2.5937), (50, 2.6916), (100, 2.7048), (500, 2.7156), (1000, 2.7169), (10000, 2.7181), (100,000,000, 2.7182818). Since $e \approx 2.71828$, the answer is yes.

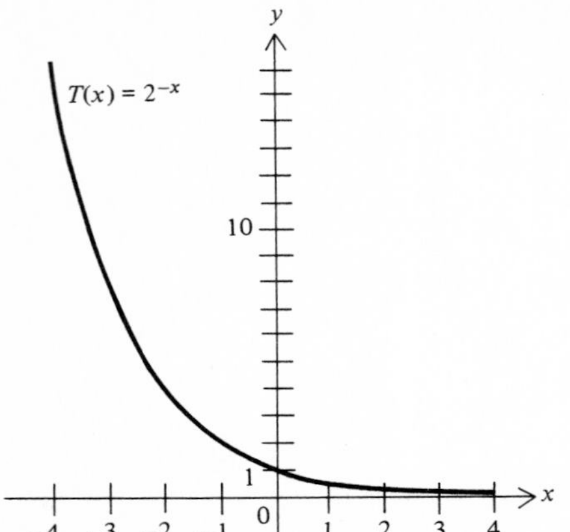

Diagram A89

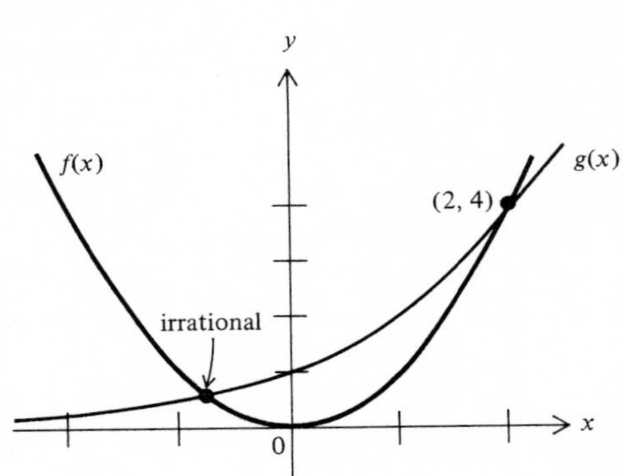

Diagram A90

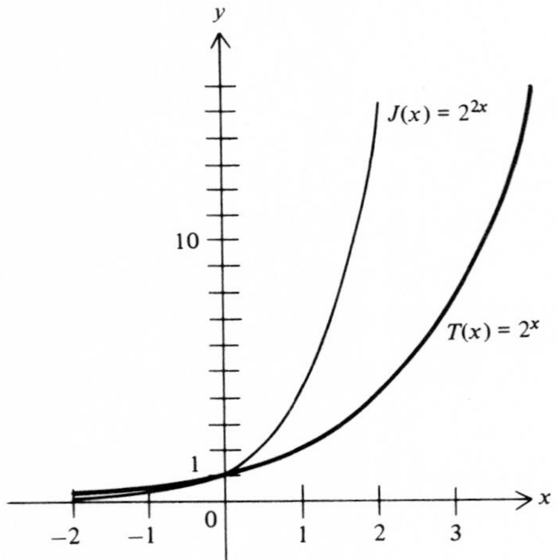

Diagram A91

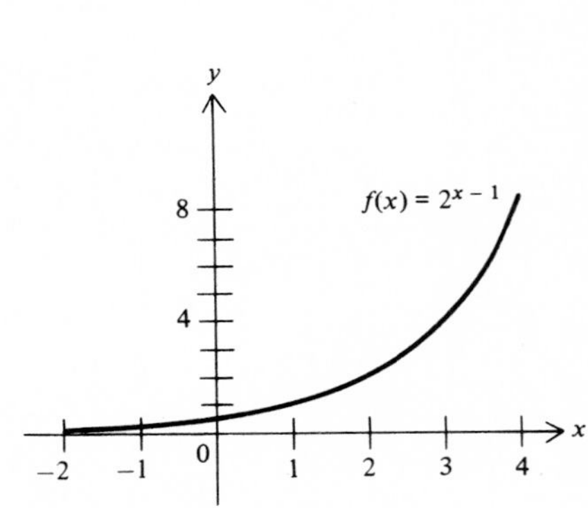

Diagram A92

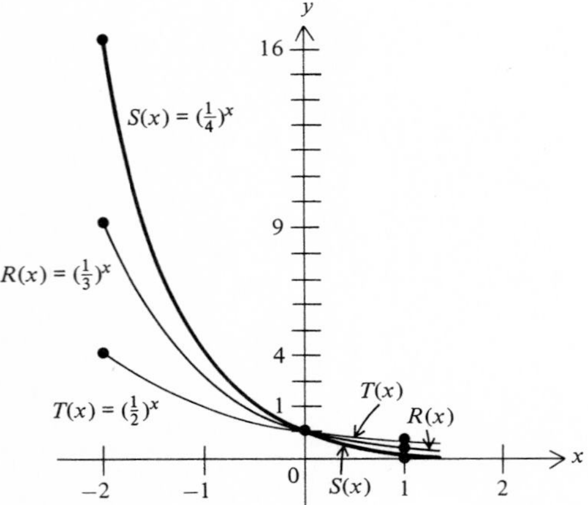

Diagram A93

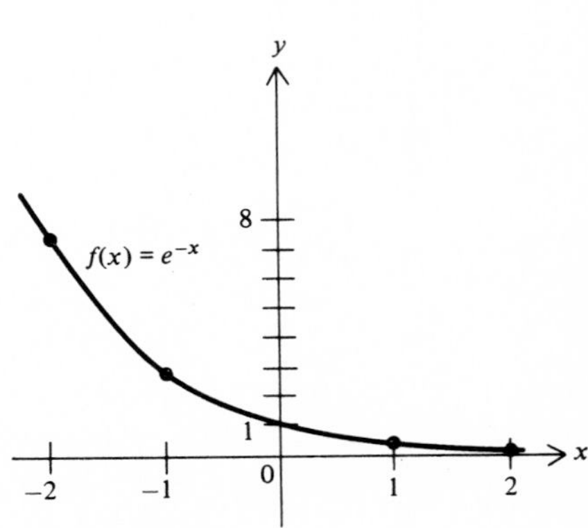

Diagram A94

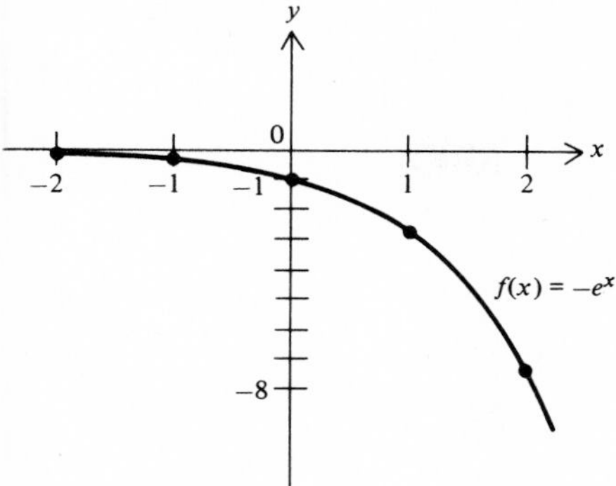

Diagram A95

Section 10.4 (Page 316)

1 $2^4 = 16$ **3** $10^1 = 10$ **5** $(\frac{2}{3})^{-2} = \frac{9}{4}$
7 $\log_4 256 = 4$ **9** $\log_8 (\frac{1}{64}) = -2$ **11** 3
13 $\frac{1}{2}$ **15** $\frac{1}{3}$ **17** Diagram A96

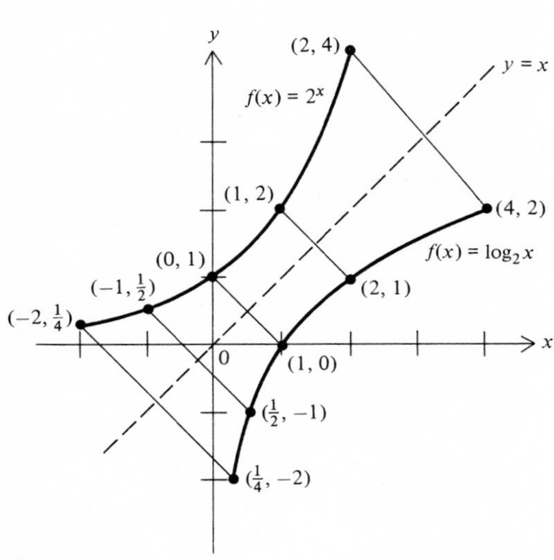

Diagram A96

19 (a) 2.8 **(b)** 1.4 **(c)** 5.8 **(d)** 0.7
21 (a) 0.3 **(b)** 0.48 **(c)** 0.6 **(d)** 0.9
23 Diagram A97 **25** Diagram A98 **27** Diagram A99
29 Exercise 23: $y = e^x$; Exercise 24: $y = (\frac{1}{3})^x$; Exercise 25: $y = 10^x$; Exercise 26: $y = 0.6^x$; Exercise 27: no inverse; Exercise 28: $y = 2^x, x \geq 0$

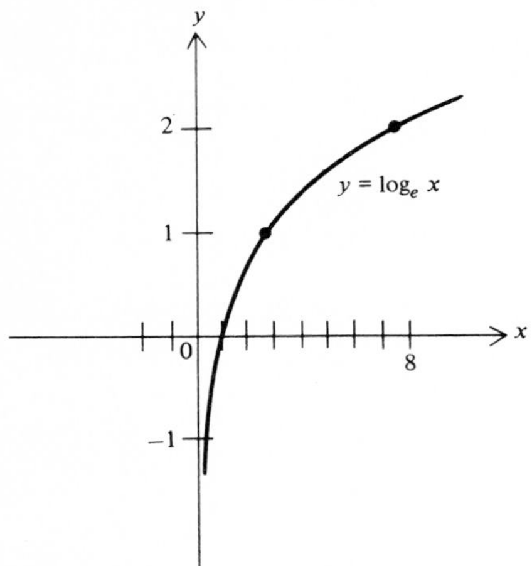

Diagram A97

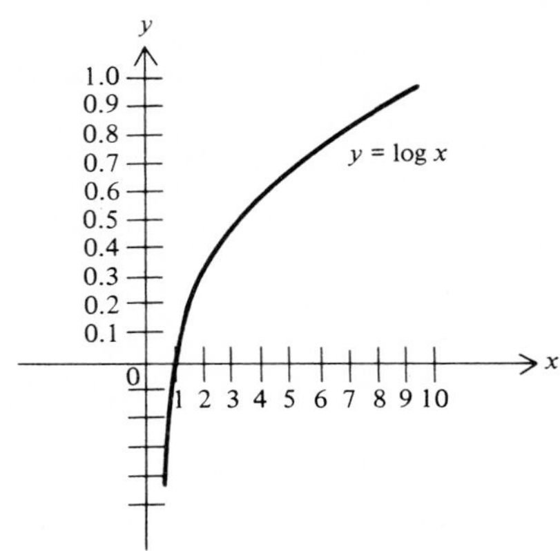

Diagram A98

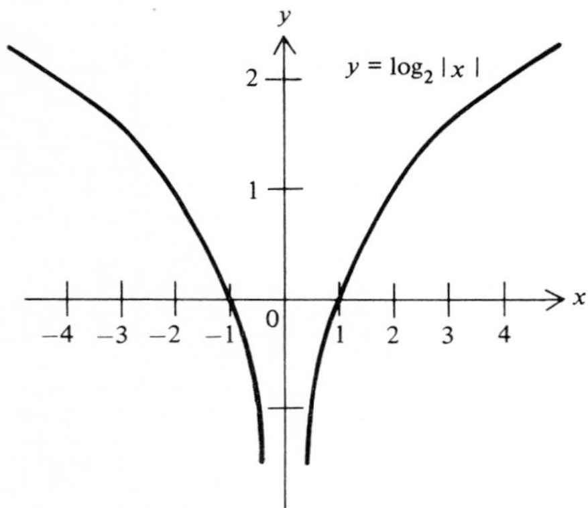

Diagram A99

Section 10.5 (Page 326)

1 Diagram A100 **3** Diagram A101 **5** Diagram A102
13 **(a)** 1.23 **(b)** 0.812
15 6.225
19 **(a)** 90.5 **(b)** 2894
21 **(a)** 9626 **(b)** 8188
23 11.7
25 200
27 8.30
29 20.5
31 711
33 0.767
35 0.255
37 0.00648
39 184
41 1.74
43 1.36
45 0.218
47 0.220
49 154
51 **(a)** moon 2.489, Jupiter 0.6145 **(b)** moon 0.1318, Jupiter 2.162
53 $r = 0.7, A = 1.54$
55 3.96
57 **(a)** 1062 **(b)** 1061
59 73.8
61 79%
63 55

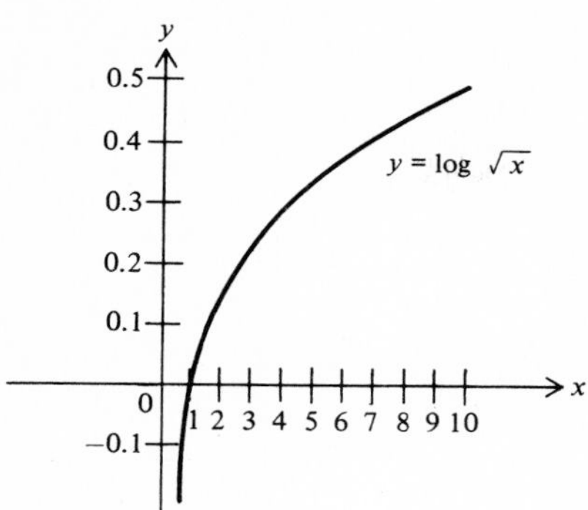

Diagram A100

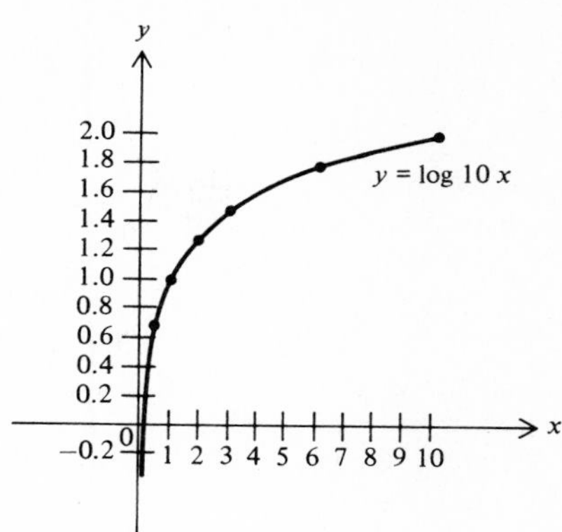

Diagram A101

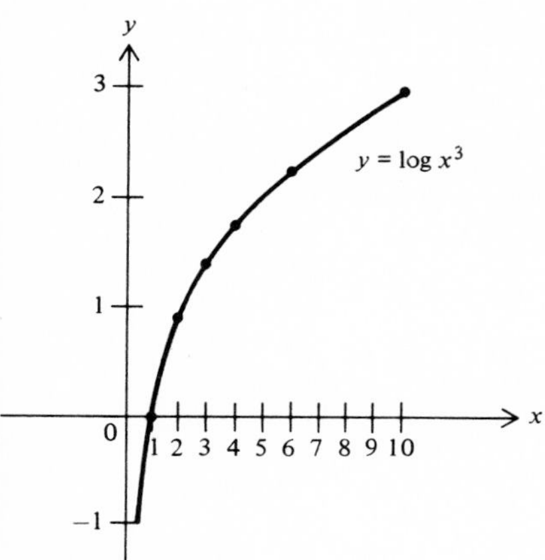

Diagram A102

CHAPTER 11

Section 11.2 (Page 335)

1 $(x-2)^2+(y-3)^2=20$, $c(2,3)$, $r=2\sqrt{5}$
3 $(x-\frac{3}{2})^2+(y+\frac{4}{3})^2=\frac{144}{36}$, $c(\frac{3}{2},-\frac{4}{3})$, $r=2$
5 locus is the point $(1,-2)$
7 $(x-\frac{1}{2})^2+(y+\frac{3}{2})^2=4$, $c(\frac{1}{2},-\frac{3}{2})$, $r=2$
9 $(4,0)$, $(-3,0)$, $(0,2)$, $(0,-6)$
13 $(3,6)$
15 **(a)** $x^2+y^2=2a^2$ **(b)** $2a^2(4-\pi)$
17 $(-\frac{1}{5},\frac{8}{5})$, $(1,-2)$
19 $4x+3y+10=0$

21 **(a)** $(-2, 4)$, $(\frac{2}{5}, \frac{4}{5})$ **(b)** at $(-2, 4)$, $2x - y + 8 = 0$, $x + 2y - 6 = 0$; at $(\frac{2}{5}, \frac{4}{5})$, $11x + 2y - 6 = 0$, $2x - 11y + 8 = 0$

25 $y = 2$, $x = 1$, $4x - 3y - 10 = 0$, $3x + 4y - 5 = 0$

Section 11.3 (Page 340)

1 $(x - \frac{3}{5})^2 + (y - \frac{16}{5})^2 = \frac{153}{5}$

3 $(x + 1)^2 + (y - 10)^2 = 25$, $c(-1, 10)$, $r = 5$

5 $(x - 3)^2 + (y + 2)^2 = 25$, $c(3, -2)$, $r = 5$

7 $x^2 + y^2 + 4x - 4y - 17 = 0$

9 $x^2 + y^2 - 12x + 4y + 23 = 0$

11 **(a)** $(\frac{3}{5}, -\frac{4}{5})$, $(3, 4)$ **(b)** $x + y - 7 = 0$, $x + 7y + 5 = 0$

13 $(x + 2)^2 + (y + 3)^2 = 32$

15 $x - y - 3 = 0$

17 $(x - \frac{5}{2})^2 + (y - \frac{1}{2})^2 = \frac{5}{2}$, $c(\frac{5}{2}, \frac{1}{2})$, $r = \sqrt{5}/2$

19 $2y = x$, $2x + y + 5 = 0$

Section 11.4 (Page 344)

1 focus: $(2, 0)$, directrix: $x = -2$

3 focus: $(-3, 0)$, directrix: $x = 3$

5 focus: $(0, -\frac{1}{2})$, directrix: $y = \frac{1}{2}$

7 $y^2 = -8x$

9 focus: $(0, \frac{1}{6})$, directrix: $y = -\frac{1}{6}$

11 $(2, 2)$, $(2, -2)$

13 **(b)** $x^2 = -4y$ **(c)** $h = 8$, $w = 4$ **(d)** 5

15 $y = 2x - 2$ is tangent to $y^2 = -16x$ at $(-1, -4)$.

17 $34\sqrt{6}$

Section 11.5 (Page 351)

1 $(\frac{3}{4}, -\frac{23}{40})$

3 $V(2, -\frac{3}{2})$, $F(2, -\frac{1}{2})$; directrix: $y = -\frac{5}{2}$

5 $V(\frac{1}{2}, -2)$, $F(-\frac{3}{2}, -2)$; directrix: $x = \frac{5}{2}$

7 $V(-\frac{3}{2}, 2)$, $F(-\frac{1}{2}, 2)$; directrix: $x = -\frac{5}{2}$

9 $V(1, -\frac{2}{3})$, $F(1, -1.042)$; directrix: $y = -0.292$

11 $V(2, -2)$, $F(2, -\frac{4}{3})$; directrix: $y = -\frac{8}{3}$

13 $V(2, -\frac{3}{2})$, $F(2, -\frac{3}{4})$; directrix: $y = -\frac{9}{4}$

15 $V(-\frac{1}{2}, 3)$, $F(\frac{5}{2}, 3)$; directrix: $x = -\frac{7}{2}$

17 $V(2, -\frac{3}{2})$, $F(\frac{13}{4}, -\frac{3}{2})$; directrix: $x = \frac{3}{4}$

19 $V(2, -1)$, $F(\frac{1}{2}, -1)$; directrix: $x = \frac{7}{2}$

23 $\left(-\dfrac{q}{2p}, r - \dfrac{q^2}{4p}\right)$

25 $(y - 3)^2 = -8(x - 2)$

27 $x^2 + 8x - 8y - 1 = 0$

29 $y = x^2$

31 $(y + \frac{1}{2}) = \frac{12}{5}(x - 2)$

33 $(-2, 4)$

35 **(a)** $y^2 - 3y - x + 1 = 0$, $V(-\frac{5}{4}, \frac{3}{2})$, $F(-1, \frac{3}{2})$; directrix: $x = -\frac{3}{2}$; **(b)** $x^2 - 3x - 3y + 2 = 0$, $V(\frac{3}{2}, -\frac{1}{12})$, $F(\frac{3}{2}, \frac{2}{3})$; directrix: $y = -\frac{5}{6}$ (Diagram A103)

37 $y^2 = -4a(x - a)$

39 **(a)** $y = \sqrt{3}x - x^2/6400$ **(b)** $6400\sqrt{3}$ **(c)** 4800

41 **(a)** $20\sqrt{3}$ **(b)** 15 and $(625\sqrt{2})/8$

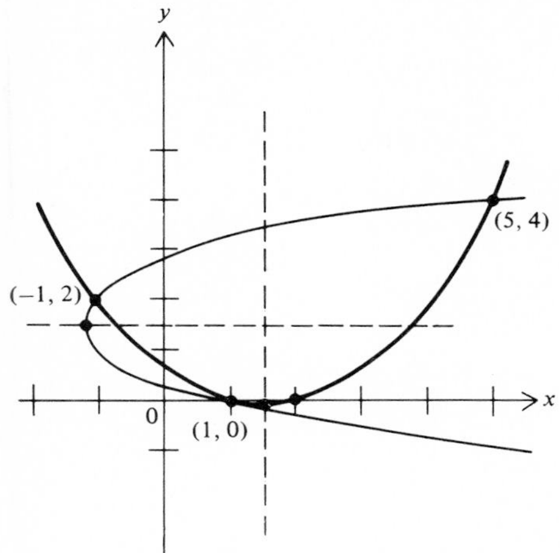

Diagram A103

CHAPTER 12

Section 12.1 (Page 360)

1 $F(4, 0), (-4, 0), V(5, 0), (-5, 0), (0, 3), (0, -3)$
3 $F(3.464, 0), (-3.464), V(4, 0), (-4, 0), (0, 2), (0, -2)$
5 $F(0.764, 0), (-0.764, 0), V(1.155, 0), (-1.155, 0), (-1.55, 0), (0,0.866), (0, -0.866)$
7 $x^2/12 + y^2/16 = 1$
9 $x + 3y - 4 = 0, 3x + y + 12 = 0; (-\frac{5}{2}, -\frac{9}{2})$
11 $x^2/80 + y^2/20 = 1$
15 **(1)** $\frac{4}{5}$; **(2)** $\sqrt{5}/3$; **(3)** $\sqrt{3}/2$; **(4)** $\sqrt{3}/3$
17 closest = 88.5 million miles, farthest = 91.5 million miles
21 **(a)** 180 **(b)** 90 **(c)** 135 **(d)** 6.4 units smaller

Section 12.2 (Page 365)

1 $C(2, 0), F(3.732, 0), (0.268, 0), V(4, 0), (0, 0), (2, -1)$
3 $c(4, -2), F(5.414, -2), (2.586, -2), V(6, -2), (2, -2), (4, -0.586), (4, -3.414)$
5 $C(-2, 3), F(-2, 5.646), (-2, 0.354), V(-2, 7), (-2, -1), (1, 3), (-5, 3)$
7 $C(6, -4), F(10.472, -4), (1.528, -4), V(12, -4), (0, -4), (6, 0), (6, -8)$
9 $C(4, 1), F(4, 3.828), (4, -1.828), V(4, -3), (4 - 2\sqrt{2}, 1), (4 + 2\sqrt{2}, 1)$
11 $C(4, -3), F(9.745, -3), (-1.745, -3), V(4, -7), (-3, -3), (11, -3)$
15 **(a)** $\dfrac{(x + 2)^2}{36} + \dfrac{(y + 1)^2}{9} = 1$ **(b)** 36 **(c)** 72 **(d)** 56.52
13 $C(4, 3), F(4, 6.464), (4, -0.464)$
17 $2x^2 + 5y^2 + 8x - 10y - 64 = 0$
19 $\dfrac{(x + 1)^2}{225} + \dfrac{(y - 2)^2}{400} = 1$

Section 12.3 (Page 371)

1 $F(5, 0)$, $(-5, 0)$, $V(4, 0)$, $(-4, 0)$; asymptotes: $y = \pm 3x/4$
3 $F(0, 6)$, $(0, -6)$, $V(0, 3)$, $(0, -3)$; asymptotes: $y = \pm 0.577x$
5 $F(2.236, 0)$, $(-2.236, 0)$, $V(1, 0)$, $(1-1, 0)$; asymptotes: $y = \pm 2x$
7 $F(0, 8)$, $(0, -8)$, $V(0, 6)$, $(0, -6)$; asymptotes: $y = \pm 3x\sqrt{\sqrt{7}}$
9 $F(2\sqrt{15}, 0)$, $(-2\sqrt{15}, 0)$, $V(6, 0)$, $(-6, 0)$; asymptotes: $y = \pm\sqrt{6}x/3$
11 $y = \pm 2x/\sqrt{5}$
13 $x^2/16 - y^2/36 = 1$
15 $x^2/16 - 9y^2/256 = 1$
17 $5y^2/36 - x^2/4 = 1$
19 $y = \pm x/2$ (for both)
21 $4y^2 - 4x^2/3 = 1$

Section 12.4 (Page 378)

1 $C(2, -3)$, $F(7, -3)$, (-3), $(-3, -3)$, $V(5, -3)$, $(-1, -3)$; asymptotes: $4x - 3y - 17 = 0$, $4x + 3y + 1 = 0$
3 $C(-5, 4)$, $F(s5, 14)$, $(-5, -6)$, $V(-5, 12)$, $(-5, -4)$; asynptotes: $4x - 3y + 32 = 0$, $4x + 3y + 8 = 0$
5 $C(-3, 1)$, $F(-3, 5)$, $(-3, -3)$, $V(-3, 3)$, $(-3, -1)$; asymptotes: $\sqrt{3x} - 3y + 3\sqrt{3} + 3 = 0$, $\sqrt{3x} + 3y + 3\sqrt{3} - 1 = 0$
7 $C(1, -2)$, $F(1, -2 \pm 2\sqrt{5})$, $V(1, 0)$, $(1, -4)$; asymptotes: $x - 2y - 5 = 0$, $x + 2y + 3 = 0$
9 $C(2, -3)$, $F(2 \pm 3\sqrt{5}, -3)$, $V(5, -3)$, $(-1, -3)$; asymptotes: $2x - y - 7 = 0$, $2x + y - 1 = 0$
11 $C(3, -2)$, $F(3 \pm 2\sqrt{5}, -2)$, $V(7, -2)$, $(-1, -2)$; asymptotes: $x - 2y - 7 = 0$, $x + 2y + 1 = 0$
13 $C(-3, 2)$, $F(-3 \pm \sqrt{61}, 2)$, $V(2, 2)$, $(-8, 2)$; asymptotes: $6x - 5y + 28 = 0$,
15 $\dfrac{(x-1)^2}{16} - \dfrac{(y-2)^2}{9} = 1$
17 $\dfrac{(x-1)^2}{9} - \dfrac{(y+1)^2}{16} = 1$

Section 12.5 (Page 392)

3 rotated π radians: $(x')^2 + (y')^2 + 4x' = 0$; rotated $\pi/2$ radians: $(x')^2 + (y')^2 + 4y' = 0$; rotated $\pi/4$ radians: $(x' - \sqrt{2})^2 + (y' + \sqrt{2})^2 = 4$
5 $\pi/8$
7 $(x')^2/12 + (y')^2/4 = 1$, $\alpha = \pi/6$ (Diagram A104)
9 $(x')^2/5 + (y')^2/9 = 1$, $\alpha = \pi/3$ (Diagram A105)
11 degenerate ellipse (no locus)
13 $y' = 4(x')^2$, $\alpha = \pi/3$ (Diagram A106)
15 degenerate parabola, $\tan 2\alpha = -\frac{12}{5}$ (two vertical lines) $x' = -5$, $x' = 1$
17 $x' = (y')^2/8$, $\tan 2\alpha = -\frac{4}{3}$ (Diagram A107)
19 $(x')^2/10 - (y')^2/10 = 1$, $\tan 2\alpha = \frac{3}{4}$ (Diagram A108)
21 **(a)** $x' = 0$ **(b)** $x' = 2$, $x' = -2$ **(c)** no locus

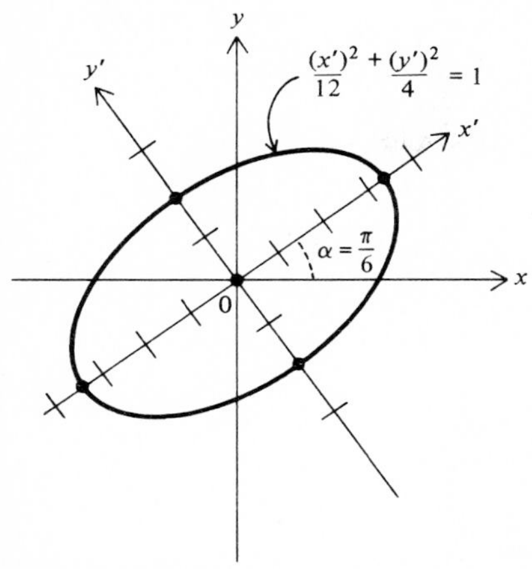

Diagram A104

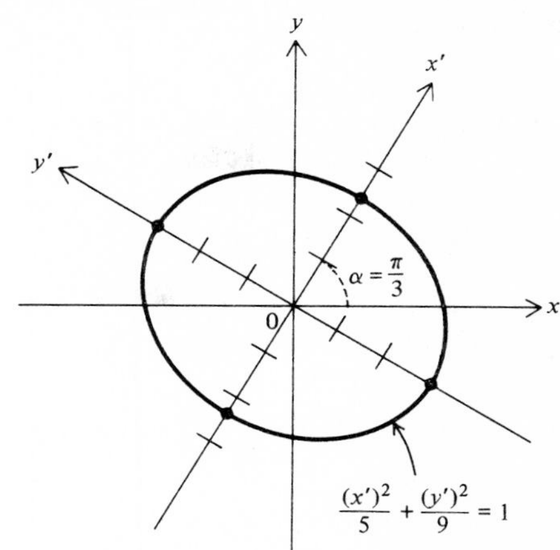

Diagram A105

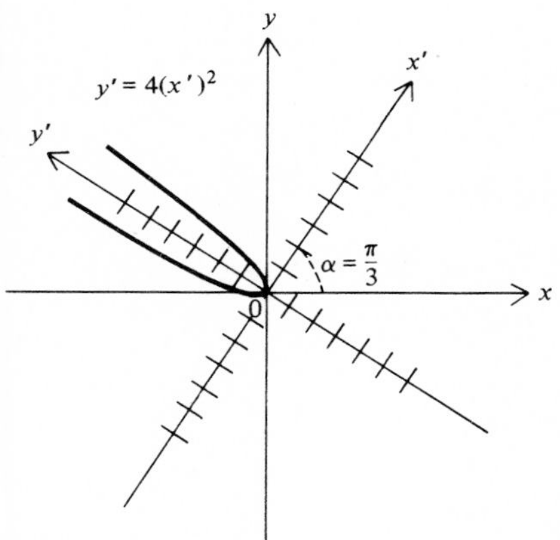

Diagram A106

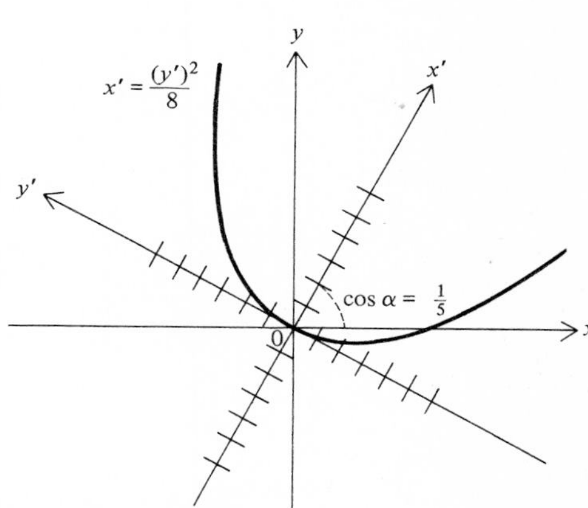

Diagram A107

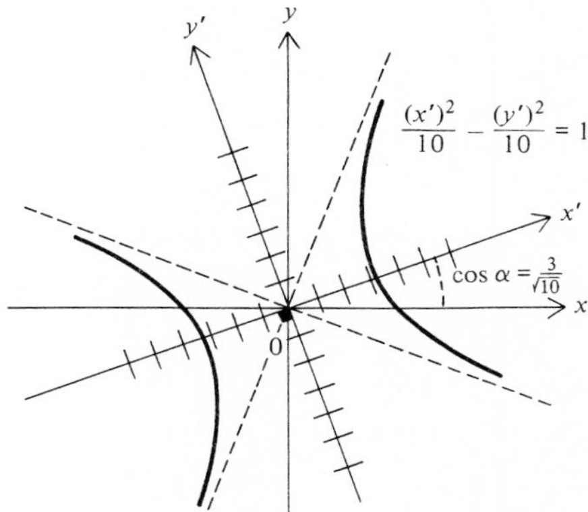

Diagram A108

Index

A

Abscissa, 3
Absolute value:
  definition of, 6
  inequalities, 101
Absolute-value function, 49
Algebraic functions, 195
Ambiguous case, 231–232
Amplitude, 276
Angle, 203
  complementary, 203
  of depression, 213
  of elevation, 213
  negative, 221
  positive, 221
  reference, 225
  standard position of, 221
Antilogarithm, 319
Asymptote:
  horizontal, 157
  of a hyperbola, 367, 369, 372, 373
  intersection with a graph, 161
  oblique, 166
  vertical, 156
Axes:
  coordinate, 1
  rotation of, 378
Axis:
  of an ellipse, 356, 362
  of a hyperbola, 367, 373
  of symmetry, 57, 61

B

Base, 301
Bearing, 216
Binary search, 143–145

C

Centroid, 34 (Prob. 22)
Characteristic, 318
Characteristic curve:
  of the cosine, 270
  of the sine, 271
Circle, 331
Circular function, 252
Circum center, 34 (Prob. 23), 337
Common domain, 183
Common logarithm tables, 398–399
Common logarithmic function, 317
Common tangents, 336 (Prob. 25)
Complementary angles, 203
Completing the square, 61, 62
Complex conjugates, 132
Complex number, 130
Complex root, 130
Composite function, 185
Compound interest, 327 (Probs. 20–24)
Conjugate axis, 368
Cosecant function, 296
Cosine function, 260
  graph of, 269

Course, 216
Cramer's rule, 30, 32
Cross-product term, 378
Crucial values, 154
Cube root, 50
  tables of, 403
Cubic polynomial:
  applications, 145
  turning point of, 146

D

Degenerate loci, 350, 379
Delian problem, 145 (Prob. 12)
Delta x, 4
Delta y, 4
Dependent variable, 41
Depressed equation, 117
Determinant:
  2 x 2, 29
  3 x 3, 31
Directed line segment, 241
Directrix, 342
Discriminant, 74
Distance:
  from a point to a line, 37
  between two points, 7, 9
Division algorithm, 112
Domain, 41, 42

E

e, 310
Eccentricity, 361
Ellipse:
  with center at $C(h, k)$, 361, 362
  definition of, 354
  equations of, with center at origin, 357
  general form of, 364
Equality:
  of functions, 46
  of ordered pairs, 3
  of polynomials, 175
Equation of a line, 17
  general form, 20
  point-slope form, 18
  slope-intercept form, 18
Euler line, 36
Exponential functions, 305–309
Exponents:
  definition of, 301
  rules for, 302

F

Factor theorem, 115
Focal chord, 345 (Prob. 12)
Focus, 342, 354, 367
Force diagram, 247–248
Frustum of a cone, 81 (Prob. 14)
Function:
  absolute value, 49
  algebraic, 195
  composite, 185
  constant, 48
  cost, 50
  cube root, 50
  definition of, 42
  equality, 46
  exponential, 305–309
  graph of, 46
  hyperbolic, 312 (Prob. 14)
  identity, 48
  inverse, 188, 290
  linear, 48
  logarithmic, 313, 317
  notation, 42
  one-to-one, 189
  periodic, 259
  polynomial, 108
  power, 48
  quadratic, 56
  rational, 153
  signum, 53 (Prob. 25)
  trigonometric, 260, 296
  winding, 252, 258
Fundamental period, 259

G

Graph of an equation, 17
Greater than, 86

H

Heron's formula, 328 (Prob. 52)
Horizontal asymptote, 157
Horizontal-line test, 189
Hyperbola:
  with center at $C(h, k)$, 372
  definition of, 367
  equation of, with center at origin, 367
Hyperbolic functions, 312 (Prob. 14)

I

Identity function, 48
Imaginary number, 71
Increments, 5
Independent variable, 41
Inequalities:
  absolute value, 101
  linear, 90
  involving products and quotients, 97–100
  properties, 86–88, 101–103
  sense, 86
Infinite series, 310
Infinity, 90, 105
Intercept form of a line, 21 (Prob. 15)
Interpolation, 320
Interval:
  closed, 90
  half-open, 90
  notation, 89, 90
  open, 89
Inverse:
  cosine, 291
  definition of, 188
  sine, 293
Irrational roots, 143

L

Law of cosines, 236
Law of sines, 230
Less than, 86
Linear asymptote, 168
Location principle, 135
Locus, 330
Logarithm:
  applications, 325
  computations, 317
  function, 313
  natural, 400–402
  tables, common, 398–399

M

Mantissa, 318
Maximum, minimum problems, 82, 146
Midpoint of a line segment, 11
Minutes, 203
Mollweides' equation, 240 (Prob. 32)
Multiple zeros, 138
Multiplicity of factors, 119

N

Nagel line, 35 (Prob. 29)
Natural logarithmic function, 317
  table of, 400–402
Newton's law of cooling, 327 (Ex. 20)
Nine-point circle, 37
Nonlinear asymptotes, 168

O

Oblique linear asymptote, 168
Oblique triangle, 230
One to one, 189

Ordered pair, 3
Ordinate, 3
Orthocenter, 35 (Prob. 24)
Orthogonality, 336 (Prob. 16)

P

Parabola:
  definition of, 342
  equations, vertex at origin, 342
  vertex at $V(h, k)$, 346
Parallel lines, 22, 23
Parallelogram law, 241
Parameter, 193
Parametric equations, 193
Partial fraction, 174
Pascal line, 35 (Prob. 30)
Period, 258, 259
  of the cosine and sine, 277
Periodic function, 259
Perpendicular bisector, 24
Perpendicular lines, 23
Phase shift, 285
Point-slope form, 18
Polynomial:
  application with such functions, 129
  equality of, 175
  functions, 108
Power functions, 48
Principle root, 303
Proper rational function, 167
Pythagorean theorem, 9, 202

Q

Quadrant, 1
Quadrantal point, 253
Quadratic equation, roots of, 65
  formula for, 68
Quadratic functions, applications of, 75–84

R

Range, 41
Rational functions, 153
Rational root theorem, 122, 127
Rectangular coordinate system, 1
Reference rectangle, 368
  right triangle, 206, 223
Remainder theorem, 114
Resultant, 241
Right triangle, 201
Rotation of axes, 378

S

Sign patterns, 96
Sine function, 260
  graph of, 269
Slope of a line, 13
Slope-intercept form, 18
Square root, table of, 403
Symmetry, 56
Synthetic division, 117
Synthetic substitution, 109–110
System of linear equations, 26
  solution to, 26

T

Transverse axis, 368
Trigonometric functions:
  cosecant, 296
  cosine, 260
  cotangent, 296
  secant, 296
  sine, 260
  tangent, 296
Trigonometric ratios, 204–205
Trigonometric tables:
  of angles, 404–411
  of real numbers, 396–397
Turning point, 58, 343

V

Variable:
dependent, 41
independent, 41
Vectors, 240
standard position of, 242
Vertex, 58
Vertical asymptote, 156
Vertical line test, 47

W

Winding function, 252
definition of, 258

Z

Zeros, 65